Operator Theory: Advances and Applications

Vol. 195

Operator Algebras, Operator Theory and Applications

18th International Workshop on Operator Theory and Applications, Potchefstroom, July 2007

J.J. Grobler
L.E. Labuschagne
M. Möller
Editors

Birkhäuser
Basel · Boston · Berlin

Editors:

J.J. Grobler
School for Computer, Statistical and
Mathematical Sciences
North-West University
2520 Potchefstroom
South Africa
e-mail: Koos.Grobler@nwu.ac.za

L.E. Labuschagne
Department of Mathematical Sciences
University of South Africa
PO Box 392
0003 UNISA
South Africa
e-mail: labusle@unisa.ac.za

M. Möller
John Knopfmacher Centre for
Applicable Analysis and Number Theory
School of Mathematics
University of the Witwatersrand
2050 WITS
South Africa
e-mail: Manfred.Moller@wits.ac.za

2000 Mathematical Subject Classification: 46-06, 47-06

Library of Congress Control Number: 2009932070

Bibliographic information published by Die Deutsche Bibliothek.
Die Deutsche Bibliothek lists this publication in the Deutsche Nationalbibliografie;
detailed bibliographic data is available in the Internet at http://dnb.ddb.de

ISBN 978-3-0346-0173-3 Birkhäuser Verlag AG, Basel - Boston - Berlin

Basel · Boston · Berlin
P.O. Box 133, CH-4010 Basel, Switzerland
Part of Springer Science+Business Media
Printed on acid-free paper produced from chlorine-free pulp. TCF∞
Printed in Germany

ISBN 978-3-0346-0173-3

e-ISBN 978-3-0346-0174-0

9 8 7 6 5 4 3 2 1

www.birkhauser.ch

Contents

Preface

The eighteenth *International Workshop on Operator Theory and Applications* (*IWOTA*) was hosted by the Unit for Business Mathematics and Informatics of the North-West University, Potchefstroom, South Africa from July 3 to 6, 2007. The conference was dedicated to Professor Joseph A. Ball on the occasion of his 60th birthday and to Professor Marinus M. Kaashoek on the occasion of his 70th birthday, and we wish to similarly honour them by dedicating this volume of the proceedings to them.

The aim of the IWOTA meetings is to bring together mathematicians working in operator theory and its applications to related fields. Each conference therefore has a unique character, depending on the related fields chosen by the local organizers. In the present case Functional Analysis and Von Neumann algebras were decided on, due to the number of people actively working in those fields in South Africa. The meetings are intended to be truly international (the seventeen previous ones were held in 11 different countries) and this conference was no exception, with participants drawn from 17 countries.

IWOTA is directed by an international steering committee of 25 members under the guidance of the president I. Gohberg (Tel Aviv), assisted by the Vice Presidents J.W. Helton (La Jolla) and M.A. Kaashoek (Amsterdam). The Proceedings of the IWOTA workshops regularly appear in the Birkhäuser series: *Operator Theory: Advances and Applications*, and we thank them for their willingness to continue this cooperation by publishing this volume.

We thank all authors who contributed to this volume. Every paper was refereed by at least two independent referees, and we thank them for adding value to these proceedings by their expert evaluations and suggestions.

The conference was sponsored by grants from the North-West University, Potchefstroom Campus, the Vrije Universiteit, Amsterdam, the National Research Foundation of South Africa, Birkhäuser Publishing Ltd, Basel, Switzerland, the Commission on Development and Exchanges of the International Mathematical Union and the Abdus Salam International Centre for Theoretical Physics, Italy. Without their assistance it would have been impossible to organize a conference so far from the European, American, and Asian centres.

Jacobus J. Grobler Louis E. Labuschagne Manfred Möller

Laudatio for M.A. (Rien) Kaashoek and J.A. (Joe) Ball

Ladies and gentlemen,

It is an honour and a pleasure to speak here on behalf of the members of the organizing committee on the achievements of the two men we honour at this conference. These are two of the great men of operator theory. Marinus Adrianus Kaashoek (better known as Rien) will turn 70 in November, Joseph Anthony Ball (better known as Joe) has turned 60 last month. Since Rien is the oldest, I will first tell you a little about his career, before turning to Joe.

Rien studied at the University of Leiden in the Netherlands, and did his PhD there in 1964 under supervision of A.C. Zaanen. As it happens, the chairman of the organizing committee of this conference, Koos Grobler, also did his PhD with Zaanen, but Rien and Koos where not in Leiden at the same time. After finishing his PhD, Rien spent a year as postdoc at UCLA. After that he was immediately appointed at the Vrije Universiteit in Amsterdam, where he has remained ever since, with the exception of several extended research visits abroad. Rien's qualities were soon recognized at the Vrije Universiteit: in 1969 he was appointed full professor at the age of 32. Officially Rien retired in 2002. However, it seems to those of us who are able to observe him from close by, that retirement has not slowed him down, quite the contrary! He quit teaching of course and he quit the management duties, but that just gives him more time to do the things that are nearest to his heart: mathematical research being perhaps one the most important activities among those, besides visiting his children and grandchildren.

In his career, Rien has been a wonderful teacher (as many of the alumni of the VU can attest), he was an efficient and very competent organizer (among other duties he was Dean of the Faculty during four years shortly before retiring), but above all he still is an outstanding researcher. Rien has had, to date (2007), 16 PhD students, with number 17 finishing this year. I am one of these PhD students, as is Gilbert Groenewald of this university. Since Mark Petersen, also of this university, is Gilbert's PhD student, Rien has a mathematical brother, a son and a grandson in Potchefstroom.

To work under Rien was a wonderful experience. There are of course the regular weekly meetings to discuss progress. There, not only does he point out how many commas you have misplaced in your poorly written first attempt at writing mathematics in English, he actually explains to you in detail how to write a good and comprehensible piece of mathematics. Also, there were the weekly seminars, where many speakers from abroad would come and give lectures, but where you, as a student, could make your first uncertain steps into presenting your mathematical results to a critical audience. I am certain that all Rien's students became better lecturers because of his kind remarks made after our lectures at the seminar.

Rien has visited here before, in November 2004, when he was main lecturer at the yearly SAMS conference. On that occasion he had his wife, Wies, with him. It seems they enjoyed their stay here very much, because at the end of the visit they said to Gilbert: "now we have to go back to that miserable weather, and endure the little food you get at expensive restaurants." (I must say that on the whole I can sympathize with that wholeheartedly!)

Rien is co-author of more than 170 papers, and the number is rapidly increasing each year. Rien is co-author of seven books, and that number is also still increasing. He has been visiting professor at universities in America, Canada and Israel, he is on the editorial board of three journals and two book series. On top of all that he found time to organize many conferences, and he is vice-president of the Steering Committee for the IWOTA conferences. All these activities did not go unnoticed. Upon his retirement Rien was made a Knight in the Royal Order of the Dutch Lion by Her Majesty the queen. But you don't have to call him "Sir Rien", as this is not the custom in The Netherlands.

Let me turn to Joe. Joe did his bachelors close to home, at Georgetown University in Washington DC, but for his masters he moved a bit further: the university of Virginia at Charlottesville, where he also did his PhD under supervision of M. Rosenblum. There he also worked with James Rovnyak. After completing his PhD he moved yet further from home: (not too far though) he was appointed at Virginia Tech, where he has stayed ever since. Well, that is to say, with the exception of several long periods as visiting professor at UCSD, working with the other vice-president of the Steering Committee of the IWOTA conferences, Bill Helton, and with the exception of a half year period at the Weizmann institute, and a semester at the MSRI in Berkeley. Joe's qualities too were quickly recognized by Virginia Tech. He was appointed full professor there at the age of 35.

I should say that to my surprise I found out that Joe has also had a short stint working for industry: he spent one summer early in his career working for Dhalgren U.S. Navy Weapons Research Lab. I am pretty sure they cannot afford him at this time!

Joe has been advisor for seven PhD students up till now, with probably some more to come. He has co-authored several books, and has also published more than 170 papers, mostly on topics involving the interplay of operator theory and mathematical systems theory. He was principal investigator for several NSF grants, among which an important grant that made the IWOTA conference in Blacksburg, at Virginia Tech possible. Joe has been and still is on the editorial board of several important mathematical journals, among them, the Proceedings of the American Mathematical Society. Joe was also recipient of an Alumni Award for Research

Excellence from Virginia Tech. The fact that Joe is really a celebrity can perhaps be illustrated best by the story he told yesterday: last year, when visiting Japan, Joe was approached by a young lady, who was carrying one of his papers, and she wanted his autograph on the paper. Joe told us that he felt like a rock-star at that moment.

To work with Joe is a very stimulating experience, as many of us can testify. It should be said that Joe is very intense as a researcher. This is shown best by the fact that at any time his brain is at least partly occupied with mathematics. During lunch the discussion is usually about mathematics, and when watching the ball games on the television he usually has pad and pen on his lap. If the game gets boring, there is always mathematics! So, whenever you want to discuss mathematics with him, he is ready to do just that! A typical afternoon of working with Joe will consist of a discussion at lunch, a short time alone in the office to rethink what Joe has said and digest it (not just the lunch!), and then a track to his office. There you will find Joe with feet up and stretched almost into the corridor, a pad on his lap and a pen in his hand, scribbling away at the same problem. You explain to him what you think, and Joe goes: yeah, aaah, well, let's see, and while he scratches his head he waves you to the blackboard to explain. Before you know it the sun is down, and you are still discussing the problem. And Joe won't rest before the problem is solved, and he will not be satisfied with half-solutions either. Fortunately, Joe has Amelia to remind him on occasion that the world does not only revolve around work and mathematics, but that there are other things of interest as well. Joe may grumble that this costs him work-time, but he will go along and enjoy himself.

Joe and Rien have a close connection, which started when Joe was visiting at the Vrije Universiteit in the early eighties. After that, they wrote 14 papers together, mostly with Israel Gohberg as co-author. Also, several of Rien's PhD students started to visit Virginia Tech for postdoc periods. Cornelis van der Mee was there, but he worked mostly with a colleague of Joe, after that I was there for a wonderful year, Gilbert was there for long time visits (with more to come), and now Rien's latest PhD student, Sanne ter Horst, will also go to work with Joe.

Rien and Joe, several people in this audience have a lot to thank both of you for, and on behalf of all of them I say from the bottom of my heart: guys, thank you for what we have learned from you.

André Ran (VU University Amsterdam) 2007

Operator Theory:
Advances and Applications, Vol. 195, 1–72

Multivariable Operator-valued Nevanlinna-Pick Interpolation: a Survey

Joseph A. Ball and Sanne ter Horst

Abstract. The theory of Nevanlinna-Pick and Carathéodory-Fejér interpolation for matrix- and operator-valued Schur class functions on the unit disk is now well established. Recent work has produced extensions of the theory to a variety of multivariable settings, including the ball and the polydisk (both commutative and noncommutative versions), as well as a time-varying analogue. Largely independent of this is the recent Nevanlinna-Pick interpolation theorem by P.S. Muhly and B. Solel for an abstract Hardy algebra set in the context of a Fock space built from a W^*-correspondence E over a W^*-algebra $\mathcal{A}$ and a $*$-representation σ of $\mathcal{A}$. In this review we provide an exposition of the Muhly-Solel interpolation theory accessible to operator theorists, and explain more fully the connections with the already existing interpolation literature. The abstract point evaluation first introduced by Muhly-Solel leads to a tensor-product type functional calculus in the main examples. A second kind of point-evaluation for the W^*-correspondence Hardy algebra, also introduced by Muhly and Solel, is here further investigated, and a Nevanlinna-Pick theorem in this setting is proved. It turns out that, when specified for examples, this alternative point-evaluation leads to an operator-argument functional calculus and corresponding Nevanlinna-Pick interpolation. We also discuss briefly several Nevanlinna-Pick interpolation results for Schur classes that do not fit into the Muhly-Solel W^*-correspondence formalism.

Mathematics Subject Classification (2000). Primary 47A57; Secondary 42A83, 47A48.

Keywords. (completely) positive kernel, (completely) positive map, left/right-tangential, operator-argument, Drury-Arveson space, Schur class, Schur-Agler class, Nevanlinna class, positive-real-odd class, Toeplitz operators, W^*-correspondence, Fock space

1. Introduction

The classical interpolation theorems of Nevanlinna [74] and Pick [79], now approaching the age of one hundred years, can be stated as follows: *Given N distinct points $\lambda_1, \dots, \lambda_N$ in the unit disk $\mathbb{D} = \{\lambda \in \mathbb{C} \colon |\lambda| < 1\}$ together with N complex numbers $w_1, \dots, w_N$, there exists a Schur class function s* (i.e., s holomorphic from the unit disk $\mathbb{D}$ to the closed disk $\overline{\mathbb{D}}$) *such that*

$$s(\lambda_i) = w_i \text{ for } i = 1, \dots, N$$

if and only if the so-called Pick matrix

$$\mathbb{P} := \left[\frac{1 - w_i \overline{w_j}}{1 - \lambda_i \overline{\lambda_j}} \right]_{i,j=1}^{N}$$

is positive semidefinite.

There is a parallel result usually attributed to Carathéodory and Fejér as well as to Schur (see [33, 34, 50, 99, 100]) for the case where one prescribes an initial segment of Taylor coefficients at the origin rather than functional values at distinct points: *Given $N+1$ complex numbers $s_0, \dots, s_N$, there exists a Schur class function s such that*

$$\frac{1}{i!} \frac{d^i s}{d\lambda^i}(0) = s_i \text{ for } i = 0, \dots, N$$

if and only if the $(N+1) \times (N+1)$ lower-triangular Toeplitz matrix

$$\begin{bmatrix} s_0 & 0 & \cdots & 0 \\ s_1 & s_0 & \ddots & \vdots \\ \vdots & \ddots & \ddots & 0 \\ s_N & s_{N-1} & \cdots & s_0 \end{bmatrix}$$

is contractive.

Over the years these results have been an inspiration for new mathematics, often interacting symbiotically with assorted engineering applications. We mention in particular that the operator-theoretic formulation of the Nevanlinna-Pick/Carathéodory-Fejér interpolation problem due to Sarason [97] led to the advance in operator theory known as commutant lifting theory introduced by Sz.-Nagy-Foias [72]; see also [51, 53]. Commutant lifting in turn provides a unifying framework for the handling of a variety of interpolation problems of Nevanlinna-Pick/Carathéodory-Fejér type for matrix- and operator-valued functions on the disk, and, more generally, for left- and right-tangential interpolation with operator-argument (**LTOA/RTOA**); cf., [53]. The operator-argument approach to interpolation emerged as one of the most popular ways of handling Nevanlinna-Pick and Carathéodory-Fejér interpolation conditions in a unified way and can be seen to be equivalent to the Sarason formulation (see [24, 21]). We mention that it is the Sarason formulation which plays a prominent role in connection with the H^∞-control theory (see, e.g., [24]) where it takes the form of a model matching problem.

Our goal here is to survey recent generalizations of the theory of Nevanlinna-Pick interpolation to several-variable contexts. In addition there has been a lot of recent work on noncommutative function theory, where one plugs in a tuple of noncommuting operators (or matrices) as the function arguments and outputs a matrix or operator. In this regard, we mention the papers [2, 3, 4, 5, 11, 17, 18, 19, 21, 28, 29, 32, 40, 48, 65, 93, 96] for the commutative setting and [20, 37, 55, 58, 83, 85, 86, 87, 88] for the noncommutative setting. We mention yet another direction having an analogue of the Schur class and of Nevanlinna-Pick interpolation, namely: the unit ball of the algebra of lower triangular matrices acting as operators on $\ell^2(\mathbb{Z})$; in the 1990s there was a lot of activity on Beurling-Lax representation and an analogue of **LTOA/RTOA** interpolation theory having connections with robust control and model reduction for time-varying systems (see [8, 23, 41, 42, 98]).

Going beyond these types of results is the recent generalized Nevanlinna-Pick interpolation theorem [69] and the generalized Schur class [70] of Muhly-Solel, where the concept of *completely positive kernel* or *completely positive map* enters into various characterizations. In particular, the Muhly-Solel result, when specialized to various specific settings, is different from the standard **LTOA** or **RTOA** formulation, in that the point-evaluation is actually with a tensor-type functional calculus. The simplest instance of this is what we call *Riesz-Dunford* interpolation, where one is given operators Z and W on a Hilbert space $\mathcal{Z}$ and one seeks a scalar Schur class function s so that $s(Z) = W$. We discuss how a solution of this problem can be had by using existing theory for **LTOA**, for both the unit disk setting and the right half-plane setting. In particular, the theorem for the right half-plane setting gives a solution which appears quite different from that obtained by Cohen-Lewkowicz [36]. We also show how the Muhly-Solel solution criterion (involving complete positivity of a kernel or of a map between C^*-algebras) can be seen to be equivalent to the criterion (involving positivity of a single block matrix) obtained from the **LTOA** theory. A similar story holds in the setting of the unit ball (commutative and noncommutative). There is an analogue of the Riesz-Dunford Nevanlinna-Pick interpolation problem in the setting of the unit ball which is handled by the Muhly-Solel theory; to the uninitiated the solution criterion looks somewhat strange since it involves *complete positivity* (in the sense of [31]) rather than merely *positivity* of a kernel, or, in another formulation, *complete positivity* of a map between C^*-algebras rather than merely positive semidefiniteness of a single block operator matrix. We resolve this situation by giving a direct proof that the two solution criteria are equivalent. In the recent paper [71] Muhly and Solel introduced a second kind of point-evaluation in the context of a generalized Poisson kernel. We develop here some further properties of this point-evaluation and derive the corresponding Nevanlinna-Pick interpolation theorem. For instance, it is shown that the Muhly-Solel Hardy algebra with respect to this point-evaluation is in general not multiplicative; a multiplicative law reminiscent of that in the operator-argument functional calculus and time-varying system theory literature is proved. This connection with time-varying systems was already hinted upon in [71]. In fact, when specified for the standard one-variable Schur class, the Nevanlinna-Pick interpolation theorem for this alternative point-evaluation gives us precisely the corresponding left-tangential operator-argument result. The precise connections with the time-varying-system interpolation theory remain to be worked out.

There is yet another generalized theory of Schur class (or, more precisely, Schur-Agler class) and Nevanlinna-Pick interpolation where one defines a Schur class starting from a family of test functions (see [44, 46, 64] and see also [6, Chapter 13] for an introduction to this approach). It is well appreciated that the Agler theory for the polydisk is an example for this theory (indeed, this is the motivating example); precise specification of what other examples can be covered, such as the higher rank graph algebras of [62] and the Hardy algebras associated with product decompositions along semigroups more general than $\mathbb{Z}$ (see [102]), is an ongoing area of investigation [15].

In this survey of interpolation problems of Nevanlinna-Pick type in a variety of settings, we discuss only criteria for existence of solutions; we do not discuss characterizations of the various Schur classes via realization as the transfer function of a conservative linear system or of the construction of solutions or parametrization of solutions of interpolation problems via linear-fractional formulas, although in the various cases often such topics are worked out in the literature. In general we do not discuss the techniques used for establishing these solution criteria; let us only mention here that, just as in the classical case, there are a variety of techniques for analyzing these types of multivariable interpolation problems. We mention specifically commutant lifting theory [80, 81, 28, 29], the "lurking isometry" method [4, 5, 28, 29, 17, 18, 19], the Fundamental-Matrix-Inequality method of Potapov [32, 21] as well as the Ball-Helton Grassmannian Kreĭn-space method [49].

The paper [21] also surveys the various types of operator-valued interpolation problems but with a focus on the Drury-Arveson Schur-multiplier class. In addition to **LTOA**, the paper [21] treats a more general version of **LTOA** (where the joint spectra of $Z^{(1)}, \dots, Z^{(N)}$ are no longer required to be in the open ball $\mathbb{B}^d$), and a still more general Abstract Interpolation Problem (see [59] for the single-variable version). These more general formalisms handle more general interpolation problems (e.g., boundary interpolation) which go beyond the original results based on the commutant lifting approach. We do not discuss these more general problems here.

We focus on interpolation theory on the unit disk and multivariable generalizations of the unit disk; this means we ignore all the activity that has been going on of late on interpolation theory for the Nevanlinna class (holomorphic functions taking the right half-plane into itself) and its multivariable generalizations. We do however give the complex disk analogue of results from [35, 36] which are set out in a real right half-plane setting.

The paper is organized as follows. In Section 2 we review Nevanlinna-Pick interpolation for the single-variable case. Perhaps new is the derivation of the result for the tangential Riesz-Dunford interpolation problem due to Rosenblum-Rovnyak [96, Section 2.3] and for the full Riesz-Dunford interpolation problem as an application of known results for **LTOA**. In the third section various extensions of the one-variable theory for functions of several variables are discussed. We consider the unit ball case (both commutative and noncommutative) in Subsections 3.1 and 3.2, and the generalization to so-called free semigroupoid algebras [66, 68, 60, 61] in Subsection 3.3 where the starting point is a directed graph (also called a quiver). New in the free semigroupoid algebra setting is a more explicit formula for the point-evaluation and of the criterion for existence of solutions of the associated Nevanlinna-Pick interpolation problem. We also work out the Nevanlinna-Pick result for a specific quiver. This example corresponds to Nevanlinna-Pick interpolation for functions in a certain subalgebra of a matrix-valued H^∞-space. In Subsection 3.4 we develop parallel results for the Schur-Agler class in the polydisk setting for both the commutative and noncommutative settings. Most of the material in Sections 2 and 3 should be well known to the experts in the interpolation theory community; the focus here is on the connections among the various results. Section 4 contains the relatively new results on a generalized Schur class and Nevanlinna-Pick interpolation obtained by P.S. Muhly and B. Solel [69, 70]. The material is presented for the expert in operator theory not familiar with such notions as "W^*-correspondence" and "Hilbert module" which are more common in the operator algebra literature. The novel part of the exposition here is to define the point-evaluation in a direct way without making explicit use of the dual correspondence. Moreover, we prove a Nevanlinna-Pick theorem for the second kind of point-evaluation for the W^*-correspondence Schur class that was recently introduced in [71]. We recover the results for the single-variable case (Section 2) and the free semigroupoid algebra setting (Subsection 3.3) as an application of the general Muhly-Solel theory; a key ingredient here is careful understanding of the connections between completely positive kernels and completely positive maps versus merely positive semidefinite operator matrices, including an application of Choi's theorem [38]. A final subsection gives some perspective on connections of the Muhly-Solel theory with the time-varying interpolation theory of the 1990s. The last section discusses the test-function approach and directions for future work.

The notation is mostly standard but we mention here a few conventions for reference. For Ω any index set and B a Banach space with norm $\|\ \|_B$, the symbol $\ell^2_B(\Omega)$ denotes the space of B-valued norm-squared summable sequences indexed by Ω:

$$\ell^2_B(\Omega) = \Big\{\xi\colon \Omega \to B\colon \sum_{\omega\in\Omega} \|\xi(\omega)\|^2_B < \infty\Big\}.$$

Most often the choice $\Omega = \mathbb{Z}$ (the integers) or $\Omega = \mathbb{Z}_+$ (the nonnegative integers) appears. For Hilbert spaces $\mathcal{U}$ and $\mathcal{Y}$ the symbol $\mathcal{L}(\mathcal{U}, \mathcal{Y})$ stands for the space of bounded linear operators from $\mathcal{U}$ into $\mathcal{Y}$. We write $H^2_{\mathcal{U}}(\mathbb{D})$ for the Hardy space of analytic functions $f : \mathbb{D} \to \mathcal{U}$ that can be extended to square integrable functions on the unit circle $\mathbb{T} := \{\lambda \in \mathbb{C}\colon |\lambda| = 1\}$. With $H^\infty_{\mathcal{L}(\mathcal{U},\mathcal{Y})}(\mathbb{D})$ we denote the Banach space of uniformly bounded analytic functions on the unit disc $\mathbb{D}$ with values in $\mathcal{L}(\mathcal{U}, \mathcal{Y})$.

2. Operator-valued Nevanlinna-Pick interpolation: the one-variable case

Let $\mathcal{U}$ and $\mathcal{Y}$ be Hilbert spaces. With $\mathcal{S}(\mathcal{U},\mathcal{Y})$ we denote the $\mathcal{L}(\mathcal{U},\mathcal{Y})$-valued Schur class, i.e., the set of all holomorphic functions S on the unit disc $\mathbb{D}$ whose values are contractive operators in $\mathcal{L}(\mathcal{U},\mathcal{Y})$:

$$\|S(\lambda)\| \leq 1 \text{ for all } \lambda \in \mathbb{D}. \tag{2.1}$$

In this section we consider variations on the classical Nevanlinna-Pick interpolation problem for functions in the Schur class $\mathcal{S}(\mathcal{U},\mathcal{Y})$.

2.1. Standard functional calculus Nevanlinna-Pick interpolation problems

The standard operator-valued Nevanlinna-Pick interpolation problem and the left- and right-tangential versions are the following problems:

(1) The Full Operator-Valued Nevanlinna-Pick (**FOV-NP**) interpolation problem: *Given $\lambda_1,\dots,\lambda_N$ in $\mathbb{D}$ and operators $W_1,\dots,W_N$ in $\mathcal{L}(\mathcal{U},\mathcal{Y})$, determine when there exists a Schur class function $S \in \mathcal{S}(\mathcal{U},\mathcal{Y})$ such that $S(\lambda_i) = W_i$ for $i = 1,\dots,N$.*

(2) The Left-Tangential Nevanlinna-Pick (**LT-NP**) interpolation problem: *Given $\lambda_1,\dots,\lambda_N$ in $\mathbb{D}$, an auxiliary Hilbert space $\mathcal{C}$ and operators $X_1,\dots,X_N$ in $\mathcal{L}(\mathcal{Y},\mathcal{C})$ and $Y_1,\dots,Y_N$ in $\mathcal{L}(\mathcal{U},\mathcal{C})$, determine when there exists a Schur class function $S \in \mathcal{S}(\mathcal{U},\mathcal{Y})$ such that $X_iS(\lambda_i) = Y_i$ for $i = 1,\dots,N$.*

(3) The Right-Tangential Nevanlinna-Pick (**RT-NP**) interpolation problem: *Given $\lambda_1,\dots,\lambda_N$ in $\mathbb{D}$, an auxiliary Hilbert space $\mathcal{C}$ and operators U_1, ..., U_N in $\mathcal{L}(\mathcal{C},\mathcal{U})$ and V_1, ..., V_N in $\mathcal{L}(\mathcal{C},\mathcal{Y})$, determine when there exists a Schur class function $S \in \mathcal{S}(\mathcal{U},\mathcal{Y})$ such that $S(\lambda_i)U_i = V_i$ for $i = 1,\dots,N$.*

Note that the **FOV-NP** interpolation problem is a particular case of the **LT-NP** interpolation problem, namely with $\mathcal{C} = \mathcal{Y}$ and $X_i = I_{\mathcal{Y}}$ for $i = 1,\dots,N$. Moreover, **LT-NP** reduces to **RT-NP**, and vice versa, as follows. A function $S : \mathbb{D} \to \mathcal{L}(\mathcal{U},\mathcal{Y})$ is in the Schur class $\mathcal{S}(\mathcal{U},\mathcal{Y})$ if and only if the function $S^\sharp : \mathbb{D} \to \mathcal{L}(\mathcal{Y},\mathcal{U})$ defined by

$$S^\sharp(\lambda) = S(\overline{\lambda})^* \quad (\lambda \in \mathbb{D}) \tag{2.2}$$

is in $\mathcal{S}(\mathcal{Y},\mathcal{U})$. In particular, S is a solution to the **LT-NP** interpolation problem, with data as above, if and only if $S^\sharp : \mathbb{D} \to \mathcal{L}(\mathcal{Y},\mathcal{U})$ is a solution to the **RT-NP** with data $\overline{\lambda}_1,\dots,\overline{\lambda}_N \in \mathbb{D}$, $X_1^*,\dots,X_N^* \in \mathcal{L}(\mathcal{C},\mathcal{Y})$ and $Y_1^*,\dots,Y_N^* \in \mathcal{L}(\mathcal{C},\mathcal{U})$.

The relations among these variations on the classical Nevanlinna-Pick interpolation problem are also exhibited in their solutions; see, e.g., [24, 51].

Theorem 2.1. *Let the data for the* **FOV-NP**, **LT-NP** *and the* **RT-NP** *interpolation problem be as given in* (1), (2) *and* (3) *above.*

1. *A solution to the* **FOV-NP** *interpolation problem exists if and only if the associated Pick matrix*

$$\mathbb{P}_{\mathrm{FOV}} := \left[\frac{I_{\mathcal{Y}} - W_i W_j^*}{1 - \lambda_i \overline{\lambda}_j} \right]_{i,j=1}^{N} \tag{2.3}$$

is positive semidefinite.

2. *A solution to the* **LT-NP** *interpolation problem exists if and only if the associated Pick matrix*

$$\mathbb{P}_{\mathrm{LT}} := \left[\frac{X_i X_j^* - Y_i Y_j^*}{1 - \lambda_i \overline{\lambda}_j} \right]_{i,j=1}^{N} \tag{2.4}$$

is positive semidefinite.

3. *A solution to the* **RT-NP** *interpolation problem exists if and only if the associated Pick matrix*

$$\mathbb{P}_{\mathrm{RT}} := \left[\frac{U_i^* U_j - V_i^* V_j}{1 - \overline{\lambda}_i \lambda_j} \right]_{i,j=1}^{N} \tag{2.5}$$

is positive semidefinite.

2.2. Operator-argument functional calculus Nevanlinna-Pick interpolation problems

Let S be a Schur class function in $\mathcal{S}(\mathcal{U},\mathcal{Y})$ with Taylor coefficients $S_0, S_1, \dots$ in $\mathcal{L}(\mathcal{U},\mathcal{Y})$, i.e.,

$$S(\lambda) = \sum_{n=0}^{\infty} \lambda^n S_n \quad (\lambda \in \mathbb{D}).$$

For an auxiliary Hilbert space $\mathcal{C}$ and operators X in $\mathcal{L}(\mathcal{Y},\mathcal{C})$ and T in $\mathcal{L}(\mathcal{C})$ with $r_{\mathrm{spec}}(T) < 1$ the left-tangential operator-argument point-evaluation $(XS)^{\wedge L}(T)$ is given by

$$(XS)^{\wedge L}(T) = \sum_{n=0}^{\infty} T^n X S_n. \tag{2.6}$$

Similarly, for an auxiliary Hilbert space $\mathcal{C}$ and operators U in $\mathcal{L}(\mathcal{C},\mathcal{U})$ and A in $\mathcal{L}(\mathcal{C})$ with $r_{\mathrm{spec}}(A) < 1$ the right-tangential operator-argument point-evaluation $(SU)^{\wedge R}(A)$ is given by

$$(SU)^{\wedge R}(A) = \sum_{n=0}^{\infty} S_n U A^n. \tag{2.7}$$

With respect to these operator-argument functional calculi we consider the following tangential Nevanlinna-Pick interpolation problems.

(1) The Left-Tangential Nevanlinna-Pick interpolation problem with Operator-Argument (**LTOA-NP**): *Given an auxiliary Hilbert space $\mathcal{C}$ together with operators $T_1, \dots, T_N$ in $\mathcal{L}(\mathcal{C})$ with $r_{\mathrm{spec}}(T_i) < 1$ for $i = 1, \dots, N$ and operators $X_1, \dots, X_N$ in $\mathcal{L}(\mathcal{Y},\mathcal{C})$ and $Y_1, \dots, Y_N$ in $\mathcal{L}(\mathcal{U},\mathcal{C})$, determine when there exists a Schur class function S in $\mathcal{S}(\mathcal{U},\mathcal{Y})$ so that $(X_i S)^{\wedge L}(T_i) = Y_i$ for $i = 1, \dots, N$.*

(2) The Right-Tangential Nevanlinna-Pick interpolation problem with Operator-Argument (**RTOA-NP**): *Given an auxiliary Hilbert space $\mathcal{C}$ together with operators $A_1, \dots, A_N$ in $\mathcal{L}(\mathcal{C})$ with $r_{\rm spec}(A_i) < 1$ for $i = 1, \dots, N$ and operators $U_1, \dots, U_N$ in $\mathcal{L}(\mathcal{C}, \mathcal{U})$ and $V_1, \dots, V_N$ in $\mathcal{L}(\mathcal{C}, \mathcal{Y})$, determine when there exists a Schur class function S in $\mathcal{S}(\mathcal{U}, \mathcal{Y})$ so that $(SU_i)^{\wedge R}(A_i) = V_i$ for $i = 1, \dots, N$.*

The **LT-NP** (**RT-NP**) interpolation problem is the special case of the **LTOA-NP** (**RTOA-NP**) interpolation problem with $T_i = \lambda_i I_{\mathcal{U}}$ ($A_i = \lambda_i I_{\mathcal{U}}$). As in the case of the standard functional calculus left- and right-tangential interpolation problems, here also **LTOA-NP** reduces to **RTOA-NP** and vice versa; again via the transformation $S \mapsto S^\sharp$ in (2.2) and a similar transformation of the data. Moreover, it suffices to consider the **LTOA-NP** interpolation problem for the case $N = 1$, since the general case, with data as above, is covered by the **LTOA-NP** interpolation problem with data

$$T = \begin{bmatrix} T_1 & & & \\ & T_2 & & \\ & & \ddots & \\ & & & T_N \end{bmatrix}, \quad X = \begin{bmatrix} X_1 \\ X_2 \\ \vdots \\ X_N \end{bmatrix} \quad \text{and} \quad Y = \begin{bmatrix} Y_1 \\ Y_2 \\ \vdots \\ Y_N \end{bmatrix};$$

see [53, Section I.3] for more details. In particular, a multipoint classical interpolation problem of Nevanlinna-Pick Carathéodory-Fejér type can be conveniently encoded as a single-point **LTOA-NP** or **RTOA-NP** problem; we refer to [24, Sections 16.8-9] for the single-variable case and [19] for multivariable examples. However, we write out the results without this reduction for better comparison with the classical case.

The solutions to the **LTOA-NP** and **RTOA-NP** interpolation problems are given in the following theorem.

Theorem 2.2. *Let the data for the* **LTOA-NP** *and* **RTOA-NP** *interpolation problems be as given in (1) and (2) above.*

1. *A solution to the* **LTOA-NP** *interpolation problem exists if and only if the associated Pick matrix*

$$\mathbb{P}_{\rm LTOA} := \left[\ \textstyle\sum_{n=0}^{\infty} T_i^n (X_i X_j^* - Y_i Y_j^*) T_j^{*n}\ \right]_{i,j=1}^N \tag{2.8}$$

is positive semidefinite.

2. *A solution to the* **RTOA-NP** *interpolation problem exists if and only if the associated Pick matrix*

$$\mathbb{P}_{\rm RTOA} := \left[\ \textstyle\sum_{n=0}^{\infty} A_i^{*n} (U_i^* U_j - V_i^* V_j) A_j^n\ \right]_{i,j=1}^N \tag{2.9}$$

is positive semidefinite.

2.3. Riesz-Dunford functional calculus Nevanlinna-Pick interpolation problems

As a third variant we consider Nevanlinna-Pick interpolation problems for Schur class functions with a Riesz-Dunford functional calculus. In this case the Hilbert spaces $\mathcal{U}$ and $\mathcal{Y}$ are both equal to $\mathbb{C}$, i.e., the Schur class functions are scalar-valued, but the unit disc is replaced by strict contractions on some fixed Hilbert space $\mathcal{Z}$. Given a Schur class function s in $\mathcal{S}(\mathbb{C},\mathbb{C})$ and an operator Z in $\mathcal{L}(\mathcal{Z})$ with spectral radius less than 1, we define $s(Z)$ via the Riesz-Dunford functional calculus:

$$s(Z) = \frac{1}{2\pi i}\int_{\rho\mathbb{T}} (\zeta I_{\mathcal{Z}} - Z)^{-1} s(\zeta)\, d\zeta = \sum_{n=0}^{\infty} s_n \cdot Z^n \text{ if } s(\lambda) = \sum_{n=0}^{\infty} s_n \lambda^n. \tag{2.10}$$

Here the multiplication $s_n \cdot Z^n$ is scalar multiplication and $\rho < 1$ is chosen large enough so that the circle $\rho\mathbb{T}$ encloses the spectrum of Z.

With respect to this functional calculus we consider the following Nevanlinna-Pick interpolation problems.

(1) The Full Riesz-Dunford Nevanlinna-Pick (**FRD-NP**) interpolation problem: *Given a Hilbert space $\mathcal{Z}$, strict contractions $Z_1, \dots, Z_N$ in $\mathcal{L}(\mathcal{Z})$ and operators $W_1, \dots, W_N$ in $\mathcal{L}(\mathcal{Z})$, determine when there exists a Schur class function s in $\mathcal{S}(\mathbb{C},\mathbb{C})$ such that $s(Z_i) = W_i$ for $i = 1, \dots, N$.*

(2) The Left-Tangential Riesz-Dunford Nevanlinna-Pick (**LTRD-NP**) interpolation problem: *Given Hilbert spaces $\mathcal{Z}$ and $\mathcal{C}$, strict contractions $Z_1, \dots, Z_N$ in $\mathcal{L}(\mathcal{Z})$ and operators $X_1, \dots, X_N, Y_1, \dots, Y_N$ in $\mathcal{L}(\mathcal{Z},\mathcal{C})$, determine when there exists a Schur class function s in $\mathcal{S}(\mathbb{C},\mathbb{C})$ such that $X_i s(Z_i) = Y_i$ for $i = 1, \dots, N$.*

(3) The Right-Tangential Riesz-Dunford Nevanlinna-Pick (**RTRD-NP**) interpolation problem: *Given Hilbert spaces $\mathcal{Z}$ and $\mathcal{C}$, strict contractions $Z_1, \dots, Z_N$ in $\mathcal{L}(\mathcal{Z})$ and operators $U_1, \dots, U_N, V_1, \dots, V_N$ in $\mathcal{L}(\mathcal{C},\mathcal{Z})$, determine when there exists a Schur class function s in $\mathcal{S}(\mathbb{C},\mathbb{C})$ such that $s(Z_i)U_i = V_i$ for $i = 1, \dots, N$.*

The **FRD-NP** interpolation problem is the special case of the **RTRD-NP** interpolation problem with $\mathcal{C} = \mathcal{Z}$ and $U_1 = \cdots = U_N = I_{\mathcal{Z}}$. Let us first consider the **RTRD-NP** interpolation problem with $\mathcal{C} = \mathbb{C}$, that is, the operators $U_1, \dots, U_N, V_1, \dots, V_N$ are vectors $u_1, \dots, u_N, v_1, \dots, v_N$ in $\mathcal{Z}$. This type of Nevanlinna-Pick interpolation problem is studied in the book [96] (see also [94, 95]), in a slightly more general form where only a local weak version of the Riesz-Dunford functional calculus is required; in particular, there it is shown how any classical Nevanlinna-Pick Carathéodory-Fejér interpolation problem can be encoded as a (even single-point) **RTRD-NP** problem. In this case $s(Z_i)u_i$ is equal to $(u_i s)^{\wedge L}(Z_i)$ as defined in (2.6), and thus, the **RTRD-NP** interpolation problem is a **LTOA-NP** interpolation problem with the same data set. If $\mathcal{C} \neq \mathbb{C}$ is finite dimensional with orthonormal basis $\{e_1, \dots, e_\kappa\}$, then the **RTRD-NP** interpolation problem can still be seen as a **LTOA-NP** interpolation problem. In this case the identity $s(Z_i)U_i = V_i$ holds if and only if

$$s(Z_i)U_ie_j = V_ie_j \text{ for } j = 1,\dots,\kappa.$$

Thus the **RTRD-NP** interpolation problem becomes a **LTOA-NP** interpolation problem with tangential interpolation conditions indexed by the Cartesian product set

$$\{1,\dots,N\} \times \{1,\dots,\kappa\}$$

and with interpolation data

$$Z_{i,j} := Z_i, \quad x_{i,j} := U_ie_j, \quad y_{i,j} := V_ie_j \text{ for } i = 1,\dots,N \text{ and } j = 1,\dots,\kappa.$$

The **LTRD-NP** interpolation problem reduces to a **RTOA-NP** interpolation problem in a similar way.

In conclusion, we have the following result.

Theorem 2.3. *Let the data for the* **FRD-NP**, **LTRD-NP** *and* **RTRD-NP** *interpolation problems be as given in (1) and (2) above with, for the* **LTRD-NP** *and* **RTRD-NP** *interpolation problems* $\dim \mathcal{C} = \kappa$ *and* $\{e_1,\dots,e_\kappa\}$ *an orthonormal basis for* $\mathcal{C}$, *and for the* **FRD-NP** *interpolation problem* $\dim \mathcal{Z} = \kappa$ *and* $\{e_1,\dots,e_\kappa\}$ *an orthonormal basis for* $\mathcal{Z}$.

1. *A solution to the* **FRD-NP** *interpolation problem exists if and only if the associated Pick matrix*
$$\mathbb{P}_{\text{FRD}} := \left[\ \textstyle\sum_{n=0}^{\infty} Z_i^n(e_{i'}e_{j'}^* - W_ie_{i'}e_{j'}^*W_j^*)Z_j^{*n}\ \right]_{(i,i'),(j,j')\in\{1,\dots,N\}\times\{1,\dots,\kappa\}} \tag{2.11}$$
is positive semidefinite.
2. *A solution to the* **LTRD-NP** *interpolation problem exists if and only if the associated Pick matrix*
$$\mathbb{P}_{\text{LTRD}} := \left[\ \textstyle\sum_{n=0}^{\infty} Z_i^{*n}(X_i^*e_{i'}e_{j'}^*X_j - Y_i^*e_{i'}e_{j'}^*Y_j)Z_j^n\ \right]_{(i,i'),(j,j')\in\{1,\dots,N\}\times\{1,\dots,\kappa\}} \tag{2.12}$$
is positive semidefinite.
3. *A solution to the* **RTRD-NP** *interpolation problem exists if and only if the associated Pick matrix*
$$\mathbb{P}_{\text{RTRD}} := \left[\ \textstyle\sum_{n=0}^{\infty} Z_i^n(U_ie_{i'}e_{j'}^*U_j^* - V_ie_{i'}e_{j'}^*V_j^*)Z_j^{*n}\ \right]_{(i,i'),(j,j')\in\{1,\dots,N\}\times\{1,\dots,\kappa\}} \tag{2.13}$$
is positive semidefinite.

The statements in Theorem 2.3 remain true for the case that $\mathcal{C}$ is a separable Hilbert space (i.e., $\kappa = \infty$). The Pick matrices $\mathbb{P}_{\text{FRD}}$, $\mathbb{P}_{\text{LTRD}}$ and $\mathbb{P}_{\text{RTRD}}$ in this case are infinite operator matrices; positivity is then to be interpreted as positivity of all $M \times M$-finite sections for each $M \in \mathbb{Z}_+$.

We wish to point out that Cohen and Lewkowicz [35, 36] have studied the Full Riesz-Dunford Nevanlinna-Pick interpolation problem in a somewhat different setting (where the unit disk is replaced by the right half-plane and where the Schur class is replaced by positive-real-odd functions – holomorphic functions f on the right half-plane such that (i) $\overline{f(\lambda)} = f(-\overline{\lambda})$ and $f(\lambda) + \overline{f(\lambda)} \geq 0$ for $\lambda + \overline{\lambda} > 0$

and (ii) f has meromorphic pseudocontinuation to the left half-plane such that $\overline{f(-\overline{\lambda})} = f(\lambda)$). Adaption of these results to our setting leads to the following refinement of part (1) of Theorem 2.3. For simplicity we consider only the case where $N = 1$; just as is the case for **LTOA-NP**, this reduction is without loss of generality.

Theorem 2.4. *Suppose that we are given the data set*

$$Z \in \mathcal{L}(\mathcal{Z}), \quad W \in \mathcal{L}(\mathcal{Z})$$

*for a Full Riesz-Dunford Nevanlinna-Pick (***FRD-NP***) interpolation problem with* $N = 1$. *Assume that* $\dim \mathcal{Z} = \kappa < \infty$ *with* $\{e_1, \dots, e_\kappa\}$ *equal to an orthonormal basis for* $\mathcal{Z}$. *Then the following are equivalent:*

1. *A solution to the* **FRD-NP** *interpolation problem exists, i.e., the Pick matrix* $\mathbb{P}_{\mathrm{FRD}}$ *given by* (2.11) (*with* $N = 1$, $Z = Z_1$, $W = W_1$)
$$\mathbb{P}_{\mathrm{FRD}} := \left[\ \sum_{n=0}^{\infty} Z^n(e_i e_j^* - W e_i e_j^* W^*) Z^{*n} \ \right]_{i,j=1,\dots,\kappa} \tag{2.14}$$
is positive semidefinite.
2. *For each vector* $u \in \mathcal{Z}$, *the* **RTRD-NP** *problem*
   find a scalar Schur-class function s so that $s(Z)u = Wu$ (*where a priori, the solution* $s = s_u$ *depends on the vector* u)
has a solution, or, equivalently, for each vector $u \in \mathcal{Z}$ *the Pick matrix* $\mathbb{P}_{\mathrm{FRD},u}$ *given by*
$$\mathbb{P}_{\mathrm{FRD},u} = \sum_{n=0}^{\infty} Z^n(uu^* - Wuu^*W^*)Z^{*n} \tag{2.15}$$
is positive-semidefinite.
3. *The Pick matrix* $\mathbb{P}_{\mathrm{FRD},u}$ *is positive-semidefinite for each vector* $u \in \mathcal{Z}$ *satisfying the additional constraint*
$$u = \arg\max_{u'} \operatorname{rank} \begin{bmatrix} u' & Zu' & \cdots & Z^{n-1}u' \end{bmatrix}, \tag{2.16}$$
i.e., for any maximally Z-controllable vector u.
4. *The matrix* W *is in the double commutant* $\{Z\}''$ *of* Z *and the Pick matrix* $\mathbb{P}_{\mathrm{FRD},u}$ (2.15) *is positive-semidefinite for some maximally* Z*-controllable vector* u (*i.e., a vector* u *satisfying* (2.16)).

Proof. Note first that the statements in condition (1) are equivalent as an application of part (1) of Theorem 2.3 and the statements in condition (2) are equivalent as an application of part (3) of Theorem 2.3. We show (1) $\Longrightarrow$ (2) $\Longleftrightarrow$ (3) and (2) $\Longrightarrow$ (4) $\Longrightarrow$ (1).

(1) $\Longrightarrow$ **(2):** Note that positive-semidefiniteness of $\mathbb{P}_{FRD}$ in (2.14) can equivalently be expressed as

$$\sum_{i,j=1}^{\kappa} u_i^* \left[\sum_{n=0}^{\infty} Z^n(e_i e_j^* - W e_i e_j^* W^*) Z^{*n} \right] u_j \geq 0 \text{ for all } u_1, \dots, u_\kappa \in \mathcal{Z}. \tag{2.17}$$

In particular, if we take u_i to be of the form $u_i = c_i u_0$ where $c_1, \dots, c_\kappa$ are scalars and $u_0 \in \mathbb{C}^\kappa$ is a fixed vector, condition (2.17) becomes

$$u_0^* \mathbb{P}_{FRD,u} u_0 \geq 0$$

where we have set $u = \sum_{i=1}^{\kappa} c_i e_i$. As u and u_0 are arbitrary, we see that we have verified condition (2).

(2) $\iff$ (3): Note that (2) $\implies$ (3) is trivial. For the converse, one can use the fact that the set of maximal Z-controllability-rank vectors u form a dense set in the set of all vectors in $\mathcal{Z}$ – for details we refer to [36].

(2) $\implies$ (4): Note that the second part of (4) follows from (2) trivially. It remains to show that (2) implies that $W \in \{Z\}''$. If we assume condition (1) of the theorem so that $W = s(Z)$ for some scalar Schur-class s, it is then trivial that $W \in \{Z\}''$; the issue is to prove that $W \in \{Z\}''$ under the seemingly weaker hypothesis (2). The following proof is adapted from a similar argument in [35] – see Remark 2.6 below for additional discussion.

We first observe that positive-semidefiniteness of the matrix $\mathbb{P}_{\mathrm{FRD},u}$ for all u is invariant under similarity transformations

$$(Z, W) \mapsto (T^{-1}ZT, T^{-1}WT).$$

More precisely, if we use the notation $\mathbb{P}_{\mathrm{FRD},u}(Z, W)$ to indicate the dependence of the matrix in (2.15) on Z, W, it is easily verified that

$$\mathbb{P}_{\mathrm{FRD},u}(T^{-1}ZT, T^{-1}WT) = T^{-1}\left(\mathbb{P}_{\mathrm{FRD},T^{-1}u}(Z, W)\right) T^{-1*} \tag{2.18}$$

from which the assertion follows.

We show next that, under hypothesis (2),

$$\textit{any } Z\textit{-invariant subspace of } \mathcal{Z} \textit{ is also } W\textit{-invariant.} \tag{2.19}$$

Indeed, by an appropriate choice of basis in $\mathcal{Z}$, we may assume that Z has the upper triangular form $Z = \left[\begin{smallmatrix} Z_{11} & Z_{12} \\ 0 & Z_{22} \end{smallmatrix}\right]$ and W is given in block 2×2 form $W = \left[\begin{smallmatrix} W_{11} & W_{12} \\ W_{21} & W_{22} \end{smallmatrix}\right]$ with respect to the same basis. Our goal is to show that $W_{21} = 0$. Toward this goal, introduce the similarity transformation $T_\epsilon = \left[\begin{smallmatrix} I & 0 \\ 0 & \epsilon I \end{smallmatrix}\right]$. By the claim in the preceding paragraph, we may replace (Z, W) by (Z_ϵ, W_ϵ) where

$$Z_\epsilon := T_\epsilon^{-1} Z T_\epsilon = \left[\begin{smallmatrix} Z_{11} & \epsilon Z_{12} \\ 0 & Z_{22} \end{smallmatrix}\right], \quad W_\epsilon := \left[\begin{smallmatrix} W_{11} & \epsilon W_{12} \\ \frac{1}{\epsilon} W_{21} & W_{22} \end{smallmatrix}\right].$$

The hypothesis (2) tells us that the matrix

$$\sum_{n=0}^{\infty} Z_\epsilon^n \left(uu^* - W_\epsilon uu^* W_\epsilon^*\right) Z_\epsilon^{*n} \geq 0 \text{ for all } \epsilon > 0. \tag{2.20}$$

As we let ϵ tend to zero, the expression (2.20) has the same asymptotics as

$$\sum_{n=0}^{\infty} Z^n (uu^* - W_\epsilon uu^* W_\epsilon^*) Z^{*n}. \tag{2.21}$$

If we take the vector u to have the form $u = \left[\begin{smallmatrix} u_1 \\ 0 \end{smallmatrix}\right]$, then the (2, 2)-block of the second term in the $n = 0$ term in (2.21) in turn has the form

$$-\begin{bmatrix} \frac{1}{\epsilon}W_{21} & W_{22} \end{bmatrix} \begin{bmatrix} u_1 \\ 0 \end{bmatrix} \begin{bmatrix} u_1^* & 0 \end{bmatrix} \begin{bmatrix} \frac{1}{\epsilon}W_{21}^* \\ W_{22}^* \end{bmatrix} = -\frac{1}{\epsilon^2} W_{21} u_1 u_1^* W_{21}^*. \tag{2.22}$$

Since the vector u_1 is arbitrary, it is clear from (2.22) that it is not possible for the quantity in (2.20) to be positive-semidefinite for all $\epsilon > 0$ if $W_{21} \neq 0$. We therefore conclude that we must have $W_{21} = 0$ and (2.19) follows.

As a consequence of (2.19), if we choose a basis for $\mathcal{Z}$ to put Z in upper triangular Jordan form, say

$$Z = \begin{bmatrix} J_{\lambda_1}^{(n_1)} & & \\ & \ddots & \\ & & J_{\lambda_k}^{(n_k)} \end{bmatrix} \text{ where } J_{\lambda_i}^{(n_i)} = \begin{bmatrix} \lambda_i & 1 & & \\ & \ddots & \ddots & \\ & & \ddots & 1 \\ & & & \lambda_i \end{bmatrix} \text{ (of size } n_i \times n_i),$$

then it follows that $W = \begin{bmatrix} W_{11} & & \\ & \ddots & \\ & & W_{kk} \end{bmatrix}$ is block diagonal with each block-diagonal entry at least in upper triangular form when expressed as a matrix in this same basis. To show that W is in the double commutant of Z, we must show that each such diagonal entry W_{ii} is in fact Toeplitz with repeated diagonal partial blocks corresponding to repeated partial Jordan blocks in $\mathcal{Z}$. As a first step we argue that each diagonal block W_{ii} is upper triangular Toeplitz.

From the fact that the pair (Z, W) satisfies condition (2) in the statement of the theorem, it follows that the pair of diagonal blocks $(J_{\lambda_i}^{(n_i)}, W_{ii})$ does as well; indeed, this can be seen simply by restricting the vectors u to be of the form $u = \begin{bmatrix} 0 & \dots & u_i^\top & \dots & 0 \end{bmatrix}^\top$. It therefore suffices to assume that the original pair (Z, W) satisfies (2) with Z given in Jordan form $Z = J_\lambda^{(n)}$. Write $Z = \lambda I + N$ where N is the nilpotent Jordan cell. Set $T_t = I - tN$ where t is a real parameter, so $T_t \in \{Z\}'$. If we define $(Z_t, W_t) = (T_t^{-1} Z T_t, T_t^{-1} W T_t)$. Then $Z_t = Z$ for all t and a simple computation gives

$$W_t = W + tC$$

where

$$C = (N + \cdots + t^{n-2}N^{n-1})W - WN - (tN + \cdots + t^{n-1}N^{n-1})WN.$$

Note that $C = 0$ exactly when W is Toeplitz. By our assumption (2) combined with (2.18), we know that $\mathbb{P}_{\mathrm{RTRD},u}(Z_t, W_t) \geq 0$ for all $t \in \mathbb{R}$, i.e., the quantity

$$\sum_{n=0}^{\infty} Z^n \left[uu^* - (W + tC)uu^*(W + tC)^*\right] Z^{*n} \tag{2.23}$$

$$= \sum_{n=0}^{\infty} Z^n (uu^* - Wuu^*W^*) Z^{*n} - \sum_{n=0}^{\infty} Z^n \left[tCuu^*W + tWuu^*C + t^2 Cuu^*C^*\right] Z^{*n}$$

is positive semidefinite for all real t and all vectors u. If C is not zero, it is clear from the last term that this is not possible. It follows that we must have $C = 0$ and consequently W is upper triangular Toeplitz as wanted.

It remains to handle the case where the eigenvalues in some of the Jordan blocks may repeat. We therefore suppose that

$$Z = \begin{bmatrix} J_\lambda^{(n_1)} & 0 \\ 0 & J_\lambda^{(n_2)} \end{bmatrix} = \begin{bmatrix} \lambda I_{n_1} + N_1 & 0 \\ 0 & \lambda I_{n_2} + N_2 \end{bmatrix}, \quad W = \begin{bmatrix} W_1 & 0 \\ 0 & W_2 \end{bmatrix}$$

where say $n_1 \geq n_2$ and N_i is the upper triangular $n_i \times n_i$ nilpotent cell for $i = 1, 2$. We let X be the $n_2 \times n_1$ matrix given by

$$X = \begin{bmatrix} I_{n_2} & 0_{n_2 \times (n_1 - n_2)} \end{bmatrix}$$

and note that X has the intertwining property

$$X J_\lambda^{(n_1)} = J_\lambda^{(n_2)} X. \tag{2.24}$$

We introduce the similarity

$$S_\epsilon = \begin{bmatrix} I_{n_1} & 0 \\ X & \epsilon I_{n_2} \end{bmatrix}$$

depending on the real nonzero parameter ϵ. From (2.24) it follows that S_ϵ commutes with Z, so $Z_\epsilon := S_\epsilon^{-1} Z S_\epsilon = Z$ for each t. To show that $W \in \{Z\}''$, it suffices to show that X commutes with W as well. In general we have

$$W_\epsilon := S_\epsilon^{-1} W S_\epsilon = \begin{bmatrix} W_1 & 0 \\ \epsilon^{-1}(W_2 X - X W_1) & W_2 \end{bmatrix}.$$

By our assumption (2) in the statement of the theorem combined with the observation (2.18), we see that we know that the matrix

$$\begin{aligned} &\sum_{n=0}^{\infty} Z^n \left(uu^* - \begin{bmatrix} W_1 & 0 \\ \epsilon^{-1}(W_2X - XW_1) & W_2 \end{bmatrix} uu^* \begin{bmatrix} W_1^* & \epsilon^{-1}(W_2X - XW_1)^* \\ 0 & W_2^* \end{bmatrix} \right) Z^{*n} \\ &= \sum_{n=0}^{\infty} Z^n (uu^* - Wuu^*W^*) Z^{*n} - \epsilon^{-1} \sum_{n=0}^{\infty} Z^n \begin{bmatrix} 0 & 0 \\ W_2X - XW_1 & 0 \end{bmatrix} uu^* W^* Z^{*n} \\ &\quad - \epsilon^{-1} \sum_{n=0}^{\infty} Z^n W uu^* \begin{bmatrix} 0 & (W_2X - XW_1)^* \\ 0 & 0 \end{bmatrix} Z^{*n} \\ &\quad - \epsilon^{-2} \sum_{n=0}^{\infty} Z^n \begin{bmatrix} 0 & 0 \\ W_2X - XW_1 & 0 \end{bmatrix} uu^* \begin{bmatrix} 0 & (W_2X - XW_1)^* \\ 0 & 0 \end{bmatrix} Z^{*n} \end{aligned} \tag{2.25}$$

is positive-semidefinite for all nonzero real numbers ϵ and all vectors u. As the vector u sweeps all possible vectors in $\mathbb{C}^\kappa$, it is clear that (2.25) cannot hold unless the commutator $W_2X - XW_1$ is zero, as was wanted

(4) $\Longrightarrow$ (1): We consider first the case where Z is non-derogatory, i.e., in the Jordan form for Z, there is associated only one Jordan cell with each eigenvalue λ of Z. It is known that Z is non-derogatory if and only if there exists a vector u such that (Z, u) is controllable, i.e., if and only if the controllability matrix appearing in (2.16) has rank equal to the dimension κ of the whole state space for any maximally Z-controllable vector u. We therefore suppose that we are given a vector u such that the pair (Z, u) is controllable and the Pick matrix $\mathbb{P}_{FRD,u}$ is positive semidefinite. Note that $\mathbb{P}_{\mathrm{FRD},u}$ is the same as the matrix $\mathbb{P}_{\mathrm{LTOA}}$ in (2.8) if we take

$$N = 1, \quad T_1 = Z, \quad X_1 = u, \quad Y_1 = Wu.$$

Hence by part (2) of Theorem 2.3, the positive-semidefiniteness of $\mathbb{P}_{FRD,u}$ implies that we can find a scalar Schur-class function s so that $s(Z)u = Wu$. As we are also assuming that W is in the second commutant of Z, we know that W and Z commute (in fact, in the non-derogatory case, the commutant and double commutant of Z are the same). We therefore have

$$s(Z)Z^k u = Z^k s(Z)u = Z^k Wu = WZ^k u$$

from which we get

$$s(Z)\begin{bmatrix} u & Zu & \cdots & Z^{\kappa-1}u \end{bmatrix} = W\begin{bmatrix} u & Zu & \cdots & Z^{\kappa-1}u \end{bmatrix}. \tag{2.26}$$

As the controllability matrix $\begin{bmatrix} u & Zu & \cdots & Z^{\kappa-1}u \end{bmatrix}$ is surjective (and in fact invertible since it is square), from (2.26) we see that s solves the FRD-NP problem $s(Z) = W$, i.e., condition (1) of the theorem holds.

Under the assumption that W is in the double commutant of Z, the derogatory case can be reduced to the non-derogatory case by working with the Jordan canonical form for Z; we refer to [36] for details. □

Remark 2.5. There is also a left-tangential version of Theorem 2.4 where conditions (2), (3) and (4) are replaced by their left versions:

2′. *For each linear functional $v \in \mathcal{L}(\mathcal{Z}, \mathbb{C})$, the* **LTRD-NP** *problem*

find a scalar Schur-class function s so that $vs(Z) = vW$ *(where a priori, the solution $s = s_v$ depends on the linear functional v)*

has a solution, or, equivalently, for each linear functional $v \in \mathcal{L}(\mathcal{Z}, \mathbb{C})$ the Pick matrix $\mathbb{P}_{v,\mathrm{FRD}}$ given by

$$\mathbb{P}_{v,\mathrm{FRD}} = \sum_{n=0}^{\infty} Z^{*n}(v^*v - W^*v^*vW)Z \tag{2.27}$$

is positive-semidefinite.

3′. *The Pick matrix $\mathbb{P}_{v,\mathrm{FRD}}$ is positive-semidefinite for each linear functional $v \in \mathcal{L}(\mathcal{Z}, \mathbb{C})$ satisfying the additional constraint*

$$v = \arg\max_{v'} \operatorname{rank} \begin{bmatrix} v' \\ v'Z \\ \vdots \\ v'Z^{n-1} \end{bmatrix}, \tag{2.28}$$

i.e., for any maximally Z-controllable linear functional v.

4′. *The operator W is in the double commutant $\{Z\}''$ of Z and the Pick matrix $\mathbb{P}_{v,\mathrm{FRD}}$ (2.27) is positive-semidefinite for some maximally Z-controllable linear functional v (i.e., a linear functional v satisfying (2.28)).*

Note that condition (1) in Theorem 2.4 implies condition (2′) trivially. The remaining implications (2′) $\Longleftrightarrow$ (3′) and (2′) $\Longrightarrow$ (4′) $\Longrightarrow$ (1) can be proved by paralleling the proof of Theorem 2.4 (with the role of vectors u acting as operators from $\mathbb{C}$ into $\mathcal{Z}$ acting on the right replaced by linear functionals v acting on the left). In this way we see that each of (2′), (3′) and (4′) is also equivalent to any of (1), (2), (3) and (4) in Theorem 2.4.

Remark 2.6. For purposes of this remark we identify $\mathcal{Z}$ with $\mathbb{C}^\kappa$, so operators Z on $\mathcal{Z}$ can be viewed concretely as $\kappa \times \kappa$ complex matrices. The paper [35] is concerned with a "Lyapunov ordering" for real symmetric matrices based on solving Lyapunov inequalities: $ZS + SZ^* \geq 0$. Theorem 2.4 is closely related to the analogue of these results with two adjustments: we work with complex Hermitian rather than real symmetric matrices and we work with Stein inequalities $S - ZSZ^* \geq 0$ rather than Lyapunov inequalities. Following [35], given a matrix Z (say of size $\kappa \times \kappa$) we give a Stein ordering on a pair of complex Hermitian $\kappa \times \kappa$ matrices Z, W as follows: $Z \leq_{\mathfrak{S}} W$ *if* $S - WSW^* \geq 0$ *whenever* $S - ZSZ^* \geq 0$. As we are assuming that Z has spectral radius less than 1, the operator $\mathcal{S}_Z \colon X \mapsto X - ZXZ^*$ is invertible as an operator on complex Hermitian $\kappa \times \kappa$ matrices with inverse $\mathcal{S}_Z^{-1}$ given explicitly as

$$\mathcal{S}_Z^{-1}(X) = \sum_{n=0}^{\infty} Z^n X Z^n.$$

Then the condition $Z \leq_{\mathfrak{S}} W$ can be expressed in operator-theoretic form as

$$\mathcal{S}_W \circ \mathcal{S}_Z^{-1}(\mathcal{P}_+) \subset \mathcal{P}_+$$

where we use the notation $\mathcal{P}_+$ to denote the positive semidefinite $\kappa \times \kappa$ matrices, or, more explicitly,

$$X \in \mathbb{C}^{\kappa\times\kappa} \text{ with } X \geq 0 \Longrightarrow \mathbb{P}_X := \sum_{n=0}^{\infty} Z^n X Z^{*n} - W\left(\sum_{n=0}^{\infty} Z^n X Z^{*n}\right) W^* \geq 0. \tag{2.29}$$

An equivalent formulation of (2.29) is that, for all positive semidefinite $\kappa \times \kappa$ matrices X and Y we have

$$\begin{aligned} &\operatorname{tr}\left(\left[\sum_{n=0}^{\infty} Z^n X Z^{*n} - W\left(\sum_{n=0}^{\infty} Z^n X Z^{*n}\right) W^*\right] Y\right) \\ &\quad = \operatorname{tr}\left(\sum_{n=0}^{\infty} [Z^{*n}(Y - W^* Y W) Z^n]\, X\right) \geq 0. \end{aligned}$$

or, equivalently,

$$\sum_{n=0}^{\infty}[Z^{*n}(Y - W^*YW)Z^n] \geq 0 \text{ for all } Y \geq 0. \tag{2.30}$$

If we restrict to matrices Y which generate extreme rays in the cone of positive semidefinite matrices, i.e., Y of the form $Y = v^*v$ with v a row vector of size $1 \times \kappa$, then we may rewrite (2.30) as

$$\sum_{n=0}^{\infty}[Z^{*n}(v^*v - W^*v^*vW)Z^n] \geq 0 \text{ for all } v \in \mathbb{C}^{1\times\kappa}$$

which we recognize as condition (2′) in Remark (2.5). Thus the Stein ordering $Z \leq_{\mathfrak{S}} W$ can be interpreted as being able to solve the **LTRD-NP** interpolation problem $vs(Z) = vW$ along every row vector v and thus is equivalent to any of the other conditions listed in Theorem 2.4 and Remark 2.5.

Alternatively, by an argument parallel to the proof of (2) $\Longrightarrow$ (4) in Theorem 2.4, one can see that $Z \leq_{\mathfrak{S}} W$ implies that $W \in \{Z\}''$, so W commutes with Z; in fact this is proved in [35] for the real symmetric case (in which case the result holds with some modifications which can be ignored in the complex case) and is the inspiration for our proof of (2) $\Longrightarrow$ (4) in Theorem 2.4. We note that to check whether $Z \leq_{\mathfrak{S}} W$ it suffices to check whether the matrix $\mathbb{P}_X$ is positive semidefinite for matrices X which generate extreme rays in $\mathcal{P}_+$, i.e., for X of the form uu^* where now we take u to be a column vector in $\mathbb{C}^\kappa$. We use the fact that Z and W commute to get

$$\begin{aligned}\mathbb{P}_{uu^*} - Z\mathbb{P}_{uu^*}Z^* &= \sum_{n=0}^{\infty} Z^n uu^* Z^{*n} - W\left(\sum_{n=0}^{\infty} Z^n uu^* Z^{*n}\right)W^* \\ &\quad - Z\left[\sum_{n=0}^{\infty} Z^n uu^* Z^{*n} - W\left(\sum_{n=0}^{\infty} Z^n uu^* Z^{*n}\right)W^*\right]Z^* \\ &= \sum_{n=0}^{\infty} Z^n uu^* Z^{*n} - \sum_{n=0}^{\infty} Z^n W uu^* W^* Z^{*n} \\ &\quad - \sum_{n=1}^{\infty} Z^n uu^* Z^{*n} - \sum_{n=1}^{\infty} Z^n W uu^* W^* Z^{*n} \\ &= uu^* - Wuu^*W^*.\end{aligned}$$

from which we see that $\mathbb{P}_{uu^*} = \mathbb{P}_{\mathrm{FRD},u}$ as given by (2.15). Thus positive semidefiniteness of $\mathbb{P}_{uu^*}$ for all vectors u is equivalent to the positive semidefiniteness of the matrix $\mathbb{P}_{\mathrm{FRD},u}$ for all vectors $u \in \mathbb{C}^\kappa$, i.e., to condition (2) in Theorem 2.4. Hence, with use of the result that $Z \leq_{\mathfrak{S}} W \Longrightarrow W \in \{Z\}''$, we have a direct verification of the equivalence of $Z \leq_{\mathfrak{S}} W$ with being able to solve the **RTRD-NP** problem $s(Z)u = Wu$ along every column vector u.

Via either of these approaches we see that we may add a fifth statement to the list of equivalent statements in Theorem 2.4 and Remark 2.5:

5. $Z \leq_{\mathfrak{S}} W$ *in the sense of Cohen-Lewkowicz* (as adapted to complex Stein inequalities as done here).

Remark 2.7. As observed in the proof of (1) $\implies$ (2), the positive semidefiniteness of the block matrix (2.14) is equivalent to condition (2.17) while (2.15) is equivalent to (2.17) holding for the vectors $u_1, \dots, u_\kappa$ of the special form

$$u_i = c_i u_0 \text{ for some scalars } c_1, \dots, c_\kappa \text{ and some fixed but arbitrary vector } u_0. \tag{2.31}$$

Since (1) and (2) in Theorem 2.4 are equivalent, we see that, conversely, the validity of (2.17) for all $u_1, \dots, u_\kappa$ of the special form (2.31) implies the validity of (2.17) for all $u_1, \dots, u_\kappa$. The following simple example shows that the reverse implication fails in general for block matrices $\mathbb{P}$ not coming from a Riesz-Dunford interpolation problem. Indeed, take $\kappa = 2$, $\mathcal{Z} = \mathbb{C}^2$ and $\mathbb{P} = \left[\begin{smallmatrix} \mathbb{P}_{11} & \mathbb{P}_{12} \\ \mathbb{P}_{21} & \mathbb{P}_{22} \end{smallmatrix}\right]$ where

$$\mathbb{P}_{11} = \begin{bmatrix} 1 & 0 \\ 0 & 0 \end{bmatrix}, \quad \mathbb{P}_{12} = \begin{bmatrix} 0 & \frac{1}{2} \\ \frac{1}{2} & 0 \end{bmatrix} = \mathbb{P}_{21}, \quad \mathbb{P}_{22} = \begin{bmatrix} 0 & 0 \\ 0 & 1 \end{bmatrix}.$$

If we choose $u_1 = \left[\begin{smallmatrix} 1 \\ 0 \end{smallmatrix}\right]$, $u_2 = \left[\begin{smallmatrix} -1 \\ 0 \end{smallmatrix}\right]$, we get $\left[\begin{smallmatrix} u_1^* & u_2^* \end{smallmatrix}\right] \mathbb{P} \left[\begin{smallmatrix} u_1 \\ u_2 \end{smallmatrix}\right] = -1$ and hence $\mathbb{P}$ is not positive definite. Nevertheless, if we restrict to u_1, u_2 of the special form (2.31), we have

$$\begin{aligned} \begin{bmatrix} \overline{c_1} u_0^* & \overline{c_2} u_0^* \end{bmatrix} \mathbb{P} \begin{bmatrix} c_1 u_0 \\ c_2 u_0 \end{bmatrix} &= u_0^* \begin{bmatrix} |c_1|^2 & \operatorname{Re} c_1 \overline{c_2} \\ \operatorname{Re} c_1 \overline{c_2} & |c_2|^2 \end{bmatrix} u_0 \\ &= \frac{1}{2} \left(\| \begin{bmatrix} \overline{c_1} & \overline{c_2} \end{bmatrix} u_0 \|^2 + \| \begin{bmatrix} c_1 & c_2 \end{bmatrix} u_0 \|^2 \right) \geq 0. \end{aligned}$$

It is a remarkable fact that (2.17) holding for $u_1, \dots, u_\kappa$ of the special form (2.31) implies that (2.17) holds for arbitrary $u_1, \dots, u_\kappa$ if $\mathbb{P}$ has the special form $\mathbb{P}_{\mathrm{FRD}}$ for some **FRD-NP** problem.

Besides the standard definition of the Schur class $\mathcal{S}(\mathcal{U}, \mathcal{Y})$ used in this section there are various other ways to define $\mathcal{S}(\mathcal{U}, \mathcal{Y})$. One of these definitions is: *A function $S : \mathbb{D} \to \mathcal{L}(\mathcal{U}, \mathcal{Y})$ is in the Schur class $\mathcal{S}(\mathcal{U}, \mathcal{Y})$ in case S defines a contractive multiplication operator M_S from the Hardy space $H^2_{\mathcal{U}}(\mathbb{D})$ into the Hardy space $H^2_{\mathcal{Y}}(\mathbb{D})$ via*

$$M_S : f(\lambda) = S(\lambda) f(\lambda) \quad (\lambda \in \mathbb{D}).$$

We also remark that the Hardy space $H^2_{\mathcal{U}}(\mathbb{D})$ is a reproducing kernel Hilbert space, whose reproducing kernel is the classical Szegő kernel

$$k(\lambda, \zeta) = \frac{1}{1 - \lambda \overline{\zeta}} \quad (\lambda, \zeta \in \mathbb{D}).$$

Another characterization of Schur class functions in $\mathcal{S}(\mathcal{U}, \mathcal{Y})$ is based on an associated kernel function: *A function $S : \mathbb{D} \to \mathcal{L}(\mathcal{U}, \mathcal{Y})$ is in the Schur class $\mathcal{S}(\mathcal{U}, \mathcal{Y})$ in case the function $k_S : \mathbb{D} \times \mathbb{D} \to \mathcal{L}(\mathcal{Y})$ given by*

$$k_S(\lambda, \zeta) = \frac{I_{\mathcal{Y}} - S(\lambda) S(\zeta)^*}{1 - \lambda \overline{\zeta}} \quad (\lambda, \zeta \in \mathbb{D})$$

is a positive kernel in the sense of Aronszajn [12] *adapted to the operator-valued setting: For any finite collection of points* $\omega_1, \dots, \omega_M \in \mathbb{D}$ *and vectors* $y_1, \dots, y_M \in \mathcal{Y}$ *it holds that*

$$\sum_{i,j=1}^{M} \langle k_S(\omega_i, \omega_j) y_j, y_i \rangle \geq 0.$$

The extensions of Nevanlinna-Pick interpolation theory for functions of several variables that we consider in the following sections are based on Schur-class functions defined as "contractive multipliers" or via "associated positive kernels", rather than as functions having contractive values as in (2.1).

3. Operator-valued Nevanlinna-Pick interpolation: the multivariable case

In this section we consider Nevanlinna-Pick interpolation problems for Schur-class function of several variables. The notion of Schur class function in these cases generalizes the definition via "contractive multipliers" or "positive kernels" mentioned at the end of Section 2, rather that the standard definition via "contractive values" (2.1) as given in Section 2.

3.1. The commutative unit ball setting

A much studied multivariable analogue of the classical Szegö kernel is the Drury-Arveson kernel k_d on the unit ball $\mathbb{B}^d = \{\lambda = (\lambda_1, \dots, \lambda_d) \in \mathbb{C}^d \colon \sum_{i=1}^d |\lambda_i|^2 < 1\}$ given by

$$k_d(\lambda, \zeta) = \frac{1}{1 - \langle \lambda, \zeta \rangle} = \frac{1}{1 - \lambda_1 \overline{\zeta}_1 - \cdots - \lambda_d \overline{\zeta}_d} \quad (\lambda, \zeta \in \mathbb{B}^d).$$

The associated reproducing kernel space $\mathcal{H}(k_d)$ is called the *Drury-Arveson space* which is the prototype for a reproducing kernel space with a *complete Pick kernel*; some of the seminal references on this topic are [47, 93, 13, 14, 5, 11, 29, 40, 48, 54, 65, 63, 84].

For a Hilbert space $\mathcal{U}$ we let $\mathcal{H}_{\mathcal{U}}(k_d)$ be the space $\mathcal{H}(k_d) \otimes \mathcal{U}$ of Drury-Arveson space functions with values in $\mathcal{U}$. It can be shown that a holomorphic function $h \colon \mathbb{B}^d \to \mathcal{U}$ with power series representation

$$h(\lambda) = \sum_{n \in \mathbb{Z}_+^d} h_n \lambda^n \quad \text{where } h_n \in \mathcal{U} \text{ for } n \in \mathbb{Z}_+^d$$

is in the vector-valued Drury-Arveson space $\mathcal{H}_{\mathcal{U}}(k_d)$ if and only if

$$\|h\|^2_{\mathcal{H}_{\mathcal{U}}(k_d)} = \sum_{n \in \mathbb{Z}^d} \frac{n!}{|n|!} \|h_n\|^2_{\mathcal{U}} < \infty.$$

Here we use standard multivariable notation: For

$$n = (n_1, \dots, n_d) \in \mathbb{Z}_+^d \quad \text{and} \quad \lambda = (\lambda_1, \dots, \lambda_d) \in \mathbb{B}^d \subset \mathbb{C}^d,$$

we set

$$\lambda^n = \lambda_1^{n_1} \cdots \lambda_d^{n_d}, \quad n! = n_1! \cdots n_d! \quad \text{and} \quad |n| = n_1 + \cdots + n_d.$$

For coefficient Hilbert spaces $\mathcal{U}$ and $\mathcal{Y}$ we define the operator-valued Drury-Arveson Schur-multiplier class $\mathcal{S}_d(\mathcal{U}, \mathcal{Y})$ to be the space of holomorphic functions $S\colon \mathbb{B}^d \to \mathcal{L}(\mathcal{U}, \mathcal{Y})$ such that the multiplication operator

$$M_S\colon f(\lambda) \mapsto S(\lambda)f(\lambda) \quad (\lambda \in \mathbb{B})$$

maps $\mathcal{H}_{\mathcal{U}}(k_d)$ contractively into $\mathcal{H}_{\mathcal{Y}}(k_d)$. We can then pose the Drury-Arveson space versions of the problems formulated in Section 2 by simply replacing the unit disk $\mathbb{D}$ by the unit ball $\mathbb{B}^d$ and the Schur class $\mathcal{S}(\mathcal{U}, \mathcal{Y})$ by the Drury-Arveson Schur-multiplier class $\mathcal{S}_d(\mathcal{U}, \mathcal{Y})$.

All results except the "right-tangential" versions extend in a natural way. The trick to reduce a right-tangential problem to a left-tangential problem via (2.2) fails to generalize to the Drury-Arveson setting; if $S \in \mathcal{S}_d(\mathcal{U}, \mathcal{Y})$ and we set $S^\sharp(\lambda) := S(\overline{\lambda})^* = S(\overline{\lambda_1}, \dots, \overline{\lambda_d})^*$ for each $\lambda \in \mathbb{B}^d$, then it is usually not the case that $S^\sharp$ is in $\mathcal{S}_d(\mathcal{Y}, \mathcal{U})$. Results on right-tangential Nevanlinna-Pick interpolation in the Drury-Arveson space exist (see [18, 19] for a general setting), but they are of the flavor of the results for Nevanlinna-Pick interpolation for the Schur-Agler class to be discussed in Subsection 3.4 below: rather than having a test with a single Pick matrix, the criterion involves being able to solve certain equations for a family of Pick matrices. This makes the theory for $\mathcal{S}_d(\mathcal{U}, \mathcal{Y})$ completely asymmetric with respect to left versus right. We summarize the results in the following theorem, with the more complicated statement for right-tangential interpolation problems omitted.

Theorem 3.1.
(1) Drury-Arveson Full Operator-Valued Nevanlinna-Pick interpolation: *Suppose that we are given points* $\lambda^{(1)}, \dots, \lambda^{(N)}$ *in* $\mathbb{B}^d$ *together with operators* $W_1, \dots, W_N$ *in* $\mathcal{L}(\mathcal{U}, \mathcal{Y})$. *Then there exists an* $S \in \mathcal{S}_d(\mathcal{U}, \mathcal{Y})$ *with* $S(\lambda^{(i)}) = W_i$ *for* $i = 1, \dots, N$ *if and only if the Pick matrix*

$$\mathbb{P}_{\mathrm{FOV}} := \left[\frac{I_{\mathcal{Y}} - W_i W_j^*}{1 - \langle \lambda^{(i)}, \lambda^{(j)} \rangle} \right]_{i,j=1}^N$$

is positive semidefinite.

(2) Drury-Arveson Left-Tangential Nevanlinna-Pick interpolation: *Suppose that we are given an auxiliary Hilbert space* $\mathcal{C}$, *points* $\lambda^{(1)}, \dots, \lambda^{(N)}$ *in* $\mathbb{B}^d$, *operators* $X_1, \dots, X_N$ *in* $\mathcal{L}(\mathcal{Y}, \mathcal{C})$ *and operators* $Y_1, \dots, Y_N$ *in* $\mathcal{L}(\mathcal{U}, \mathcal{C})$. *Then there exists an* $S \in \mathcal{S}_d(\mathcal{U}, \mathcal{Y})$ *such that* $X_i S(\lambda^{(i)}) = Y_i$ *for* $i = 1, \dots, N$ *if and only if the Pick matrix*

$$\mathbb{P}_{\mathrm{LT}} := \left[\frac{X_i X_j^* - Y_i Y_j^*}{1 - \langle \lambda^{(i)}, \lambda^{(j)} \rangle} \right]_{i,j=1}^N$$

is positive semidefinite.

(3) Drury-Arveson Left-Tangential Nevanlinna-Pick interpolation with Operator-Argument: *Suppose that we are given an auxiliary Hilbert space $\mathcal{C}$ together with commutative d-tuples*

$$Z^{(1)} = (Z_1^{(1)}, \ldots, Z_d^{(1)}), \ldots, Z^{(N)} = (Z_1^{(N)}, \ldots, Z_d^{(N)}) \in \mathcal{L}(\mathcal{C})^d,$$

i.e., $Z_k^{(i)} \in \mathcal{L}(\mathcal{C})$ for $i = 1, \ldots, N$ and $k = 1, \ldots, d$ and for each fixed i, the operators $Z_1^{(i)}, \ldots, Z_d^{(i)}$ commute pairwise, with the property that each d-tuple $Z^{(i)}$ has joint spectrum contained in $\mathbb{B}^d$. Assume in addition that we are given operators $X_1, \ldots, X_N$ in $\mathcal{L}(\mathcal{Y}, \mathcal{C})$ and operators $Y_1, \ldots, Y_N$ in $\mathcal{L}(\mathcal{U}, \mathcal{C})$. Then there is an $S \in \mathcal{S}_d(\mathcal{U}, \mathcal{Y})$ so that

$$(X_i S)^{\wedge L}(Z^{(i)}) := \sum_{n \in \mathbb{Z}_+^d} (Z^{(i)})^n X_i S_n = Y_i \quad \text{for} \quad i = 1, \ldots, N$$

if and only if the associated Pick matrix

$$\mathbb{P}_{\mathrm{LTOA}} := \left[\sum_{n \in \mathbb{Z}_+^d} (Z^{(i)})^n (X_i X_j^* - Y_i Y_j^*)(Z^{(j)})^{n*} \right]_{i,j=1}^N$$

is positive semidefinite. Here we use the multivariable notation: $Z^n = Z_1^{n_1} \cdots Z_d^{n_d}$ if $Z = (Z_1, \ldots, Z_d) \in \mathcal{L}(\mathcal{C})^d$ and $n = (n_1, \ldots, n_d) \in \mathbb{Z}_+^d$.

(4) Full Drury-Arveson Riesz-Dunford Nevanlinna Pick interpolation: *Suppose that we are given commutative d-tuples $Z^{(1)}, \ldots, Z^{(N)}$ as in statement (3) above, acting on a separable auxiliary Hilbert space $\mathcal{Z}$ with orthonormal basis $\{e_1, \ldots, e_\kappa\}$ (with possibly $\kappa = \infty$). We also assume that we are given operators $W_1, \ldots, W_N$ in $\mathcal{L}(\mathcal{Z})$. Then there exists a scalar Drury-Arveson Schur class function $s \in \mathcal{S}_d(\mathbb{C}, \mathbb{C})$ so that*

$$s(Z^{(i)}) := \sum_{n \in \mathbb{Z}_+^d} s_n (Z^{(i)})^n = W_i \quad \text{for} \quad i = 1, \ldots, N$$

if and only if the Pick matrix

$$\mathbb{P}_{\mathrm{FRD}} := \left[\sum_{N \in \mathbb{Z}_+^d} (Z^{(i)})^n (e_{i'} e_{j'}^* - W_i e_{i'} e_{j'}^* W_j^*)(Z^{(j)})^{n*} \right]_{(i,i'),(j,j') \in \{1,\ldots,N\} \times \{1,\ldots,\kappa\}}$$

is positive semidefinite.

Proof. Statement (1) **FOV**-interpolation) was obtained in [11] and [40] via the method of descending from the corresponding result for the noncommutative case (see Theorem 3.2 below). Similarly the second and third statements (**LT** and LTOA interpolation) can be obtained as a consequence of the corresponding noncommutative result in [83] and [85]. Alternatively, one can obtain the results directly without reference to the noncommutative theory, as is done in [5, 29, 17]. The fourth statement (**RD**-interpolation) follows from the result on the **LTOA**-problem in the same way as was done for the single-variable case in Section 2. We note also that **FOV** is a special case of **LT** and that **LT** is a special case of **LTOA** just as in the single-variable case. We mention that an analogue of the Rosenblum-Rovnyak theory for the tangential Riesz-Dunford interpolation for this setting appears in [83], again via the connection with the noncommutative theory. □

3.2. The noncommutative unit ball setting

There is also a noncommutative version of the Drury-Arveson Schur class; see [85, 30]. To describe this Schur class, let $\{1,\dots,d\}$ be an alphabet consisting of d letters and let $\mathcal{F}_d$ be the associated free semigroup generated by the letters $1,\dots,d$ consisting of all words γ of the form $\gamma = i_N \cdots i_1$, where each $i_k \in \{1,\dots,d\}$ and where $N = 1,2,\dots$. For $\gamma = i_N \cdots i_1 \in \mathcal{F}_d$ we set $|\gamma| := N$ to be the *length* of the word γ. Multiplication of two words $\gamma = i_N \cdots i_1$ and $\gamma' = j_{N'} \cdots j_1$ is defined via concatenation:

$$\gamma\gamma' = i_N \cdots i_1 j_{N'} \cdots j_1.$$

The empty word $\emptyset$ is included in $\mathcal{F}_d$ and acts as the unit element for this multiplication; by definition $|\emptyset| = 0$. We set

$$\gamma\gamma'^{-1} = \begin{cases} \gamma'' & \text{if there is a } \gamma'' \in \mathcal{F}_d \text{ so that } \gamma = \gamma''\gamma', \\ \text{undefined} & \text{otherwise.} \end{cases}$$

For a Hilbert space $\mathcal{U}$, the associated *Fock space* $\ell^2_{\mathcal{U}}(\mathcal{F}_d)$ is the $\mathcal{U}$-valued ℓ^2 space indexed by the free semigroup $\mathcal{F}_d$:

$$\ell^2_{\mathcal{U}}(\mathcal{F}_d) = \Big\{ u \colon \mathcal{F}_d \to \mathcal{U} \colon \sum_{\gamma \in \mathcal{F}_d} \|u(\gamma)\|^2_{\mathcal{U}} < \infty \Big\}.$$

Given two coefficient Hilbert spaces $\mathcal{U}$ and $\mathcal{Y}$ we define the *noncommutative d-variable Schur class* $\mathcal{S}_{nc,d}(\mathcal{U},\mathcal{Y})$ to be the set consisting of all formal power series $s(z) = \sum_{\gamma \in \mathcal{F}_d} S_\gamma z^\gamma$ in noncommuting indeterminates $z = (z_1,\dots,z_d)$ (where we think of $z^\gamma = z_{i_N} \cdots z_{i_1}$ if $\gamma = i_N \cdots i_1$) with coefficients $S_\gamma \in \mathcal{L}(\mathcal{U},\mathcal{Y})$ such that the associated Toeplitz matrix

$$R = \left[S_{\gamma\gamma'^{-1}} \right]_{\gamma,\gamma' \in \mathcal{F}_d}, \text{ where we set } S_{\text{undefined}} = 0,$$

defines a contraction operator from $\ell^2_{\mathcal{U}}(\mathcal{F}_d)$ into $\ell^2_{\mathcal{Y}}(\mathcal{F}_d)$.

There has been a variety of interpolation results for this setting, for instance Sarason type, Rosenblum-Rovnyak or tangential Riesz-Dunford type and left-tangential operator-argument type; we refer to [83, 40, 11, 85, 86]. As already mentioned, from results for the noncommutative Schur class $\mathcal{S}_{nc,d}(\mathcal{U},\mathcal{Y})$ one can arrive at interpolation results for the Drury-Arveson space as in Theorem 3.1 by abelianizing the variables (see [13, 40, 11, 22]). We state here only the analogues of left-tangential operator-argument Nevanlinna-Pick and full Riesz-Dunford Nevanlinna-Pick interpolation for this noncommutative setting. The right-tangential versions again do not reduce to the left-tangential problems due to the same phenomenon discussed for the commutative case. In the noncommutative setting we distinguish between the *Riesz-Dunford functional calculus* which uses

$$f(Z) := \sum_{\gamma \in \mathcal{F}_d} f_\gamma Z^\gamma$$

where we set $Z^\gamma = Z_{i_N} \cdots Z_{i_1}$ if $\gamma = i_N \cdots i_1$, and the *transposed Riesz-Dunford functional calculus* which uses

$$f^\top(Z) := \sum_{\gamma \in \mathcal{F}_d} f_\gamma Z^{\gamma^\top}$$

where $\gamma^\top = i_1, \dots, i_N$ if $\gamma = i_N \cdots i_1$ and $Z^{\gamma^\top} = Z_{i_1} \cdots Z_{i_N}$ if $Z = (Z_1, \dots, Z_d)$.

Theorem 3.2.
(1) Free-semigroup algebra Left-Tangential Nevanlinna-Pick interpolation with Operator-Argument: *Suppose that we are given a coefficient Hilbert space $\mathcal{C}$ and a collection*

$$Z^{(1)} = (Z_1^{(1)}, \dots, Z_d^{(1)}), \dots, Z^{(N)} = (Z_1^{(N)}, \dots, Z_d^{(N)})$$

of (not necessarily commutative) d-tuples of operators on $\mathcal{C}$ such that, for each fixed $i = 1, \dots, N$, the block row matrix $\left[\, Z_1^{(i)} \cdots Z_d^{(i)} \,\right]$ defines a strict contraction operator from $\mathcal{C}^d = \oplus_{k=1}^d \mathcal{C}$ into $\mathcal{C}$. In addition, suppose that we are given operators $X_1, \dots, X_N$ in $\mathcal{L}(\mathcal{Y}, \mathcal{C})$ and operators $Y_1, \dots, Y_N$ in $\mathcal{L}(\mathcal{U}, \mathcal{C})$. Then there exists a formal power series $S(z) = \sum_{\gamma \in \mathcal{F}_d} S_\gamma z^\gamma$ in the noncommutative Schur class $\mathcal{S}_{nc,d}(\mathcal{U}, \mathcal{Y})$ such that

$$(X_i S)^{\wedge L}(Z^{(i)}) := \sum_{\gamma \in \mathcal{F}_d} (Z^{(i)})^{\gamma^\top} X_i S_\gamma = W_i \quad \textit{for} \quad i = 1, \dots, N$$

if and only if the associated Pick matrix

$$\mathbb{P}_{\mathrm{LTOA}} := \left[\sum_{\gamma \in \mathcal{F}_d} (Z^{(i)})^\gamma (X_i X_j^* - Y_i Y_j^*)(Z^{(j)})^{\gamma *} \right]_{i,j=1}^N$$

is positive semidefinite.

(2) Free-semigroup algebra full transposed Riesz-Dunford Nevanlinna-Pick interpolation: *Suppose that we are given d-tuples of operators $Z^{(1)}, \dots, Z^{(N)}$ as in statement (1) above, acting on a separable auxiliary Hilbert space $\mathcal{Z}$ with orthonormal basis $\{e_1, \dots, e_\kappa\}$ (with possibly $\kappa = \infty$). Suppose also that we are given operators $W_1, \dots, W_N$ in $\mathcal{L}(\mathcal{Z})$. Then there exists a formal power series $s(z) = \sum_{\gamma \in \mathcal{F}_d} s_\gamma z^\gamma$ in the scalar noncommutative Schur class $\mathcal{S}_{nc,d}(\mathbb{C}, \mathbb{C})$ satisfying the transposed Riesz-Dunford interpolation conditions*

$$s^\top(Z^{(i)}) := \sum_{\gamma \in \mathcal{F}_d} s_\gamma (Z^{(i)})^{\gamma^\top} = W_i \quad \textit{for} \quad i = 1, \dots, N \tag{3.1}$$

if and only if the associated Pick matrix

$$\mathbb{P}_{\mathrm{FRD}} := \left[\sum_{\gamma \in \mathcal{F}_d} (Z^{(i)})^{\gamma} (e_{i'} e_{j'}^* - W_i e_{i'} e_{j'}^* W_j^*)(Z^{(j)})^{\gamma *} \right]_{(i,i'),(j,j') \in \{1,\dots,N\} \times \{1,\dots,\kappa\}} \tag{3.2}$$

is positive semidefinite.

(3) Free-semigroup algebra full Riesz-Dunford Nevanlinna-Pick interpolation: *Suppose that we are given d-tuples of operators $Z^{(1)}, \dots, Z^{(N)}$ as in statement* (2) *above acting on a finite-dimensional auxiliary Hilbert space $\mathcal{Z}$ with orthonormal basis $\{e_1, \dots, e_\kappa\}$ along with operators $W_1, \dots, W_N$ in $\mathcal{L}(\mathcal{Z})$. Then there exists a formal power series $s(z) = \sum_{\gamma \in \mathcal{F}_d} s_\gamma z^\gamma$ in the scalar noncommutative Schur class $\mathcal{S}_{nc,d}(\mathbb{C}, \mathbb{C})$ satisfying the Riesz-Dunford-type interpolation conditions*

$$s(Z^{(i)}) := \sum_{\gamma \in \mathcal{F}_d} s_\gamma (Z^{(i)})^\gamma = W_i \quad \text{for} \quad i = 1, \dots, N \tag{3.3}$$

if and only if the associated Pick matrix

$$\mathbb{P}_{\mathrm{FRD}^*} := \left[\sum_{\gamma \in \mathcal{F}_d} (Z^{(i)})^{\gamma *} (e_{i'} e_{j'}^* - W_i^* e_{i'} e_{j'}^* W_j)(Z^{(j)})^{\gamma} \right]_{(i,i'),(j,j') \in \{1,\dots,N\} \times \{1,\dots,\kappa\}} \tag{3.4}$$

is positive semidefinite.

Proof. The first statement can be found in [85] as well as in [20]. A tangential version of the second statement (i.e., tangential Riesz-Dunford interpolation) analogous to the Rosenblum-Rovnyak theory for the single-variable case is given in [83]. The full transpose Riesz-Dunford interpolation result in the second statement follows from the first statement on **LTOA** in the same way as in Section 2 for the single-variable case.

We reduce the third statement to the second as follows. By taking adjoints, we may rewrite (3.1) as

$$\sum_{\gamma \in \mathcal{F}_d} \overline{s}_\gamma (Z^{(i)})^{\gamma^*} = \sum_{\gamma \in \mathcal{F}_d} \overline{s}_\gamma (Z^{(i)*})^{\gamma^\top} = W_i^* \tag{3.5}$$

where we set

$$Z^{(i)*} := (Z_1^{(i)*}, \dots, Z_d^{(i)*}) \text{ for } i = 1, \dots, N.$$

Note that

$$\tau \colon \{h_\gamma\}_{\gamma \in \mathcal{F}_d} \mapsto \{\overline{h}_\gamma\}_{\gamma \in \mathcal{F}_d}$$

is a conjugation (conjugate-linear norm-preserving involution) on $\ell^2(\mathcal{F}_d)$ and, if $R = \left[s_{\gamma\gamma'^{-1}} \right]_{\gamma,\gamma' \in \mathcal{F}_d}$ is the Toeplitz operator associated with the formal power series $s(z) = \sum_{\gamma \in \mathcal{F}_d} s_\gamma z^\gamma$ acting on $\ell^2(\mathcal{F}_d)$, then $\tau \circ R \circ \tau = \overline{R}$ where $\overline{R} = \left[\overline{s}_{\gamma\gamma'^{-1}} \right]_{\gamma,\gamma' \in \mathcal{F}_d}$. We conclude that $s(z) = \sum_{\gamma \in \mathcal{F}_d} s_\gamma z^\gamma$ is in the noncommutative Schur-multiplier class $\mathcal{S}_{nc,d}$ if and only if $\overline{s}(z) = \sum_{\gamma \in \mathcal{F}_d} \overline{s}_\gamma z^\gamma$ is in $\mathcal{S}_{nc,d}$. Thus there is an $s(z) = \sum_{\gamma \in \mathcal{F}_d} s_\gamma z^\gamma$ in $\mathcal{S}_{nc,d}$ satisfying (3.5) if and only if there is another $s \in \mathcal{S}_{nc,d}$ (namely, $\overline{s}$) satisfying

$$s^\top(Z^{(i)*}) := \sum_{\gamma \in \mathcal{F}_d} s_\gamma (Z^{(i)*})^{\gamma^\top} = W_i^* \text{ for } i = 1, \dots, N. \tag{3.6}$$

By Part (2) of Theorem 3.2, this last condition is equivalent to the positive-semidefiniteness of the matrix $\mathbb{P}_{\mathrm{FRD}^*}$ in Part (3). This completes the proof of Theorem 3.2. □

3.3. Nevanlinna-Pick interpolation for Toeplitz algebras associated with directed graphs

This example is discussed in [69, Example 4.3 and pages 52–53] and [70, Section 5]. Here we work out the example, in particular, the nature of the point-evaluation and the Nevanlinna-Pick theorem, in more detail. The papers [66, 68, 101] introduced this class of algebras and studied the uniform closure of the algebra (called the quiver algebra) generated by creation operators. The associated weak-$*$ closed Toeplitz algebra obtained here is also known as a *free semigroupoid algebra*; the papers [60, 61] give concrete representations (some as explicit matrix-function algebras) arising from particular choices of the quiver (i.e., directed graph).

Quivers. Formally a *quiver* G is a quadruple $\{Q_0, Q_1, s, r\}$ that consists of two finite sets Q_0 and Q_1 and two maps s and r from Q_1 to Q_0. We think of the elements of Q_0 as vertices and those of Q_1 as arrows; for any $\alpha \in Q_1$ we think of α as an arrow from $s(\alpha)$ to $r(\alpha)$. It is possible to work with infinite sets Q_0 and Q_1 in which case they need to be equipped with a topology [60, 61], but we shall not consider that case here. With the quiver G we associate the *transposed quiver* $\widetilde{G} = \{Q_0, Q_1, r, s\}$ (with respect to G), i.e., where the source and range maps are interchanged. As a particular example, the reader is invited to take $Q_0 = \{v_0\}$ (i.e., the set of vertices is a singleton) and $Q_1 = \{\alpha_1, \dots, \alpha_d\}$ with $s(\alpha_k) = r(\alpha_k) = v_0$ for $k = 1, \dots, d$. The present example then collapses to the free semigroup algebra example $\mathcal{L}_d$ discussed in Subsection 3.2. The difference between the standard and the transposed Riesz-Dunford functional calculus depends on whether one starts with the quiver G or its reverse $\widetilde{G}$.

Paths. For each nonnegative integer n we write Q_n for the paths of length n and Γ for the collection $\cup_{n=0}^\infty Q_n$ of all finite paths of whatever length. Thus a $\gamma \in Q_n$ is an n-tuple $(\alpha_n, \dots, \alpha_1)$ consisting of arrows $\alpha_k \in Q_1$ with the property that $r(\alpha_k) = s(\alpha_{k+1})$ for $k = 1, \dots, n-1$. In that case we write $s_n(\gamma)$ for $s(\alpha_1)$ and $r_n(\gamma)$ for $r(\alpha_n)$. Note that for $n = 1$ this definition is consistent with that of Q_1 if the elements of Q_1 are seen as paths of length 1. For $n = 0$ we can view the elements of Q_0 as paths of length 0, with for any $v \in Q_0$, $r_0(v) = s_0(v) = v$. If the length n of γ in Γ is not specified, we write simply $s(\gamma)$ and $r(\gamma)$ rather than $s_n(\gamma)$ and $r_n(\gamma)$. The set of paths Γ associated with the quiver G forms a semigroupoid when multiplication is defined via concatenation: For

$$\gamma = (\alpha_n, \dots, \alpha_1), \gamma' = (\alpha'_m, \dots, \alpha'_1) \in \Gamma \text{ with } s(\gamma) = r(\gamma')$$

we set

$$\gamma \cdot \gamma' := (\alpha_n, \dots, \alpha_1, \alpha'_m, \dots, \alpha'_1).$$

Inversion is then given by

$$\gamma \cdot \gamma'^{-1} = \begin{cases} \gamma'' & \text{if } \gamma'' \in \Gamma \text{ so that } \gamma = \gamma'' \cdot \gamma', \\ \text{undefined} & \text{otherwise.} \end{cases} \tag{3.7}$$

The Fock space and Toeplitz algebra associated with the quiver G. For a Hilbert space $\mathcal{U}$ with direct sum decomposition $\mathcal{U} = \oplus_{v \in Q_0} \mathcal{U}_v$ we write, with some abuse of notation, $\ell^2_{\mathcal{U}}(\Gamma)$ for the space $\oplus_{\gamma \in \Gamma} \mathcal{U}_{r(\gamma)}$; the $\mathcal{U}$-valued *Fock space defined by the quiver* G.

Given two coefficient Hilbert spaces $\mathcal{U} = \oplus_{v \in Q_0} \mathcal{U}_v$ and $\mathcal{Y} = \oplus_{v \in Q_0} \mathcal{Y}_v$, the associated Banach space of Toeplitz operators $\mathfrak{L}_\Gamma(\mathcal{U}, \mathcal{Y})$ consists of operators R from $\ell^2_{\mathcal{U}}(\Gamma)$ to $\ell^2_{\mathcal{Y}}(\Gamma)$ and hence can be given in terms of an infinite operator-matrix with rows and columns indexed by Γ:

$$R = [R_{\gamma,\gamma'}]_{\gamma,\gamma' \in \Gamma} \quad \text{with} \quad R_{\gamma,\gamma'} \in \mathcal{L}(\mathcal{U}_{r(\gamma')}, \mathcal{Y}_{r(\gamma)}).$$

The Toeplitz structure for this setting means that the matrix entries $R_{\gamma,\gamma'}$ are completely determined from the particular entries $R_\gamma := R_{\gamma, s(\gamma)}$ according to the rule

$$R_{\gamma,\gamma'} = R_{\gamma\gamma'^{-1}, r(\gamma')}$$

where we take $R_{\text{undefined},v} = 0$ for each $v \in Q_0$.

If $\mathcal{U} = \mathcal{Y}$, then we write $\mathfrak{L}_\Gamma(\mathcal{U})$ instead of $\mathfrak{L}_\Gamma(\mathcal{U}, \mathcal{U})$, and $\mathfrak{L}_\Gamma(\mathcal{U})$ is an algebra that we call the *Toeplitz algebra* associated with G and $\mathcal{U}$. In the special case that $\mathcal{U}_v = \mathcal{Y}_v = \mathbb{C}$ for all $v \in Q_0$, this Toeplitz algebra $\mathfrak{L}_\Gamma$ is otherwise known as the (weak-$*$ closed) *path algebra* corresponding to the quiver G. The algebra $\mathfrak{L}_\Gamma$ also appears as the weak-$*$ closed unital subalgebra of $\mathcal{L}(\ell^2_{\mathbb{C}}(\Gamma))$ generated by the creation operators C_α, for $\alpha \in Q_1$, given by

$$C_\alpha(\oplus_{\gamma \in \Gamma} f_\gamma) = \oplus_{\gamma' \in \Gamma} \tilde{f}_{\gamma'} \text{ where } \tilde{f}_{\gamma'} = \begin{cases} f_\gamma & \text{if } \gamma \in \Gamma \text{ so that } \alpha = \gamma' \cdot \gamma^{-1}, \\ 0 & \text{otherwise,} \end{cases}$$

or, more succinctly, where

$$\tilde{f}_{\gamma'} = f_{\alpha^{-1}\gamma'}$$

and we use the analogue of the convention (3.7) for the case where the inverse path is on the left.

The Schur class associated with the quiver G. Given a quiver G and coefficient spaces $\mathcal{U}$ and $\mathcal{Y}$ we define the noncommutative Schur class $\mathcal{S}_G(\mathcal{U}, \mathcal{Y})$ to be the set of formal power series $S(z) = \sum_{\gamma \in \Gamma} S_\gamma z^\gamma$ in noncommutative indeterminates $z = (z_\alpha \colon \alpha \in Q_1)$ (where $z^\gamma = z_{\alpha_n} \cdots z_{\alpha_1}$ in case $\gamma = (\alpha_n, \dots, \alpha_1) \in \Gamma$) such that the sequence of Taylor coefficients $S_\gamma \in \mathcal{L}(\mathcal{U}_{s(\gamma)}, \mathcal{Y}_{r(\gamma)})$ defines an element $S = [\ S_{\gamma,\gamma'}\ ]_{\gamma,\gamma' \in \Gamma}$ (with $S_{\gamma,\gamma'} = S_{\gamma\gamma'^{-1}}$ and $S_{\text{undefined}} = 0$) in $\mathfrak{L}_\Gamma(\mathcal{U}, \mathcal{Y})$ of norm at most one. One can also define a commutative analog of the Schur class $\mathcal{S}_G(\mathcal{U}, \mathcal{Y})$, but we will not develop this here.

Now assume we are also given an auxiliary Hilbert space $\mathcal{Z}$ with direct sum decomposition $\mathcal{Z} = \oplus_{v\in Q_0}\mathcal{Z}_v$. We then define the generalized unit disc $\mathbb{D}_{G,\mathcal{Z}}$ associated with $\mathcal{Z}$ and the quiver G to be the set of tuples of operators of the form $(Z_\alpha \in \mathcal{L}(\mathcal{Z}_{s(\alpha)}, \mathcal{Z}_{r(\alpha)})\colon \alpha \in Q_1)$ such that for each $v \in Q_0$ the row matrix formed by all Z_α with $r(\alpha) = v$ is a strict contraction:

$$Z_v := \left[\begin{array}{c} Z_\alpha \end{array}\right]_{\substack{\alpha\in Q_1 \\ r(\alpha)=v}} : \oplus_{\substack{\alpha\in Q_1 \\ r(\alpha)=v}} \mathcal{Z}_{s(\alpha)} \to \mathcal{Z}_v \text{ satisfies } \|Z_v\| < 1.$$

In other words, the operator matrix

$$\left[\begin{array}{c} Z_{v,\alpha} \end{array}\right]_{\substack{v\in Q_0 \\ \alpha\in Q_1}} : \oplus_{\alpha\in Q_1}\mathcal{Z}_{s(\alpha)} \to \mathcal{Z} \text{ with } Z_{v,\alpha} = \begin{cases} Z_\alpha & \text{if } v = r(\alpha), \\ 0 & \text{otherwise,} \end{cases}$$

is a strict contraction. For $Z \in \mathbb{D}_{G,\mathcal{Z}}$ given by the tuple $(Z_\alpha\colon \alpha \in Q_1)$ and $\gamma = (\alpha_n, \ldots, \alpha_1) \in \Gamma$ we introduce the notation Z^γ for the operator

$$Z^\gamma = Z_{\alpha_n} \cdots Z_{\alpha_1} : \mathcal{Z}_{s_n(\gamma)} \to \mathcal{Z}_{r_n(\gamma)}. \tag{3.8}$$

In the sequel we shall use the abbreviations: $\mathcal{R}_v = \mathcal{U}_v \otimes \mathcal{Z}_v$ and $\mathcal{Q}_v = \mathcal{Y}_v \otimes \mathcal{Z}_v$ for each $v \in Q_0$, and $\mathcal{R} = \oplus_{v\in Q_0}\mathcal{R}_v$ and $\mathcal{Q} = \oplus_{v\in Q_0}\mathcal{Q}_v$.

Given a Schur class function S with Taylor coefficients $\{S_\gamma\colon \gamma \in \Gamma\}$ and $Z \in \mathbb{D}_{G,\mathcal{Z}}$ we define the value of S at Z to be given by the tensor functional-calculus:

$$S(Z) = \sum_{\gamma\in\Gamma} i_{\mathcal{Q}_{r(\gamma)}}(S_\gamma \otimes Z^\gamma) i^*_{\mathcal{R}_{s(\gamma)}} \in \mathcal{L}(\mathcal{R}, \mathcal{Q}). \tag{3.9}$$

Here we use the standard notation that for a subspace $\mathcal{V}$ of a Hilbert space $\mathcal{W}$ we write $i_\mathcal{V}$ for the canonical embedding of $\mathcal{V}$ into $\mathcal{W}$.

With respect to this tensor-product point-evaluation we consider what we will call the *Quiver Left-Tangential Tensor functional calculus Nevanlinna-Pick interpolation problem* (**QLTT-NP**): *Given a data set*

$$\mathfrak{D}: \ Z^{(1)}, \ldots, Z^{(N)} \in \mathbb{D}_{G,\mathcal{Z}}, \ X_1, \ldots, X_N \in \mathcal{L}(\mathcal{Q}, \mathcal{C}), \ Y_1, \ldots, Y_N \in \mathcal{L}(\mathcal{R}, \mathcal{C}),$$

where $\mathcal{C}$ is an auxiliary Hilbert space, determine when there exists a Schur class function S in $\mathcal{S}_G(\mathcal{U}, \mathcal{Y})$ that satisfies

$$X_i S(Z^{(i)}) = Y_i \quad \textit{for} \quad i = 1, \ldots, N.$$

The Riesz-Dunford functional calculus for this setting is then just the tensor functional calculus for the special case that the coefficient spaces $\mathcal{U}$ and $\mathcal{Y}$ are both equal to $\oplus_{v\in Q_0}\mathbb{C}$. Hence $\mathcal{R} = \mathcal{Q} = \mathcal{Z}$ and for $Z \in \mathbb{D}_{G,\mathcal{Z}}$ and a Schur class function s in $\mathcal{S}_G$ with Taylor coefficients $\{s_\gamma\colon \gamma \in \Gamma\}$ the value of s at Z is given by

$$s(Z) = \sum_{\gamma\in\Gamma} s_\gamma \cdot i_{\mathcal{Z}_{r(\gamma)}} Z^\gamma i^*_{\mathcal{Z}_{s(\gamma)}} \in \mathcal{L}(\mathcal{Z}).$$

The corresponding Nevanlinna-Pick problem is referred to as the *Quiver Left-Tangential Riesz-Dunford Nevanlinna-Pick interpolation problem* (**QLTRD-NP**).

We can also define an operator-argument functional-calculus for the Schur class $\mathcal{S}_G(\mathcal{U},\mathcal{Y})$. For this purpose, assume we have another Hilbert space $\mathcal{X}$, again admitting an orthogonal direct sum decomposition of the form $\mathcal{X} = \oplus_{v\in Q_0}\mathcal{X}_v$. The points in this case come from the generalized disk $\mathbb{D}_{\widetilde{G},\mathcal{X}}$, where $\widetilde{G}$ is the transposed quiver of G. Thus a $T \in \mathbb{D}_{\widetilde{G},\mathcal{X}}$ corresponds to a tuple $(T_\alpha \in \mathcal{L}(\mathcal{X}_{r(\alpha)},\mathcal{X}_{s(\alpha)})\colon \alpha \in Q_1)$ with the property that the operator matrix

$$\left[\ T_{v,\alpha}\ \right]_{\substack{v\in Q_0\\ \alpha\in Q_1}} : \oplus_{\alpha\in Q_1}\mathcal{X}_{r(\alpha)} \to \mathcal{X} \text{ with } T_{v,\alpha} = \begin{cases} T_\alpha & \text{if } v = s(\alpha),\\ 0 & \text{otherwise}\end{cases}$$

is a strict contraction. Given a $T = (T_\alpha\colon \alpha\in Q_1) \in \mathbb{D}_{\widetilde{G},\mathcal{X}}$ and $\gamma = (\alpha_n,\dots,\alpha_1) \in \Gamma$, then $\gamma^\top := (\alpha_1,\dots,\alpha_n) \in \widetilde{\Gamma}$ and we set

$$T^{\gamma^\top} = T_{\alpha_1}\cdots T_{\alpha_n} : \mathcal{X}_{r(\gamma)} \to \mathcal{X}_{s(\gamma)}. \tag{3.10}$$

Then for a Schur class function $S \in \mathcal{S}_G(\mathcal{U},\mathcal{Y})$, a $T \in \mathbb{D}_{\widetilde{G},\mathcal{X}}$ and a block diagonal operator $X = \operatorname{diag}_{v\in Q_0}(X_v)$, with $X_v \in \mathcal{L}(\mathcal{Y}_v,\mathcal{X}_v)$, we define the left-tangential operator-argument point-evaluation $(XS)^{\wedge L}(T)$ by

$$(XS)^{\wedge L}(T) = \sum_{\gamma\in\Gamma} i_{\mathcal{X}_{s(\gamma)}} T^{\gamma^\top} X_{r(\gamma)} S_\gamma i^*_{\mathcal{U}_{s(\gamma)}} = \sum_{\widetilde{\gamma}\in\widetilde{\Gamma}} i_{\mathcal{X}_{r(\widetilde{\gamma})}} T^{\widetilde{\gamma}} X_{s(\widetilde{\gamma})} S_{\widetilde{\gamma}^\top} i^*_{\mathcal{U}_{r(\widetilde{\gamma})}} \in \mathcal{L}(\mathcal{U},\mathcal{X}).$$

Notice that $(XS)^{\wedge L}(T)$ is a block diagonal operator in $\mathcal{L}(\mathcal{U},\mathcal{X})$, that is, the operator $(XS)^{\wedge L}(T)$ maps $\mathcal{U}_v$ into $\mathcal{X}_v$ for each $v \in Q_0$.

We then consider the *quiver Left-Tangential Nevanlinna-Pick interpolation problem with Operator Argument* (**QLTOA-NP**): *Given a data set*

$$T^{(i)} \in \mathbb{D}_{\widetilde{G},\mathcal{X}},\ X^{(i)} = \operatorname{diag}_{v\in Q_0}(X_v^{(i)}),\ Y^{(i)} = \operatorname{diag}_{v\in Q_0}(Y_v^{(i)}),\ \textit{for } i = 1,\dots,N,$$

where $X_v^{(i)} \in \mathcal{L}(\mathcal{Y}_v,\mathcal{X}_v)$ *and* $Y_v^{(i)} \in \mathcal{L}(\mathcal{U}_v,\mathcal{X}_v)$, *determine when there exists a Schur class function* S *in* $\mathcal{S}_G(\mathcal{U},\mathcal{Y})$ *that satisfies*

$$(X^{(i)}S)^{\wedge L}(T^{(i)}) = Y^{(i)} \quad \textit{for} \quad i = 1,\dots,N.$$

The solutions to these interpolation problems are given in the following theorem. The proofs of these statements are given in Subsection 4.9 below.

Theorem 3.3. *Let the data for the* **QLTT-NP**, **QLTRD-NP** *and* **QLTOA-NP** *problems be as given above.*

1. *Assume that* $\mathcal{Z}$ *is separable, and that* $\{e_1^{(v)},\dots,e_{\kappa_v}^{(v)}\}$ *is an orthonormal basis for* $\mathcal{Z}_v$ *for each* $v \in Q_0$ *(with possibly* $\kappa_v = \infty$*). Then a solution to the* **QLTT-NP** *interpolation problem exists if and only if for each* $v \in Q_0$ *the associated Pick matrix* $\mathbb{P}^{(v)}_{QLTT} \in \mathcal{L}(\mathcal{C})^{\kappa_v N\times\kappa_v N}$ *given by*

$$\left[\sum_{\gamma\in\Gamma, s(\gamma)=v} \begin{matrix} X_i i_{\mathcal{Q}_{r(\gamma)}}\left(I_{\mathcal{Y}_{r(\gamma)}} \otimes (Z^{(i)})^\gamma e_{i'}^{(v)} e_{j'}^{(v)*} (Z^{(j)})^{\gamma*}\right) i^*_{\mathcal{Q}_{r(\gamma)}} X_j^* + \\ -Y_i i_{\mathcal{Q}_{r(\gamma)}}\left(I_{\mathcal{U}_{r(\gamma)}} \otimes (Z^{(i)})^\gamma e_{i'}^{(v)} e_{j'}^{(v)*} (Z^{(j)})^{\gamma*}\right) i^*_{\mathcal{Q}_{r(\gamma)}} Y_j^* \end{matrix}\right]_{(i,i'),(j,j')}$$

where (i,i') and (j,j') range over $\{1,\dots,N\}\times\{1,\dots,\kappa_v\}$, is positive semidefinite.

2. *Assume that $\mathcal{C}$ is separable, and that $\{e_1,\dots,e_\kappa\}$ is an orthonormal basis for $\mathcal{C}$ (with possibly $\kappa=\infty$). Then a solution to the* **QLTRD-NP** *interpolation problem exists if and only if the Pick matrix $\mathbb{P}_{\text{QLTRD}}\in\mathcal{L}(\mathcal{Z})^{\kappa N\times\kappa N}$, for which the entry corresponding to the pairs $(i,i'),(j,j',)\in\{1,\dots,N\}\times\{1,\dots,\kappa\}$ is given by*

$$\left[\ \mathbb{P}_{\text{QLTRD}}\ \right]_{(i,i'),(j,j')} = \sum_{\gamma\in\Gamma} i_{\mathcal{Z}_{s(\gamma)}}(Z^{(i)})^{\gamma*}i^*_{\mathcal{Z}_{r(\gamma)}}(X_i^*e_{i'}e_{j'}^*X_j - Y_i^*e_{i'}e_{j'}^*Y_j)i_{\mathcal{Z}_{r(\gamma)}}(Z^{(j)})^{\gamma}i^*_{\mathcal{Z}_{s(\gamma)}}$$

is positive semidefinite.

3. *A solution to the* **QLTOA-NP** *interpolation problem exists if and only if the Pick matrix*

$$\mathbb{P}_{\text{QLTOA}} = \left[\ \sum_{\gamma\in\Gamma} i_{\mathcal{X}_{s(\gamma)}}(T^{(i)})^{\gamma}(X^{(i)}_{r(\gamma)}X^{(j)*}_{r(\gamma)} - Y^{(i)}_{r(\gamma)}Y^{(j)*}_{r(\gamma)})(T^{(j)})^{\gamma*}i^*_{\mathcal{X}_{s(\gamma)}}\ \right]_{i,j=1}^{N}$$

is positive semidefinite.

We conclude this subsection with an example for a concrete quiver; an example also considered in [60]. Let $G=\{Q_0,Q_1,s,r\}$ be the quiver with two vertices $Q_0=\{a,b\}$, two arrows $Q_1=\{\alpha,\beta\}$, and source and range map given by

$$s(\alpha)=r(\alpha)=a,\quad s(\beta)=a\quad\text{and}\quad r(\beta)=b.$$

With the vertices a and b we associate Hilbert spaces $\mathcal{A}$ and $\mathcal{B}$, and we consider the Toeplitz algebra $\mathfrak{L}_\Gamma(\mathcal{A}\oplus\mathcal{B})$. Here Γ is the path semigroupoid of G which equals

$$\Gamma=\{\alpha^n,\beta\alpha^n,b\colon n\in\mathbb{Z}_+\},$$

where α^n and $\beta\alpha^n$ are abbreviations for $(\alpha,\dots,\alpha)$ (with length n) and $(\beta,\alpha,\dots,\alpha)$ (with length $n+1$), respectively, and $\alpha^0=a$. The elements in the Toeplitz algebra are then given by infinite tuples

$$R=(V_n,W_n,B_0\colon n\in\mathbb{Z}_+,V_n\in\mathcal{L}(\mathcal{A}),W_n\in\mathcal{L}(\mathcal{A},\mathcal{B}),B_0\in\mathcal{L}(\mathcal{B}))\tag{3.11}$$

with the property that the infinite operator matrix

$$\begin{bmatrix} V_0 & 0 & 0 & 0 & 0 & 0 & \cdots\\ 0 & B_0 & 0 & 0 & 0 & 0 & \cdots\\ V_1 & 0 & V_0 & 0 & 0 & 0 & \cdots\\ W_0 & 0 & 0 & B_0 & 0 & 0 & \cdots\\ V_2 & 0 & V_1 & 0 & V_0 & 0 & \cdots\\ W_1 & 0 & W_0 & 0 & 0 & B_0 & \\ \vdots & \vdots & \vdots & \vdots & \vdots & & \ddots \end{bmatrix}\tag{3.12}$$

defines a bounded operator on $\ell^2_{\mathcal{A}\oplus\mathcal{B}}(\mathbb{Z}_+)$; the norm of R in $\mathfrak{L}_\Gamma(\mathcal{A}\oplus\mathcal{B},\mathcal{A}\oplus\mathcal{B})$ is equal to the operator norm of the operator matrix in $\mathcal{L}(\ell^2_{\mathcal{A}\oplus\mathcal{B}}(\mathbb{Z}_+))$. After rearranging rows and columns it follows that the operator norm of (3.12) is the same as that of

$$\begin{bmatrix} B_0 & 0 & 0 & 0 & 0 & 0 & \cdots \\ 0 & V_0 & 0 & 0 & 0 & 0 & \cdots \\ 0 & W_0 & B_0 & 0 & 0 & 0 & \cdots \\ 0 & V_1 & 0 & V_0 & 0 & 0 & \cdots \\ 0 & W_1 & 0 & W_0 & B_0 & 0 & \cdots \\ 0 & V_2 & 0 & V_1 & 0 & V_0 & \\ \vdots & \vdots & \vdots & \vdots & \vdots & \vdots & \ddots \end{bmatrix}$$

as an element in $\mathcal{L}(\mathcal{B} \oplus \ell^2_{\mathcal{A}\oplus\mathcal{B}}(\mathbb{Z}_+))$, and one can even leave out the first row and column. Thus R can be identified with the functions $V \in H^\infty_{\mathcal{L}(\mathcal{A})}(\mathbb{D})$ and $W \in H^\infty_{\mathcal{L}(\mathcal{A},\mathcal{B})}(\mathbb{D})$ given by

$$V(\lambda) = \sum_{n=0}^{\infty} \lambda^n V_n, \quad W(\lambda) = \sum_{n=0}^{\infty} \lambda^n W_n, \tag{3.13}$$

and the constant function with value B_0, which we also denote by B_0. The norm of R is then equal to the norm of the multiplication operator

$$\begin{bmatrix} M_V & 0 \\ M_W & M_{B_0} \end{bmatrix} \text{ on } \begin{bmatrix} H^2_{\mathcal{A}}(\mathbb{D}) \\ H^2_{\mathcal{B}}(\mathbb{D}) \end{bmatrix}, \tag{3.14}$$

where M_V, M_W and M_{B_0} denote the multiplication operators for the functions V, W and B_0, respectively. The Toeplitz algebra $\mathfrak{L}_\Gamma(\mathcal{A} \oplus \mathcal{B})$ can thus be identified with the algebra

$$\begin{bmatrix} H^\infty_{\mathcal{A}}(\mathbb{D}) & 0 \\ H^\infty_{\mathcal{L}(\mathcal{A},\mathcal{B})}(\mathbb{D}) & \mathcal{L}(\mathcal{B}) \end{bmatrix},$$

(with $\mathcal{L}(\mathcal{B})$ identified with the space of constant functions), which is easily seen to be isometrically isomorphic to the algebra

$$\begin{bmatrix} H^\infty_{\mathcal{A}}(\mathbb{D}) & 0 \\ H^\infty_{\mathcal{L}(\mathcal{A},\mathcal{B}),0}(\mathbb{D}) & \mathcal{L}(\mathcal{B}) \end{bmatrix}.$$

Here $H^\infty_{\mathcal{L}(\mathcal{A},\mathcal{B}),0}(\mathbb{D})$ is the Banach space of functions $W \in H^\infty_{\mathcal{L}(\mathcal{A},\mathcal{B})}(\mathbb{D})$ with $W(0) = 0$. The identification with the latter algebra was already obtained in [60].

We first consider Nevanlinna-Pick interpolation for the Riesz-Dunford functional calculus, i.e., when $\mathcal{A} = \mathcal{B} = \mathbb{C}$. Let $\mathcal{Z} = \mathcal{Z}_a \oplus \mathcal{Z}_b$ be an auxiliary Hilbert space. The generalized unit disk $\mathbb{D}_{G,\mathcal{Z}}$ then consists of all pairs of operators (Z_α, Z_β) with $Z_\alpha \in \mathcal{L}(\mathcal{Z}_a)$ and $Z_\beta \in \mathcal{L}(\mathcal{Z}_a, \mathcal{Z}_b)$ such that $\left[\begin{smallmatrix} Z_\alpha \\ Z_b \end{smallmatrix}\right]$ is a strict contraction. Given a point (Z_α, Z_β) in $\mathbb{D}_{G,\mathcal{Z}}$ and an

$$R = (v_n, w_n, b_0 \in \mathbb{C} \colon n \in \mathbb{Z}_+) \in \mathfrak{L}_\Gamma,$$

the point evaluation of R in (Z_α, Z_β) is given by

$$R(Z_\alpha, Z_\beta) = \begin{bmatrix} \sum_{n=0}^\infty v_n Z_\alpha^n & 0 \\ Z_\beta \sum_{n=0}^\infty w_n Z_\alpha^n & b_0 I_{\mathcal{Z}_b} \end{bmatrix} \text{ on } \begin{bmatrix} \mathcal{Z}_a \\ \mathcal{Z}_b \end{bmatrix}.$$

Assume we are given data for the **QLTRD-NP** problem:

$$(Z_\alpha^{(i)}, Z_\beta^{(i)}) \in \mathbb{D}_{G,\mathcal{Z}},\ X^{(i)} = \begin{bmatrix} X_a^{(i)} & X_b^{(i)} \end{bmatrix},\ Y^{(i)} = \begin{bmatrix} Y_{a^{(i)}} & Y_b^{(i)} \end{bmatrix} \text{ for } i = 1, \dots, N,$$

with $X^{(i)}, Y^{(i)} \in \mathcal{L}(\mathcal{Z}, \mathcal{C})$, where $\mathcal{C}$ is some separable Hilbert space with orthonormal basis $\{e_1, \dots, e_\kappa\}$. It then follows from Theorem 3.3 that there exists an $S \in \mathfrak{L}_\Gamma$ with $\|S\| \leq 1$ such that

$$X^{(i)} S(Z_\alpha^{(i)}, Z_\beta^{(i)}) = Y^{(i)} \text{ for } i = 1, \dots, N$$

if and only if the Pick matrices $\mathbb{P}^{(1)}_{\text{QLTRD}}$ and $\mathbb{P}^{(2)}_{\text{QLTRD}}$, with $\mathbb{P}^{(1)}_{\text{QLTRD}}$ given by

$$\left[\sum_{n=0}^{\infty} (Z_\alpha^{(i)})^{n*} \begin{pmatrix} X_a^{(i)*} e_{i'} e_{j'}^* X_a^{(j)} + Z_\beta^{(i)*} X_b^{(i)*} e_{i'} e_{j'}^* X_b^{(j)} Z_\beta^{(j)} + \\ -Y_a^{(i)*} e_{i'} e_{j'}^* Y_a^{(j)} - Z_\beta^{(i)*} Y_b^{(i)*} e_{i'} e_{j'}^* Y_b^{(j)} Z_\beta^{(j)} \end{pmatrix} (Z_\alpha^{(j)})^n \right]_{(i,i'),(j,j')}$$

and

$$\mathbb{P}^{(2)}_{\text{QLTRD}} = \left[X_b^{(i)*} e_{i'} e_{j'}^* X_b^{(j)} - Y_b^{(i)*} e_{i'} e_{j'}^* Y_b^{(j)} \right]_{(i,i'),(j,j')},$$

are both positive semidefinite. The range of the pairs (i, i') and (j, j') in the definition of the Pick matrices is $\{1, \dots, N\} \times \{1, \dots, \kappa\}$.

Now assume that $S = (v_n, w_n, b_0 \in \mathbb{C} \colon n \in \mathbb{Z}_+) \in \mathfrak{L}_\Gamma$ is a solution. Then b_0 necessarily satisfies $|b_0| \leq 1$ and $b_0 X_b^{(i)} = Y_b^{(i)}$ for $i = 1, \dots, N$. The existence of a number b_0 with these properties turns out to be equivalent to the positive semidefiniteness of the pick matrix $\mathbb{P}^{(2)}_{\text{QLTRD}}$, which is the content of the following lemma.

Lemma 3.4. *For $i = 1, \dots, N$ let $X_i, Y_i \in \mathcal{L}(\mathcal{H}, \mathcal{K})$, where $\mathcal{H}$ and $\mathcal{K}$ are Hilbert spaces and $\mathcal{H}$ is separable with orthonormal basis $\{e_1, \dots, e_\kappa\}$. Then*

$$\left[X_i e_{i'} e_{j'} X_j^* - Y_i e_{i'} e_{j'} Y_j^* \right]_{(i,i'),(j,j') \in \{1,\dots,N\} \times \{1,\dots,\kappa\}} \tag{3.15}$$

is a positive semidefinite if and only if there exist a $\delta \in \mathbb{C}$ with $|\delta| \leq 1$ such that $\delta X_i = Y_i$ for all $i = 1, \dots, N$.

Proof. Set

$$X = \begin{bmatrix} X_1 \\ \vdots \\ X_N \end{bmatrix} \quad \text{and} \quad Y = \begin{bmatrix} Y_1 \\ \vdots \\ Y_N \end{bmatrix}.$$

Then (3.15) is unitarily equivalent, via a permutation matrix, to

$$\text{col}_{i' \in \{1,\dots,\kappa\}}(X e_{i'})(\text{col}_{i' \in \{1,\dots,\kappa\}}(X e_{i'}))^* - \text{col}_{i' \in \{1,\dots,\kappa\}}(Y e_{i'})(\text{col}_{i' \in \{1,\dots,\kappa\}}(Y e_{i'}))^*. \tag{3.16}$$

Thus positive semidefiniteness of (3.15) corresponds to positive semidefiniteness of (3.16). It follows right away from the Douglas factorization lemma [43] that (3.16) being positive semidefinite is equivalent to the existence of a $\delta \in \mathbb{C}$ with $|\delta| \leq 1$ such that $\delta \text{col}_{i' \in \{1,\dots,\kappa\}}(X e_{i'}) = \text{col}_{i' \in \{1,\dots,\kappa\}}(Y e_{i'})$, which is the same as $\delta X_i = Y_i$ for all $i = 1, \dots, N$. □

Next assume that both Pick matrices $\mathbb{P}^{(1)}_{\rm QLTRD}$ and $\mathbb{P}^{(2)}_{\rm QLTRD}$ are positive semidefinite. According to Lemma 3.4 there exists a b_0 with $|b_0| \leq 1$ such that $b_0 X_b^{(i)} = Y_b^{(i)}$ for $i = 1, \dots, N$. Using this number b_0 we can rewrite $\mathbb{P}^{(2)}_{\rm QLTRD}$ as

$$\left[\sum_{n=0}^{\infty} (Z_\alpha^{(i)})^{n*} \begin{pmatrix} X_a^{(i)*} e_{i'} e_{j'}^* X_a^{(j)} - Y_a^{(i)*} e_{i'} e_{j'}^* Y_a^{(j)} + \\ \quad +(1-|b_0|^2) Z_\beta^{(i)*} X_b^{(i)*} e_{i'} e_{j'}^* X_b^{(j)} Z_\beta^{(j)} \end{pmatrix} (Z_\alpha^{(j)})^n \right]_{(i,i'),(j,j')},$$

where again the range of the pairs (i, i') and (j, j') in the definition of the Pick matrices is $\{1, \dots, N\} \times \{1, \dots, \kappa\}$. We then distinguish between two case, namely (1) $|b_0| = 1$ and (2) $|b_0| < 1$. In case $|b_0| = 1$, for $S = (v_n, w_n, b_0 \in \mathbb{C} \colon n \in \mathbb{Z}_+) \in \mathfrak{L}_\Gamma$ to be a solution it is necessary that $w_n = 0$ for all $n \in \mathbb{Z}_+$ due to the norm constraint $\|S\| \leq 1$. On the other hand, the Pick matrix $\mathbb{P}^{(1)}_{\rm QLTRD}$ reduces to

$$\left[\sum_{n=0}^{\infty} (Z_\alpha^{(i)})^{n*} \left(X_a^{(i)*} e_{i'} e_{j'}^* X_a^{(j)} - Y_a^{(i)*} e_{i'} e_{j'}^* Y_a^{(j)}\right) (Z_\alpha^{(j)})^n \right]_{(i,i'),(j,j') \in \{1,\dots,N\} \times \{1,\dots,\kappa\}}$$

which is a Pick matrix of the form appearing in Part 2 of Theorem 2.3. In fact, it is the pick matrix for the **LTRD-NP** problem for the scalar-valued Schur class $\mathcal{S}$ with data

$$Z_i = Z_\alpha^{(i)},\ X_i = X_a^{(i)},\ Y_i = Y_a^{(i)}, \quad i = 1, \dots, N.$$

Applying Theorem 2.3 to this data set we obtain a $v \in \mathcal{S}$ with $X_a^{(i)} v(Z_\alpha^{(i)}) = Y_a^{(i)}$ for $i = 1, \dots, N$. Let $v_0, v_1, \dots$ be the Taylor coefficients of v. It is then not difficult to see that $S = (v_n, w_n, b_0 \in \mathbb{C} \colon n \in \mathbb{Z}_+,\ w_n = 0)$ is a solution. The case that $|b_0| < 1$ does not reduce to a **LTRD-NP** problem, but rather to a left-tangential tensor functional-calculus Nevanlinna-Pick (**LTT-NP**) problem which is a type of problem we discuss in Subsection 4.8 below. The problem can then be solved directly using some of the techniques developed there, but we will not work out the details here.

Next we specify the left-tangential operator-argument Nevanlinna-Pick result for our example. Let $\mathcal{X} = \mathcal{X}_\alpha \oplus \mathcal{X}_\beta$ be a given Hilbert space. in this case, points are elements of the generalized disk $\mathbb{D}_{\widetilde{G},\mathcal{X}}$ (with $\widetilde{G}$ the transposed quiver of G), which corresponds to the set of pairs (T_α, T_β) with $T_\alpha \in \mathcal{L}(\mathcal{X}_a, \mathcal{X}_a)$ and $T_\beta \in \mathcal{L}(\mathcal{X}_b, \mathcal{X}_a)$ such that the row operator $\begin{bmatrix} T_\alpha & T_\beta \end{bmatrix}$ is a strict contraction. Now let $R \in \mathfrak{L}_\Gamma(\mathcal{A} \oplus \mathcal{B})$ be given by (3.11) and $(T_\alpha, T_\beta) \in \mathbb{D}_{\widetilde{G},\mathcal{X}}$. If in addition we are also given a block diagonal operator $X = \operatorname{diag}(X_a, X_b)$ with $X_a \in \mathcal{L}(\mathcal{A}, \mathcal{X}_a)$ and $X_b \in \mathcal{L}(\mathcal{B}, \mathcal{X}_b)$, then the left-tangential operator-argument point evaluation $(XR)^{\wedge L}(T_\alpha, T_\beta)$ is given by

$$(XR)^{\wedge L}(T_\alpha, T_\beta) = \begin{bmatrix} \displaystyle\sum_{n=0}^{\infty} T_\alpha^n (X_a V_n + T_\beta X_b W_n) & 0 \\ 0 & X_b B_0 \end{bmatrix}.$$

The data for the **QLTOA-NP** problem in then given by

$$(T_\alpha^{(i)}, T_\beta^{(i)}) \in \mathbb{D}_{\widetilde{G},\mathcal{X}},\ X^{(i)} = \operatorname{diag}(X_a^{(i)}, X_b^{(i)}),\ Y^{(i)} = \operatorname{diag}(Y_a^{(i)}, Y_b^{(i)}) \text{ for } i = 1, \dots, N$$

with $X_a^{(i)}, Y_a^{(i)} \in \mathcal{L}(\mathcal{A}, \mathcal{X}_a)$ and $X_b^{(i)}, Y_b^{(i)} \in \mathcal{L}(\mathcal{B}, \mathcal{X}_b)$, and it follows from Part 3 of Theorem 3.3 that there exists an $S \in \mathfrak{L}_\Gamma(\mathcal{A} \oplus \mathcal{B})$ with $\|S\| \leq 1$ such that

$$(X^{(i)}S)^{\wedge L}(T_\alpha^{(i)}, T_\beta^{(i)}) = Y^{(i)} \quad \text{for} \quad i = 1, \dots, N \tag{3.17}$$

if and only if the Pick matrix $\mathbb{P}_{\text{QLTOA}}$ in Part 3 of Theorem 3.3 is positive semidefinite. In this case, after rearranging columns and rows, $\mathbb{P}_{\text{QLTOA}}$ can be identified with

$$\mathbb{P}_{\text{QLTOA}} = \begin{bmatrix} \mathbb{P}^{(1)}_{\text{QLTOA}} & 0 \\ 0 & \mathbb{P}^{(2)}_{\text{QLTOA}} \end{bmatrix}$$

where $\mathbb{P}^{(1)}_{\text{QLTOA}}$ is the Pick matrix given by

$$\left[\sum_{n=0}^{\infty}(T_\alpha^{(i)})^n(X_a^{(i)}X_a^{(j)*}+T_\beta^{(i)}X_b^{(i)}X_b^{(j)*}T_\beta^{(j)*}-Y_a^{(i)}Y_a^{(j)*}-T_\beta^{(i)}Y_b^{(i)}Y_b^{(j)*}T_\beta^{(j)*})(T_\alpha^{(j)})^{n*}\right]_{i,j=1}^{N}$$

and

$$\mathbb{P}^{(2)}_{\text{QLTOA}} = \left[\; X_b^{(i)}X_b^{(j)*} - Y_b^{(i)}Y_b^{(j)*} \;\right]_{i,j=1}^{N}.$$

To see the sufficiency of this Pick matrix criterion, assume that $\mathbb{P}_{\text{QLTOA}}$ is positive semidefinite, and thus, equivalently, that $\mathbb{P}^{(1)}_{\text{QLTOA}}$ and $\mathbb{P}^{(2)}_{\text{QLTOA}}$ are positive semidefinite. Notice that $\mathbb{P}^{(2)}_{\text{QLTOA}}$ can also be written as

$$\begin{bmatrix} X_b^{(1)} \\ \vdots \\ X_b^{(N)} \end{bmatrix}\begin{bmatrix} X_b^{(1)*} & \cdots & X_b^{(N)*} \end{bmatrix} - \begin{bmatrix} Y_b^{(1)} \\ \vdots \\ Y_b^{(N)} \end{bmatrix}\begin{bmatrix} Y_b^{(1)*} & \cdots & Y_b^{(N)*} \end{bmatrix}.$$

Hence the positive semidefiniteness of $\mathbb{P}^{(2)}_{\text{QLTOA}}$, again using Douglas factorization lemma, corresponds to the existence of a contraction $B_0 \in \mathcal{L}(\mathcal{B})$ with

$$X_b^{(i)}B_0 = Y_b^{(i)} \quad \text{for} \quad i = 1, \dots, N.$$

Let $D_{B_0^*}$ denote the defect operator of B_0^*, that is, $D_{B_0^*}$ is the positive square root of $I_{\mathcal{B}} - B_0B_0^*$. We can then rewrite the first Pick matrix $\mathbb{P}^{(1)}_{\text{QLTOA}}$ as

$$\left[\sum_{n=0}^{\infty}(T_\alpha^{(i)})^n(X_a^{(i)}X_a^{(j)*} + T_\beta^{(i)}X_b^{(i)}D^2_{B_0^*}X_b^{(j)*}T_\beta^{(j)*} - Y_a^{(i)}Y_a^{(j)*})(T_\alpha^{(j)})^{n*}\right]_{i,j=1}^{N}.$$

In this form $\mathbb{P}^{(1)}_{\text{QLTOA}}$ is a Pick matrix of the type appearing in Part 1 of Theorem 2.2. In fact, it is the Pick matrix for the **LTOA-NP** problem for functions from the Schur class $\mathcal{S}(\mathcal{A}, \mathcal{A} \oplus \mathcal{B})$ with data

$$T_i = T_\alpha^{(i)},\; X_i = \begin{bmatrix} X_a^{(i)} & T_\beta^{(i)}X_b^{(i)}D_{B_0^*} \end{bmatrix},\; Y_i = Y_a^{(i)}, \quad \text{for } i = 1, \dots, N. \tag{3.18}$$

Applying Theorem 2.2 to this data set, we obtain a function

$$H = \begin{bmatrix} V \\ \widetilde{W} \end{bmatrix} \in \mathcal{S}(\mathcal{A}, \mathcal{A} \oplus \mathcal{B}) \tag{3.19}$$

with

$$(X_iH)^{\wedge L}(T_i) = Y_i \quad \text{for} \quad i = 1, \ldots, N.$$

In other words, the Taylor coefficients $V_0, V_1, \ldots$ of V and $\widetilde{W}_0, \widetilde{W}_1, \ldots$ of $\widetilde{W}$ satisfy

$$\sum_{n=0}^{\infty} (T_\alpha^{(i)})^n (X_a^{(i)} V_n + T_\beta^{(i)} X_b^{(i)} D_{B_0^*} \widetilde{W}_n) = Y_a^{(i)} \text{ for } i = 1, \ldots, N.$$

Now set $W_n = D_{B_0^*} \widetilde{W}_n$ for each $n \in \mathbb{Z}_+$. It follows that $S = (V_n, W_n, B_0 \colon n \in \mathbb{Z}_+)$ is in $\mathfrak{L}_\Gamma(\mathcal{A} \oplus \mathcal{B})$ and satisfies the interpolation conditions (3.17), and it is not difficult to see that $\|S\| \leq 1$. Thus S is a solution to the **QLTOA-NP** problem for the quiver G considered in this example. It is also possible to provide a direct proof of the necessity of the Pick matrix condition; we leave the details as an exercise for the interested reader.

3.4. The polydisk setting: commutative and noncommutative

For the setting of the polydisk $\mathbb{D}^d = \{\lambda = (\lambda_1, \ldots, \lambda_d) \colon |\lambda_k| < 1 \text{ for } k = 1, \ldots, d\}$, the results concerning Nevanlinna-Pick-like interpolation are of a different flavor. We define what is now called the d-variable *Schur-Agler class* $\mathcal{SA}_d$ to consist of those holomorphic complex-valued functions $s(\lambda) = \sum_{n \in \mathbb{Z}_+^d} s_n \lambda^n$ on $\mathbb{D}^d$ with the property that, for every commutative d-tuple $Z = (Z_1, \ldots, Z_d)$ of strict contraction operators ($\|Z_k\| < 1$ for $k = 1, \ldots, d$) on a Hilbert space $\mathcal{K}$, it happens that the resulting operator

$$s(Z) = \sum_{n \in \mathbb{Z}_+^d} s_n Z^n$$

(d-variable Riesz-Dunford functional calculus) has $\|s(Z)\| \leq 1$. The special choice $Z = (\lambda_1 I_\mathcal{K}, \ldots, \lambda_d I_\mathcal{K})$ with $\lambda = (\lambda_1, \ldots, \lambda_d) \in \mathbb{D}^d$ shows that the Schur-Agler class is a subset of the Schur class (defined to be the set of holomorphic functions mapping $\mathbb{D}^d$ into the closed unit disk $\overline{\mathbb{D}}$). The converse holds for the cases $d = 1$ and $d = 2$ as a consequence of the von Neumann inequality holding for these two cases, as can be seen from the Sz.-Nagy dilation theorem for the case $d = 1$ and from the Andô dilation theorem for the case $d = 2$ – see, e.g., [6].

In fact, as a consequence of the Drury-von Neumann inequality/dilation theorem for commutative row contractions $Z = (Z_1, \ldots, Z_d)$ [47] and the Popescu-von Neumann inequality/dilation theorem [82], a Schur-Agler type characterization also holds for the Drury-Arveson Schur-multiplier class $\mathcal{S}_d$ and the free-semigroup algebra $\mathcal{L}_d$: *a holomorphic function s on the ball $\mathbb{B}^d$ is in the Drury-Arveson Schur-multiplier class $\mathcal{S}_d$ if and only if $\|s(Z)\| \leq 1$ for all commutative d-tuples $Z = (Z_1, \ldots, Z_d)$ for which the block row matrix $\mathbf{Z} = \begin{bmatrix} Z_1 & \cdots & Z_d \end{bmatrix}$ is a strict contraction*, and, similarly, *a formal power series $s(z) = \sum_{\gamma \in \mathcal{F}_d} s_\gamma z^\gamma$ is in the noncommutative Schur-multiplier class $\mathcal{S}_{nc,d}$ if and only if $s(Z) = \sum_{\gamma \in \mathcal{F}_d} s_\gamma Z^\gamma$ has $\|s(Z)\| \leq 1$ for any (not necessarily commutative) d-tuple $Z = (Z_1, \ldots, Z_d)$ for which the row matrix $\mathbf{Z} := \begin{bmatrix} Z_1 & \cdots & Z_d \end{bmatrix}$ is a strict contraction.*

The main result on interpolation for the Schur-Agler class is the following result of Agler.

Theorem 3.5. (*See* [2, 3, 4, 6].) *Suppose that we are given a subset X of $\mathbb{D}^d$ and a function $f\colon X \to \mathbb{C}$. Then there exists a function $s\colon \mathbb{D}^d \to \mathbb{C}$ in the Schur-Agler class $\mathcal{SA}_d$ such that*

$$s|_X = f$$

if and only if there exist d positive kernels $K_1, \dots, K_d$ on $X \times X$ so that

$$1 - f(\lambda)\overline{f(\zeta)} = \sum_{k=1}^{d} (1 - \lambda_k \overline{\zeta_k}) K_k(\lambda, \zeta) \text{ for all } \lambda, \zeta \in X. \tag{3.20}$$

We mention that an independent proof valid for a more general function-algebra setting has been given by Paulsen [76].

Given two Hilbert spaces $\mathcal{U}$ and $\mathcal{Y}$, the operator-valued version $\mathcal{SA}_d(\mathcal{U}, \mathcal{Y})$ of the Schur-Agler class can be defined via tensor functional calculus as follows. Given a holomorphic function $S(\lambda) = \sum_{n \in \mathbb{Z}_+^d} S_n \lambda^n$ with coefficients (and hence also values) in $\mathcal{L}(\mathcal{U}, \mathcal{Y})$ and given a commutative d-tuple $Z := (Z_1, \dots, Z_d)$ of operators on another auxiliary Hilbert space $\mathcal{K}$ such that each Z_k is a strict contraction for $k = 1, \dots, d$, we define $S(Z)$ via

$$S(Z) = \sum_{n \in \mathbb{Z}_+^d} S_n \otimes Z^n \in \mathcal{L}(\mathcal{U} \otimes \mathcal{K}, \mathcal{Y} \otimes \mathcal{K}),$$

using the (commutative) multivariable notation for Z^n defined in Part 3 of Theorem 3.1. We then say that $S \in \mathcal{SA}_d(\mathcal{U}, \mathcal{Y})$ if $\|S(Z)\| \leq 1$ whenever $Z = (Z_1, \dots, Z_d)$ is a commutative d-tuple of strict contraction operators on $\mathcal{K}$. We will not state the results precisely here, but rather merely mention that the extensions to left- and right-tangential interpolation have been given in [28]. The theory can be generalized to an arbitrary domain $\mathcal{D}$ with polynomial-matrix defining function ($\mathcal{D} = \mathcal{D}_Q = \{\lambda \in \mathbb{C}^d \colon \|Q(\lambda)\| < 1\}$ for a fixed matrix polynomial $Q(\lambda)$) – see [10, 18]; note that the polydisk corresponds to the case

$$Q(\lambda) = \begin{bmatrix} \lambda_1 & & \\ & \ddots & \\ & & \lambda_d \end{bmatrix}.$$

In this general setting the analysis of the left- and right-tangential Nevanlinna-Pick problem with operator-argument (using an adaptation of the Taylor-Vasilescu functional calculus) has been worked out in [19].

A noncommutative version of the Schur-Agler class can be defined as follows. Given coefficient Hilbert spaces $\mathcal{U}$ and $\mathcal{Y}$ and a formal power series S of the form $S(z) = \sum_{\gamma \in \mathcal{F}_d} S_\gamma z^\gamma$, where for each γ in the free semigroup $\mathcal{F}_d$ defined in Subsection 3.2 the coefficient S_γ is in $\mathcal{L}(\mathcal{U}, \mathcal{Y})$, we say that S is in the noncommutative operator-valued d-variable Schur-Agler class $\mathcal{SA}_{nc,d}(\mathcal{U},\mathcal{Y})$ if, for any (not necessarily commutative) d-tuple $Z=(Z_1,\dots,Z_d)$ of strict contraction operators, it is the case that

$$S(Z)=\sum_{\gamma\in\mathcal{F}_d} S_\gamma\otimes Z^\gamma$$

has $\|S(Z)\|\le 1$. Here we use the noncommutative multivariable notation of Subsection 3.1: $Z^\gamma=Z_{i_N}\cdots Z_{i_1}$ if $\gamma=i_N\dots i_1$. It is one of the results of [9] that in fact one need only check $\|S(Z)\|\le 1$ for $Z=(Z_1,\dots,Z_d)$ a d-tuple of matrices of finite size $\kappa\times\kappa$ for arbitrary $\kappa=1,2,\dots$.

We then have the following result for left-tangential Nevanlinna-Pick interpolation with operator-argument on the noncommutative polydisk.

Theorem 3.6. (1) Left-Tangential interpolation with Operator-Argument for the noncommutative polydisk: *Suppose that we are given N d-tuples of (not necessarily commutative) strict contraction operators*

$$T^{(1)}=(T_1^{(1)},\dots,T_d^{(1)}),\dots,T^{(N)}=(T_1^{(N)},\dots,T_d^{(N)})$$

on a Hilbert space $\mathcal{C}$ together with operators $X_1,\dots,X_N$ in $\mathcal{L}(\mathcal{Y},\mathcal{C})$ and $Y_1,\dots,Y_N$ in $\mathcal{L}(\mathcal{U},\mathcal{C})$. Then there exists a formal power series $S(z)=\sum_{\gamma\in\mathcal{F}_d}S_\gamma z^\gamma$ in the noncommutative d-variable operator-valued Schur class $\mathcal{S}_{nc,d}(\mathcal{U},\mathcal{Y})$ such that

$$(X_iS)^{\wedge L}(T^{(i)}):=\sum_{\gamma\in\mathcal{F}_d}(T^{(i)})^{\gamma^\top}X_iS_\gamma=Y_i \text{ for } i=1,\dots,N$$

if and only if there exist d positive semidefinite block matrices $K_1,\dots,K_d$ with entries in $\mathcal{L}(\mathcal{C})$ of the form

$$K_k=[K_k(i,j)]_{i,j=1,\dots,N}$$

so that the noncommutative Agler decomposition

$$X_iX_j^*-Y_iY_j^*=\sum_{k=1}^d\left(K_k(i,j)-T_k^{(i)}K_k(i,j)(T_k^{(j)})^*\right)$$

holds.

(2) Riesz-Dunford interpolation for the noncommutative polydisk. *Suppose that $\mathcal{Z}$ is a separable Hilbert space with orthonormal basis $\{e_1,\dots,e_\kappa\}$ (with possibly $\kappa=\infty$) and that we are given N (not necessarily commutative) d-tuples of strictly contractive operators $Z^{(1)}=(Z_1^{(1)},\dots,Z_d^{(1)})$, …, $Z^{(N)}=(Z_1^{(N)},\dots,Z_d^{(N)})$ in $\mathcal{L}(\mathcal{Z})$. Assume also that we are given operators $W_1,\dots,W_N$ in $\mathcal{L}(\mathcal{Z})$. Then there exists a formal power series $s(z)=\sum_{\gamma\in\mathcal{F}_d}s_\gamma z^\gamma$ in the scalar noncommutative Schur-Agler class $\mathcal{SA}_{nc,d}$ such that*

$$s(Z^{(i)}):=\sum_{\gamma\in\mathcal{F}_d}s_\gamma(Z^{(i)})^{\gamma^\top}=W_i \text{ for } i=1,\dots,N$$

if and only if there exist d positive semidefinite operator matrices $K_1, \dots, K_d$ *of size* $(N \cdot \kappa) \times (N \cdot \kappa)$ *with entries in* $\mathcal{L}(\mathcal{Z})$ *written in the form*

$$K_k = [K_k((i,i'), (j,j'))]_{(i,i'),(j,j') \in \{1,\dots,N\} \times \{1,\dots,\kappa\}}$$

so that the following noncommutative Agler decomposition holds:

$$e_{i'}e_{j'}^* - W_i e_{i'} e_{j'}^* W_j^* = \sum_{k=1}^{d} \left(K_k\left((i,i'),(j,j')\right) - Z_k^{(i)} K_k\left((i,i'),(j,j')\right) (Z_k^{(j)})^* \right).$$

Proof. Statement (1) is a particular case of [20, Theorem 7.9]. Statement (2) then follows from statement (1) in the same way as was done for the single-variable case in Section 2. □

As an example we next discuss how the Schur interpolation problem for the noncommutative polydisk can be handled as an application of Theorem 3.6. We first say that a subset Γ of $\mathcal{F}_d$ is *lower inclusive* if, whenever $\gamma \in \Gamma$ and γ factors as $\gamma = \gamma'\gamma''$, then also $\gamma' \in \Gamma$. The Carathéodory-Fejér interpolation problem for the noncommutative polydisk can be formulated as follows: *Given a lower inclusive subset* Γ *of* $\mathcal{F}_d$ *and given a collection of operators* $\{F_\gamma \colon \gamma \in \Gamma\} \subset \mathcal{L}(\mathcal{U}, \mathcal{Y})$, *find a formal power series* $S(z) = \sum_{\gamma \in \mathcal{F}_d} S_\gamma z^\gamma$ *in the noncommutative operator-valued Schur class* $\mathcal{SA}_{nc,d}(\mathcal{U}, \mathcal{Y})$ *such that*

$$S_\gamma = F_\gamma \text{ for } \gamma \in \Gamma.$$

In [20, Section 7.4] it is shown how to choose the data set

$$\mathfrak{D}: \quad T^{(1)}, \dots, T^{(N)}, \quad X_1, \dots, X_d, \quad Y_1, \dots, Y_d$$

so that the associated left-tangential interpolation problem with operator-argument handled by statement (1) in Theorem 3.6 is equivalent to the Carathéodory-Fejér interpolation problem for the noncommutative polydisk. We note that this Carathéodory-Fejér problem was handled directly earlier in [57].

Just as was the case for the commutative case, the theory for the noncommutative polydisk can be extended to more general noncommutative domains. This is done in [26] for noncommutative operator domains with a certain type of linear defining function Q: $Z = (Z_1, \dots, Z_d) \in \mathcal{D}_Q$ if $\|Q(Z)\| < 1$ where $Q(z) = Q_1 z_1 + \dots + Q_d z_d$ and $Q(Z) = Q_1 \otimes Z_1 + \dots + Q_d \otimes Z_d$. The results on Left (and/or Right) Tangential interpolation with Operator-Argument in [20] are actually given for this level of generality.

4. W^*-correspondence Nevanlinna-Pick theorems

4.1. Preliminaries

Let $\mathcal{A}$ and $\mathcal{B}$ be C^*-algebras and E a linear space. We say that E is an $(\mathcal{A}, \mathcal{B})$-*correspondence* when E is a bi-module with respect to a given right $\mathcal{B}$-action and a left $\mathcal{A}$-action, and E is endowed with a $\mathcal{B}$-valued inner product $\langle\ ,\ \rangle_E$ satisfying the following axioms: For any $\lambda, \mu \in \mathbb{C}$, $\xi, \eta, \zeta \in E$, $a \in \mathcal{A}$ and $b \in \mathcal{B}$

1. $\langle \lambda\xi + \mu\zeta, \eta\rangle_E = \lambda\langle\xi,\eta\rangle_E + \mu\langle\zeta,\eta\rangle_E$;
2. $\langle \xi \cdot b, \eta\rangle_E = \langle\xi,\eta\rangle_E b$;
3. $\langle a\cdot\xi, \eta\rangle_E = \langle\xi, a^*\cdot\eta\rangle_E$;
4. $\langle\xi,\eta\rangle_E^* = \langle\eta,\xi\rangle_E$;
5. $\langle\xi,\xi\rangle_E \geq 0$ (in $\mathcal{B}$);
6. $\langle\xi,\xi\rangle_E = 0$ implies that $\xi = 0$;

and such that E is a Banach space with respect to the norm $\|\ \|_E$ defined by

$$\|\xi\|_E = \|\langle\xi,\xi\rangle_E\|_{\mathcal{B}}^{\frac{1}{2}} \quad (\xi \in E),$$

where $\|\ \|_{\mathcal{B}}$ denotes the norm of $\mathcal{B}$. We also impose that

$$(\lambda\xi)\cdot b = \xi\cdot(\lambda b) \quad \text{and} \quad (\lambda a)\cdot\xi = a\cdot(\lambda\xi) \quad (\lambda\in\mathbb{C}, a\in\mathcal{A}, b\in\mathcal{B}, \xi\in E).$$

In practice we usually write $\langle\ ,\ \rangle$ and $\|\ \|$ for the inner product and norm on E, and in case $\mathcal{A} = \mathcal{B}$ we say that E is an $\mathcal{A}$-correspondence.

Given two $(\mathcal{A},\mathcal{B})$-correspondences E and F, the set of bounded linear operators from E to F is denoted by $\mathcal{L}(E,F)$. It may happen that a $T\in\mathcal{L}(E,F)$ is not adjointable, i.e., it is not necessarily the case that

$$\langle T\xi,\gamma\rangle = \langle\xi, T^*\gamma\rangle \quad (\xi\in E, \gamma\in F)$$

for some $T^*\in\mathcal{L}(F,E)$. We will write $\mathcal{L}^a(E,F)$ for the set of adjointable operators in $\mathcal{L}(E,F)$. As usual we have the abbreviations $\mathcal{L}(E)$ and $\mathcal{L}^a(E)$ in case $F = E$.

The third inner-product axiom implies that the left $\mathcal{A}$-action can be identified with a $*$-homomorphism φ of $\mathcal{A}$ into the C^*-algebra $\mathcal{L}^a(E)$. In case this $*$-homomorphism φ is specified we will occasionally write $\varphi(a)\xi$ instead of $a\cdot\xi$.

Furthermore, an operator $T\in\mathcal{L}(E,F)$ is said to be a *right $\mathcal{B}$-module map* if

$$T(\xi\cdot b) = T(\xi)\cdot b \quad (\xi\in E, b\in\mathcal{B}),$$

and a *left $\mathcal{A}$-module map* whenever

$$T(a\cdot\xi) = a\cdot T(\xi) \quad (\xi\in E, a\in\mathcal{A}).$$

It is easily checked that an adjointable map $T\in\mathcal{L}^a(E,F)$ is automatically a right module map. Occasionally we leave out the $\mathcal{B}$ and $\mathcal{A}$, and just say left or right module map. In case T is both a left and right module map we also say that T is a *bi-module map*. Notice that the product of two left (right) module maps is again a left (right) module map, and the adjoint of an adjointable left (right) module map, is also a left (right) module map.

We will have a need for various constructions which create new correspondences out of given correspondences.

Given two $(\mathcal{A},\mathcal{B})$-correspondences E and F, we define the *direct-sum correspondence* $E\oplus F$ to be the direct-sum vector space $E\oplus F$ together with the diagonal left $\mathcal{A}$-action and right $\mathcal{B}$-action and the direct-sum $\mathcal{B}$-valued inner-product defined by setting for each $\xi,\xi'\in E$, $\gamma,\gamma'\in F$, $a\in\mathcal{A}$ and $b\in\mathcal{B}$:

$$a\cdot(\xi\oplus\gamma) = (a\cdot\xi)\oplus(a\cdot\gamma), \qquad (\xi\oplus\gamma)\cdot b = (\xi\cdot b)\oplus(\gamma\cdot b),$$
$$\langle\xi\oplus\gamma, \xi'\oplus\gamma'\rangle_{E\oplus F} = \langle\xi,\xi'\rangle_E + \langle\gamma,\gamma'\rangle_F.$$

Bounded linear operators between direct-sum correspondences admit operator matrix decompositions in precisely the same way as in the Hilbert space case ($\mathcal{B} = \mathbb{C}$), while adjointability and the left and right module map property of such an operator corresponds to the operators in the decomposition being adjointable, or left or right module maps, respectively.

Now suppose that we are given three C^*-algebras $\mathcal{A}, \mathcal{B}$ and $\mathcal{C}$ together with an $(\mathcal{A}, \mathcal{B})$-correspondence E and a $(\mathcal{B}, \mathcal{C})$-correspondence F. Then we define the *tensor-product correspondence* $E \otimes F$ to be the completion of the linear span of all tensors $\xi \otimes \gamma$ (with $\xi \in E$ and $\gamma \in F$) subject to the identification

$$(\xi \cdot b) \otimes \gamma = \xi \otimes (b \cdot \gamma) \quad (\xi \in E, \gamma \in F, b \in \mathcal{B}), \tag{4.1}$$

with left $\mathcal{A}$-action, right $\mathcal{C}$-action and the $\mathcal{C}$-valued inner-product defined by setting for each $\xi, \xi' \in E$, $\gamma, \gamma' \in F$, $a \in \mathcal{A}$ and $c \in \mathcal{C}$:

$$a \cdot (\xi \otimes \gamma) = (a \cdot \xi) \otimes \gamma, \qquad (\xi \otimes \gamma) \cdot c = \xi \otimes (\gamma \cdot c),$$
$$\langle \xi \otimes \gamma, \xi' \otimes \gamma' \rangle_{E \otimes F} = \langle \langle \xi, \xi' \rangle_E \cdot \gamma, \gamma' \rangle_F.$$

In case the left action on F is given by the $*$-homomorphism φ we occasionally emphasize this by writing $E \otimes_\varphi F$ for $E \otimes F$.

It is more complicated to characterize the bounded linear operators between tensor-product correspondences. One way to construct such operators is as follows. Let E and E' be $(\mathcal{A}, \mathcal{B})$-correspondences and F and F' $(\mathcal{B}, \mathcal{C})$-correspondences, for C^*-algebras $\mathcal{A}$, $\mathcal{B}$ and $\mathcal{C}$. Furthermore, let $X \in \mathcal{L}(E, E')$ be a right module map and $Y \in \mathcal{L}(F, F')$ a left module map. Then we write $X \otimes Y$ for the operator in $\mathcal{L}(E \otimes F, E' \otimes F')$ which is determined by

$$X \otimes Y(\xi \otimes \gamma) = (X\xi) \otimes (Y\gamma) \quad (\xi \otimes \gamma \in E \otimes F). \tag{4.2}$$

The module map properties are needed to guarantee that the balancing in the tensor-product (see (4.1)) is respected by the operator $X \otimes Y$.

If, in addition, X is also a left module map, then $X \otimes Y$ is a left module map, while Y also being a right module map guarantees that $X \otimes Y$ is a right module map. Moreover, $X \otimes Y$ is adjointable in case X and Y are both adjointable operators, with $(X \otimes Y)^* = X^* \otimes Y^*$.

Notice that the left action on $E \otimes F$ can now be written as $a \mapsto \varphi(a) \otimes I_F \in \mathcal{L}^a(E \otimes F)$, where $I_F \in \mathcal{L}^a(F)$ is the identity operator on F.

4.2. The Fock space $\mathcal{F}^2(E)$ and the Toeplitz algebra $\mathcal{F}^\infty(E)$

In this section we shall consider the situation where $\mathcal{A} = \mathcal{B}$, i.e, E is an $\mathcal{A}$-correspondence. We also restrict our attention to the case where $\mathcal{A}$ is a von Neumann algebra and let E be an $\mathcal{A}$-W^*-correspondence. This means that E is an $\mathcal{A}$-correspondence which is also *self-dual* in the sense that any right $\mathcal{A}$-module map $\rho \colon E \to \mathcal{A}$ is given by taking the inner-product against some element e_ρ of E:

$$\rho(e) = \langle e, e_\rho \rangle_E \in \mathcal{A}. \tag{4.3}$$

It is easily seen that such maps are adjointable with adjoint $\rho^* \colon \mathcal{A} \to E$ given by

$$\rho^*(a) = e_\rho \cdot a \tag{4.4}$$

and hence also *any* right module map $\nu\colon \mathcal{A} \to E$ has the form $\nu = \rho^*$ as in (4.4). Moreover, the space $\mathcal{L}^a(E)$ of adjointable operators on the W^*-correspondence E is in fact a W^*-algebra, i.e., it is the abstract version of a von Neumann algebra with an ultra-weak topology (see [69]).

Since E is an $\mathcal{A}$-correspondence, we may define the self-tensor-product $E^{\otimes 2} = E \otimes E$ to get another $\mathcal{A}$-correspondence, and, inductively, an $\mathcal{A}$-correspondence $E^{\otimes n} = E \otimes (E^{\otimes(n-1)})$ for each $n = 1, 2, \dots$. If we use $a \mapsto \varphi(a)$ to denote the left $\mathcal{A}$-action $\varphi(a)e = a \cdot e$ on E, we denote the left $\mathcal{A}$-action on $E^{\otimes n}$ by $\varphi^{(n)}$:

$$\varphi^{(n)}(a)\colon \xi_n \otimes \xi_{n-1} \otimes \cdots \otimes \xi_1 \mapsto (\varphi(a)\xi_n) \otimes \xi_{n-1} \otimes \cdots \otimes \xi_1.$$

Note that, using the notation in (4.2), we may write $\varphi^{(n)}(a) = \varphi(a) \otimes I_{E^{\otimes n-1}}$. We formally set $E^{\otimes 0} = \mathcal{A}$. Then the *Fock space* $\mathcal{F}^2(E)$ is defined to be

$$\mathcal{F}^2(E) = \oplus_{n=0}^{\infty} E^{\otimes n} \tag{4.5}$$

and is also an $\mathcal{A}$-W^*-correspondence. The left $\mathcal{A}$-action on $\mathcal{F}^2(E)$ is given by φ_∞:

$$\varphi_\infty(a)\colon\ \oplus_{n=0}^{\infty} \xi^{(n)} \mapsto \oplus_{n=0}^{\infty}(\varphi^{(n)}(a)\xi^{(n)}) \text{ for } \oplus_{n=0}^{\infty} \xi^{(n)} \in \oplus_{n=0}^{\infty} E^{\otimes n}, \tag{4.6}$$

or, more succinctly,

$$\varphi_\infty(a) = \operatorname{diag}(a, \varphi^{(1)}(a), \varphi^{(2)}(a), \dots).$$

In addition to the operators $\varphi_\infty(a) \in \mathcal{L}^a(\mathcal{F}^2(E))$, we introduce the so-called *creation operators* on $\mathcal{F}^2(E)$ given, for each $\xi \in E$, by the subdiagonal (or shift) block matrix

$$T_\xi = \begin{bmatrix} 0 & 0 & 0 & \cdots \\ T_\xi^{(0)} & 0 & 0 & \cdots \\ 0 & T_\xi^{(1)} & 0 & \cdots \\ \vdots & \ddots & \ddots & \ddots \end{bmatrix}$$

where the block entry $T_\xi^{(n)}\colon E^{\otimes n} \to E^{\otimes n+1}$ is given by

$$T_\xi^{(n)}\colon \xi_n \otimes \cdots \otimes \xi_1 \mapsto \xi \otimes \xi_n \otimes \cdots \otimes \xi_1.$$

The operator T_ξ is also in $\mathcal{L}^a(\mathcal{F}^2(E))$. In summary, both T_ξ and $\varphi_\infty(a)$ are right $\mathcal{A}$-module maps with respect to the right $\mathcal{A}$-action on $\mathcal{F}^2(E)$ for each $\xi \in E$ and $a \in \mathcal{A}$. Moreover, one easily checks that

$$\varphi_\infty(a)T_\xi = T_{a\xi} = T_{\varphi(a)\xi} \quad \text{and} \quad T_\xi\varphi_\infty(a) = T_{\xi a} \quad \text{for each } a \in \mathcal{A} \text{ and } \xi \in E.$$

We let $\mathcal{F}^\infty(E)$ denote the *Toeplitz algebra* equal to the weak-$*$ closed algebra generated by the linear span of the collection of operators

$$\{\varphi_\infty(a),\ T_\xi\colon a \in \mathcal{A} \text{ and } \xi \in E\}$$

in the W^*-algebra $\mathcal{L}^a(\mathcal{F}(E))$. The justification for the term "Toeplitz algebra" comes from the following proposition, which is a variation on Proposition 4.2 in [16].

Proposition 4.1. *If $R \in \mathcal{L}^a(\mathcal{F}^2(E))$ is in the Toeplitz algebra $\mathcal{F}^\infty(E)$ with matrix representation*

$$R = [R_{i,j}]_{i,j=0,1,2,\dots} \tag{4.7}$$

where $R_{i,j} \in \mathcal{L}^a(E^{\otimes j}, E^{\otimes i})$, then there exists a sequence $\xi^{(0)}, \xi^{(1)}, \xi^{(2)}, \dots$ with $\xi^{(n)} \in E^{\otimes n}$ such that

$$R_{i,j} = \begin{cases} 0 & \text{if } i < j, \\ T^{(0)}_{\xi^{(i-j)}} \otimes I_{E^{\otimes j}} & \text{if } i \geq j, \end{cases} \tag{4.8}$$

where for $n = 0, 1, 2, \dots$ the operator $T^{(0)}_{\xi^{(n)}} \in \mathcal{L}^a(\mathcal{A}, E^{\otimes n})$ is given by

$$T^{(0)}_{\xi^{(n)}} a = \xi^{(n)} a \quad (a \in \mathcal{A}). \tag{4.9}$$

In particular, R is completely determined by the entries of its first column. Conversely, if $\xi^{(0)}, \xi^{(1)}, \xi^{(2)}, \dots$ is a sequence with $\xi^{(n)} \in E^{\otimes n}$ such that the infinite operator matrix R given by (4.7), (4.8) *and* (4.9) *induces a bounded operator on $\mathcal{F}^2(E)$, then R is in $\mathcal{F}^\infty(E)$.*

Proof. Let $\mathcal{F}^{\infty\prime}(E)$ be the set of all adjointable operators on $\mathcal{F}^2(E)$ with matrix representation $R = [R_{i,j}]_{i,j=0,1,\dots}$ given by (4.8) and (4.9). Note first that $\mathcal{F}^{\infty\prime}(E)$ is an algebra; indeed if S and R in $\mathcal{F}^{\infty\prime}(E)$ are given by the sequences $\xi^{(0)}, \xi^{(1)}, \xi^{(2)}, \dots$ and $\zeta^{(0)}, \zeta^{(1)}, \zeta^{(2)}, \dots$ with $\xi^{(n)}, \zeta^{(n)} \in E^{\otimes n}$, respectively, then it is straightforward to check that SR is the element of $\mathcal{F}^{\infty\prime}(E)$ given by the sequence $\rho^{(0)}, \rho^{(1)}, \rho^{(2)}, \dots$ with $\rho^{(n)} \in E^{\otimes n}$ equal to

$$\rho^{(n)} = \sum_{k=0}^{n} \xi^{(k)} \otimes \zeta^{(n-k)}.$$

To show that $\mathcal{F}^\infty(E) \subset \mathcal{F}^{\infty\prime}(E)$, it therefore suffices to check that (1) each of the generators $\varphi_\infty(a)$ and T_ξ (for $a \in \mathcal{A}$ and $\xi \in E$) is in $\mathcal{F}^{\infty\prime}(E)$, (2) that $\mathcal{F}^{\infty\prime}(E)$ is closed under addition, and that (3) $\mathcal{F}^{\infty\prime}(E)$ is weak-$*$ closed. These verifications are straightforward and are left to the reader.

Conversely, to show that $\mathcal{F}^{\infty\prime}(E) \subset \mathcal{F}^\infty(E)$, since $\mathcal{F}^\infty(E)$ is weak-$*$ closed by definition, it suffices to consider that case of an operator $R \in \mathcal{F}^{\infty\prime}(E)$ with support only on a subdiagonal: $R_{i,j} = 0$ unless $i - j = k$ for some $k \geq 0$. Moreover, we can restrict to the case that $R_{k,0} \in \mathcal{L}^a(\mathcal{A}, E^{\otimes k})$ is defined by some pure tensor $\xi_k \otimes \cdots \otimes \xi_1$ in $E^{\otimes k}$ via $R_{k,0}a = \xi_k \otimes \cdots \otimes \xi_2 \otimes (\xi_1 a)$ for $a \in \mathcal{A}$. To see that this is the case, first note that, since $E^{\otimes k}$ is an $\mathcal{A}$-W^*-correspondence, $R_{k,0}$ is of the form $R_{k,0}a = \xi^{(k)}a$ for some $\xi^{(k)} \in E^{\otimes k}$. The claim then follows since any element $\xi^{(k)} \in E^{\otimes k}$ can be approximated by linear combinations of pure tensors in $E^{\otimes k}$, and because $\mathcal{F}^\infty(E)$ is weak-$*$ closed. Finally, assume that R is only supported on the kth diagonal and that $R_{k,0}$ is defined by the pure tensor $\xi_k \otimes \cdots \otimes \xi_1 \in E^{\otimes k}$ as described above. It is then straightforward to check that $R = T_{\xi_k} \cdots T_{\xi_1}$. Thus R is in $\mathcal{F}^\infty(E)$. □

4.3. Correspondence-representation pairs and their dual

In addition to the von Neumann algebra $\mathcal{A}$ and the $\mathcal{A}$-correspondence E, suppose that we are also given an auxiliary Hilbert space $\mathcal{E}$ and a representation (meaning a nondegenerate $*$-homomorphism) $\sigma\colon \mathcal{A} \to \mathcal{L}(\mathcal{E})$; as this will be the setting for much of the analysis to follow, we refer to such a pair (E, σ) as a *correspondence-representation pair.* We further assume that σ is faithful (injective) and normal (σ-weakly continuous). Then the Hilbert space $\mathcal{E}$ equipped with σ becomes an $(\mathcal{A}, \mathbb{C})$-correspondence with left $\mathcal{A}$-action given by σ:

$$a \cdot y = \sigma(a)y \text{ for all } a \in \mathcal{A} \text{ and } y \in \mathcal{E}.$$

Thus we can form the tensor-product $(\mathcal{A}, \mathbb{C})$-correspondence $E \otimes_\sigma \mathcal{E}$. With E^σ we denote the set of all bounded linear operators $\mu\colon \mathcal{E} \to E \otimes_\sigma \mathcal{E}$ which are also left $\mathcal{A}$-module maps:

$$E^\sigma = \{\mu\colon \mathcal{E} \to E \otimes_\sigma \mathcal{E}\colon \mu\sigma(a) = (\varphi(a) \otimes I_\mathcal{E})\mu\}. \tag{4.10}$$

It turns out that E^σ is itself a W^*-correspondence (the correspondence *dual* to E^σ; see [69, Section 3]), not over $\mathcal{A}$ but over the W^*-algebra

$$\sigma(\mathcal{A})' = \{b \in \mathcal{L}(\mathcal{E})\colon b\sigma(a) = \sigma(a)b \text{ for all } a \in \mathcal{A}\} \subset \mathcal{L}(\mathcal{E})$$

(the *commutant* of the image $\sigma(\mathcal{A})$ of the representation σ in $\mathcal{L}(\mathcal{E})$) with left and right $\sigma(\mathcal{A})'$-action and $\sigma(\mathcal{A})'$-valued inner-product $\langle \cdot, \cdot \rangle_{E^\sigma}$ given by

$$\mu \cdot b = \mu b, \quad b \cdot \mu = (I_\mathcal{E} \otimes b)\mu,$$
$$\langle \mu, \nu \rangle_{E^\sigma} = \nu^* \mu.$$

The intertwining relations of elements μ and ν in E^σ with $\sigma(a)$ and $a \otimes I_\mathcal{E}$ for $a \in \mathcal{A}$ imply that $\nu^*\mu$ is indeed in $\sigma(\mathcal{A})'$.

Notice that elements $b \in \sigma(\mathcal{A})'$ and $\mu \in E^\sigma$ define operators on the $(\mathcal{A}, \mathbb{C})$-correspondence

$$\mathcal{F}^2(E, \sigma) := \mathcal{F}^2(E) \otimes_\sigma \mathcal{E} = \oplus_{n=0}^\infty E^{\otimes n} \otimes_\sigma \mathcal{E}$$

via the operator matrices

$$I_{\mathcal{F}^2(E)} \otimes b = \operatorname{diag}(b, I_E \otimes b, I_{E^{\otimes 2}} \otimes b, \dots)$$

and

$$I_{\mathcal{F}^2(E)} \otimes \mu = \begin{bmatrix} 0 & 0 & 0 & 0 & \cdots \\ \mu & 0 & 0 & 0 & \cdots \\ 0 & I_E \otimes \mu & 0 & 0 & \cdots \\ 0 & 0 & I_{E^{\otimes 2}} \otimes \mu & 0 & \cdots \\ \vdots & \vdots & \ddots & \ddots & \ddots \end{bmatrix}.$$

Here, in interpreting $I_{E^{\otimes n}} \otimes \mu$, we use the identification

$$E^{\otimes n} \otimes_\sigma \mathcal{E} = E^{\otimes n-1} \otimes (E \otimes_\sigma \mathcal{E}). \tag{4.11}$$

The operators $I_{E^{\otimes n}} \otimes b$ and $I_{E^{\otimes n}} \otimes \mu$ are well defined because b and μ are left $\mathcal{A}$-module maps.

The following theorem provides another way of characterizing the elements of $\mathcal{F}^\infty(E)$ using the operators $I_{\mathcal{F}^2(E)} \otimes b$ and $I_{\mathcal{F}^2(E)} \otimes \mu$. The result follows directly from the combination of Theorems 3.9 and 3.10 in [69].

Proposition 4.2. *Given an operator $X \in \mathcal{L}(\mathcal{F}^2(E,\sigma))$, there exists an $R \in \mathcal{F}^\infty(E)$ so that $X = R \otimes I_\mathcal{E}$ if and only if X commutes with $I_{\mathcal{F}^2(E)} \otimes b$ and $I_{\mathcal{F}^2(E)} \otimes \mu$ for each $b \in \sigma(\mathcal{A})'$ and $\mu \in E^\sigma$. Moreover, if $X = R \otimes I_\mathcal{E}$ for some $R \in \mathcal{F}^\infty(E)$, then $\|X\| = \|R\|$.*

We can then prove the following concrete version of the C^*-correspondence commutant lifting theorem [67, Theorem 4.4].

Theorem 4.3. *Given subspaces $\mathcal{M}$ and $\mathcal{N}$ of $\mathcal{F}^2(E,\sigma)$ that are both invariant under $I_{\mathcal{F}^2(E)} \otimes b$ and $I_{\mathcal{F}^2(E)} \otimes \mu^*$ for all $b \in \sigma(\mathcal{A})'$ and $\mu \in E^\sigma$, and a contractive operator X from $\mathcal{M}$ into $\mathcal{N}$ such that*

$$X^*(I_{\mathcal{F}^2(E)} \otimes b)|_\mathcal{N} = (I_{\mathcal{F}^2(E)} \otimes b)X^* \quad \text{and} \quad X^*(I_{\mathcal{F}^2(E)} \otimes \mu^*)|_\mathcal{N} = (I_{\mathcal{F}^2(E)} \otimes \mu^*)X^*,$$

for all $b \in \sigma(\mathcal{A})'$ and $\mu \in E^\sigma$, there exists an $S \in \mathcal{F}^\infty(E)$ with $\|S\| \le 1$ so that $(S^* \otimes I_\mathcal{E})\mathcal{N} \subset \mathcal{M}$ and $X = P_\mathcal{N}(S \otimes I_\mathcal{E})|_\mathcal{M}$.

Proof. Using the C^*-correspondence commutant lifting theorem [67, Theorem 4.4] and the intertwining relations of X it follows that X can be lifted to a contractive operator Y on $\mathcal{F}^2(E,\sigma)$ that commutes with $I_{\mathcal{F}^2(E)} \otimes b$ and $I_{\mathcal{F}^2(E)} \otimes \mu$ for each $b \in \sigma(\mathcal{A})'$ and $\mu \in E^\sigma$, with the property that $Y^*\mathcal{N} \subset \mathcal{M}$ and $X = P_\mathcal{N} Y|_\mathcal{M}$. The claim then follows immediately with Proposition 4.2. □

4.4. The generalized disk $\mathbb{D}((E^\sigma)^*)$ and the first Muhly-Solel point-evaluation

For the definition of point-evaluations to follow, however, the important object is $(E^\sigma)^*$, the set of adjoints of elements of E^σ (which are also left $\mathcal{A}$-module maps):

$$(E^\sigma)^* = \{\eta : E \otimes_\sigma \mathcal{E} \to \mathcal{E} : \eta^* \in E^\sigma\}. \tag{4.12}$$

For a given $\eta \in (E^\sigma)^*$ and a positive integer n, we may define the generalized power $\eta^n : E^{\otimes n} \otimes_\sigma \mathcal{E} \to \mathcal{E}$ by

$$\eta^n = \eta(I_E \otimes \eta) \cdots (I_{E^{\otimes n-1}} \otimes \eta)$$

where we again use the identification (4.11) in this definition. We also set $\eta^0 = I_\mathcal{E} \in \mathcal{L}(\mathcal{E})$. Again the fact that η is a left $\mathcal{A}$-module map ensures that $I_{E^{\otimes k}} \otimes \eta$ is a well-defined operator in $\mathcal{L}(E^{\otimes k+1} \otimes_\sigma \mathcal{E}, E^{\otimes k} \otimes_\sigma \mathcal{E})$. The defining $\mathcal{A}$-module property of η in (4.12) then extends to the generalized powers η^n in the form

$$\eta^n(\varphi^{(n)}(a) \otimes I_\mathcal{E}) = \sigma(a)\eta^n, \tag{4.13}$$

i.e., η^n is also an $\mathcal{A}$-module map.

Denote by $\mathbb{D}((E^\sigma)^*)$ the set of strictly contractive elements of $(E^\sigma)^*$:

$$\mathbb{D}((E^\sigma)^*) = \{\eta \in (E^\sigma)^* : \|\eta\| < 1\}.$$

We then consider elements $R \in \mathcal{F}^\infty(E)$ as functions $\widehat{R}$ on $\mathbb{D}((E^\sigma)^*)$ with values in $\mathcal{L}(\mathcal{E})$ according to the formula for the *first Muhly-Solel point evaluation*

$$\widehat{R}(\eta) = \sum_{n=0}^{\infty} \eta^n (R_{n,0} \otimes I_{\mathcal{E}}) \text{ where } R = [R_{i,j}] \in \mathcal{F}^\infty(E). \tag{4.14}$$

From the facts that $\|\eta\| < 1$ and $\|R_{n,0}\| \leq M < \infty$ (since R is bounded on $\mathcal{F}^2(E)$), one can see that the series in (4.14) converges in operator norm.

Remark 4.4. The definition of the first Muhly-Solel point-evaluation in [69, 70, 71] is actually more elaborate. There it is observed that an element $\eta \in \overline{\mathbb{D}}((E^\sigma)^*)$ induces an ultraweakly continuous completely contractive bimodule map $\widehat{\eta}\colon E \to \mathcal{L}(\mathcal{E})$ via the formula

$$\widehat{\eta}(\xi)e = \eta(\xi \otimes e) \text{ for } \xi \in E,\ e \in \mathcal{E}. \tag{4.15}$$

The bimodule property means that

$$\widehat{\eta}(\varphi(a)\xi b) = \sigma(a)\widehat{\eta}(\xi)\sigma(b) \text{ for all } a, b \in \mathcal{A},\ \xi \in E.$$

Conversely, if $S\colon E \to \mathcal{L}(\mathcal{E})$ is an ultraweakly continuous completely contractive bimodule map, then the same formula (4.15) turned around

$$\widetilde{S}(\xi \otimes e) = S(\xi)e \tag{4.16}$$

can be used to define an element $\widetilde{S}$ of the closed generalized disk $\overline{\mathbb{D}}((E^\sigma)^*)$. The pair $(\widehat{\eta}, \sigma)$ (or (S, σ)) is said to be a *covariant representation* of the correspondence E. Given a covariant representation (S, σ), in case $\|S\| < 1$ (or, equivalently, $\|\widetilde{S}\| < 1$), there is an associated completely contractive representation $S \times \sigma$ of the Toeplitz algebra $\mathcal{F}^\infty(E)$ defined on generators via

$$\begin{aligned} (S \times \sigma)(\varphi_\infty(a)) &= \sigma(a), \\ (S \times \sigma)(T_\xi) &= S(\xi). \end{aligned} \tag{4.17}$$

Conversely, if ρ is any ultraweakly continuous completely contractive $\mathcal{L}(\mathcal{E})$-valued representation of $\mathcal{F}^\infty(E)$, the same formulas (4.17) can be turned around

$$\begin{aligned} \sigma(a) &= \rho(\varphi_\infty(a)), \\ S(\xi) &= \rho(T_\xi) \end{aligned} \tag{4.18}$$

to define a covariant representation (S, σ) of E. Given any covariant representation (S, σ) of E, the formulas (4.17) can always be extended to define a representation of the *uniform closure* $\mathcal{T}^\infty(E)$ of the algebra generated by $\varphi_\infty(a)$ and T_ξ ($a \in \mathcal{A}$ and $\xi \in E$). It is an unsolved problem in the theory to identify which covariant representations (S, σ) have the property that (4.17) can be extended to define an ultraweakly continuous completely contractive representation of the weak-$*$ closed Toeplitz algebra $\mathcal{F}^\infty(E)$.

If we fix a representation $\sigma\colon \mathcal{A} \to \mathcal{L}(\mathcal{E})$, and let $\eta \in \mathbb{D}((E^\sigma)^*)$ and $R \in \mathcal{F}^\infty(E)$, then the value $\widehat{R}(\eta)$ of R at η is defined to be the element of $\mathcal{L}(\mathcal{E})$ assigned to R by the representation $\widehat{\eta} \times \sigma$:

$$\widehat{R}(\eta) = (\widehat{\eta} \times \sigma)(R).$$

In the end this definition agrees with the definition (4.14). The formula (4.14) for the point-evaluation is called the *Cauchy transform* in [69].

Following [70] (see also [16]), given a correspondence-representation pair (E, σ), we define the *Schur class* $\mathcal{S}_{E,\sigma}$ to be the class of all functions $S\colon \mathbb{D}((E^\sigma)^*) \to \mathcal{L}(\mathcal{E})$ which can be expressed in the form $S(\eta) = \widehat{R}(\eta)$ for some $R \in \mathcal{F}^\infty(E)$ with $\|R\| \leq 1$. The associated left-tangential Nevanlinna-Pick interpolation problem is: *Given a subset* Ω *of* $\mathbb{D}((E^\sigma)^*)$ *and two functions* $F\colon \Omega \to \mathcal{L}(\mathcal{E})$ *and* $G\colon \Omega \to \mathcal{L}(\mathcal{E})$, *determine when there exists a function* S *in the Schur class* $\mathcal{S}_{E,\sigma}$ *so that* $GS|_\Omega = F$. We shall explain the solution to this problem due to Muhly-Solel [69] in Subsection 4.6. For the moment we note the following two extreme cases:

(I) $\Omega = \mathbb{D}((E^\sigma)^*)$ **and** $G(\eta) = I_{\mathcal{E}}$. In this case the function F is defined on all of $\mathbb{D}((E^\sigma)^*)$ and we seek a test to decide if $F \in \mathcal{S}_{E,\sigma}$.

(II) Ω **finite.** In this case Ω consists of finitely many points, say $\eta_1, \dots, \eta_N \in \mathbb{D}((E^\sigma)^*)$, and we are given operators $Y_1 = F(\eta_1), \dots, Y_N = F(\eta_N)$ and $Y_1 = F(\eta_1), \dots, Y_N = F(\eta_N)$ in $\mathcal{L}(\mathcal{E})$. We seek a test to decide if there is a function S (or, more ambitiously, a description of all such functions S) in the Schur class $\mathcal{S}_{E,\sigma}$ satisfying the interpolation conditions

$$X_j S(\eta_j) = Y_j \text{ for } j = 1, \dots, N.$$

4.5. Positive and completely positive kernels/maps

The solution of the Nevanlinna-Pick interpolation problem in [69] involves the notion of a completely positive map while the characterization of the Schur class $\mathcal{S}_{E,\sigma}$ in [70] involves the notion of *completely positive kernel* introduced in [31].

In the following discussion $\mathcal{A}$, $\mathcal{B}$ and $\mathcal{C}$ are C^*-algebras (in particular, possibly W^*-algebras) and Ω is a set. We begin with the notion of *positive kernel* which goes back at least to Aronszajn [12]; a function $K\colon \Omega \times \Omega \to \mathcal{C}$ is said to be a *positive kernel* if

$$\sum_{i,j=1}^{n} c_i^* K(\omega_i, \omega_j) c_j \geq 0$$

for all choices of $\omega_1, \dots, \omega_n \in \Omega$ and $c_1, \dots, c_n \in \mathcal{C}$ for $n = 1, 2, \dots$. Equivalently, for every choice of n points $\omega_1, \dots, \omega_n \in \Omega$, the matrix

$$[K(\omega_i, \omega_j)]_{i,j=1,\dots,n}$$

is a positive element of $\mathcal{C}^{n \times n}$ where $n = 1, 2, \dots$. Given two C^*-algebras $\mathcal{A}$ and $\mathcal{B}$, a map $\varphi\colon \mathcal{A} \to \mathcal{B}$ is said to be a *positive map* if

$$a \geq 0 \text{ in } \mathcal{A} \Longrightarrow \varphi(a) \geq 0 \text{ in } \mathcal{B}.$$

Such a map $\varphi\colon \mathcal{A} \to \mathcal{B}$ is said to be *completely positive* if

$$\begin{bmatrix} a_{11} & \dots & a_{1n} \\ \vdots & & \vdots \\ a_{n1} & \dots & a_{nn} \end{bmatrix} \succeq 0 \text{ in } \mathcal{A}^{n\times n} \Longrightarrow \begin{bmatrix} \varphi(a_{11}) & \dots & \varphi(a_{1n}) \\ \vdots & & \vdots \\ \varphi(a_{n1}) & \dots & \varphi(a_{nn}) \end{bmatrix} \succeq 0 \text{ in } \mathcal{B}^{n\times n} \quad (4.19)$$

for every $n = 1, 2, \dots$. If the implication (4.19) holds for a fixed n in $\mathbb{Z}_+$, then we say that φ is *n-positive.* In case $\mathcal{A} = \mathbb{C}^{N\times N}$, a result of Choi [38] (see also [77, Theorem 3.14]) says that $\varphi\colon \mathcal{A} \to \mathcal{B}$ *is completely positive if and only if the single block matrix*

$$\begin{bmatrix} \varphi(e_{11}) & \dots & \varphi(e_{1N}) \\ \vdots & & \vdots \\ \varphi(e_{N1}) & \dots & \varphi(e_{NN}) \end{bmatrix}$$

is positive in $\mathcal{B}^{N\times N}$, where e_{ij} are the standard matrix units

$$[e_{ij}]_{\alpha,\beta} = \delta_{i,\alpha}\delta_{j,\beta}$$

in $\mathbb{C}^{N\times N}$ (here we make use of the Kronecker delta – $\delta_{i,j} = 1$ for $i = j$ and $\delta_{i,j} = 0$ for $i \neq j$). There is an alternative characterization in terms of positive kernels (see [103]): *A map* $\varphi\colon \mathcal{A} \to \mathcal{B}$ *is completely positive if and only if the kernel* $k_\varphi\colon \mathcal{A} \times \mathcal{A} \to \mathcal{B}$ *defined by*

$$k_\varphi(a, a') = \varphi(a^* a')$$

is a positive kernel. Barreto-Bhat-Liebscher-Skeide in [31] (with motivation from quantum physics which need not concern us here) combined these notions as follows: we say that a kernel $\mathbb{K}\colon \Omega \times \Omega \to \mathcal{L}(\mathcal{A}, \mathcal{B})$ is *completely positive* if the associated kernel $k\colon (\Omega \times \mathcal{A}) \times (\Omega \times \mathcal{A}) \to \mathcal{B}$ given by

$$k((\omega, a), (\omega', a')) = \mathbb{K}(\omega, \omega')[a^* a']$$

is a positive kernel, i.e., if, for all choices $(\omega_1, a_1), \dots, (\omega_n, a_n)$ in $\Omega \times \mathcal{A}$ and for all choices $b_1, \dots, b_n$ in $\mathcal{B}$ it is the case that

$$\sum_{i,j=1}^{n} b_i^* \mathbb{K}(\omega_i, \omega_j)[a_i^* a_j] b_j \geq 0 \text{ in } \mathcal{B} \quad (4.20)$$

for $n = 1, 2, \dots$. Note that the notion of completely positive kernel contains the notions of positive kernel and of completely positive map as special cases: in case the set Ω consists of a single point, then $\mathbb{K}$ can be considered simply as a linear map from $\mathcal{A}$ to $\mathcal{B}$, and the condition that $\mathbb{K}$ be a completely positive kernel collapses to the condition that $\mathbb{K}$ be a completely positive map from $\mathcal{A}$ to $\mathcal{B}$ by the positive-kernel formulation of completely positive map mentioned above. On the other extreme, if $\mathcal{A}$ is just the complex numbers $\mathbb{C}$, then the condition that $\mathbb{K}$ be completely positive collapses to the condition that the kernel $k(\cdot,\cdot) := \mathbb{K}(\cdot,\cdot)[1]$ be a positive kernel.

A number of equivalent characterizations of complete positivity for a kernel $\mathbb{K}\colon \Omega \times \Omega \to \mathcal{L}(\mathcal{A}, \mathcal{B})$ is given in [31]. Let us mention some of these which will be convenient for our analysis of various generalized Nevanlinna-Pick interpolation problems.

Proposition 4.5. *Suppose that* $\mathbb{K}$ *is a function from* $\Omega \times \Omega$ *into* $\mathcal{L}(\mathcal{A}, \mathcal{B})$. *Then the following are equivalent:*

1. $\mathbb{K}$ *is a completely positive kernel, i.e., the kernel* $k\colon (\Omega \times \mathcal{A}) \times (\Omega \times \mathcal{A}) \to \mathcal{B}$ *given by*
$$k((\omega, a), (\omega', a')) = \mathbb{K}(\omega, \omega')[a^* a']$$
is a positive kernel.
2. *For every choice of* n *points* $\omega_1, \dots, \omega_n$ *in* Ω, *the map* $\varphi_{\omega_1,\dots,\omega_n}\colon \mathcal{A}^{n\times n} \to \mathcal{B}^{n\times n}$ *given by*
$$\varphi_{\omega_1,\dots,\omega_n}([a_{ij}]_{i,j=1,\dots,n}) = [\mathbb{K}(\omega_i, \omega_j)[a_{ij}]]_{i,j=1,\dots,n} \tag{4.21}$$
is a completely positive map for any $n = 1, 2, \dots$.
3. *For every choice of* n *points* $\omega_1, \dots, \omega_n$ *in* Ω, *the map* $\varphi_{\omega_1,\dots,\omega_n}\colon \mathcal{A}^{n\times n} \to \mathcal{B}^{n\times n}$ *given by* (4.21) *is a positive map for any* $n = 1, 2, \dots$.

Proof. The equivalence of (1), (2) and (3) correspond to the equivalence of (1), (4) and (5) in Lemma 3.2.1 from [31]. □

An important example of a completely positive kernel for our purposes here appearing implicitly in the work of Muhly-Solel [69, 70] and developed explicitly in [16] is the Szegő kernel associated with a correspondence-representation pair (E, σ) and defined as follows. We suppose that we are given an $\mathcal{A}$-W^*-correspondence E together with a representation $\sigma\colon \mathcal{A} \to \mathcal{L}(\mathcal{E})$ for a Hilbert space $\mathcal{E}$. We then define the associated *Szegő kernel*
$$\mathbb{K}_{E,\sigma}\colon \mathbb{D}((E^\sigma)^*) \times \mathbb{D}((E^\sigma)^*) \to \mathcal{L}(\sigma(\mathcal{A})', \sigma(\mathcal{A})')$$
by
$$\mathbb{K}_{E,\sigma}(\eta, \zeta)[b] = \sum_{n=0}^{\infty} \eta^n (I_{E^{\otimes n}} \otimes b) \zeta^{n*}. \tag{4.22}$$
Then it can be seen that $\mathbb{K}_{E,\sigma}$ is a completely positive kernel and in fact is the reproducing kernel for a reproducing kernel correspondence $H^2(E, \sigma)$ whose elements are $\mathcal{E}$-valued functions on the generalized disk $\mathbb{D}((E^\sigma)^*)$ which also carries a representation ι_∞ of $\sigma(\mathcal{A})'$. This construction is based on yet another and perhaps the most fundamental characterization of completely positive kernel which is not mentioned in Proposition 4.5, namely: $\mathbb{K}$ *is completely positive if and only if* $\mathbb{K}$ *has a* Kolmogorov decomposition *in the sense of* [31]; the reproducing-kernel correspondence associated with the completely positive kernel can then be taken to be the middle space in its Kolmogorov decomposition and the Schur class $\mathcal{S}_{E,\sigma}$ can alternatively be characterized as the space of contractive-multiplier $\sigma(\mathcal{A})'$-module maps acting on $H^2(E, \sigma)$. For further details on these ideas, we refer to [31].

In short there are at least two points of view to the Schur class $\mathcal{S}_{E,\sigma}$. The first is the connection with the representation theory for $\mathcal{F}^\infty(E)$ as developed in the work of Muhly-Solel [69, 70, 71], while the second is the view of the Schur class as contractive-multiplier module maps from [16]. Our purpose here is to ignore all this finer structure and use the most direct definition (4.14) of the point-evaluation to make explicit how the Muhly-Solel Nevanlinna-Pick interpolation theorem hooks up with the existing literature on assorted generalizations (in particular, to multivariable contexts) of Nevanlinna-Pick interpolation.

4.6. A W^*-correspondence Nevanlinna-Pick interpolation theorem

The solution to the W^*-correspondence version of the Nevanlinna-Pick interpolation problem is as follows. In the statement of the result we make use of the Szegő kernel $\mathbb{K}_{E,\sigma}$ defined as in (4.22).

Theorem 4.6. *Suppose that we are given a correspondence-representation pair (E,σ) together with a subset Ω of $\mathbb{D}((E^\sigma)^*)$ and two functions $F\colon \Omega \to \mathcal{L}(\mathcal{E})$ and $G\colon \Omega \to \mathcal{L}(\mathcal{E})$. Then the following are equivalent.*

1. *There exists a function $S\colon \mathbb{D}((E^\sigma)^*) \to \mathcal{L}(\mathcal{E})$ in the Schur class $\mathcal{S}_{E,\sigma}$ such that*
$$GS|_\Omega = F.$$
2. *The kernel $\mathbb{K}_F\colon \Omega \times \Omega \to \mathcal{L}(\sigma(\mathcal{A})', \mathcal{L}(\mathcal{E}))$ given by*
$$\mathbb{K}_F(\eta,\eta')[b] = G(\eta)\mathbb{K}_{E,\sigma}(\eta,\eta')[b]G(\zeta)^* - F(\eta)\mathbb{K}_{E,\sigma}(\eta,\zeta)[b]F(\zeta)^*$$
is completely positive.
3. *For each choice of $\eta_1,\dots,\eta_n \in \omega$, the map $\varphi_{\eta_1,\dots,\eta_n}$ from $(\sigma(\mathcal{A})')^{n\times n}$ to $\mathcal{L}(\mathcal{E})^{n\times n}$ given by*
$$\begin{aligned}&\varphi_{\eta_1,\dots,\eta_n}\left([b_{ij}]_{i,j=1}^n\right)\\ &\quad = \left[G(\eta_i)\mathbb{K}_{E,\sigma}(\eta_i,\eta_j)[b_{ij}]G(\eta_j)^* - F(\eta_i)\mathbb{K}_{E,\sigma}(\eta_i,\eta_j)[b_{ij}]F(\eta_j)^*\right]_{i,j=1}^n\end{aligned}$$
is a completely positive (or even just a positive) map for any $n = 1,2,3,\dots$.

Remark 4.7. In case Ω is finite, say $\Omega = \{\eta_1,\dots,\eta_N\}$, and $X_1 = G(\eta_1),\dots,X_N = G(\eta_N)$ and $Y_1 = F(\eta_1),\dots,Y_N = F(\eta_N)$, we can always assume n in condition (3) to be of the form $n = kN$ for some positive integer k and the sequence of points from X to be the sequence $\eta_1,\dots,\eta_N$ repeated k times. It then follows that condition (3) can be written as: *The map $\varphi : \sigma(\mathcal{A})'^{N\times N} \to \mathcal{L}(\mathcal{E})^{N\times N}$ given by*
$$\varphi\left(\left[\, b_{i,j}\,\right]_{i,j=1}^N\right) = \left[X_i\mathbb{K}_{E,\sigma}(\eta_i,\eta_j)[b_{ij}]X_j^* - Y_i\mathbb{K}_{E,\sigma}(\eta_i,\eta_j)[b_{ij}]Y_j^*\right]_{i,j=1}^N \tag{4.23}$$
is a completely positive map.

Proof of Theorem 4.6. Note that the equivalence of (2) and (3) is just a particular case of Proposition 4.5. The equivalence of (1) and (3) for the case that Ω is finite is given in [69] (see also Remark 4.7). For the case where $\Omega = \mathbb{D}((E^\sigma)^*)$ and $G(\eta) = I_\mathcal{E}$ for each $\eta \in \Omega$ the equivalence of (1) and (2) is given in [70] (see also [16] for the reproducing-kernel point of view). The case of a general Ω can be seen to follow from the case of a finite Ω via a standard weak-$*$ compactness argument; one way to organize this argument is as an application of Kurosh's Theorem (see [6, page 30]). □

4.7. The second Muhly-Solel point-evaluation and associated Nevanlinna-Pick theorem

In the recent paper [71] Muhly and Solel introduced a Poisson kernel, and applied this object to define a second point-evaluation for elements of the Toeplitz algebra $H^\infty(E)$. As pointed out in [71], this point-evaluation has certain characteristics that resemble those of the point-evaluation used in discrete-time time-varying interpolation and system theory, as developed in the 1990s; cf. [23, 8, 53]. In [52] it was observed that many time-varying interpolation problems can be recast as classical interpolation problems with an operator argument. In this section we prove a Nevanlinna-Pick interpolation theorem for the alternative point-evaluation of [71]; in the examples to follow (see Subsections 4.8 and 4.9 below) we show that this Nevanlinna-Pick theorem indeed corresponds to the operator-argument versions in the settings considered there.

For the second point-evaluation, the points are formed by pairs (ζ, a), where a is from the W^*-algebra $\mathcal{A}$ and ζ is from the set

$$\mathbb{D}(E^*) := \{\zeta \colon \zeta^* \in E, \ \|\zeta^*\|_E < 1\}.$$

Given such a pair (ζ, a) we set

$$\zeta^*_{(n)} := \zeta^* \otimes \cdots \otimes \zeta^* \in E^{\otimes n} \text{ for } n = 1, 2, \ldots,$$

and define

$$W_{\zeta^*, a^*} = \varphi_\infty(a^*) \begin{bmatrix} 1_{\mathcal{A}} \\ T^{(0)}_{\zeta^*_{(1)}} \\ T^{(0)}_{\zeta^*_{(2)}} \\ \vdots \end{bmatrix} \in \mathcal{L}^a(\mathcal{A}, \mathcal{F}^2(E)),$$

where $\varphi_\infty(a^*)$ and $T^{(0)}_{\zeta^*_{(n)}}$ are given by (4.6) and (4.9), respectively. The Poisson kernel defined in [71] is the special case of W_{ζ^*,a^*} with $a^* = T^{(0)*}_{\zeta^*} T^{(0)}_{\zeta^*}$. In that case W_{ζ^*,a^*} is well defined even if $\|\zeta^*\| = 1$, and W_{ζ^*,a^*} is a contractive operator. It is easy to see that for each $\zeta \in \mathbb{D}(E^*)$ and $a \in \mathcal{A}$ the operator W_{ζ^*,a^*} still defines a bounded operator, but not necessarily contractive.

For an $R \in \mathcal{F}^\infty(E)$ we define the evaluation of R in a point (ζ, a) from $\mathbb{D}(E^*) \times \mathcal{A}$ by

$$\widehat{R}(\zeta, a) = W^*_{\zeta^*,a^*} R|_{\mathcal{A}} \in \mathcal{L}^a(\mathcal{A}) = \mathcal{A}.$$

Here we used the fact that for any W^*-algebra $\mathcal{A}$, considered in the standard way as a W^*-correspondence over itself, we can identify $\mathcal{L}^a(\mathcal{A})$ with the C^*-algebra $\mathcal{A}$ itself. If $R \in \mathcal{F}^\infty(E)$ is given by a sequence of elements of $E^{\otimes n}$, $n = 0, 1, \ldots$, as in Proposition 4.1, then $\widehat{R}(\zeta, a)$ can be written more concretely.

Proposition 4.8. *Let $R \in \mathcal{F}^\infty(E)$ be given by (4.7)–(4.9) with $\xi^{(n)} \in E^{\otimes n}$ for $n = 0, 1, 2, \ldots$. Then for any $(\zeta, a) \in \mathbb{D}(E^*) \times \mathcal{A}$ we have*

$$\widehat{R}(\zeta, a) = \sum_{n=0}^{\infty} \langle a\xi^{(n)}, \zeta_{(n)}^* \rangle.$$

Proof. The statement follows directly from the fact that for any $a' \in \mathcal{A}$ we have

$$\begin{aligned}\widehat{R}(\zeta, a)a' &= \sum_{n=0}^{\infty} T_{\zeta_{(n)}^*}^{(0)*} \varphi_n(a) T_{\xi^{(n)}}^{(0)} a' = \sum_{n=0}^{\infty} T_{\zeta_{(n)}^*}^{(0)*} a \cdot \xi^{(n)} \cdot a' \\ &= \sum_{n=0}^{\infty} \langle a \cdot \xi^{(n)} \cdot a', \zeta_{(n)}^* \rangle = \sum_{n=0}^{\infty} \langle a \cdot \xi^{(n)}, \zeta_{(n)}^* \rangle a'. \qquad \square\end{aligned}$$

The following lemma will be useful in the sequel.

Lemma 4.9. *For any $(\zeta, a) \in \mathbb{D}(E^*) \times \mathcal{A}$ and $R \in \mathcal{F}^\infty(E)$ we have*

$$W_{\zeta^*, a^*}^* R = W_{\zeta^*, \widehat{R}(\zeta,a)^*}^*. \tag{4.24}$$

Proof. Let $(\zeta, a) \in \mathbb{D}(E^*) \times \mathcal{A}$ and $R \in \mathcal{F}^\infty(E)$. Let $\xi^{(n)} \in E^{\otimes n}$, for $n = 0, 1, 2, \ldots$, such that R is given by (4.7)–(4.9). First observe that for $n, k = 0, 1, 2, \ldots$ and for any $\rho \in E^{\otimes k}$ we have

$$\begin{aligned} T_{\zeta_{(n+k)}^*}^{(0)*} \varphi_{n+k}(a)(T_{\xi^{(n)}}^{(0)} \otimes I_{E^{\otimes k}})\rho &= T_{\zeta_{(n+k)}^*}^{(0)*} (a \cdot \xi^{(n)}) \otimes \rho \\ &= \langle (a \cdot \xi^{(n)}) \otimes \rho, \zeta_{(n+k)}^* \rangle \\ &= \langle (a \cdot \xi^{(n)}) \otimes \rho, \zeta_{(n)}^* \otimes \zeta_{(k)}^* \rangle \\ &= \langle \langle a \cdot \xi^{(n)}, \zeta_{(n)}^* \rangle \cdot \rho, \zeta_{(k)}^* \rangle \\ &= T_{\zeta_{(k)}^*}^{(0)*} \varphi_k(\langle a \cdot \xi^{(n)}, \zeta_{(n)}^* \rangle)\rho. \end{aligned}$$

It then follows that the kth entry in the infinite bock row matrix $W_{\zeta^*, a^*}^* R$ is equal to

$$\begin{aligned} \sum_{n=0}^{\infty} T_{\zeta_{(n+k)}^*}^{(0)*} \varphi_{n+k}(a)(T_{\xi^{(n)}}^{(0)} \otimes I_{E^{\otimes k}}) &= \sum_{n=0}^{\infty} T_{\zeta_{(k)}^*}^{(0)*} \varphi_k(\langle a \cdot \xi^{(n)}, \zeta_{(n)}^* \rangle) \\ &= T_{\zeta_{(k)}^*}^{(0)*} \varphi_k \left(\sum_{n=0}^{\infty} \langle a \cdot \xi^{(n)}, \zeta_{(n)}^* \rangle \right) \\ &= T_{\zeta_{(k)}^*}^{(0)*} \varphi_k(\widehat{R}(\zeta, a)). \end{aligned}$$

This proves our claim. $\square$

One application of Lemma 4.9 is the following multiplicative law for elements in the Toeplitz algebra $\mathcal{F}^\infty(E)$ with respect to the point-evaluation considered in the present subsection.

Proposition 4.10. *Let $R, S \in \mathcal{F}^\infty(E)$ and $(\zeta, a) \in \mathbb{D}(E^*) \times \mathcal{A}$. Then*
$$(\widehat{RS})(\zeta, a) = \widehat{S}(\zeta, \widehat{R}(\zeta, a)).$$

Proof. It follows from Lemma 4.9 that
$$(\widehat{RS})(\zeta, a) = W^*_{\zeta, a^*} RS|_{\mathcal{A}} = W^*_{\zeta, \widehat{R}(\zeta,a)^*} S|_{\mathcal{A}} = \widehat{S}(\zeta, \widehat{R}(\zeta, a)). \qquad \square$$

It follows, in particular, from this proposition that the alternative point-evaluation is not multiplicative; unlike the first Muhly-Solel point-evaluation discussed in Subsection 4.4; cf., [16, Proposition 4.4]. The multiplicative law obtained in this case shows more resemblance to that appearing in the operator-argument functional calculus; cf., formula I.2.7 in [53].

We now prove the following Nevanlinna-Pick interpolation theorem.

Theorem 4.11. *Suppose that we are given $(\zeta_1, a_1), \dots, (\zeta_N, a_N) \in \mathbb{D}(E^*) \times \mathcal{A}$ and $w_1, \dots, w_N \in \mathcal{A}$. Then there exists an $S \in \mathcal{F}^\infty(E)$ with $\|S\| \leq 1$ and $\widehat{S}(\zeta_k, a_k) = w_k$ for $k = 1, \dots, N$ if and only if the operator matrix*
$$\left[\sum_{n=0}^{\infty} \langle (a_i a_j^* - w_i w_j^*) \zeta^*_{j\,(n)}, \zeta^*_{i\,(n)} \rangle_{E^{\otimes n}} \right]_{i,j=0}^{N} \tag{4.25}$$
is a positive element of $\mathcal{A}^{N \times N}$.

Proof. Assume we have an $S \in \mathcal{F}^\infty(E)$ with $\|S\| \leq 1$ and $\widehat{S}(\zeta_k, a_k) = w_k$ for $k = 1, \dots, N$. Then, by Lemma 4.9, $W^*_{\zeta_k^*, a_k^*} S = W^*_{\zeta_k^*, \widehat{S}(\zeta_k, a_k)^*} = W^*_{\zeta_k^*, w_k^*}$ for $k = 1, \dots, N$, and thus
$$S^* \begin{bmatrix} W_{\zeta_1^*, a_1^*} & \cdots & W_{\zeta_N^*, a_N^*} \end{bmatrix} = \begin{bmatrix} W_{\zeta_1^*, w_1^*} & \cdots & W_{\zeta_N^*, w_N^*} \end{bmatrix}.$$
Since $\|S\| \leq 1$, this implies that
$$\begin{bmatrix} W^*_{\zeta_1^*, a_1^*} \\ \vdots \\ W^*_{\zeta_N^*, a_N^*} \end{bmatrix} \begin{bmatrix} W_{\zeta_1^*, a_1^*} \cdots W_{\zeta_N^*, a_N^*} \end{bmatrix} - \begin{bmatrix} W^*_{\zeta_1^*, w_1^*} \\ \vdots \\ W^*_{\zeta_N^*, w_N^*} \end{bmatrix} \begin{bmatrix} W_{\zeta_1^*, w_1^*} \cdots W_{\zeta_N^*, w_N^*} \end{bmatrix}$$
$$= \begin{bmatrix} W^*_{\zeta_1^*, a_1^*} \\ \vdots \\ W^*_{\zeta_N^*, a_N^*} \end{bmatrix} (I - SS^*) \begin{bmatrix} W_{\zeta_1^*, a_1^*} \cdots W_{\zeta_N^*, a_N^*} \end{bmatrix} \succeq 0 \quad (\text{in } \mathcal{A}^{N \times N}). \tag{4.26}$$
Next observe that for any $(\zeta, a), (\zeta', a') \in \mathbb{D}(E^*) \times \mathcal{A}$ we have
$$W^*_{\zeta^*, a^*} W_{\zeta'^*, a'^*} = \sum_{n=0}^{\infty} \langle a a'^* \zeta'^*_{(n)}, \zeta^*_{(n)} \rangle \in \mathcal{A} = \mathcal{L}^a(\mathcal{A}).$$
From this computation it follows that the left-hand side of the identity in (4.26) is equal to the Pick matrix (4.25). Thus (4.25) is a positive element of $\mathcal{A}^{N \times N}$.

Conversely, assume that (4.25) is positive in $\mathcal{A}^{N\times N}$. Let $\sigma : \mathcal{A} \to \mathcal{L}(\mathcal{E})$ be a faithful (i.e., injective) representation of $\mathcal{A}$ into $\mathcal{L}(\mathcal{E})$ for some Hilbert space $\mathcal{E}$. Fix $(\zeta, a) \in \mathbb{D}(E^*) \times \mathcal{A}$. For notational convenience we set $\widetilde{W}_{\zeta^*,a^*} = W_{\zeta^*,a^*} \otimes I_{\mathcal{E}} \in \mathcal{L}^a(\mathcal{E}, \mathcal{F}^2(E,\sigma))$, where $\mathcal{F}^2(E,\sigma) := \mathcal{F}^2(E) \otimes_\sigma \mathcal{E}$. We claim that for each $b \in \sigma(\mathcal{A})'$ and $\mu \in E^\sigma$ we have

$$\begin{aligned} \widetilde{W}^*_{\zeta^*,a^*}(I_{\mathcal{F}^2(E)} \otimes b) &= b\widetilde{W}^*_{\zeta^*,a^*}, \\ \widetilde{W}^*_{\zeta^*,a^*}(I_{\mathcal{F}^2(E)} \otimes \mu) &= ((T^{(0)*}_{\zeta^*} \otimes I_{\mathcal{E}})\mu)\widetilde{W}^*_{\zeta^*,a^*}. \end{aligned} \tag{4.27}$$

The first identity follows directly from the computation

$$\widetilde{W}^*_{\zeta^*,a^*}(I_{\mathcal{F}^2(E)} \otimes b) = (W^*_{\zeta^*,a^*} \otimes I_{\mathcal{E}})(I_{\mathcal{F}^2(E)} \otimes b) = (I_{\mathcal{A}} \otimes b)(W^*_{\zeta^*,a^*} \otimes I_{\mathcal{E}}) = b\widetilde{W}^*_{\zeta^*,a^*}.$$

Here we used the fact that $\mathcal{A} \otimes_\sigma \mathcal{E}$ can be identified with $\mathcal{E}$. To see that the second identity in (4.27) holds we show that the nth entry

$$(T^{(0)*}_{\zeta^*_{(n+1)}} \varphi_{n+1}(a) \otimes I_{\mathcal{E}})(I_{E^{\otimes n}} \otimes \mu) : E^{\otimes n} \otimes_\sigma \mathcal{E} \to \mathcal{A}$$

in the infinite block row matrix $\widetilde{W}^*_{\zeta^*,a^*}(I_{\mathcal{F}^2(E)} \otimes \mu)$ is equal to

$$((T^{(0)*}_{\zeta^*} \otimes I_{\mathcal{E}})\mu)(T^{(0)*}_{\zeta^*_{(n)}} \varphi_n(a) \otimes I_{\mathcal{E}}).$$

It suffices to check the equality for elements from $E^{\otimes n} \otimes_\sigma \mathcal{E}$ of the form $\xi^{(n)} \otimes e$ with $\xi^{(n)} \in E^{\otimes n}$ and $e \in \mathcal{E}$. For such elements $\xi^{(n)} \otimes e \in E^{\otimes n} \otimes_\sigma \mathcal{E}$ and for each $e' \in \mathcal{E}$ we obtain that the

$$\begin{aligned} &\langle (T^{(0)*}_{\zeta^*_{(n+1)}} \varphi_{n+1}(a) \otimes I_{\mathcal{E}})(I_{E^{\otimes n}} \otimes \mu)(\xi^{(n)} \otimes e), e' \rangle_{\mathcal{E}} \\ &\quad = \langle (a \cdot \xi^{(n)}) \otimes \mu e, \zeta^*_{(n+1)} \otimes e' \rangle = \langle \langle a \cdot \xi^{(n)}, \zeta^*_{(n)} \rangle \cdot \mu e, \zeta^* \otimes e' \rangle \\ &\quad = \langle \mu\sigma(\langle a \cdot \xi^{(n)}, \zeta^*_{(n)} \rangle) e, \zeta^* \otimes e' \rangle = \langle ((T^{(0)*}_{\zeta^*_{(1)}} \otimes I_{\mathcal{E}})\mu)\sigma(T^{(0)*}_{\zeta^*_{(n)}} \varphi_n(a)\xi^{(n)})e, e' \rangle \\ &\quad = \langle ((T^{(0)*}_{\zeta^*_{(1)}} \otimes I_{\mathcal{E}})\mu)(T^{(0)*}_{\zeta^*_{(n)}} \varphi_n(a) \otimes I_{\mathcal{E}})(\xi^{(n)} \otimes e), e' \rangle. \end{aligned}$$

Thus our claim follows. Now set

$$\mathcal{H}_{\zeta,a} := \overline{\operatorname{Im} \widetilde{W}_{\zeta,a}} \subset \mathcal{F}^2(E,\sigma) := \mathcal{F}^2(E) \otimes_\sigma \mathcal{E}.$$

Using the intertwining relations (4.27) we see that $\mathcal{H}_{\zeta,a}$ is an invariant subspace for the operators $I_{\mathcal{F}^2(E)} \otimes \mu^*$ and $I_{\mathcal{F}^2(E)} \otimes b$ for each $\mu \in E^\sigma$ and $b \in \sigma(\mathcal{A})'$.

Identifying $\mathcal{A}$ with its image in $\mathcal{L}(\mathcal{E})$ under the representation σ, we see that the Pick matrix (4.25) defines a positive semidefinite element in $\mathcal{L}(\mathcal{E})^{N\times N}$. Moreover, from the first part of the proof we see that, equivalently, the left-hand side of the identity in (4.26), with $W^*_{\zeta^*_k,a^*_k}$ replaced by $\widetilde{W}^*_{\zeta^*_k,a^*_k}$, defines positive semidefinite element in $\mathcal{L}(\mathcal{E})^{N\times N}$. Thus, by the Douglas factorization lemma [43], we can define a contraction

$$V : \mathcal{M} \to \mathcal{N} \quad \text{by} \quad \widetilde{W}_{\zeta^*_k,a^*_k} V = \widetilde{W}^*_{\zeta^*_k,w^*_k} \text{ for } k = 1,\ldots,N,$$

where we set

$$\mathcal{M} := \bigvee_{k=1}^{N} \operatorname{Im} \widetilde{W}_{\zeta_k^*, w_k^*} \quad \text{and} \quad \mathcal{N} := \bigvee_{k=1}^{N} \operatorname{Im} \widetilde{W}_{\zeta_k^*, a_k^*}.$$

Now, for $b \in \sigma(\mathcal{A})'$ and $\mu \in E^\sigma$, let T_b, $\widetilde{T}_b$ and T_μ, $\widetilde{T}_\mu$ be the compressions of $I_{\mathcal{F}^2(E)} \otimes b$ and $I_{\mathcal{F}^2(E)} \otimes \mu$ to $\mathcal{M}$ and $\mathcal{N}$, respectively. Then, from the intertwining relations (4.27) it follows for $k = 1, \ldots, N$ that

$$\widetilde{W}^*_{\zeta_k^*, a_k^*} T_b V = b \widetilde{W}^*_{\zeta_k^*, a_k^*} V = b \widetilde{W}^*_{\zeta_k^*, w_k^*} = \widetilde{W}^*_{\zeta_k^*, w_k^*} \widetilde{T}_b = \widetilde{W}^*_{\zeta_k^*, a_k^*} V \widetilde{T}_b,$$

and with a similar computation $\widetilde{W}^*_{\zeta_k^*, a_k^*} T_\mu V = \widetilde{W}^*_{\zeta_k^*, a_k^*} V \widetilde{T}_\mu$. Since these identities hold for all $k = 1, \ldots, N$, we obtain that V intertwines T_b with $\widetilde{T}_b$ and T_μ with $\widetilde{T}_\mu$ for each $b \in \sigma(\mathcal{A})'$ and $\mu \in E^\sigma$. It then follows from Theorem 4.3 that there exists an $S \in \mathcal{F}^\infty(E)$ with $\|S\| \leq 1$, such that $(S^* \otimes I_\mathcal{E})\mathcal{N} \subset \mathcal{M}$ and $V = P_\mathcal{N}(S \otimes I_\mathcal{E})|_\mathcal{M}$. Hence for $k = 1, \ldots, N$ we have

$$\begin{aligned} \sigma(\widehat{S}(\zeta_k, a_k)) &= \widehat{S}(\zeta_k^*, a_k) \otimes I_\mathcal{E} = (W^*_{\zeta_k^*, a_k^*} S|_\mathcal{A}) \otimes I_\mathcal{E} = \widetilde{W}^*_{\zeta_k^*, a_k^*} (S \otimes I_\mathcal{E})|_\mathcal{E} \\ &= \widetilde{W}^*_{\zeta_k^*, w_k^*}|_\mathcal{E} = \sigma(w_k). \end{aligned}$$

Thus $\widehat{S}(\zeta_k, a_k) = w_k$ for $k = 1, \ldots, N$, because σ is injective. □

4.8. Example: Operator-valued Nevanlinna-Pick interpolation on the unit disk $\mathbb{D}$

As a first example we consider the case where $E = \mathcal{A} = \mathcal{L}(\mathcal{V})$ for $\mathcal{V}$ some Hilbert space. Then E is a W^*-$\mathcal{A}$-correspondence with left and right actions given by the usual left and right multiplication in $\mathcal{L}(\mathcal{V})$ and with $\mathcal{L}(\mathcal{V})$-valued inner product given by

$$\langle R, Q \rangle = Q^* R \text{ for all } R, Q \in \mathcal{L}(\mathcal{V}).$$

The preliminaries of this case have been spelled out in Subsection 6.1 of [16]. We recall here the results needed in the sequel. The balancing (4.1) in the tensor products causes the tensor spaces $E^{\otimes n}$ to reduce to $E = \mathcal{L}(\mathcal{V})$ via the identification

$$R_n \otimes \cdots \otimes R_1 = I_\mathcal{V} \otimes \cdots \otimes I_\mathcal{V} \otimes (R_n \cdots R_1) \equiv R_n \cdots R_1. \tag{4.28}$$

Then the Fock space $\mathcal{F}^2(E)$ is equal to $\mathcal{L}(\mathcal{V}, \ell^2_\mathcal{V}(\mathbb{Z}_+))$, and the Toeplitz algebra $\mathcal{F}^\infty(E)$ is the algebra of block Toeplitz matrices

$$R = \begin{bmatrix} R_0 & 0 & 0 & \cdots \\ R_1 & R_0 & 0 & \cdots \\ R_2 & R_1 & R_0 & \ddots \\ \vdots & \ddots & \ddots & \ddots \end{bmatrix}, \quad R_j \in \mathcal{L}(\mathcal{V}) \tag{4.29}$$

that induce bounded operators on $\mathcal{L}(\mathcal{V}, \ell^2_{\mathcal{V}}(\mathbb{Z}_+))$ via matrix multiplication: if we write $X \in \mathcal{L}(\mathcal{V}, \ell^2_{\mathcal{V}}(\mathbb{Z}_+))$ in block column-matrix form

$$X = \begin{bmatrix} X_1 \\ X_2 \\ \vdots \end{bmatrix} \text{ with } X_k \in \mathcal{L}(\mathcal{V}) \text{ for } k = 1, 2, \dots,$$

then $R\colon X \mapsto R \cdot X$ where the matrix for R is given by (4.29). By the closed-graph theorem, for any such fixed X, $R \cdot X \in \mathcal{L}(\mathcal{V}, \ell^2_{\mathcal{V}}(\mathbb{Z}_+))$ if and only if $(R \cdot X) \cdot v = R \cdot (X \cdot v) \in \ell^2_{\mathcal{V}}(\mathbb{Z}_+)$ for each $v \in \mathcal{V}$. As X and v are arbitrary, an equivalent condition is that $R \cdot w \in \ell^2_{\mathcal{V}}(\mathbb{Z}_+)$ for each $w \in \ell^2_{\mathcal{V}}(\mathbb{Z}_+)$, i.e., (again by the closed-graph theorem) R is bounded when viewed as an ordinary Toeplitz matrix acting on $\ell^2_{\mathcal{V}}(\mathbb{Z}_+)$. Furthermore one can check that the norm of R viewed as an element of $\mathcal{L}(\mathcal{L}(\mathcal{V}, \ell^2_{\mathcal{V}}(\mathbb{Z}_+)))$ is equal to its norm when viewed as an element of $\mathcal{L}(\ell^2_{\mathcal{V}}(\mathbb{Z}_+))$. Thus a sequence of coefficients $\{R_n\}_{n \in \mathbb{Z}}$ with $R_n \in \mathcal{L}(\mathcal{V})$ and $R_n = 0$ for $n < 0$ gives rise to an element $[R_{i-j}]_{i,j \in \mathbb{Z}_+}$ in $\mathcal{F}^\infty(E)$ exactly when $R(\lambda) := \sum_{n=0}^{\infty} R_n \lambda^n$ is in the operator-valued Schur class $\mathcal{S}(\mathcal{V})$.

Let $\mathcal{G}$ be another Hilbert space. We take $\mathcal{E}$ to be the tensor product Hilbert space $\mathcal{V} \otimes \mathcal{G}$, and define a representation $\sigma : \mathcal{A} \to \mathcal{L}(\mathcal{E})$ by

$$\sigma(R) = R \otimes I_{\mathcal{G}} \text{ for } R \in \mathcal{L}(\mathcal{V}) = \mathcal{A}.$$

Then $(E \otimes_\sigma \mathcal{E})^*$ can be identified with $\mathcal{E}$ via the identification

$$R \otimes_\sigma (v \otimes g) = I_{\mathcal{U}} \otimes_\sigma ((Rv) \otimes g) \equiv (Rv) \otimes g \text{ for } R \in \mathcal{L}(\mathcal{V}),\ v \in \mathcal{V},\ g \in \mathcal{G}.$$

We identify E^σ as

$$E^\sigma = \{\eta \in \mathcal{L}(\mathcal{E})\colon \eta(R \otimes I_{\mathcal{G}}) = (R \otimes I_{\mathcal{G}})\eta \text{ for all } R \in \mathcal{L}(\mathcal{V})\} = I_{\mathcal{V}} \otimes \mathcal{L}(\mathcal{G}),$$

and hence the generalized disk $\mathbb{D}((E^\sigma)^*)$ can be identified with the set of strict contraction operators on $\mathcal{G}$. For $\eta \in \mathbb{D}((E^\sigma)^*)$ identified with a strict contraction operator $Z \in \mathcal{L}(\mathcal{G})$ and for $R = [R_{i-j}]_{i,j \in \mathbb{Z}_+} \in \mathcal{F}^\infty(E)$, one can work out that the associated first point-evaluation $\widehat{R}(Z)$ is given by a tensor functional-calculus

$$\widehat{R}(Z) = \sum_{n=0}^{\infty} R_n \otimes Z^n.$$

Thus, the correspondence Nevanlinna-Pick interpolation problem for this setting becomes: *Given strict contraction operators $Z_1, \dots, Z_N$ in $\mathcal{L}(\mathcal{G})$ and operators $X_1, \dots, X_N$ and $Y_1, \dots, Y_N$ in $\mathcal{L}(\mathcal{V} \otimes \mathcal{G})$, find a Schur class function $S(\lambda) = \sum_{n=0}^{\infty} S_n \lambda^n \in \mathcal{S}(\mathcal{V})$ so that*

$$X_i S(Z_i) := X_i \sum_{n=0}^{\infty} S_n \otimes Z_i^n = Y_i \text{ for } i = 1, \dots, n. \tag{4.30}$$

We call this the *left-tangential tensor functional-calculus Nevanlinna-Pick interpolation problem* (**LTT-NP**). Note that this contains the **LT-NP** (take $\mathcal{G} = \mathbb{C}$) and

LTRD-NP (take $\mathcal{V} = \mathbb{C}$) problems discussed in Subsections 2.1 and 2.3, respectively, as special cases. The solution to the **LTT-NP** problem is readily obtained as a direct application of Theorem 4.6.

Theorem 4.12. *Suppose that we are given the data set*

$$\mathfrak{D}: \quad Z_1, \dots, Z_N \in \mathcal{L}(\mathcal{G}), \quad X_1, \dots, X_N, Y_1, \dots, Y_N \in \mathcal{L}(\mathcal{V} \otimes \mathcal{G})$$

for an **LTT-NP** *problem. Then a solution $S \in \mathcal{S}(\mathcal{V})$ exists if and only if any of the following equivalent conditions holds:*

(1) *The kernel $\mathbb{K}_{\mathfrak{D}}$ mapping $\{1, \dots, N\} \times \{1, \dots, N\}$ into $\mathcal{L}^a(\mathcal{L}(\mathcal{G}), \mathcal{L}(\mathcal{V} \otimes \mathcal{G}))$ given by*

$$\mathbb{K}_{\mathfrak{D}}(i,j)[B] = \sum_{n=0}^{\infty} X_i \left(I_{\mathcal{V}} \otimes \left(Z_i^n B Z_j^{*n}\right)\right) X_j^* - Y_i \left(I_{\mathcal{V}} \otimes \left(Z_i^n B Z_j^{*n}\right)\right) Y_j^*$$

is completely positive.

(2) *The map $\varphi\colon \mathcal{L}(\mathcal{G})^{N \times N} \to \mathcal{L}(\mathcal{V} \otimes \mathcal{G})^{N \times N}$ given by*

$$\varphi\left(\left[B_{ij}\right]_{i,j=1,\dots,N}\right) = \tag{4.31}$$
$$= \left[\sum_{n=0}^{\infty} X_i \left(I_{\mathcal{V}} \otimes \left(Z_i^n B_{ij} Z_j^{*n}\right)\right) X_j^* - Y_i \left(I_{\mathcal{V}} \otimes \left(Z_i^n B_{ij} Z_j^{*n}\right)\right) Y_j^*\right]_{i,j=1}^{N}$$

is a completely positive map.

Note that criteria of Theorem 4.12 give seemingly different criteria than the ones obtained in Section 2. We now show how, after some reductions, one can see directly the equivalence of the criteria in Theorem 4.12 with the Pick-matrix criteria of Section 2.

First assume that $\mathcal{V}$ and $\mathcal{G}$ admit direct sum decompositions

$$\mathcal{V} = \mathcal{U} \oplus \mathcal{Y} \oplus \mathcal{C} \quad \text{and} \quad \mathcal{G} = \mathcal{Z} \oplus \mathcal{D},$$

and that the operators in the data set $\mathfrak{D}$ are actually operators

$$Z_i \in \mathcal{L}(\mathcal{Z}), \quad X_i \in \mathcal{L}(\mathcal{Y} \otimes \mathcal{Z}, \mathcal{C} \otimes \mathcal{D}), \quad Y_i \in \mathcal{L}(\mathcal{U} \otimes \mathcal{Z}, \mathcal{C} \otimes \mathcal{D}) \text{ for } i = 1, \dots, N,$$

identified with operators in $\mathcal{L}(\mathcal{G})$ and $\mathcal{L}(\mathcal{E})$, respectively, by adding zero-operators in the operator matrix decompositions. A solution to the "non-square" **LTT-NP** problem: *Given strict contraction operators $Z_1, \dots, Z_N$ in $\mathcal{L}(\mathcal{Z})$, and operators $X_1, \dots, X_N \in \mathcal{L}(\mathcal{Y} \otimes \mathcal{Z}, \mathcal{C} \otimes \mathcal{D})$ and $Y_1, \dots, Y_N \in \mathcal{L}(\mathcal{U} \otimes \mathcal{Z}, \mathcal{C} \otimes \mathcal{D})$, find a Schur class function $S(\lambda) = \sum_{n=0}^{\infty} S_n \lambda^n \in \mathcal{S}(\mathcal{U}, \mathcal{Y})$ satisfying* (4.30), can then be obtained from a solution $\tilde{S} \in \mathcal{S}(\mathcal{E})$ just by assigning $S(\lambda)$ to be the compression of $\tilde{S}(\lambda)$ to $\mathcal{L}(\mathcal{U}, \mathcal{Y})$. Conversely, a solution $S \in \mathcal{L}(\mathcal{U}, \mathcal{Y})$ to the "non-square" problem defines a solution $\tilde{S} \in \mathcal{L}(\mathcal{E})$ to the "square" problem by embedding the values $S(\lambda) \in \mathcal{L}(\mathcal{U}, \mathcal{Y})$ into $\mathcal{L}(\mathcal{E})$. It thus follows that a solution to the "non-square" problem exists if and only if the map φ in criterion 2 of Theorem 4.12 (which now can also be seen as a map from $\mathcal{L}(\mathcal{Z})^{N \times N}$ into $\mathcal{L}(\mathcal{Z} \otimes \mathcal{D})^{N \times N}$) is completely positive.

Transforming a result from a "square" Nevanlinna-Pick problem to one for Schur class functions that take "non-square" values can even be performed on the level of W^*-correspondence Nevanlinna-Pick interpolation using techniques developed in [70].

Standard functional calculus Nevanlinna-Pick interpolation. Assume that $\mathcal{G} = \mathbb{C}$. Hence the data for our Nevanlinna-Pick interpolation is of the **LT-NP** form:

$$\mathfrak{D}: \ \lambda_1, \dots, \lambda_N \in \mathbb{D}, \ X_1, \dots, X_N \in \mathcal{L}(\mathcal{Y}, \mathcal{C}), \ Y_1, \dots, Y_N \in \mathcal{L}(\mathcal{U}, \mathcal{C}).$$

In that case, the kernel $\mathbb{K}_{\mathfrak{D}} : \{1, \dots, N\} \times \{1, \dots, N\} \to \mathcal{L}(\mathbb{C}, \mathcal{L}(\mathcal{L}(\mathcal{V})))$ reduces to

$$\mathbb{K}_{\mathfrak{D}}(i,j)[c] = c \cdot \frac{X_i X_j^* - Y_i Y_i^*}{i - \lambda_i \overline{\lambda}_j}.$$

It is then straightforward to see that complete positivity of the kernel $\mathbb{K}_{\mathfrak{D}}$ collapses to positive semidefiniteness of the standard Pick matrix

$$\left[\frac{X_i X_j^* - Y_i Y_i^*}{i - \lambda_i \overline{\lambda}_j} \right]_{j,i=1}^N.$$

Thus we recover the criterion of Theorem 2.1 (Part 2).

Riesz-Dunford functional calculus Nevanlinna-Pick interpolation. Next we consider the case that $\mathcal{V} = \mathbb{C}$. In that case the data for our Nevanlinna-Pick interpolation problem takes the **LTRD-NP** form:

$$\mathfrak{D}: \ Z_1, \dots, Z_N \in \mathcal{L}(\mathcal{Z}), \ X_1, \dots, X_N, Y_1, \dots, Y_N \in \mathcal{L}(\mathcal{Z}, \mathcal{C}).$$

To see that Theorem 4.12 also contains the result of Theorem 2.3 (Part 2), it is convenient to introduce a third criterion.

Theorem 4.13. *In case $\mathcal{V} = \mathbb{C}$, in addition to the two conditions* (1) *and* (2) *in Theorem* 4.12, *a third condition equivalent to the existence of an $S \in \mathcal{S}(\mathbb{C}, \mathbb{C})$ that satisfies* (4.30) *is that the map φ_* from $\mathcal{L}(\mathcal{C})^{N \times N}$ to $\mathcal{L}(\mathcal{C})^{N \times N}$ given by*

$$\varphi_* \left([C_{ij}]_{i,j=1}^N \right) = \left[\sum_{n=0}^{\infty} Z_i^{*n} (X_i^* C_{ij} X_j - Y_i^* C_{ij} Y_j) Z_j^n \right]_{i,j=1}^N \tag{4.32}$$

be a completely positive map.

Proof. We prove that positivity of φ is equivalent to positivity of φ_*. The proof of the equivalence for k-positivity with $k \in \mathbb{Z}_+$ arbitrary (and thus for complete positivity) goes analogously. The map φ in (4.31) being positive implies that

$$\sum_{i,j=1}^N \operatorname{trace} \left(C_{ji} \left(\sum_{n=0}^{\infty} X_i Z_i^n B_{ij} Z_j^{*n} X_j^* - Y_i Z_i^n B_{ij} Z_j^{*n} Y_j^* \right) \right) \geq 0 \tag{4.33}$$

for all $\mathbf{B} = [B_{ij}]_{i,j=1}^N \succeq 0$ and $\mathbf{C} = [C_{ij}]_{i,j=1}^N \succeq 0$ in $\mathcal{L}(\mathcal{G})^{N \times N}$ with $\mathbf{B}$ also in the trace class. Using the invariance of the trace under cyclic permutations, we may rewrite (4.33) as

$$\sum_{i,j=1}^{N} \operatorname{trace}\left(\left(\sum_{n=0}^{\infty} Z_j^{*n}(X_j^* C_{ji} X_i - Y_j^* C_{ji} Y_i) Z_i^n\right) B_{ij}\right) \geq 0. \tag{4.34}$$

From this it follows that φ_* is a positive map. To see the converse implication one just reverses the above computation but now with $\mathbf{C}$ in the trace class. □

In order to transform the completely positive map criterion of Theorem 4.13 into one of checking positivity of a (possibly infinite) operator matrix we need the following extension of Choi's theorem [38].

Theorem 4.14. *A weak-$*$ continuous map $\psi : \mathcal{L}(\mathcal{H}) \to \mathcal{A}$, where $\mathcal{H}$ a separable Hilbert space with orthonormal basis $\{e_1, \dots, e_\kappa\}$ (with possibly $\kappa = \infty$) and $\mathcal{A}$ is a C^*-algebra, is completely positive if and only if the (possibly infinite) block matrix*

$$\left[\ \psi(e_i e_j^*)\ \right]_{i,j=1}^{\kappa} \tag{4.35}$$

is a positive element of $\mathcal{A}^{\kappa\times\kappa}$ (i.e., if $\kappa = \infty$, then all M-finite sections are positive in $\mathcal{A}^{M\times M}$ for each $M \in \mathbb{Z}_+$).

Proof. If $\kappa < \infty$, then the statement is just Choi's theorem [38]. So assume that $\kappa = \infty$. Complete positivity of ψ means that for each $N \in \mathbb{Z}_+$ the map

$$\left[\ B_{i,j}\ \right]_{i,j=1}^{N} \mapsto \left[\ \psi(B_{i,j})\ \right]_{i,j=1}^{N}$$

from $\mathcal{L}(\mathcal{H})^{N\times N}$ into $\mathcal{A}^{N\times N}$ is positive. In particular, using $B_{i,j} = e_i e_j^*$ it follows directly that ψ being completely positive implies that the block matrix (4.35) is a positive element of $\mathcal{A}^{\kappa\times\kappa}$.

Now assume that the block matrix (4.35) is positive. For each $M \in \mathbb{Z}_+$, let P_M denote the projection onto $\mathcal{H}_M := \operatorname{span}\{e_1, \dots, e_M\}$. Using Choi's theorem and the fact that the M-finite section $\left[\psi(e_i e_j^*)\right]_{i,j=1}^{M}$ of (4.35) is a positive element of $\mathcal{A}^{M\times M}$ we obtain that the map $\tilde{B} \mapsto \psi(P_M \tilde{B} P_M)$ from $\mathcal{H}_M$ into $\mathcal{A}$ is completely positive. Fix a $N \in \mathbb{Z}_+$ and a positive $\left[B_{i,j}\right]_{i,j=1}^{N}$ in $\mathcal{L}(\mathcal{H})^{N\times N}$. Then for each $M \in \mathbb{Z}_+$ we have $\left[\psi(P_M B_{i,j} P_M)\right]_{i,j=1}^{N} \succeq 0$. Moreover, $P_M B_{i,j} P_M$ converges to $B_{i,j}$ in the weak-$*$ topology as $M \to \infty$, and thus, by hypothesis $\psi(P_M B_{i,j} P_M) \to \psi(B_{i,j})$ in the weak-$*$ topology as $M \to \infty$ for all $i, j = 1, \dots, N$. This implies that

$$\left[\ \psi(P_M B_{i,j} P_M)\ \right]_{i,j=1}^{N} \to \left[\ \psi(B_{i,j})\ \right]_{i,j=1}^{N} \text{ weak-}* \text{ as } M \to \infty.$$

Since positivity is preserved under weak-$*$ convergence it follows that $\left[\psi(B_{i,j})\right]_{i,j=1}^{N}$ is a positive element of $\mathcal{A}^{N\times N}$, and thus, because N was chosen arbitrarily, that ψ is a completely positive map. □

Now assume that $\mathcal{C}$ is a separable Hilbert space with orthonormal basis $\{e_1, \dots, e_\kappa\}$. It then follows from Theorem 4.14 that complete positivity of the maps φ_* given by (4.32) reduces to positivity of the Pick matrix

$$\left[\ \sum_{n=0}^{\infty} Z_i^{*n}(X_i^* e_{i'} e_{j'}^* X_j - Y_i^* e_{i'} e_{j'}^* Y_j) Z_j^n\ \right]_{(i,i'),(j,j')\in\{1,\dots,N\}\times\{1,\dots,\kappa\}}.$$

Hence we recover Part 2 of Theorem 2.3. Indeed, the one thing left to verify is that the map φ_* is weak-$*$ continuous, which we leave as an exercise to the reader.

Operator-argument functional calculus Nevanlinna-Pick interpolation. Finally, we show that the second point-evaluation for $\mathcal{F}^\infty(E)$ discussed in Subsection 4.7 gives us the operator-argument Nevanlinna-Pick theorem for the operator-valued Schur class. To see that this is the case, note that the points are elements of

$$\mathbb{D}(E^*) \times \mathcal{A} = \{T \in \mathcal{L}(\mathcal{V}) \colon \|T\| < 1\} \times \mathcal{L}(\mathcal{V}).$$

For each $\zeta \in \mathbb{D}(E^*)$ identified with a strict contraction operator $T \in \mathcal{L}(\mathcal{V})$ the element $\zeta_{(n)}^* = \zeta^* \otimes \cdots \otimes \zeta^*$ of $E^{\otimes n}$ corresponds to T^{*n} (via the identification (4.28)), and the operator $T^{(0)}_{\zeta^*_{(n)}}$ in (4.9) is just multiplication with T^{*n}. Thus, we obtain that for $(T, X) \in \mathbb{D}(E^*) \times \mathcal{A}$ and $R = \left[\ R_{i-j}\ \right]_{i,j\in\mathbb{Z}_+} \in \mathcal{F}^\infty(E)$ the value of R in (T, X) is equal to the left-tangential point-evaluation:

$$\widehat{R}(T, X) = \sum_{n=0}^{\infty} T^n X R_n = (XR)^{\wedge L}(T),$$

where we also use R to indicate the Schur-class function in $\mathcal{S}(\mathcal{V})$ corresponding to $R \in \mathcal{F}^\infty(E)$. So, in this case, the Nevanlinna-Pick problem with data $\zeta_k = T_k \in \mathbb{D}(E^*)$, $a_k = X_k, w_k = Y_k \in \mathcal{A}$, for $k = 1, \dots, N$, considered in Subsection 4.7 is the **LTOA** problem of Subsection 2.2. One easily computes that the operator matrix (4.25) reduces to the Pick matrix $\mathbb{P}_{\text{LTOA}}$ in (2.8). Thus, Theorem 4.11 here gives us precisely Part 1 of Theorem 2.2 (for the case $\mathcal{U} = \mathcal{Y} = \mathcal{C}$; the non-square case can be obtained as explained above).

4.9. Example: Nevanlinna-Pick interpolation for Toeplitz algebras associated with directed graphs

Next, following Section 5 in [70], we show how the Toeplitz algebra associated with a directed graph/quiver can be seen as an example of the W^*-correspondence formalism, and we prove Theorem 3.3. We follow the notation and terminology used in Subsection 3.3. Let $G = \{Q_0, Q_1, s, r\}$ be a quiver and $\mathcal{V}$ a given Hilbert space that admits an orthogonal decomposition $\mathcal{V} = \oplus_{v\in Q_0} \mathcal{V}_v$. For each nonnegative integer n we associate with Q_n, the set of paths of length n, the space $C_{G,\mathcal{V}}(Q_n)$ of continuous $\mathcal{L}(\mathcal{V})$-valued functions f on Q_n, where $f(\gamma)$ maps $\mathcal{V}_{s(\gamma)}$ into $\mathcal{V}_{r(\gamma)}$ and $f(\gamma)|_{\mathcal{V}^\perp_{s(\gamma)}} = 0$ for each $\gamma \in Q_n$; of course the continuity is automatic as Q_n is a finite set with the discrete topology – this notation is used for consistency with more general settings where Q_n is a more general topological space as in [60, 61]. Usually we will just write $C_G(Q_n)$, rather than $C_{G,\mathcal{V}}(Q_n)$, for notational convenience.

The space $C_G(Q_n)$ can be seen as a W^*-$C_G(Q_0)$-correspondence with left and right multiplication and $C_G(Q_0)$-valued inner-product given by:

$$\begin{aligned} (f\cdot\xi)(\gamma) &= f(r_n(\gamma))\xi(\gamma) \\ (\xi\cdot f)(\gamma) &= \xi(\gamma)f(s_n(\gamma)) \\ \langle \xi,\eta\rangle(v) &= \sum_{s_n(\gamma')=v} \overline{\eta(\gamma')}\xi(\gamma') \end{aligned} \qquad (\xi,\eta\in C_G(Q_n), f\in C_G(Q_0), \gamma\in Q_n, v\in Q_0).$$

Tensoring $C_G(Q_n)$ with $C_G(Q_m)$ gives $C_G(Q_{n+m})$, hence, in particular, we obtain for each $n\in\mathbb{Z}_+$ that $C_G(Q_n) = C_G(Q_1)^{\otimes n}$. More precisely, for $\xi_n,\dots,\xi_1\in C_G(Q_1)$ we can identify $\xi_1\otimes\cdots\otimes\xi_n\in C_G(Q_1)^{\otimes n}$ with the element $\xi^{(n)}\in C_G(Q_n)$ given by

$$\xi^{(n)}(\gamma) := \xi_n(\alpha_n)\cdots\xi_1(\alpha_1) \quad \text{for} \quad \gamma=(\alpha_n,\dots,\alpha_1)\in Q_n. \tag{4.36}$$

To see that this is the case, note that, rather than viewing elements of $C_G(Q_n)$ as functions, they are also given by tuples $F=(F_\gamma\in\mathcal{L}(\mathcal{V}_{s(\gamma)},\mathcal{V}_{r(\gamma)})\colon \gamma\in Q_n)$ that we usually identify with operator matrices

$$\mathbf{F} = \left[\ F_{\gamma,v}\ \right]_{\gamma\in Q_n, v\in Q_0} : \bigoplus_{v\in Q_0}\mathcal{V}_v \to \bigoplus_{\gamma\in Q_n}\mathcal{V}_{r(\gamma)}, \text{ where } F_{\gamma,v} = \begin{cases} F_\gamma & \text{if } s_n(\gamma)=v, \\ 0 & \text{otherwise.}\end{cases} \tag{4.37}$$

The advantage of the operator matrix representation is that for $F,F'\in C_G(Q_n)$ the norm $\|F\|_{C_G(Q_n)}$ is equal to the operator norm of $\mathbf{F}$, and $\langle F,F'\rangle$ can be identified with $\mathbf{F}'^*\mathbf{F}$; the operator matrix $\mathbf{F}'^*\mathbf{F}\in\mathcal{L}(\mathcal{V})$ is block diagonal and, for each $v\in Q_0$, the diagonal entry from $\mathcal{L}(\mathcal{V}_v)$ corresponds to the value of $\langle F,F'\rangle(v)$. Note that $C_G(Q_0)$ corresponds to the W^*-algebra of block diagonal operators $\operatorname{diag}_{v\in Q_0}(\Lambda_v)$ on $\mathcal{V}=\oplus_{v\in Q_0}\mathcal{V}_v$. Moreover, the operator $T_F^{(0)}$ defined by (4.9) mapping $C_G(Q_0)$ (block diagonal operators) into $C_G(Q_n)$ can be identified with multiplication with $\mathbf{F}$. Subject to this identification, the operator $\mathbf{F}\otimes I_{C_G(Q_1)}$ from $C_G(Q_1)$ to $C_G(Q_{n+1})$ corresponds to multiplication with the block operator matrix

$$\mathbf{F}\otimes I_{C_G(Q_1)} = \left[\ F_{\gamma,\alpha}\ \right]_{\gamma\in Q_{n+1},\alpha\in Q_1}, \text{ where } F_{\gamma,\alpha} = \begin{cases} F_{\gamma'} & \text{if } \gamma=(\gamma',\alpha), \\ 0 & \text{otherwise.}\end{cases}$$

Analogously, one obtains formulas for $\mathbf{F}\otimes I_{C_G(Q_1)^{\otimes n}} = \mathbf{F}\otimes I_{C_G(Q_n)}$ for each $n\in\mathbb{Z}_+$. Taking a product of such operators $\mathbf{F_1}, \mathbf{F_2}\otimes I_{C_G(Q_1)},\dots,\mathbf{F_n}\otimes I_{C_G(Q_1)^{\otimes n}}$, with $F_1,F_2,\dots,F_n\in C_G(Q_1)$, we finally arrive at (4.36).

We now take E to be the W^*-$C_G(Q_0)$-correspondence $C_G(Q_1)$. Elements of the Fock space $\mathcal{F}^2(C_G(Q_1)) = \oplus_{n\in\mathbb{Z}_+}C_G(Q_n)$ are then given by tuples $F=(F_\gamma\in\mathcal{L}(\mathcal{V}_{s(\gamma)},\mathcal{V}_{r(\gamma)})\colon \gamma\in\Gamma)$, where Γ is the collection of all finite paths of whatever length, such that the operator matrix

$$\mathbf{F} = \left[\ F_{\gamma,v}\ \right]_{v\in Q_0,\gamma\in\Gamma}$$

is in $\mathcal{L}(\mathcal{V},\oplus_{\gamma\in\Gamma}\mathcal{V}_{r(\gamma)})$ (i.e., bounded). Here $F_{\gamma,v}$ is as defined in (4.37). As in Subsection 3.3 we denote the Hilbert space $\oplus_{\gamma\in\Gamma}\mathcal{V}_{r(\gamma)}$ by $\ell^2_{\mathcal{V}}(\Gamma)$. The Toeplitz algebra $\mathcal{F}^\infty(E)$, seen as a subalgebra of $\mathcal{L}(\ell^2_{\mathcal{V}}(\Gamma))$, is then precisely the Toeplitz algebra $\mathfrak{L}_\Gamma(\mathcal{V},\mathcal{V})$ defined in Subsection 3.3. That is, an element R in $\mathcal{F}^\infty(E)$ is given by a tuple $R = (R_\gamma \in \mathcal{L}(\mathcal{V}_{s(\gamma)}, \mathcal{V}_{r(\gamma)})\colon \gamma \in \Gamma)$ with the property that the operator matrix

$$\mathbf{R} = \left[\ R_{\gamma,\gamma'}\ \right]_{\gamma,\gamma'\in\Gamma}, \text{ where } R_{\gamma,\gamma'} = R_{\gamma\gamma'^{-1}} \text{ (with } R_{\text{undefined}} = 0),$$

induces a bounded operator on $\ell^2_{\mathcal{V}}(\Gamma)$. Here $\gamma\gamma'^{-1}$ is defined by (3.7).

In addition to $\mathcal{G}$ we assume that we are given a Hilbert space $\mathcal{G}$ that also has an orthogonal-sum decomposition of the form $\mathcal{G} = \oplus_{v\in Q_0}\mathcal{G}_v$. We then set $\mathcal{E}_v = \mathcal{V}_v \otimes \mathcal{G}_v$ for each $v \in Q_0$ and $\mathcal{E} = \oplus_{v\in Q_0}\mathcal{E}_v$. Now let $\sigma : C_G(Q_0) \to \mathcal{L}(\mathcal{E})$ be the representation of $C_G(Q_0)$ given by

$$\sigma(\operatorname{diag}_{v\in Q_0}(A_v)) = \operatorname{diag}_{v\in Q_0}(A_v \otimes I_{\mathcal{G}_v}).$$

Then, for each $n \in \mathbb{Z}_+$, we have $C_G(Q_n) \otimes_\sigma \mathcal{E} = \oplus_{\gamma\in Q_n}(\mathcal{V}_{r_n(\gamma)} \otimes \mathcal{G}_{s_n(\gamma)})$ and thus

$$\mathcal{F}^2(E,\sigma) := \mathcal{F}^2(E) \otimes_\sigma \mathcal{E} = \oplus_{\gamma\in\Gamma}(\mathcal{V}_{r(\gamma)} \otimes \mathcal{G}_{s(\gamma)}).$$

It is straightforward that

$$\sigma(C_G(Q_0))' = \left\{\operatorname{diag}_{v\in Q_0}(I_{\mathcal{V}_v} \otimes B_v)\colon B_v \in \mathcal{L}(\mathcal{G}_v)\right\}.$$

Moreover, the space $C_G(Q_1)^\sigma$ consists of those operator matrices mapping $\mathcal{E} = \oplus_{v\in Q_0}(\mathcal{V}_v \otimes \mathcal{G}_v)$ into $C_G(Q_1) \otimes \mathcal{E} = \oplus_{\alpha\in Q_1}(\mathcal{V}_{r(\alpha)} \otimes \mathcal{G}_{s(\alpha)})$ that are of the form

$$\left[\ I_{\mathcal{V}_{r(\alpha)}} \otimes K_{\alpha,v}\ \right]_{\alpha\in Q_1, v\in Q_0}$$

with $K_{\alpha,v} \in \mathcal{L}(\mathcal{G}_{r(\alpha)}, \mathcal{G}_{s(\alpha)})$ and $K_{\alpha,v} = 0$ in case $r(\alpha) \neq v$. Leaving out the identity operators $I_{\mathcal{V}_v}$, we can identify $\sigma(C_G(Q_0))'$ with $C_{\widetilde{G},\mathcal{G}}(Q_0)$ and $C_G(Q_1)^\sigma$ with $C_{\widetilde{G},\mathcal{G}}(Q_1)$, where $\widetilde{G} = \{Q_0, Q_1, r, s\}$ is the transposed quiver of G (i.e., with the source and range maps interchanged). Note that the generalized disk $\mathbb{D}((C_G(Q_1)^\sigma)^*)$ of strictly contractive operators with adjoint in $C_G(Q_1)^\sigma$ can be identified with the set $\mathbb{D}_{G,\mathcal{G}}$ defined in Subsection 3.3. Next observe that for $Z = (Z_\alpha \in \mathcal{L}(\mathcal{G}_{s(\alpha)}, \mathcal{G}_{r(\alpha)})\colon \alpha \in Q_1) \in \mathbb{D}_{G,\mathcal{G}}$, the nth generalized power Z^n of Z then corresponds to the tuple $(Z_\gamma \in \mathcal{L}(\mathcal{G}_{s_n(\gamma)}, \mathcal{G}_{r_n(\gamma)})\colon \gamma \in Q_n)$ with $Z_\gamma = Z^\gamma$, where Z^γ is defined by (3.8).

Now let $R = (R_\gamma\colon \gamma \in \Gamma) \in \mathcal{F}^\infty(E) = \mathfrak{L}_\Gamma(\mathcal{V},\mathcal{V})$ and $Z = (Z_\alpha\colon \alpha \in Q_1) \in \mathbb{D}_{G,\mathcal{G}}$. It then follows that the first Muhly-Solel point-evaluation $\widehat{R}(Z)$ of R in Z is given by the tensor-product functional-calculus for $\mathfrak{L}_\Gamma(\mathcal{V},\mathcal{V})$:

$$\widehat{R}(Z) = \sum_{\gamma\in\Gamma} i_{\mathcal{V}_{r(\gamma)}\otimes\mathcal{G}_{r(\gamma)}}(R_\gamma \otimes Z^\gamma) i^*_{\mathcal{V}_{s(\gamma)}\otimes\mathcal{G}_{s(\gamma)}},$$

where, as in Subsection 3.3, we use the general notation: for a subspace $\mathcal{H}$ of a Hilbert space $\mathcal{K}$ we write $i_{\mathcal{H}}$ for the canonical embedding of $\mathcal{H}$ into $\mathcal{K}$. Thus the W^*-correspondence Schur class $\mathcal{S}(E,\sigma)$ corresponds to the free semigroupoid algebra Schur class $\mathcal{S}_G(\mathcal{V},\mathcal{V})$ in combination with the point-evaluation in the generalized disk $\mathbb{D}_{G,\mathcal{G}}$.

The W^*-correspondence Nevanlinna-Pick problem in this case thus turns out to be the left-tangential tensor-product functional-calculus free semigroupoid algebra Nevanlinna-Pick problem (**QLTT-NP**): *Given a data set*

$$\mathfrak{D} : Z^{(1)}, \dots, Z^{(N)} \in \mathbb{D}_{G,\mathcal{G}},\ X_1, \dots, X_N, Y_1, \dots, Y_N \in \mathcal{L}(\mathcal{E}), \tag{4.38}$$

determine when there exists a Schur class function $S \in \mathcal{S}_G(\mathcal{V}, \mathcal{V})$ such that

$$X_i S(Z^{(i)}) = Y_i \quad \textit{for} \quad i = 1, \dots, N. \tag{4.39}$$

As in the "unit disk" example of Subsection 4.8, one can deduce from the solution to this problem the analogous result for the "non-square" case considered in Subsection 3.3, but we will not work out those details here.

In order to state the solution we remark that the Szegő kernel (4.22) specified for this setting has the form

$$\mathbb{K}_{C_G(Q_1),\sigma} \colon \mathbb{D}_{G,\mathcal{G}} \times \mathbb{D}_{G,\mathcal{G}} \to \mathcal{L}^a(C_{\widetilde{G}}(Q_0), \mathcal{L}(\mathcal{E}))$$

and can be written as

$$\mathbb{K}_{C_G(Q_1),\sigma}(Z, Z')[B] = \sum_{\gamma \in \Gamma} i_{\mathcal{E}_{r(\gamma)}} \left(I_{\mathcal{V}_{r(\gamma)}} \otimes Z^{\gamma} i^*_{\mathcal{G}_{s(\gamma)}} B i_{\mathcal{G}_{s(\gamma)}} (Z'^{\gamma})^* \right) i^*_{\mathcal{E}_{r(\gamma)}}. \tag{4.40}$$

An application of the general Theorem 4.6 then leads to the following solution of the problem.

Theorem 4.15. *Suppose we are given data as in* (4.38). *Then there exists a solution $S \in \mathcal{S}_G(\mathcal{V}, \mathcal{V})$ to the* **QLTT-NP** *interpolation problem if and only if one of the following equivalent conditions holds:*

1. *the kernel $\mathbb{K}_{\mathfrak{D}} \colon \{1, \dots, N\} \times \{1, \dots, N\} \to \mathcal{L}^a(C_{\widetilde{G}}(Q_0), \mathcal{L}(\mathcal{E}))$ given by*
$$\mathbb{K}_{\mathfrak{D}}(i,j)[B] = X_i \mathbb{K}_{C_G(Q_1),\sigma}(Z^{(i)}, Z^{(j)})[B] X_j^* - Y_i K_{C_G(Q_1),\sigma}(Z^{(i)}, Z^{(j)})[B] Y_j^*$$
is a completely positive kernel.
2. *the map φ from $C_{\widetilde{G}}(Q_0)^{N\times N}$ to $\mathcal{L}(\mathcal{E})^{N \times N}$ given by*
$$\varphi\left([B_{ij}]_{i,j=1}^N\right) = \left[\ \mathbb{K}_{\mathfrak{D}}(i,j)[B_{i,j}]\ \right]_{i,j=1}^N$$
is completely positive.

As a consequence of the extension of Choi's theorem (see Theorem 4.14) we obtain the following third condition.

Theorem 4.16. *Assume that $\mathcal{V}$ is separable, and that for each $v \in Q_0$ we have an orthonormal basis $\{e_1^{(v)}, \dots, e_{\kappa_v}^{(v)}\}$ for $\mathcal{V}_v$. Then, in addition to the two conditions* (1) *and* (2) *in Theorem* 4.15, *a third condition equivalent to the existence of an $S \in \mathcal{S}_G(\mathcal{V}, \mathcal{V})$ that satisfies* (4.39) *is that for each $v \in Q_0$ the operator matrix*

$$\left[\ \mathbb{K}_{\mathfrak{D}}(i,j)[e_{i'}^{(v)} e_{j'}^{(v)*}]\ \right]_{(i,i'),(j,j') \in \{1,\dots,N\} \times \{1,\dots,\kappa_v\}} \in \mathcal{L}(\mathcal{E})^{\kappa_v N \times \kappa_v N} \tag{4.41}$$

is positive. Here $\mathbb{K}_{\mathfrak{D}}$ is the kernel defined in Part 1 *of Theorem* 4.15.

If one writes out the definition of the kernel $\mathbb{K}_{\mathfrak{D}}$ (and that of $\mathbb{K}_{C_G(Q_1),\sigma}$) the operator matrix (4.41) turns out to be exactly the Pick matrix $\mathbb{P}_{QLTT}$ of Part 1 of Theorem 3.3. Thus we obtain the first solution criterion of Theorem 3.3.

In order to prove Theorem 4.16 it is convenient to first prove two lemmas.

Lemma 4.17. *Let $\mathcal{H}$ be a Hilbert space with orthogonal sum decomposition $\mathcal{H} = \oplus_{i=1}^N \mathcal{H}_i$. Then the map $\psi : \mathcal{L}(\mathcal{H}) \to \mathcal{L}(\mathcal{H})$ defined by*

$$\psi([\ B_{i,j}\ ]_{i,j=1}^N) = \operatorname{diag}_{i=1}^N (B_{i,i}),$$

where $[\, B_{i,j} \,]_{i,j=1}^N \in \mathcal{L}(\mathcal{H})$ is an operator matrix with $B_{i,j} \in \mathcal{L}(\mathcal{H}_j, \mathcal{H}_i)$, is a completely positive map.

More generally, the statement remains true for any conditional expectation operator $\psi : \mathcal{A} \to \mathcal{A}$, where $\mathcal{A}$ is an arbitrary C^*-algebra [104] (see also [105]), of which the map ψ in Lemma 4.17 is just a particular example. We give the following independent proof.

Proof of Lemma 4.17. It is immediately clear that ψ is positive. Let $M \in \mathbb{Z}_+$. Assume that we have an operator matrix

$$\mathbb{B} = \left[[B_{\alpha,i;\beta,j}]_{i,j=1,\dots,N}\right]_{\alpha,\beta=1,\dots,M}$$

that is a positive element of $\mathcal{L}(\oplus_{\alpha=1}^M \mathcal{H})$, where each $B_{\alpha,i;\beta,j} \in \mathcal{L}(\mathcal{H}_j, \mathcal{H}_i)$. The assumption that $\mathbb{B}$ is positive implies that the matrix

$$\widetilde{\mathbb{B}}_i = [B_{\alpha,i;\beta,i}]_{\alpha,\beta=1,\dots,M}$$

is positive for each $i = 1, \dots, N$, since $\widetilde{\mathbb{B}}_i$ is a principal submatrix of $\mathbb{B}$. On the other hand by definition we have

$$\psi(\mathbb{B}) = \left[\operatorname{diag}_{i=1}^N [B_{\alpha,i;\beta,i}]\right]_{\alpha,\beta=1,\dots,M}.$$

which is unitarily equivalent via a permutation matrix to

$$\operatorname{diag}_{i=1}^N \left([B_{\alpha,i;\beta,i}]_{\alpha,\beta=1,\dots,M}\right) = \operatorname{diag}_{i=1}^N (\widetilde{\mathbb{B}}_i).$$

Thus positivity of $\mathbb{B}$ implies positivity of $\psi(\mathbb{B})$ as required. □

Now observe that we can extend the Szegő kernel $\mathbb{K}_{C_G(Q_1),\sigma}$ to a kernel $\overline{\mathbb{K}}_{C_G(Q_1),\sigma}$ of the form

$$\overline{\mathbb{K}}_{C_G(Q_1),\sigma} \colon \mathbb{D}_{G,\mathcal{G}} \times \mathbb{D}_{G,\mathcal{G}} \to \mathcal{L}(\mathcal{L}(\mathcal{G}), \mathcal{L}(\mathcal{E}))$$

using the same formula, i.e., in the right-hand side of (4.40) we allow B to be in $\mathcal{L}(\mathcal{G})$ rather than just a block diagonal operator. We can then also extend the kernel $\mathbb{K}_{\mathfrak{D}}$ in condition (1) and the map φ in condition (2) of Theorem 4.15 to a kernel $\overline{\mathbb{K}}_{\mathfrak{D}}$ and a map $\overline{\varphi}$ of the form

$$\overline{\mathbb{K}}_{\mathfrak{D}} : \{1, \dots, N\} \times \{1, \dots, N\} \to \mathcal{L}^a(\mathcal{L}(\mathcal{G}), \mathcal{L}(\mathcal{E})), \quad \overline{\varphi} : \mathcal{L}(\mathcal{G})^{N \times N} \to \mathcal{L}(\mathcal{E})^{N \times N},$$

simply by replacing $\mathbb{K}_{C_G(Q_1),\sigma}$ by $\overline{\mathbb{K}}_{C_G(Q_1),\sigma}$ in the definitions of $\overline{\mathbb{K}}_{\mathfrak{D}}$ and φ.

Lemma 4.18. *The map φ defined in Part 2 of Theorem 4.15 is completely positive if and only if the map $\overline{\varphi}$ is completely positive.*

Proof. Let ψ be the completely positive map of Lemma 4.17 with $\mathcal{H}$ replaced by $\mathcal{G}$, relative to the orthogonal decomposition $\mathcal{G} = \oplus_{v\in Q_0}\mathcal{G}_v$. Notice that complete positivity of $\overline{\varphi}$ automatically implies complete positivity of φ. To see that the converse is also true, observe that for any $B \in \mathcal{L}(\mathcal{G})$ and any $Z, Z' \in \mathbb{D}_{G,\mathcal{G}}$ we have $\overline{\mathbb{K}}_{C_G(Q_1),\sigma}(Z,Z')[B] = \mathbb{K}_{C_G(Q_1),\sigma}(Z,Z')[\psi(B)]$, and thus also $\overline{\varphi}[[\,B_{i,j}\,]_{i,j=1}^N] = \varphi[[\,\psi(B_{i,j})\,]_{i,j=1}^N]$. Hence the converse statement follows from Lemma 4.17 and the fact that compositions of completely positive maps are again completely positive maps. □

Proof of Theorem 4.16. Let $\{e_1,\dots,e_\kappa\}$ be a reordering of the orthonormal basis $\{e_i^{(v)} : v \in Q_0,\ i = 1,\dots,\kappa_v\}$ of $\mathcal{V}$. Since $\overline{\varphi}$ is defined on $\mathcal{L}(\mathcal{V})^{N\times N} = \mathcal{L}(\mathcal{V}^N)$, we can apply Theorem 4.14 to $\overline{\varphi}$, obtaining that $\overline{\varphi}$ is completely positive if and only the operator matrix

$$\left[\ \overline{\mathbb{K}}_{\mathfrak{D}}(i,j)[e_{i'}e_{j'}^*]\ \right]_{(i,i'),(j,j')\in\{1,\dots,N\}\times\{1,\dots,\kappa\}} \tag{4.42}$$

is positive. Next observe that $\overline{\mathbb{K}}_{\mathfrak{D}}(i,j)[e_{i'}e_{j'}^*] = \mathbb{K}_{\mathfrak{D}}(i,j)[e_{i'}e_{j'}^*]$ in case $e_{i'}, e_{j'} \in \{e_1^{(v)},\dots,e_{\kappa_v}^{(v)}\}$ for some $v \in Q_0$, and that $\overline{\mathbb{K}}_{\mathfrak{D}}(i,j)[e_{i'}e_{j'}^*] = 0$ otherwise. Thus, after a reordering in the basis $\{e_1,\dots,e_\kappa\}$, we can identify the operator matrix (4.42) with the block diagonal operator matrix with the operator matrices (4.41) on the diagonal. This proves our claim. □

We now show how the solution criterion obtained above can be used to derive the results for the Riesz-Dunford and operator-argument functional calculus listed in Subsection 3.3.

Riesz-Dunford functional calculus Nevanlinna-Pick interpolation. The Riesz-Dunford functional calculus is the special case of the tensor functional calculus with $\mathcal{V}_v = \mathbb{C}$ for each $v \in Q_0$, i.e., $\mathcal{E} = \mathcal{G}$. In this case, the same argument as used in the proof of Theorem 4.13 (i.e., invariance under cyclic permutations of the trace) now applied to the map $\overline{\varphi}$, in combination with Theorem 4.14, gives us the following result.

Theorem 4.19. *In case $\mathcal{V}_v = \mathbb{C}$ for each $v \in Q_0$, then there exists an $S \in \mathcal{S}_G(\mathbb{C},\mathbb{C})$ that satisfies* (4.39) *if and only if the map φ_* from $\mathcal{L}(\mathcal{G})^{N\times N}$ to $\mathcal{L}(\mathcal{G})^{N\times N}$ given by*

$$\varphi_*\left([C_{ij}]_{i,j=1}^N\right) \tag{4.43}$$
$$= \left[\ \sum_{\gamma\in\Gamma} i_{\mathcal{G}_{s(\gamma)}}((Z^{(i)})^\gamma)^* i_{\mathcal{G}_{r(\gamma)}}^* (X_i^* C_{i,j} X_j - Y_i^* C_{i,j} Y_j) i_{\mathcal{G}_{r(\gamma)}} (Z^{(j)})^\gamma i_{\mathcal{G}_{s(\gamma)}}^*\ \right]_{i,j=1}^N$$

is a completely positive map. If $\mathcal{G}$ is separable and $\{e_1,\dots,e_\kappa\}$ is an orthonormal basis for $\mathcal{G}$, then complete positivity of φ_* is equivalent to positivity of the operator matrix $\mathbb{P}_{\mathrm{QLTRD}} \in \mathcal{L}(\mathcal{G})_{\kappa N\times\kappa N}$ for which the entry corresponding to the pairs

$(i,i'),(j,j') \in \{1,\dots,N\} \times \{1,\dots,\kappa\}$ *is given by*

$$\begin{aligned} &\left[\ \mathbb{P}_{\mathrm{QLTRD}}\ \right]_{(i,i'),(j,j')} \qquad (4.44)\\ &= \sum_{\gamma\in\Gamma} i_{\mathcal{G}_{s(\gamma)}}((Z^{(i)})^{\gamma})^{*} i^{*}_{\mathcal{G}_{r(\gamma)}}(X_i^{*}e_ie_{j'}^{*}X_j - Y_i^{*}e_ie_{j'}^{*}Y_j)i_{\mathcal{G}_{r(\gamma)}}(Z^{(j)})^{\gamma} i^{*}_{\mathcal{G}_{s(\gamma)}}. \end{aligned}$$

The definition of $\mathbb{P}_{\mathrm{QLTRD}}$ in Theorem 4.16 is the same as that in Theorem 3.3. Thus we obtain the second statement of Theorem 3.3.

Operator-argument functional calculus Nevanlinna-Pick interpolation. Note that in the case of the second Muhly-Solel point-evaluation of Subsection 4.7, specified for the setting considered here, the points are pairs $(T,A) \in \mathbb{D}(C_G(Q_1)^*) \times C_G(Q_0)$, i.e., T corresponds to a tuple $(T_\alpha \in \mathcal{L}(\mathcal{V}_{r(\alpha)}, \mathcal{V}_{s(\alpha)})\colon \alpha \in Q_1)$ so that $F = (T_\alpha^*\colon \alpha \in Q_1)$ is in $C_G(Q_1)$ and the operator matrix $\mathbf{F}$ corresponding to F is a strict contraction. Thus T is an element of the set $\mathbb{D}_{\widetilde{G},\mathcal{V}}$ defined in Subsection 3.3. The element $T^*_{(n)} = T^* \otimes \cdots \otimes T^* = F \otimes \cdots \otimes F \in C_G(Q_n)$ is then given by the tuple $(T^*_\gamma\colon \gamma \in Q_n)$ with $T_\gamma = T^\gamma$, following the notation (3.10), and the operator $T^{(0)}_{T^*_{(n)}}$ (as in (4.9)) corresponds to multiplication with the operator matrix $\mathbf{T}^*_{(\mathbf{n})}$ associated with $T^*_{(n)}$.

It is then not difficult to see that the evaluation of an element $R = (R_\gamma\colon \gamma \in \Gamma)$ of the Toeplitz algebra $\mathcal{F}^\infty(E) = \mathfrak{L}_\Gamma(\mathcal{V},\mathcal{V})$ in a pair $(T,A) \in C_G(Q_0) \times \mathbb{D}_{G,\mathbf{OA}}$ is given by the left-tangential operator-argument functional calculus:

$$\hat{R}(T,A) = \sum_{\gamma\in\Gamma} i_{s(\gamma)} T^{\gamma^{\top}} A_{r(\gamma)} R_\gamma i^*_{s(\gamma)},$$

where $A = \operatorname{diag}_{v\in Q_0}(A_v)$ and $T = (T_\alpha\colon \alpha \in Q_1)$. The corresponding Nevanlinna-Pick problem of Subsection 4.7 thus turns out to be the **QLTOA-NP** problem considered in Subsection 3.3, and one easily sees that the Pick matrix criterion of Theorem 4.11 is exactly the Pick matrix criterion of Part 3 of Theorem 3.3.

4.10. Still more examples

There are still more seemingly different examples of generalized Nevanlinna-Pick interpolation covered by the Muhly-Solel correspondence-representation formalism. We mention in particular the semicrossed product algebras of Peters [78] (see Example 2.6 in [67]). A particular instance of this setup yields as the Toeplitz algebra $\mathcal{F}^\infty(E)$ the algebra of operators on $\ell^2(\mathbb{Z})$ having lower-triangular matrix representation with points equal to strictly contractive bilateral weighted shift operators on $\ell^2(\mathbb{Z})$; this algebra can also be seen as the Toeplitz algebra associated with the infinite quiver:

$$\begin{aligned} &Q_0 = \{v_k\colon k\in\mathbb{Z}\}, \quad Q_1 = \{\alpha_k\colon k \in \mathbb{Z}\},\\ &s(\alpha_k) = v_k, \quad r(\alpha_k) = v_{k+1}. \end{aligned}$$

The point-evaluation $\widehat{R}(\eta)$ in this context has a neat interpretation in terms of the realization of the lower triangular operator R as the input-output operator for a conservative time-varying linear system

$$\Sigma\colon \begin{cases} x(n+1) &= A(n)x(n) + B(n)u(n) \\ y(n) &= C(n)x(n) + D(n)u(n) \end{cases}$$

(see [16, Section 6.3]). In order to recover the time-varying interpolation theory related to time-varying H^∞-control and model reduction carried out in the 1990s [8, 23, 41, 98, 42], one needs to work with the second Muhly-Solel point-evaluation specified for this setting. We leave details to another occasion.

5. More general Schur classes

For all the classes of Schur functions discussed to this point with the exception of the (commutative and noncommutative) polydisk examples in Subsection 3.4, the Schur class can be isometrically identified with the space of contractive multipliers between two reproducing kernel Hilbert spaces. Although there are some issues with different kinds of point-evaluations, one can say that the most general of these is the Muhly-Solel correspondence-representation setup in Section 4, and that all these are tied down to the setting of a *complete Pick kernel* (see [6]). An alternative noncommutative version of a complete Pick kernel with a rather complete theory of Nevanlinna-Pick interpolation has been worked out by Popescu in [90]. There are a number of generalized Hardy algebras which go beyond these limitations. We mention a few:

Powers of complete Pick kernels: As mentioned above, there is a good theory of transfer-function realization and Nevanlinna-Pick interpolation for the generalized Schur class defined to be the set of contractive multipliers on a reproducing Hilbert space $\mathcal{H}(k)$ with complete Pick kernel (see [5, 29, 6]). More generally, one can consider the generalized Schur-Agler class associated with the set of powers $k(z,w)^q$, $m = 1, \dots, m$ for some integer m, of some complete Pick kernel k (see [91] for the commutative setting and [89] for the noncommutative setting). While there is a good operator-model and dilation theory as well as a von Neumann inequality, there does not appear to be any approach to transfer-function realization or Nevanlinna-Pick interpolation for this generalized Schur-Agler class.

Higher-rank graph algebras: These include direct products of the graph algebras considered in Subsection 3.3 above and much more – see [62]. Whether these more general algebras are of interest for robust control theory, as those coming from SNMLs (see [27]), remains to be seen. For this general setting it remains to work out the nature of possible point-evaluations, the Schur class and a Schur-Agler type interpolation theory.

Hardy algebras associated with product systems over more general semigroups: We mention that the Fock space $\mathcal{F}^2(E)$ of Section 4 is a product decomposition over the semigroup $\mathbb{Z}_+$. Similar constructions but over more general semigroups, such as $\mathbb{Z}_+^n$, pick up the higher-rank graphs of [62] as examples. Here also it remains to work out the point-evaluations, Schur class and interpolation theory. These results should include the commutative and noncommutative interpolation theory for the polydisk discussed in Subsection 3.4.

Schur classes based on a family of test functions: In this approach we assume that we are given a set X and a family Ψ of functions on X. We then say that a positive kernel $k\colon X \times X \to \mathbb{C}$ is *admissible* whenever $k_\psi(x,x') := (1 - \psi(x)\overline{\psi(x')})k(x,x')$ is also a positive kernel; we denote the class of all admissible kernels by $\mathcal{K}_\Psi$. A function φ on X is then said to be in the Schur-Agler class $\mathcal{SA}_\Psi$ associated with Ψ if $k_\varphi(x,x') := (1 - \varphi(x)\overline{\varphi(x')})k(x,x')$ is a positive kernel whenever $k \in \mathcal{K}_\Psi$. If one takes $X = \mathbb{D}^d$ and $\Psi = \{\psi_k(\lambda) := \lambda_k\colon k = 1,\dots,d\}$, then the associated Schur-Agler class $\mathcal{SA}_\Psi$ is just the Schur-Agler class $\mathcal{SA}_d$ defined in Subsection 3.4. In addition to the Schur-Agler class on the polydisk as discussed in Subsection 3.4, it has been known since the paper of Abrahamse [1] that the Schur class over a finitely-connected planar domain fits into this framework, but with an infinite family of test functions. It turns out that many of the original ideas of Agler giving rise to transfer-function realization and Nevanlinna-Pick interpolation theorems via Agler decompositions go through for this general setting (see [39, 46, 64]); in the case of finitely-connected planar domains, one even gets a continuous analogue of the Agler decomposition (3.20) (see [45]). The paper [44] handles a more general scenario where the underlying set is a semigroupoid (satisfying some additional hypotheses) and pointwise multiplication of functions is replaced by semigroupoid convolution. The setup gives a unified formalism (an alternative to the **LTOA/RTOA-NP** or **RTRD-NP** setup discussed in Section 2) for the simultaneous encoding of interpolation problems of Nevanlinna-Pick and of Carathéodory-Fejér type inspired by the ideas of Jury [56]. For this class to include examples of interest (e.g., the Hardy algebras associated with the higher-rank graph algebras mentioned above, the (unit balls of) Toeplitz algebras $\mathcal{F}^\infty(E)$ appearing in Section 4, Schur classes associated with a complete Pick kernel [5, 29, 90], as well as the more general commutative Schur-Agler classes in [19, 20] and noncommutative Schur-Agler classes in [26]), one must identify the appropriate collection Ψ of test functions to get started. To handle these examples, the theory from [44, 46, 64] must be extended to handle matrix- or operator-valued test functions. Even after this is done, it appears that something more must be incorporated in the test-function approach in order to handle the Muhly-Solel tensor-type point-evaluation. Work has begun on finding a single formalism containing all these examples as special cases (see [15]).

Acknowledgment

We thank David Sherman for the reference [104], Nir Cohen for informative correspondence regarding Theorem 2.4, Michael Dritschel, Tirtha Bhattacharyya and Chris Todd for suggesting the example in Remark 2.7, and Vladimir Bolotnikov and Gilbert Groenewald for their useful comments and suggestions.

References

[1] M.B. Abrahamse, The Pick interpolation theorem for finitely connected domains, *Michigan Math. J.* **26** (1979), 195–203.

[2] J. Agler, Some interpolation theorems of Nevanlinna-Pick type, Preprint, 1988.

[3] J. Agler, On the representation of certain holomorphic functions defined on a polydisk, in *Topics in Operator Theory: Ernst D. Hellinger Memorial Volume* (L. de Branges, I. Gohberg and J. Rovnyak, ed.), pp. 47–66, **OT48**, Birkhäuser, Basel-Berlin-Boston, 1990.

[4] J. Agler and J.E. McCarthy, Nevanlinna-Pick interpolation on the bidisk, *J. reine angew. Math.* **506** (1999), 191–204.

[5] J. Agler and J.E. McCarthy, Complete Nevanlinna-Pick kernels, *J. Funct. Anal.*, **175** (2000), 111–124.

[6] J. Agler and J.E. McCarthy, *Pick Interpolation and Hilbert Function Spaces*, Graduate Studies in Mathematics Volume **44**, Amer. Math. Soc., Providence, 2002.

[7] D. Alpay, J.A. Ball, I. Gohberg and L. Rodman, J-unitary preserving automorphisms of rational matrix functions: state space theory, interpolation and factorization, *Linear Algebra Appl.* **197-198** (1994), 531–566.

[8] D. Alpay, P. Dewilde and H. Dym, Lossless inverse scattering and reproducing kernels for upper triangular operators, in: *Extensions and Interpolation of Linear Operators and Matrix Functions* (I. Gohberg, ed.), pp. 61–133, **OT 47**, Birkhäuser, Basel-Berlin-Boston, 1990.

[9] D. Alpay and D.S. Kalyuzhnyĭ-Verbovetzkiĭ, Matrix J-unitary non-commutative rational formal power series, in: *The State Space Method Generalizations and Applications* (D. Alpay and I. Gohberg, ed.), pp. 49–113, **OT 161**, Birkhäuser, Basel-Berlin-Boston, 2006.

[10] C.-G. Ambrozie and D. Timotin, A von Neumann type inequality for certain domains in $\mathbb{C}^n$, *Proc. Amer. Math. Soc.* **131** (2003), 859–869.

[11] A. Arias and G. Popescu, Noncommutative interpolation and Poisson transforms, *Israel J. Math.* **115** (2000), 205–234.

[12] N. Aronszajn, Theory of reproducing kernels, *Trans. Amer. Math. Soc.* **68** (1950), 337–404.

[13] W. Arveson, Subalgebras of C^* algebras III: Multivariable operator theory, *Acta Math.* **181** (1998), 159–228.

[14] W. Arveson, The curvature invariant of a Hilbert module over $\mathbb{C}[z_1, \dots, z_d]$, *J. reine angew. Math.* **522** (2000), 173–226.

[15] J.A. Ball, T. Bhattacharyya, M.A. Dritschel, S. ter Horst and C.S. Todd, The Schur-Agler class in a correspondence/test-function setting: transfer-function realization, in preparation.

[16] J.A. Ball, A. Biswas, Q. Fang and S. ter Horst, Multivariable generalizations of the Schur class: positive kernel characterization and transfer function realization, in: *Recent Advances in Operator Theory and Applications*, pp. 17–79, **OT 187**, Birkhäuser, Basel, 2008.

[17] J.A. Ball and V. Bolotnikov, On a bitangential interpolation problem for contractive-valued functions on the unit ball, *Linear Algebra Appl.* **353** (2002), 107–147.

[18] J.A. Ball and V. Bolotnikov, Realization and interpolation for Schur-Agler class functions on domains with matrix polynomial defining function in $\mathbb{C}^n$, *J. Funct. Anal.* **213** (2004), 45–87.

[19] J.A. Ball and V. Bolotnikov, Nevanlinna-Pick interpolation for Schur-Agler class functions on domains with matrix polynomial defining function in $\mathbb{C}^n$, *New York J. Math.* **11** (2005), 247–290.

[20] J.A. Ball and V. Bolotnikov, Interpolation in the noncommutative Schur-Agler class, *J. Operator Theory* **58** (2007), 83–126.

[21] J.A. Ball and V. Bolotnikov, Interpolation problems for Schur multipliers on the Drury-Arveson space: from Nevanlinna-Pick to Abstract Interpolation Problem, Integral Equations Operator Theory **62** (2008), 301–349.

[22] J.A. Ball, V. Bolotnikov and Q. Fang, Multivariable backward-shift-invariant subspaces and observability operators, *Multidimens. Syst. Signal Process.* **18** (2007), 191–248.

[23] J.A. Ball, I. Gohberg and M.A. Kaashoek, Nevanlinna-Pick interpolation for time-varying input-output maps: The discrete case, in: *Time-Variant Systems and Interpolation* (I. Gohberg, ed.), pp. 1–51, **OT 56**, Birkhäuser, Basel-Berlin-Boston, 1992.

[24] J.A. Ball, I. Gohberg and L. Rodman, *Interpolation of Rational Matrix Functions*, **OT45**, Birkhäuser, Basel-Berlin-Boston, 1990.

[25] J.A. Ball, I. Gohberg and L. Rodman, Two-sided tangential interpolation of real rational matrix functions, in: *New Aspects in Interpolation and Completion Theories*, pp. 73–102, **OT 64**, Birkhäuser, Basel-Berlin-Boston, 1993.

[26] J.A. Ball, G. Groenewald and T. Malakorn, Conservative structured noncommutative multidimensional linear systems, in: *The State Space Method Generalizations and Applications* (D. Alpay and I. Gohberg, ed.), pp. 179–223, **OT 161**, Birkhäuser, Basel-Berlin-Boston, 2006.

[27] J.A. Ball, G. Groenewald and T. Malakorn, Bounded real lemma for structured noncommutative multidimensional linear systems and robust control, *Multidimens. Syst. Signal Process.* **17** (2006), 119–150.

[28] J.A. Ball and T.T. Trent, Unitary colligations, reproducing kernel Hilbert spaces and Nevanlinna-Pick interpolation in several variables, *J. Funct. Anal.* **157** (1998), 1–61.

[29] J.A. Ball, T. T. Trent and V. Vinnikov, Interpolation and commutant lifting for multipliers on reproducing kernel Hilbert spaces, in: *Operator Theory and Analysis* (H. Bart, I. Gohberg and A.C.M. Ran, ed.), pp. 89–138, **OT 122**, Birkhäuser, Basel, 2001.

[30] J.A. Ball and V. Vinnikov, *Victor Lax-Phillips scattering and conservative linear systems: a Cuntz-algebra multidimensional setting*, Mem. Amer. Math. Soc. No. **868**, American Mathematical Society, Providence, 2006.

[31] S. Barreto, B.V.R. Bhat, V. Liebscher and M. Skeide, Type I product systems of Hilbert modules, *J. Funct. Anal.* **212** (2004), 121–181.

[32] V. Bolotnikov, Interpolation for multipliers on reproducing kernel Hilbert spaces, *Proc. Amer. Math. Soc.* **131** (2003), 1373–1383.

[33] C. Carathéodory, Über den Variabilitätsbereich der Koefficienten von Potenzreihen, die gegebene Werte nicht annehmen, *Mat. Ann.* **64** (1907), 95–115.

[34] C. Carathéodory and L. Fejér, Über den Zusammenhang der Extreme von harmonischen Funktionen mit ihren Koeffizienten und über den Picard-Landauschen Satz, *Rend. Circ. Mat. Palermo* **32** (1911), 193–217.

[35] N. Cohen and I. Lewkowicz, The Lyapunov order for real matrices, *Linear Algebra Appl.* **430** (2009), 1849–1866.

[36] N. Cohen and I. Lewkowicz, Nevanlinna-Pick interpolation: a matrix-theoretic approach, preprint, 2008.

[37] T. Constantinescu and J.L. Johnson, A note on noncommutative interpolation, *Canad. Math. Bull.* **46** (2003), 59–70.

[38] M.D. Choi, Completely positive linear maps on complex matrices, *Linear Algebra Appl.* **10** (1975), 285–290.

[39] B.J. Cole and J. Wermer, Pick interpolation, von Neumann inequalities, and hyperconvex sets, in *Complex Potential Theory (Montreal, PQ,* 1993) (P.J. Gauthier and G. Sabidussi, ed.) pp. 89–129, NATO Advanced Scientific Instruments Series. C Mathematics and Physics Science, vol. **439**, Kluwer Academic Publisher, Dordrecht, 1994.

[40] K. Davidson and D. Pitts, Nevanlinna-Pick interpolation for noncommutative analytic Toeplitz algebras, *Integral Equations Operator Theory* **31** (1998), 321–337.

[41] P. Dewilde and H. Dym, Interpolation for upper triangular operators, in: *Time-Variant Systems and Interpolation* (I. Gohberg, ed.), pp. 153–260, **OT 56**, Birkhäuser, Basel-Berlin-Boston, 1992.

[42] P. Dewilde and A.-J. van der Veen, *Time-Varying Systems and Computation*, Kluwer, Boston, 1998.

[43] R.G. Douglas, On majorization, factorization, and range inclusion of operators on Hilbert space, *Proc. Amer. Math. Soc.* **17** (1966), 413–415.

[44] M.A. Dritschel, S.A.M. Marcantognini and S. McCullough, Interpolation in semigroupoid algebras, *J. reine angew. Math.* **606** (2007), 1–40.

[45] M.A. Dritschel and S. McCullough, The failure of rational dilation on a triply connected domain, *J. Amer. Math. Soc.* **18** (2005), 873–918.

[46] M.A. Dritschel and S. McCullough, Test functions, kernels, realizations and interpolation, in: *Operator Theory, Structured Matrices, and Dilations. Tiberiu Constantinescu Memorial volume* (M. Bakonyi, A. Gheondea, M. Putinar and J. Rovnyak, ed.), pp. 153–179, Theta Foundation, Bucharest, 2007.

[47] S.W. Drury, A generalization of von Neumann's inequality to the complex ball, *Proc. Amer. Math. Soc.* **68** (1978), 300–304.

[48] J. Eschmeier and M. Putinar, Spherical contractions and interpolation problems on the unit ball, *J. reine angew. Math.* **542** (2002), 219–236.

[49] Q. Fang, *Multivariable Interpolation Problems*, Ph.D. dissertation, Virginia Tech, 2008.

[50] L. Fejér, Die Abschätzung eines Polynoms in einem Intervalle, wenn Schranken für seine Werte und ersten Ableitungswerte in einzelnen Punkten des Intervalles gegeben sind, und ihre Anwendung auf die Konvergenzfrage Hermitescher Interpolationsreihen, *Math. Zeitschrift* **32** (1930) no. 1, 426–457.

[51] C. Foias and A. Frazho, *The Commutant Lifting Approach to Interpolation Problems*, **OT44**, Birkhäuser, Basel-Berlin-Boston, 1990.

[52] C. Foias, A. Frazho, I. Gohberg, and M.A. Kaashoek, Discrete time-variant interpolation as classical interpolation with an operator argument, *Integral Equations Operator Theory* **26** (1996), 371–403.

[53] C. Foias, A. Frazho, I. Gohberg, and M.A. Kaashoek, *Metric Constrained Interpolation, Commutant Lifting and Systems*, **OT100**, Birkhäuser, Basel-Berlin-Boston, 1998.

[54] D.C. Greene, S. Richter, and C. Sundberg, *The structure of inner multipliers on spaces with complete Nevanlinna-Pick kernels*, J. Funct. Anal. **194** (2002), 311–331.

[55] W.J. Helton, S.A. McCullough and V. Vinnikov, Noncommutative convexity arises from linear matrix inequalities, *J. Funct. Anal.* **240** (2006), 105–191.

[56] M. Jury, *Matrix Products and Interpolation Problems in Hilbert Function Spaces*, Ph.D. dissertation, Washington University at St. Louis, 2002.

[57] D.S. Kalyuzhnyĭ-Verbovetzkiĭ, Carathéodory interpolation on the non-commutative polydisk, *J. Funct. Anal.* **229** (2005), 241–276.

[58] D.S. Kalyuzhnyĭ-Verbovetzkiĭ and Victor Vinnikov, Singularities of rational functions and minimal factorizations: The noncommutative and the commutative setting, *Linear Algebra Appl.* **430** (1009), 869–889.

[59] V. Katsnelson, A. Kheifets and P. Yudiskiĭ, An abstract interpolation problem and extension theory of isometric operators, in: *Operators in Spaces of Functions and Problems in Function Theory* (V.A. Marchenko, ed.), pp. 83–96, **146**, Naukova Dumka, Kiev, 1987; English Transl. in: *Topics in Interpolation Theory* (H. Dym, B. Fritzsche, V. Katsnelson and B. Kirstein, ed.), pp. 283–298, **OT 95**, Birkhäuser, Basel-Berlin-Boston, 1997.

[60] D.W. Kribs and S.C. Power, Free semigroupoid algebras, *J. Ramanujan Math. Soc.* **19** (2004), 117–159.

[61] D.W. Kribs and S.C. Power, Partly free algebras from directed graphs, in: *Current trends in operator theory and its applications*, pp. 373–385, **OT149**, Birkhäuser, Basel-Berlin-Boston, 2004.

[62] D.W. Kribs and S.C. Power, The analytic algebras of higher rank graphs, *Math. Proc. R. Ir. Acad.* **106A** (2006), 199–218.

[63] S. McCullough, The local de Branges-Rovnyak construction and complete Nevanlinna-Pick kernels, in: *Algebraic Methods in Operator Theory* (R. Curto and P.E.T. Jorgensen, eds.), Birkhäuser, Basel-Berlin-Boston, 1994, 15–24.

[64] S. McCullough and S. Sultanic, Ersatz commutant lifting with test functions, *Complex Anal. Oper. Theory* **1** (2007), 581–620.

[65] S. McCullough and T.T. Trent, Invariant subspaces and Nevanlinna-Pick kernels, *J. Funct. Anal.* **178** (2000), 226–249.

[66] P. Muhly, A finite-dimensional introduction to operator algebras, in: *Operator algebras and applications (Samos,* 1996), pp. 313–354, *NATO Adv. Sci. Inst. Ser. C Math. Phys. Sci.*, **495**, Kluwer Acad. Publ., Dordrecht, 1997.

[67] P.S. Muhly and B. Solel, Tensor algebras over C^*-correspondences: representations, dilations and C^*-envelopes, *J. Funct. Anal.* **158** (1998), 389–457.

[68] P.S. Muhly and B. Solel, Tensor algebras, induced representations, and the Wold decomposition, *Canadian J. Math.* **51** (1999), 850–880.

[69] P.S. Muhly and B. Solel, Hardy algebras, W^*-correspondences and interpolation theory, *Math. Annalen* **330** (2004), 353–415.

[70] P.S. Muhly and B. Solel, Schur class operator functions and automorphisms of Hardy algebras, *Documenta Math.* **13** (2008), 365–411.

[71] P.S. Muhly and B. Solel, The Poisson kernel for Hardy algebras, *Complex Anal. Oper. Theory* **3** (2009), 221–242.

[72] B. Sz.-Nagy and C. Foias, Dilation des commutants d'opérateurs, *C. R. Acad. Sci. Paris, série A,* **266** (1968), 493–495.

[73] M. Nakamura, M. Takesaki, and H. Umegaki, A remark on the expectations of operator algebras, *Kōdai Math. Sem. Rep.* **12** (1960), 82–90.

[74] R. Nevanlinna, Über beschränkte Funktionen, die in gegebenen Punkten vorgeschriebene Werte annehmen, *Ann. Acad. Sci. Fenn. Ser. A* **13** (1919), no. 1.

[75] R. Nevanlinna, Über beschränkte Funktionen, *Ann. Acad. Sci. Fenn. Ser. A* **32** (1929), no. 7.

[76] V.I. Paulsen, Matrix-valued interpolation and hyperconvex sets, *Integral Equations and Operator Theory* **41** (2001), 38–62.

[77] V.I. Paulsen, *Completely Bounded Maps and Operator Algebras*, Cambridge Studies in Advances Mathematics **76**, Cambridge University Press, Cambridge, 2002.

[78] J. Peters, Semi-crossed products of C^*-algebras, *J. Funct. Anal.* **59** (1984), 498–534.

[79] G. Pick, Über die Beschränkungen analytischer Funktionen, welche durch vorgegebene Funktionswerte bewirkt werden, *Math. Ann.* **77** (1916), 7–23.

[80] G. Popescu, Isometric dilations for infinite sequences of noncommuting operators, *Trans. Amer. Math. Soc.* **316** (1989), 523–536.

[81] G. Popescu, Characteristic functions for infinite sequences of noncommuting operators, *J. Operator Theory* **22** (1989), 51–71.

[82] G. Popescu, Von Neumann inequality for $(B(H)^n)_1$, *Math. Scand.* **68** (1991), 292–304.

[83] G. Popescu, Interpolation problems in several variables, *J. Math. Anal. Appl.* **227** (1998), 227–250.

[84] G. Popescu, Poisson transforms on some C^*-algebras generated by isometries, *J. Funct. Anal.* **161** (1999), 27–61.

[85] G. Popescu, Multivariable Nehari problem and interpolation, *J. Funct. Anal.* **200** (2003), 536–581.

[86] G. Popescu, *Entropy and Multivariable Interpolation*, Mem. Amer. Math. Soc. No. **868**, American Mathematical Society, Providence, 2006.

[87] G. Popescu, Free holomorphic functions on the unit ball of $B(\mathcal{H})^n$, *J. Funct. Anal.* **241** (2006), 268–333.

[88] G. Popescu, Free holomorphic functions and interpolation, *Math. Ann.* **342** (2008), 1–30.

[89] G. Popescu, Noncommutative Berezin transforms and multivariable operator model theory, *J. Funct. Anal.* **254** (2008), 1003–1057.

[90] G. Popescu, *Unitary invariants in multivariable operator theory*, Mem. Amer. Math. Soc. Nr. 941, American Mathematical Society, Providence, 2009.

[91] S. Pott, Standard models under polynomial positivity conditions, *J. Operator Theory* **41** (1999), 365–389.

[92] S.C. Power, Classifying higher rank analytic Toeplitz algebras, *New York J. Math.* **13** (2007), 271–298.

[93] P. Quiggen, For which reproducing kernel Hilbert spaces is Pick's theorem true?, *Integral Equations Operator Theory* **16** (1993), 244–266.

[94] M. Rosenblum and J. Rovnyak, An operator-theoretic approach to theorems of the Pick-Nevanlinna and Loewner types.I, *Integral Equations and Operator Theory* **3** (1980), 408–436.

[95] M. Rosenblum and J. Rovnyak, An operator-theoretic approach to theorems of the Pick-Nevanlinna and Loewner types.II, *Integral Equations and Operator Theory* **5** (1982), 870–887.

[96] M. Rosenblum and J. Rovnyak, *Hardy Classes and Operator Theory*, Oxford University Press, Oxford, 1985; reprinted: Dover Publications, Mineola, NY, 1997.

[97] D. Sarason, *Generalized interpolation in* H^∞, Trans. Amer. Math. Soc. **127** (1967), 179–203.

[98] A.H. Sayed, T. Constantinescu and T. Kailath, Time-variant displacement structure and interpolation problems, *IEEE Trans. Automat. Control* **39** (1994), 960–976.

[99] I. Schur, Über Potenzreihen, die im Innern des Einheitskreises beschränkt sind I, *J. reine und angew. Mathematik* **147** (1917), 205–232.

[100] I. Schur, Über Potenzreihen, die im Innern des Einheitskreises beschränkt sind II, *J. reine und angew. Mathematik* **148** (1918), 122–145.

[101] B. Solel, You can see the arrows in a quiver operator algebra, *J. Aust. Math. Soc.* **77** (2004), 111–122.

[102] B. Solel, Representations of product systems over semigroups and dilations of commuting CP maps, *J. Funct. Anal.* **235** (2006), 593–618.

[103] W.S. Stinespring, Positive functions on C^*-algebras, *Proc. Amer. Math. Soc.* **6** (1955), 211–216.

[104] E. Størmer, Positive linear maps of C^*-algebras, in: *Foundations of quantum mechanics and ordered linear spaces* (*Advanced Study Inst., Marburg,* 1973), pp. 85–106, Lecture Notes in Phys., Vol. 29, Springer, Berlin, 1974.

[105] J. Tomiyama, On the projection of norm one in W^*-algebras. III, Tohoku Math. J. (1) **11** (1959), 125–129.

Joseph A. Ball
Department of Mathematics
Virginia Tech, Blacksburg
VA 24061-0123, USA
e-mail: `ball@math.vt.edu`

Sanne ter Horst
Utrecht University
Budapestlaan 6
NL-3584 CD Utrecht, The Netherlands
e-mail: `s.terhorst@uu.nl`

Operator Theory:
Advances and Applications, Vol. 195, 73–82

On the Unit Ball of Operator-valued H^2-functions

Oscar Blasco

Abstract. Let X be a complex Banach space and let $H^2(\mathbb{D}, X)$ denote the space of X-valued analytic functions in the unit disc such that

$$\sup_{0<r<1} \int_0^{2\pi} \|F(re^{it})\|^2 \frac{dt}{2\pi} < \infty.$$

It is shown that a function F belongs to the unit ball of $H^2(\mathbb{D}, X)$ if and only if there exist $f \in H^\infty(\mathbb{D}, X)$ and $\phi \in H^\infty(\mathbb{D})$ such that

$$\|f(z)\|^2 + |\phi(z)|^2 \leq 1 \quad \text{and} \quad F(z) = \frac{f(z)}{1 - z\phi(z)}$$

for $|z| < 1$.

Mathematics Subject Classification (2000). Primary 46E40; Secondary 42B30.

Keywords. Operator-valued Hardy spaces, factorization results, unit ball

1. Introduction

Let $1 \leq p \leq \infty$, $\mathbb{T} = \{z \in \mathbb{C} : |z| = 1\}$, $\mathbb{D} = \{z \in \mathbb{C} : |z| < 1\}$, let X be a complex Banach space and denote by $H^p(\mathbb{T}, X)$, $1 \leq p \leq \infty$, the set of functions $f \in L^p(\mathbb{T}, X)$ such that $\hat{f}(n) = 0$ for $n < 0$, and by $H^p(\mathbb{D}, X)$ the set of holomorphic functions $F : \mathbb{D} \to X$ such that, for $1 \leq p < \infty$,

$$\|F\|_{H^p(\mathbb{D},X)} = \sup_{0<r<1} \left(\int_0^{2\pi} \|F(re^{it})\|^p \frac{dt}{2\pi} \right)^{1/p} < \infty \tag{1}$$

and

$$\|F\|_{H^\infty(\mathbb{D},X)} = \sup_{|z|<1} \|F(z)\| < \infty.$$

It is elementary to see that $H^p(\mathbb{T}, X) \subseteq H^p(\mathbb{D}, X)$ by means of the Poisson integral. That is to say, if $f \in H^p(\mathbb{T}, X)$ then $F(re^{it}) = P_r * f(e^{it}) \in H^p(\mathbb{D}, X)$ and

---

Partially supported by Proyecto MTM2005-08350-C03-03 and MTM2008-04594/MTM..

with the same norm. Actually $H^p(\mathbb{T}, X)$ can be identified with the closure of the polynomials in $H^p(\mathbb{D}, X)$ for $1 \le p < \infty$. It is a well-known fact that for $X = \mathbb{C}$ the spaces $H^p(\mathbb{T}) = H^p(\mathbb{D})$ for $1 \le p \le \infty$ due to the fact that functions in $H^p(\mathbb{D})$ do have non-tangential boundary values almost everywhere which belong to $H^p(\mathbb{T})$. The reader should be aware that this is not longer true for infinite-dimensional Banach spaces. Spaces such that any function in $H^\infty(\mathbb{D}, X)$ has radial boundary limits almost everywhere are said to have the *analytic Radon Nikodym property* (ARNP for short). This property was first considered by A.V. Bukhvalov and A.A. Danilevich (see [3] or [2]) and it was shown to be equivalent to the fact that functions in $H^p(\mathbb{D}, X)$ for some (and equivalently for all) $1 \le p < \infty$ have radial boundary limits almost everywhere. A typical example failing ARNP is $X = c_0$. It is well known and easy to see that any reflexive Banach space has the ARNP. We just mention that the fact that a Hilbert space $\mathcal{H}$ has the ARNP follows, among many other ways, from the description of functions $F \in H^2(\mathbb{D}, \mathcal{H})$ as $F(z) = \sum_{n=0}^\infty h_n z^n$ where $h_n \in \mathcal{H}$ and

$$\|F\|_{H^2(\mathbb{T},\mathcal{H})} = \|F\|_{H^2(\mathbb{D},\mathcal{H})} = \left(\sum_{n=0}^\infty \|h_n\|^2\right)^{1/2}.$$

In this paper we deal with the following factorization result due D. Sarason for scalar-valued functions.

Theorem 1. (See [9, page 490].) *Let h be an analytic function in the unit disc $\mathbb{D}$. The following are equivalent:*

(i) $h \in H^2(\mathbb{D})$ *and* $\|h\|_{H^2(\mathbb{D})} \le 1$.

(ii) *There exist $\phi, \psi \in H^\infty(\mathbb{D})$ such that $|\phi(z)|^2 + |\psi(z)|^2 \le 1$ and*

$$h(z) = \frac{\phi(z)}{1 - z\psi(z)}.$$

This gives a very interesting and useful factorization of functions in the unit ball of H^2, that is any $h \in H^2$ and $\|h\|_{H^2} \le 1$ can be written as $h(z) = \frac{f(z)}{1-zg(z)}$ where $z \to (f(z), g(z))$ belongs to the unit ball of $H^\infty(\mathbb{D}, \mathbb{C}^2)$.

Sarason's result was extended to matrix-valued functions by D. Alpay, V. Bolotnikov and Y. Peretz in [1]. Let us set the notation to establish the result. Given $p, q \in \mathbb{N}$ the authors denoted by $H_2^{p\times q}$ the Hilbert space of $\mathbb{C}^{p\times q}$-valued functions with H^2 entries endowed with the inner product

$$\langle f, g\rangle = \int_0^{2\pi} tr(g(e^{it})^* f(e^{it}))\frac{dt}{2\pi},$$

and denoted by $H_2^{p\times q}(I_q)$ the set of functions $f \in H_2^{p\times q}$ such that

$$\sum_{k=0}^\infty f_k^* f_k \le I_q$$

where I_q is the identity operator on $\mathbb{C}^q$ and $f(z) = \sum_{k=0}^\infty f_k z^k$ with $f_k \in \mathbb{C}^{p\times q}$.

Theorem 2. (See [1, Theorem 2.2].) *Let f be a $\mathbb{C}^{p\times q}$-valued function analytic in $\mathbb{D}$. Then f belongs to $H_2^{p\times q}(I_q)$ if and only if it can be written as*

$$f(z) = s_1(z)(I_q - zs_2(z))^{-1}$$

for some Schur function $S(z) = (s_1(z), s_2(z))$, that is s_1 and s_2 are $\mathbb{C}^{p\times q}$-valued and $\mathbb{C}^{q\times q}$-valued analytic functions in $\mathbb{D}$ with

$$\|s_1(z)\|^2 + \|s_2(z)\|^2 \leq 1, \quad |z| < 1,$$

and conversely.

A further generalization of Theorem 2 to the infinite-dimensional case is due to A.E. Frazho, S. ter Horst, M.A. Kaashoek (see [5, 6]).

Given two Hilbert spaces $\mathcal{H}$ and $\mathcal{H}'$ the authors denote by $\mathbf{S}(\mathcal{H},\mathcal{H}')$ the Schur class consisting in those $\mathcal{L}(\mathcal{H},\mathcal{H}')$-valued analytic functions S in the unit disc such that $S(z)$ are contractions and denote by $H^2_{\text{ball}}(\mathcal{L}(\mathcal{H},\mathcal{H}'))$ the set of $\mathcal{L}(\mathcal{H},\mathcal{H}')$-valued analytic functions F such that $F(z)h \in H^2(\mathbb{D},\mathcal{H}')$ for any $h \in \mathcal{H}$ and $\|F(z)h\|_{H^2(\mathcal{H}')} \leq \|h\|$.

Theorem 3. (See [5, Corollary 0.3].) *Let F be an $\mathcal{L}(\mathcal{H},\mathcal{H}')$-valued analytic function in the unit disc $\mathbb{D}$. Then $F \in H^2_{\text{ball}}(\mathcal{L}(\mathcal{H},\mathcal{H}'))$ if and only if there exist $S = (s_1, s_2) \in \mathbf{S}(\mathcal{H},\mathcal{H}' \bigoplus \mathcal{H})$ such that*

$$F(z) = s_1(z)(I_{\mathcal{H}} - zs_2(z))^{-1}.$$

The reader is referred to the papers [1, 9, 5, 6] for the use of these theorems in interpolation theory, computing exposed points and connections with the relaxed commutant lifting problem.

The aim of this paper is to get the analogue to Theorem 3 for functions in $H^2(\mathbb{D},\mathcal{L}(\mathcal{H},\mathcal{H}'))$. It will be shown that if

$$F \in H^2(\mathbb{D},\mathcal{L}(\mathcal{H},\mathcal{H}')) \quad \text{and} \quad \|F\|_{H^2(\mathbb{D},\mathcal{L}(\mathcal{H},\mathcal{H}'))} \leq 1$$

then the operator-valued function $s_2(z)$ can be chosen to be $\phi(z)I_{\mathcal{H}}$ for some scalar-valued function ϕ.

Namely we shall show the following results:

Theorem 4. *Let $\mathcal{H}$ and $\mathcal{H}'$ be complex Hilbert spaces. The following are equivalent*

(i) *$F \in H^2(\mathbb{D},\mathcal{L}(\mathcal{H},\mathcal{H}'))$ (respectively $F \in H^2(\mathbb{T},\mathcal{L}(\mathcal{H},\mathcal{H}'))$) satisfies that $\|F\|_{H^2(\mathbb{D},\mathcal{L}(\mathcal{H},\mathcal{H}'))} \leq 1$.*

(ii) *There exist $f \in H^\infty(\mathbb{D},\mathcal{L}(\mathcal{H},\mathcal{H}'))$ (respectively $f \in H^\infty(\mathbb{T},\mathcal{L}(\mathcal{H},\mathcal{H}'))$) and $\phi \in H^\infty(\mathbb{D})$ such that*

$$\|f(z)\|^2 + |\phi(z)|^2 \leq 1 \quad \text{and} \quad F(z) = \frac{f(z)}{1 - z\phi(z)}.$$

2. Vector-valued approach

In this note we are concerned with the case $X = \mathcal{L}(\mathcal{H}, \mathcal{H}')$ the space of bounded linear operator between two Hilbert spaces $\mathcal{H}$ and $\mathcal{H}'$. To handle this situation, given two Banach spaces X_1 and X_2, we can consider not only $H^p(\mathbb{T}, \mathcal{L}(X_1, X_2))$ and $H^p(\mathbb{D}, \mathcal{L}(X_1, X_2))$ (which turn out to be different), but we can also consider $H^p_{\text{strong}}(\mathbb{T}, \mathcal{L}(X_1, X_2))$ (respectively, $H^p_{\text{weak}}(\mathbb{T}, \mathcal{L}(X_1, X_2))$) the set of functions $f : \mathbb{T} \to \mathcal{L}(X_1, X_2)$ such that $e^{it} \to f(e^{it})(x_1) \in H^p(\mathbb{T}, X_2)$ (respectively $e^{it} \to \langle f(e^{it})(x_1), x_2^* \rangle \in H^p(\mathbb{T})$) for all $x_1 \in X_1$ (respectively $x_1 \in X_1$ and $x_2^* \in X_2^*$) and we write

$$\|f\|_{H^p_{\text{strong}}(\mathbb{T}, \mathcal{L}(X_1, X_2))} = \sup_{\|x_1\|=1} \|f(\cdot)(x_1)\|_{L^p(\mathbb{T}, X_2)}. \tag{2}$$

(respectively

$$\|f\|_{H^p_{\text{weak}}(\mathbb{T}, \mathcal{L}(X_1, X_2))} = \sup_{\|x_1\|=1, \|x_2^*\|=1} \|\langle f(\cdot)(x_1), x_2^* \rangle\|_{L^p(\mathbb{T})}.) \tag{3}$$

Similarly we can define $H^p_{\text{strong}}(\mathbb{D}, \mathcal{L}(X_1, X_2))$ and $H^p_{\text{weak}}(\mathbb{D}, \mathcal{L}(X_1, X_2))$.
Of course $H^p_{\text{weak}}(\mathbb{T}, \mathcal{L}(X_1, X_2)) = H^p_{\text{weak}}(\mathbb{D}, \mathcal{L}(X_1, X_2))$. However

$$H^p_{\text{strong}}(\mathbb{T}, \mathcal{L}(X_1, X_2)) \subsetneq H^p_{\text{strong}}(\mathbb{D}, \mathcal{L}(X_1, X_2))$$

as it is shown by taking $X_1 = \mathbb{C}$ and $X_2 = X$ not having the analytic Radon-Nikodym property.

It is also elementary to see that, for $1 \leq p \leq \infty$,

$$H^p(\mathbb{D}, \mathcal{L}(X_1, X_2)) \subseteq H^p_{\text{strong}}(\mathbb{D}, \mathcal{L}(X_1, X_2)) \subseteq H^p_{\text{weak}}(\mathbb{D}, \mathcal{L}(X_1, X_2)) \tag{4}$$

with continuous inclusions. Actually in the case $p = \infty$, taking into account that

$$\|T\| = \sup_{\|x_1\|=1} \|T(x_1)\| = \sup_{\|x_1\|=1, \|x_2^*\|=1} |\langle T(x_1), x_2^* \rangle|$$

for $T \in \mathcal{L}(X_1, X_2)$, one can conclude that

$$H^\infty(\mathbb{D}, \mathcal{L}(X_1, X_2)) = H^\infty_{\text{strong}}(\mathbb{D}, \mathcal{L}(X_1, X_2)) = H^\infty_{\text{weak}}(\mathbb{D}, \mathcal{L}(X_1, X_2)). \tag{5}$$

Let us see the expressions of the norms in these different spaces in the particular case $X = \mathcal{L}(\mathcal{H}, \mathcal{H}')$ where $\mathcal{H}$ and $\mathcal{H}'$ are Hilbert spaces.

Let $F : \mathbb{D} \to \mathcal{L}(\mathcal{H}, \mathcal{H}')$ be given by $F(z) = \sum_{k=0}^\infty T_k z^k$ where $T_k \in \mathcal{L}(\mathcal{H}, \mathcal{H}')$.
$F \in H^2(\mathbb{D}, \mathcal{L}(\mathcal{H}, \mathcal{H}'))$ if and only if

$$\|F\|_{H^2(\mathbb{D}, \mathcal{L}(\mathcal{H}, \mathcal{H}'))} = \sup_{0<r<1} \left(\int_0^{2\pi} \|\sum_{k=0}^\infty T_k r^k e^{ikt}\|^2_{\mathcal{L}(\mathcal{H}, \mathcal{H}')} \frac{dt}{2\pi} \right)^{1/2} < \infty. \tag{6}$$

$F \in H^2_{\text{strong}}(\mathbb{D}, \mathcal{L}(\mathcal{H}, \mathcal{H}'))$ if and only if

$$\|F\|_{H^2_{\text{strong}}(\mathbb{D}, \mathcal{L}(\mathcal{H}, \mathcal{H}'))} = \sup_{\|h\|=1} \left(\sum_{k=0}^\infty \|T_k(h)\|^2 \right)^{1/2} < \infty. \tag{7}$$

$F \in H^2_{\text{weak}}(\mathbb{D}, \mathcal{L}(\mathcal{H}, \mathcal{H}'))$ if and only if

$$\|F\|_{H^2_{\text{weak}}(\mathbb{D},\mathcal{L}(\mathcal{H},\mathcal{H}'))} = \sup_{\|h\|=1,\|h'\|=1} \left(\sum_{k=0}^{\infty} |\langle T_k(h), h' \rangle|^2 \right)^{1/2} < \infty. \tag{8}$$

Let us describe $H_2^{p\times q}$, $H_2^{p\times q}(I_q)$ and $H^2_{\text{ball}}(\mathcal{L}(\mathcal{H}, \mathcal{H}'))$ in terms of our vector-valued Hardy spaces. Let us observe that

$$\begin{aligned} \|f\|^2_{H_2^{p\times q}} &= \int_0^{2\pi} \operatorname{tr}(f(e^{it})^* f(e^{it})) \frac{dt}{2\pi} = \operatorname{tr}\left(\int_0^{2\pi} f(e^{it})^* f(e^{it}) \frac{dt}{2\pi} \right) \\ &= \operatorname{tr}\left(\sum_{k=0}^{\infty} f_k^* f_k \right) = \sum_{k=0}^{\infty} \sum_{i=1}^{q} \sum_{j=1}^{p} (f_k^*)_{i,j} (f_k)_{j,i} \\ &= \sum_{k=0}^{\infty} \sum_{i=1}^{q} \sum_{j=1}^{p} |(f_k)_{i,j}|^2 = \sum_{k=0}^{\infty} \sum_{i=1}^{q} \sum_{j=1}^{p} |\langle f_k(e_i), e'_j \rangle|^2, \end{aligned}$$

where e_i and e'_j stand for the canonical basis of $\mathbb{C}^q$ and $\mathbb{C}^p$ respectively. Hence

$$\|f\|^2_{H_2^{p\times q}} = \sup_{\|h\|_{\mathbb{C}^q}=1,\|h'\|_{\mathbb{C}^p}=1} \sum_{k=0}^{\infty} |\langle f_k(h), h' \rangle|^2.$$

This shows that $H_2^{p\times q}$ is nothing else than $H^2_{\text{weak}}(\mathbb{D}, \mathcal{L}(\mathbb{C}^q, \mathbb{C}^p))$.

Observe now that $H^2_{\text{strong}}(\mathbb{D}, \mathcal{L}(\mathbb{C}^q, \mathbb{C}^p))$ is now defined by the condition

$$\sup_{\|h\|_{\mathbb{C}^q}=1} \sum_{k=0}^{\infty} \|f_k(h)\|^2 < \infty,$$

which is equivalent to

$$\sup_{\|h\|_{\mathbb{C}^q}=1} \sum_{k=0}^{\infty} |\langle f_k^* f_k(h), h \rangle| < \infty.$$

In particular $H_2^{p\times q}(I_q)$ coincides with the unit ball of $H^2_{\text{strong}}(\mathbb{D}, \mathcal{L}(\mathbb{C}^q, \mathbb{C}^p))$. This point of view also shows that $f \in H^{p\times q}(I_q)$ if and only if the operator of multiplication by f is a contraction from $\mathbb{C}^q$ to $H^2(\mathbb{D}, \mathbb{C}^p)$ (see [1, Theorem 2.1]).

It is well known that a $\mathbb{C}^{p\times q}$-valued analytic function in $\mathbb{D}$ is a Schur function is equivalent to the operator of multiplication by the function is a contraction from $H^2(\mathbb{D}, \mathbb{C}^q)$ to $H^2(\mathbb{D}, \mathbb{C}^p)$. This, in our terminology, corresponds to the fact that elements in the unit ball of $H^\infty_{\text{strong}}(\mathbb{D}, \mathcal{L}(\mathbb{C}^q, \mathbb{C}^p))$ define contractions from $H^2(\mathbb{D}, \mathbb{C}^q)$ to $H^2(\mathbb{D}, \mathbb{C}^p)$ via multiplication and clearly exhibits that the difference between the Schur class and $H_2^{p\times q}(I_q)$ is nothing else but the difference between $H^\infty_{\text{strong}}(\mathbb{D}, \mathcal{L}(\mathbb{C}^q, \mathbb{C}^p))$ and $H^2_{\text{strong}}(\mathbb{D}, \mathcal{L}(\mathbb{C}^q, \mathbb{C}^p))$.

It is clear from the definition that $H^2_{\text{ball}}(\mathcal{L}(\mathcal{H}, \mathcal{H}'))$ is the unit ball of $H^2_{\text{strong}}(\mathcal{L}(\mathcal{H}, \mathcal{H}'))$ and, in particular, $H^2_{\text{ball}}(\mathbb{C}^q, \mathbb{C}^p) = H_2^{p\times q}(I_q)$.

Hence Theorem 3 can be formulated as a factorization result for the unit ball of $H^2_{\text{strong}}(\mathbb{D}, \mathcal{L}(\mathcal{H}, \mathcal{H}'))$ or, equivalently, of $H^2_{\text{strong}}(\mathbb{T}, \mathcal{L}(\mathcal{H}, \mathcal{H}'))$. We shall analyze the situation for the cases $H^2(\mathbb{D}, \mathcal{L}(\mathcal{H}, \mathcal{H}'))$ and $H^2(\mathbb{T}, \mathcal{L}(\mathcal{H}, \mathcal{H}'))$.

Let us mention that for infinite-dimensional Hilbert spaces the Hardy spaces considered above are different.

Proposition 1. *Let $\mathcal{H}$ and $\mathcal{H}'$ be separable infinite-dimensional Hilbert spaces. Then*

$$H^2(\mathbb{T}, \mathcal{L}(\mathcal{H}, \mathcal{H}')) \subsetneq H^2(\mathbb{D}, \mathcal{L}(\mathcal{H}, \mathcal{H}')).$$

Proof. Let (e_k) and (e'_k) be orthonormal basis of $\mathcal{H}$ and $\mathcal{H}'$, and write $e_k \otimes e'_k \in \mathcal{L}(\mathcal{H}, \mathcal{H}')$ for the rank 1 operator given by $h \to \langle h, e_k\rangle e'_k$. Define $T_k = e_k \otimes e'_k$ and $F(z) = \sum_{k=0}^{\infty}(e_k \otimes e'_k)z^k$. Therefore

$$F(z)(h) = \sum_k \langle h, e_k\rangle z^k e'_k.$$

It follows that $\|F(z)\|_{\mathcal{L}(\mathcal{H},\mathcal{H}')} = \sup_{n\geq 0} |z^n| = 1$. This shows that F is bounded and therefore $F \in H^2(\mathbb{D}, \mathcal{L}(\mathcal{H}, \mathcal{H}'))$.

To see that $F \notin H^2(\mathbb{T}, \mathcal{L}(\mathcal{H}, \mathcal{H}'))$ note that if there exists $\lim_{r\to 1} F(re^{it}) = F(e^{it})$ in $\mathcal{L}(\mathcal{H}, \mathcal{H}')$ for some $t \in [0, 2\pi)$ then $F(e^{it})(h) = \sum_k \langle h, e_k\rangle e^{ik\theta} e'_k$. On the other hand $F(z)$ is compact for any $z \in \mathbb{D}$ (because $(z^n) \in c_0$ for $|z| < 1$) but $F(e^{it})$ is not compact. □

Proposition 2. *Let $\mathcal{H}$ and $\mathcal{H}'$ be separable infinite-dimensional Hilbert spaces. Then*

$$H^2(\mathbb{D}, \mathcal{L}(\mathcal{H}, \mathcal{H}')) \subsetneq H^2_{\text{strong}}(\mathbb{D}, \mathcal{L}(\mathcal{H}, \mathcal{H}')) \subsetneq H^2_{\text{weak}}(\mathbb{D}, \mathcal{L}(\mathcal{H}, \mathcal{H}')).$$

Proof. Fix $h' \in \mathcal{H}'$ with $\|h'\|_{\mathcal{H}'} = 1$ and (e_k) an orthonormal basis of $\mathcal{H}$. Define $T_k = e_k \otimes h'$ and $F(z) = \sum_{k=0}^{\infty}(e_k \otimes h')z^k = (\sum_{k=0}^{\infty} e_k z^k) \otimes h'$. Therefore

$$F(z)(h) = \left(\sum_{k=0}^{\infty} \langle e_k, h\rangle z^k\right) h'$$

for any $h \in \mathcal{H}$. This implies $\|F(z)\|_{\mathcal{L}(\mathcal{H},\mathcal{H}')} = \frac{1}{(1-|z|^2)^{1/2}}$ and $\|F(z)(h)\|_{\mathcal{H}'} = |\sum_{k=0}^{\infty}\langle e_k, h\rangle z^k|$. Hence $F \in H^2_{\text{strong}}(\mathbb{D}, \mathcal{L}(\mathcal{H}, \mathcal{H}'))$ but $F \notin H^2(\mathbb{D}, \mathcal{L}(\mathcal{H}, \mathcal{H}'))$.

Fix $h_0 \in \mathcal{H}$ with $\|h_0\|_{\mathcal{H}} = 1$ and (e'_k) an orthonormal basis of $\mathcal{H}'$. Define $T_k = h_0 \otimes e'_k$ and $F(z) = \sum_{k=0}^{\infty}(h_0 \otimes e'_k)z^k = h_0 \otimes (\sum_{k=0}^{\infty} z^k e'_k)$. Therefore

$$F(z)(h) = \langle h_0, h\rangle \left(\sum_{k=0}^{\infty} z^k e'_k\right)$$

for any $h \in \mathcal{H}$. This gives $\|F(z)(h)\|_{\mathcal{H}'} = |\langle h_0, h\rangle| \frac{1}{(1-|z|^2)^{1/2}}$ and $|\langle F(z)(h), h'\rangle| = |\langle h_0, h\rangle| |\sum_{k=0}^{\infty}\langle e'_k, h'\rangle z^k)|$. From that one concludes that $F \in H^2_{\text{weak}}(\mathbb{D}, \mathcal{L}(\mathcal{H}, \mathcal{H}'))$ but $F \notin H^2_{\text{strong}}(\mathbb{D}, \mathcal{L}(\mathcal{H}, \mathcal{H}'))$. □

Before embarking in the proof of the main results let me point out the following general observation.

Proposition 3. *Let $1 \leq p < \infty$, X and Y be Banach spaces, A be a Banach algebra and $\mathcal{B} : X \times A \to Y$ a bounded bilinear map. If $f \in H^\infty(\mathbb{D}, X)$ and $g \in H^\infty(\mathbb{D}, A)$ such that*

$$\|f(z)\|_X^p + \|g(z)\|_A^p \leq 1. \tag{9}$$

Then

$$F(z) = \mathcal{B}(f(z), (1 - zg(z))^{-1}) \in H^p(\mathbb{D}, Y)$$

and

$$\|F\|_{H^p(\mathbb{D},Y)} \leq \frac{\|\mathcal{B}\| \|f\|_{H^\infty(\mathbb{D},X)}}{1 - \|g\|_{H^p(\mathbb{D},A)}}.$$

Proof. From (9) one gets $\|f\|^p_{H^p(\mathbb{D},X)} + \|g\|^p_{H^p(\mathbb{D},A)} \leq 1$. Now we can assume that $\|g\|_{H^p(\mathbb{D},A)} < 1$, otherwise $F = 0$.

We first observe that $\|g^n\|_{H^p(\mathbb{D},A)} \leq \|g\|^n_{H^p(\mathbb{D},A)}$. Indeed,

$$\begin{aligned}\int_0^{2\pi} \|g(re^{it})^n\|_A^p \frac{dt}{2\pi} &\leq \int_0^{2\pi} \|g(re^{it})\|_A^{pn} \frac{dt}{2\pi} \\ &\leq \|g\|^{p(n-1)}_{H^\infty(\mathbb{D},A)} \int_0^{2\pi} \|g(re^{it})\|_A^p \frac{dt}{2\pi} \leq \|g\|^{pn}_{H^p(\mathbb{D},A)}.\end{aligned}$$

Hence $z \to \sum_{n=0}^\infty z^n g(z)^n$ defines an absolutely convergent series in $H^p(\mathbb{D}, A)$ and we denote by $(1 - zg(z))^{-1}$ its sum which satisfies

$$\|(1 - zg(z))^{-1}\|_{H^p(\mathbb{D},A)} \leq \frac{1}{1 - \|g\|_{H^p(\mathbb{D},A)}}.$$

Define $F(z) = \mathcal{B}(f(z), (1 - zg(z))^{-1})$. It is elementary to see that F is a Y-valued analytic function. Moreover

$$\begin{aligned}\int_0^{2\pi} \|F(re^{it})\|_Y^p \frac{dt}{2\pi} &\leq \int_0^{2\pi} \|\mathcal{B}\|^p \|f(re^{it})\|_X^p \|(1 - re^{it} g(re^{it}))^{-1}\|_A^p \frac{dt}{2\pi} \\ &\leq \|\mathcal{B}\|^p \|f\|^p_{H^\infty(\mathbb{D},X)} \|(1 - zg(z))^{-1}\|^p_{H^p(\mathbb{D},A)} \leq \frac{\|\mathcal{B}\|^p \|f\|^p_{H^\infty(\mathbb{D},X)}}{(1 - \|g\|_{H^p(\mathbb{D},A)})^p}.\end{aligned}$$ □

When applying Proposition 3 for $p = 2$, $X = Y = \mathcal{L}(\mathcal{H}, \mathcal{H}')$, $A = \mathcal{L}(\mathcal{H}, \mathcal{H})$ and $\mathcal{B} : X \times A \to Y$ given by $B(T, S) = TS$ one realizes that the point in the reverse implication of Theorem 3 is actually to get that F belongs to the unit ball of $H^2_{\text{strong}}(\mathbb{D}, \mathcal{L}(\mathcal{H}, \mathcal{H}'))$ because the function $F \in H^2(\mathbb{D}, \mathcal{L}(\mathcal{H}, \mathcal{H}'))$ but with norm $\frac{1}{1 - \|g\|_{H^2(\mathbb{D},\mathcal{L}(\mathcal{H},\mathcal{H}))}} \geq 1$.

3. Proof of the main theorem

We shall see that there is nothing special in the space $\mathcal{L}(\mathcal{H},\mathcal{H}')$ and that the same factorization holds for a general Banach space X. We start working with analytic functions that can be approached by polynomials in $H^2(\mathbb{D},X)$, that is to say functions with boundary values belonging to the space $H^2(\mathbb{T},X)$. This actually allows us to give an alternative proof of Theorem 1 for $X=\mathbb{C}$. We use the same notation for the function defined at the boundary and its Poisson extension to the disc.

Theorem 5. *Let X be a complex Banach space. The following are equivalent:*

(i) $F\in H^2(\mathbb{T},X)$ *and* $\|F\|_{H^2(\mathbb{T},X)}\leq 1$

(ii) *There exist* $f\in H^\infty(\mathbb{T},X)$ *and* $\phi\in H^\infty(\mathbb{T})$ *such that for any* $|z|<1$,

$$\|f(z)\|^2+|\phi(z)|^2\leq 1 \quad \text{and} \quad F(z)=\frac{f(z)}{1-z\phi(z)}.$$

Proof. (i) $\Longrightarrow$ (ii) Assume $F\in H^2(\mathbb{T},X)$ and $\|F\|_{H^2(\mathbb{T},X)}=1$. Then $u(z)=\|F(z)\|^2$ is a subharmonic function in $\mathbb{D}$ and $\int_0^{2\pi}u(e^{it})\frac{dt}{2\pi}=1$. Consider the harmonic function $U(z)$ given by the Poisson extension of $u(e^{it})=\lim u(re^{it})$ and write $U=\Re(k)$ where k is holomorphic which takes values in $\{\Re w\geq 0\}$ and $k(0)=1$. One has that $\int_0^{2\pi}U(e^{it})\frac{dt}{2\pi}=\int_0^{2\pi}u(e^{it})\frac{dt}{2\pi}=1$ and $\|F(z)\|^2\leq\Re(k(z))$ for $|z|<1$.

Define $G(z)=\frac{1-k(z)}{1+k(z)}$, that is $k(z)=\frac{1+G(z)}{1-G(z)}$ and $\Re(k(z))=\frac{1-|G(z)|^2}{|1-G(z)|^2}$. Hence

$$\|F(z)\|^2\leq\frac{1-|G(z)|^2}{|1-G(z)|^2}.$$

Clearly $G\in\mathcal{H}(\mathbb{D})$, $|G(z)|<1$ and $G(0)=0$. So we have $G(z)=z\phi(z)$ for some $\phi\in\mathcal{H}(\mathbb{D})$. Define now $f(z)=F(z)(1-z\phi(z))$.

Obviously $f\in\mathcal{H}(\mathbb{D},X)$. Let us see that $\|f(z)\|^2+|\phi(z)|^2\leq 1$ for $z\in\mathbb{D}$. Indeed, on the boundary

$$\begin{aligned}\|f(e^{it})\|^2+|\phi(e^{it})|^2&=\|F(e^{it})(1-e^{it}\phi(e^{it}))\|^2+|\phi(e^{it})|^2\\&\leq(1-|\phi(e^{it})|^2)+|\phi(e^{it})|^2=1.\end{aligned}$$

Consider now the Banach space $X_1=X\otimes_2\mathbb{C}$ and look at the function $h(e^{it})=(f(e^{it}),\phi(e^{it}))$. We have shown that h belongs to the unit ball of $H^\infty(\mathbb{T},X_1)$. Therefore $h(z)=(f(z),\phi(z))\in H^\infty(\mathbb{D},X_1)$ and $\|f(z)\|^2+|\phi(z)|^2\leq 1$ for any $z\in\mathbb{D}$.

(ii) $\Longrightarrow$ (i) Assume that there exist $f\in H^\infty(\mathbb{T},X)$ and $\phi\in H^\infty$ such that $\|f(z)\|^2+|\phi(z)|^2\leq 1$ and

$$F(z)=\frac{f(z)}{1-z\phi(z)}.$$

Let us show that $F \in H^2(\mathbb{T}, X)$ with norm bounded by 1. Indeed,

$$\begin{aligned}\int_0^{2\pi} \|F(e^{it})\|^2 \frac{dt}{2\pi} &= \int_0^{2\pi} \frac{\|f(e^{it})\|^2}{|1-e^{it}\phi(e^{it})|^2}\frac{dt}{2\pi} \leq \int \frac{1-|e^{it}\phi(e^{it})|^2}{|1-e^{it}\phi(e^{it})|^2} dt \\ &= \lim_{r\to 1}\int_0^{2\pi} \frac{1-|re^{it}\phi(re^{it})|^2}{|1-re^{it}\phi(re^{it})|^2}\frac{dt}{2\pi} \\ &= \lim_{r\to 1} \Re\left(\int_0^{2\pi} \frac{1+re^{it}\phi(re^{it})}{1-re^{it}\phi(re^{it})}\right)\frac{dt}{2\pi} \\ &= \Re\left(\lim_{r\to 1}\int_0^{2\pi} \frac{1+re^{it}\phi(re^{it})}{1-re^{it}\phi(re^{it})}\frac{dt}{2\pi}\right) = 1.\end{aligned}$$

□

In order to extend the result for functions in $H^2(\mathbb{D}, X)$ we are going to use the following extension of a classical factorization for vector-valued functions. The reader is also referred to the work by G. Pisier [8] (and references thereby) for other factorization results of vector-valued analytic functions which depend on the geometry of the space X.

Lemma 1. (See [2] or [7], Theorem 2.10.) *Let $1 \leq p < \infty$ and $F \in H^p(\mathbb{D}, X)$. There exist $F_1 \in H^\infty(\mathbb{D}, X)$ with $\|F_1\|_{H^\infty(\mathbb{D},X)} = 1$ and $\phi \in H^p(\mathbb{D})$ with $\|\phi\|_{H^p(\mathbb{D})} = \|F\|_{H^p(\mathbb{D},X)}$ such that $F(z) = \phi(z)F_1(z)$.*

Theorem 6. *Let X be a complex Banach space $F \in \mathcal{H}(\mathbb{D}, X)$. The following are equivalent:*

(i) *$F \in H^2(\mathbb{D}, X)$ and $\|F\|_{H^2(\mathbb{D},X)} \leq 1$.*
(ii) *There exist $f \in H^\infty(\mathbb{D}, X)$ and $\phi \in H^\infty(\mathbb{D})$ such that*

$$\|f(z)\|^2 + |\phi(z)|^2 \leq 1, \quad |z| < 1$$

and

$$F(z) = \frac{f(z)}{1 - z\phi(z)}.$$

Proof. (i) $\Longrightarrow$ (ii) Use Lemma 1 to find $F_1 \in H^\infty(\mathbb{D}, X)$ with $\|F_1\|_{H^\infty(\mathbb{D},X)} = 1$ and $\phi \in H^2(\mathbb{D})$ with $\|\phi\|_{H^2(\mathbb{D})} = \|F\|_{H^2(\mathbb{D},X)}$ such that $F(z) = \phi(z)F_1(z)$.

Apply Sarason's result (or Theorem 5 for $X = \mathbb{C}$) to obtain $\phi_1, \phi_2 \in H^\infty(\mathbb{D})$ with $|\phi_1(z)| + |\phi_2(z)| \leq 1$ and $\phi(z) = \frac{\phi_1(z)}{1-z\phi_2(z)}$. Define $f(z) = F_1(z)\phi_1(z)$ and $g(z) = \phi_2(z)$ to get this implication.

(ii) $\Longrightarrow$ (i) Assume now that there exist $f \in H^\infty(\mathbb{D}, X)$ and $g \in H^\infty(\mathbb{D})$ such that $\|f(z)\|^2 + |g(z)|^2 \leq 1$ and

$$F(z) = \frac{f(z)}{1 - zg(z)}.$$

Arguing as above we have

$$\begin{aligned}\int \|F(re^{it})\|^2 dt &= \int \frac{\|f(re^{it})\|^2}{|1 - re^{it}g(re^{it})|^2} dt \\ &\leq \int \frac{1 - |g(re^{it})|^2}{|1 - re^{it}g(re^{it})|^2} dt \\ &\leq \int \frac{1 - |re^{it}g(re^{it})|^2}{|1 - re^{it}g(re^{it})|^2} dt \\ &= \Re\left(\int \frac{1 + re^{it}g(re^{it})}{1 - re^{it}g(re^{it})}\right) dt \\ &= 1.\end{aligned}$$

□

Acknowledgement

I would like to thank Prof. Kaashoek for fruitful discussions during our common stay in the 18th IWOTA meeting in Potchestroom in 2007.

References

[1] D. Alpay, V. Bolotnikokv, Y. Peretz, *On the tangential interpolation problem for H^2 functions.* Trans. Amer. Math. Soc. 347, (1995) 675–686.

[2] A.V. Bukhvalov, *On the analytic Radon-Kikodym property.* Function Spaces, Proc. Second Internat. conf. Poznana 1989, Teubner-Texte zur Math. 120, (1991) 211–228.

[3] A.V. Bukhvalov, A.A. Danilevich, *Boundary properties of analytic functions with values in Banach spaces.* Mat. Zametki 31 (1982), 203–214 (in Russian).

[4] J. Diestel, J.J. Uhl, *Vector Measures*, American Mathematical Society, Mathematical Surveys, Number 15 (1977).

[5] A.E. Frazho, S. ter Horst, M.A. Kaashoek, *Coupling and relaxed commutant lifting.* Integral Equations and Operator Theory 54, no. 1, (2006), 33–67.

[6] A.E. Frazho, S. ter Horst, M.A. Kaashoek, *All solutions to the relaxed commutant lifting problem.* Acta Sci. Math. (Szeged) 72 (2006), no. 1-2, 299–318.

[7] A. Michalak, *Translations of functions in vector Hardy classes in the unit disk.* Dissertationes Mathematicae 359 (1996).

[8] G. Pisier, *Factorization of operator-valued analytic functions.* Adv. Math. 93 (1992), no. 1, 61–125.

[9] D. Sarason, *Exposed points in H^1.* The Gohberg anniversary collection, Vol. II, OT, 41, Birkhäuser Verlag, Basel, 1989, 485–496.

Oscar Blasco
Departamento de Análisis Matemático
Universidad de Valencia
E-46100 Burjassot, Valencia, Spain
e-mail: oblasco@uv.es

Operator Theory:
Advances and Applications, Vol. 195, 83–101

On an Interpolation Problem for Generalized Schur Functions

Vladimir Bolotnikov

Dedicated to Professor Joseph Ball on occasion of his 60th birthday

Abstract. The nondegenerate Nevanlinna-Pick-Carathéodory-Fejér interpolation problem with finitely many interpolation conditions always has infinitely many solutions in a generalized Schur class $\mathcal{S}_\kappa$ for every $\kappa \geq \kappa_{\min}$ where the integer $\kappa_{\min}$ equals the number of negative eigenvalues of the Pick matrix associated to the problem and completely determined by interpolation data. A linear fractional description of all $\mathcal{S}_{\kappa_{\min}}$ solutions of the (nondegenerate) problem is well known. In this paper, we present a similar result for an arbitrary $\kappa \geq \kappa_{\min}$.

Mathematics Subject Classification (2000). 30E05.

Keywords. Generalized Schur functions, Nevanlinna-Pick interpolation problem.

1. Introduction

Let $\mathcal{S}$ stand for the Schur class of analytic functions mapping the unit disk $\mathbb{D}$ into $\overline{\mathbb{D}}$ and let $\mathcal{B}_\kappa$ be the set of finite Blaschke products of degree κ. We denote by $\mathcal{S}_\kappa$ the *generalized Schur class* of meromorphic functions of the form

$$f(z) = \frac{s(z)}{b(z)}, \tag{1.1}$$

where $s \in \mathcal{S}$ and $b \in \mathcal{B}_\kappa$ do not have common zeros (in particular, $\mathcal{S}_0 = \mathcal{S}$). Formula (1.1) is called the Krein-Langer representation of a generalized Schur function f; the entries s and b are determined by f uniquely up to a unimodular constant. Via nontangential boundary limits, the $\mathcal{S}_\kappa$-functions can be identified with the functions from the unit ball of $L^\infty(\mathbb{T})$ which admit meromorphic continuation inside the unit disk with total pole multiplicity equal κ. On the other hand, the $\mathcal{S}_\kappa$-functions f can be characterized as meromorphic functions on $\mathbb{D}$ for which the associated kernel

$$K_f(z,\zeta) := \frac{1-f(z)\overline{f(\zeta)}}{1-z\bar{\zeta}} \tag{1.2}$$

has κ negative squares on $\rho(f)$, the domain of analyticity of f: $\mathrm{sq}_-(K_f)=\kappa$.

The classes $\mathcal{S}_\kappa$ were thoroughly studied in [9, 10], the major interpolation results for $\mathcal{S}_\kappa$-functions can be found in [1, 2, 3, 4]. The objective of this paper is the Nevanlinna-Pick-Carathéodory-Fejér interpolation problem which will be denoted by $\mathbf{IP}_\kappa$ and which consists of the following:

$\mathbf{IP}_\kappa$: *Given an integer* $\kappa \geq 0$, *distinct points* $z_1,\dots,z_k \in \mathbb{D}$, *a tuple* $\mathbf{n} = (n_1,\dots,n_k) \in \mathbb{N}^k$ *and* $|\mathbf{n}| := n_1+\dots+n_k$ *complex numbers* $f_{i,j}$ $(0 \leq j \leq n_i - 1;\ 1 \leq i \leq k)$, *find all functions* $f \in \mathcal{S}_\kappa$ *(if exist) which are analytic at* z_i *and satisfy*

$$f^{(j)}(z_i) = j!\, f_{i,j} \qquad (i=1,\dots,k;\ j=0,\dots,n_i-1). \tag{1.3}$$

Necessary and sufficient conditions for the $\mathbf{IP}_\kappa$ to have a solution can be given in terms of the Pick matrix of the problem which is determined from interpolation data as follows. Let $J_n(z)$ denote the $n\times n$ Jordan block with the number z on the main diagonal and let E_n stand for the column vector of the height n with the first coordinate equals one and other coordinates equal zero:

$$J_n(z) = \begin{bmatrix} z & 1 & & 0 \\ 0 & z & \ddots & \\ \vdots & \ddots & \ddots & 1 \\ 0 & \dots & 0 & z \end{bmatrix}, \quad E_n = \begin{bmatrix} 1 \\ 0 \\ \vdots \\ 0 \end{bmatrix}.$$

Associated with the tuples $\mathbf{z} = (z_1,\dots,z_k)$ and $\mathbf{n} = (n_1,\dots,n_k)$ are the matrices

$$T = \begin{bmatrix} J_{n_1}(z_1) & & \\ & \ddots & \\ & & J_{n_k}(z_k) \end{bmatrix}, \quad E = \begin{bmatrix} E_{n_1} \\ \vdots \\ E_{n_k} \end{bmatrix}, \tag{1.4}$$

and we arrange the rest of data in the column-vector

$$C = \begin{bmatrix} C_1 \\ \vdots \\ C_k \end{bmatrix}, \quad \text{where} \quad C_i = \begin{bmatrix} f_{i,0} \\ \vdots \\ f_{i,n_i-1} \end{bmatrix}. \tag{1.5}$$

Since all the eigenvalues of T fall inside the unit disk, the Stein equation

$$P - TPT^* = EE^* - CC^* \tag{1.6}$$

has a unique solution P which is defined via the converging series

$$P = \sum_{j=0}^{\infty} T^j (EE^* - CC^*) T^{*j} \tag{1.7}$$

and which is called the *Pick matrix* of the problem $\mathbf{IP}_\kappa$. A necessary condition for the $\mathbf{IP}_\kappa$ to have a solution can be obtained as follows. Given an $f \in \mathcal{S}_\kappa$, an integer

$k \geq 0$ and two k-tuples $\mathbf{z} = (z_1, \ldots, z_k) \in \rho(f)^k$ and $\mathbf{n} = (n_1, \ldots, n_k) \in \mathbb{N}^k$, we define the column-vector

$$M_{\mathbf{n}}(f, \mathbf{z}) = \begin{bmatrix} M_{n_1}(f; z_1) \\ \vdots \\ M_{n_k}(f; z_k) \end{bmatrix}, \quad \text{where} \quad M_{n_i}(f; z_i) = \begin{bmatrix} f(z_i) \\ \frac{f'(z_i)}{1!} \\ \vdots \\ \frac{f^{(n_i-1)}(z_i)}{(n_i-1)!} \end{bmatrix} \tag{1.8}$$

and the $|\mathbf{n}| \times |\mathbf{n}|$ *Schwarz-Pick matrix*

$$P_{\mathbf{n}}(f; \mathbf{z}) = -\frac{1}{4\pi^2} \int_\Gamma \int_\Gamma (\xi - T)^{-1} E K_f(\xi, \omega) E^* (\bar{\omega} - T^*)^{-1} \, d\xi d\bar{\omega}, \tag{1.9}$$

where T and E are given in (1.4) and where $\Gamma \in \mathbb{D}$ is any contour enclosing the points $z_1, \ldots, z_k$ and such that $\overline{\operatorname{Int}\Gamma} \subset \rho(f)$. Since $\operatorname{sq}_-(K_f) = \kappa$, the standard approximation arguments show that the matrix $P_{\mathbf{n}}(f; \mathbf{z})$ has at most κ negative eigenvalues: $\operatorname{sq}_-(P_{\mathbf{n}}(f; \mathbf{z})) \leq \kappa$. Furthermore, equality

$$M_{\mathbf{n}}(f, \mathbf{z}) = \frac{1}{2\pi i} \int_\Gamma (\xi - T)^{-1} E f(\xi) \, d\xi \tag{1.10}$$

follows from definitions (1.8) and (1.4) by residue calculus; using this equality, one can readily check that the matrix $P_{\mathbf{n}}(f; \mathbf{z})$ defined in (1.9) satisfies the Stein equation

$$P_{\mathbf{n}}(f; \mathbf{z}) - T P_{\mathbf{n}}(f; \mathbf{z}) T^* = EE^* - M_{\mathbf{n}}(f; \mathbf{z}) M_{\mathbf{n}}(f; \mathbf{z})^*.$$

Now we observe that for every solution f of the problem $\mathbf{IP}_\kappa$, the Schwarz-Pick matrix $P_{\mathbf{n}}(f; \mathbf{z})$ is equal to P, the Pick matrix of the problem (indeed, if f satisfies interpolation conditions (1.3), it follows from (1.5) and (1.10) that $M_{\mathbf{n}}(f; \mathbf{z}) = C$; thus P and $P_{\mathbf{n}}(f; \mathbf{z})$ satisfy the same Stein equation which in turn, has a unique solution). In particular, if P has more than κ negative eigenvalues, the problem $\mathbf{IP}_\kappa$ has no solutions. Thus, condition $\kappa \geq \operatorname{sq}_-(P)$ is necessary for the $\mathbf{IP}_\kappa$ to have a solution. On the other hand, if

$$\kappa \geq \operatorname{sq}_-(P) \quad \text{and} \quad \det P \neq 0, \tag{1.11}$$

then the problem $\mathbf{IP}_\kappa$ has infinitely many solutions, which are parametrized by a linear fractional transformation.

Theorem 1.1. *Let the Pick matrix P of the $\mathbf{IP}_\kappa$ meet conditions* (1.11) *and let* $\Theta = \begin{bmatrix} \Theta_{11} & \Theta_{12} \\ \Theta_{21} & \Theta_{22} \end{bmatrix}$ *be the 2×2 rational matrix-valued function defined by*

$$\Theta(z) = I_2 + (z-1) \begin{bmatrix} E^* \\ C^* \end{bmatrix} (I - zT^*)^{-1} P^{-1} (I - T)^{-1} \begin{bmatrix} E & -C \end{bmatrix}. \tag{1.12}$$

Then all solutions f of the $\mathbf{IP}_\kappa$ are parametrized by the linear fractional formula

$$f(z) = \frac{\Theta_{11}(z) S(z) + \Theta_{12}(z) B(z)}{\Theta_{21}(z) S(z) + \Theta_{22}(z) B(z)}, \tag{1.13}$$

where the parameters $S \in \mathcal{S}$ *and* $B \in \mathcal{B}_{\kappa-\mathrm{sq}_-(P)}$ *do not have common zeros and satisfy conditions*

$$\Theta_{21}(z_i)S(z_i) + \Theta_{22}(z_i)B(z_i) \neq 0 \quad (i = 1, \dots, k). \tag{1.14}$$

Equivalently,

$$f(z) = \mathbf{T}_\Theta[\mathcal{E}] := \frac{\Theta_{11}(z)\mathcal{E}(z) + \Theta_{12}(z)}{\Theta_{21}(z)\mathcal{E}(z) + \Theta_{22}(z)}, \tag{1.15}$$

where the parameter $\mathcal{E} \in \mathcal{S}_{\kappa-\mathrm{sq}_-(P)}$ *satisfies*

$$\Theta_{21}(z_i)\mathcal{E}(z_i) + \Theta_{22}(z_i) \neq 0 \quad (i = 1, \dots, k) \tag{1.16}$$

or has a pole at z_i *in case* $\Theta_{21}(z_i) \neq \Theta_{22}(z_i) = 0$.

Equivalence of descriptions (1.13) and (1.15) is established via the Krein-Langer representation $\mathcal{E} = \frac{S}{B}$ of the function $\mathcal{E} \in \mathcal{S}_{\kappa-\mathrm{sq}_-(P)}$. If κ is minimally possible (i.e., if $\kappa = \mathrm{sq}_-(P)$), then the parameter $\mathcal{E}$ in (1.15) runs through the Schur class $\mathcal{S}$; this result can be found in [2, 3, 4]. A somewhat new point presented here is that in case $\kappa > \mathrm{sq}_-(P)$, some solutions of the problem may arise via formula (1.15) from parameters which are not analytic at interpolation nodes. We illustrate this possibility by a numerical example.

Example 1.2. Let $z_1 = 0$, $z_2 = 1/2$, $f_1 = 1$ and $f_2 = 1/2$ so that the interpolation conditions are

$$f(0) = 1 \quad \text{and} \quad f(1/2) = 1/2. \tag{1.17}$$

The Pick matrix $P = \begin{bmatrix} 0 & 1/2 \\ 1/2 & 1 \end{bmatrix}$ of the problem has one negative and one positive eigenvalues and thus, the problem $\mathbf{IP}_\kappa$ has a solution if and only if $\kappa \geq 1$. Furthermore, substituting

$$P^{-1} = \begin{bmatrix} -4 & 2 \\ 2 & 0 \end{bmatrix}, \quad T = \begin{bmatrix} 0 & 0 \\ 0 & \frac{1}{2} \end{bmatrix}, \quad E = \begin{bmatrix} 1 \\ 1 \end{bmatrix}, \quad C = \begin{bmatrix} 1 \\ \frac{1}{2} \end{bmatrix}$$

into (1.12) gives

$$\Theta(z) = \frac{1}{2-z} \begin{bmatrix} 3z - 2 & 2z(1-z) \\ 2(z-1) & z(3-2z) \end{bmatrix}.$$

By Theorem 1.1, all solutions of the problem $\mathbf{IP}_1$ with interpolation conditions (1.17) are parametrized by the formula

$$f(z) = \frac{(3z-2)\mathcal{E}(z) + 2z(1-z)}{2(z-1)\mathcal{E}(z) + z(3-2z)}, \tag{1.18}$$

where $\mathcal{E}$ belongs to $\mathcal{S}_0$ and satisfies

$$\mathcal{E}(0) \neq 0 \quad \text{and} \quad \mathcal{E}(1/2) \neq 1. \tag{1.19}$$

To get all solutions of the problem $\mathbf{IP}_2$ with interpolation conditions (1.17), we use the formula (1.18) with parameters $\mathcal{E} \in \mathcal{S}_1$ which are analytic at $z_1 = 0$ and $z_2 = 1/2$ and match constraints (1.19). However, since $\Theta_{22}(0) = 0$ and $\Theta_{21}(0) = -2 \neq 0$, Theorem 1.1 asserts that any function $\mathcal{E} \in \mathcal{S}_1$ which has a pole at $z_1 = 0$

and meets the second constraint in (1.19), also leads via (1.18) to a solution f of the problem.

Our interest to the "non-minimal" problem $\mathbf{IP}_\kappa$ (where $\kappa > \mathrm{sq}_-(P)$) is motivated by the following reason: if the Pick matrix P of the problem $\mathbf{IP}_\kappa$ is singular, then the minimally possible κ for which the $\mathbf{IP}_\kappa$ has a solution, may be greater than $\mathrm{sq}_-(P)$ (see, e.g., [11]). The description of all solutions for such a degenerate problem can be reduced to a family of nondegenerate "non-minimal" problems at which point Theorem 1.1 will come into play (in [6] this was done in the context of the Nevanlinna-Pick interpolation problem; the general case will be presented in a follow-up paper). The proof of Theorem 1.1 is presented in Section 2. In Section 3 we will discuss the divisor-remainder formulation of the problem $\mathbf{IP}_\kappa$ (also considered in [2, 3, 4] for $\kappa = \mathrm{sq}_-(P)$).

2. Proof of Theorem 1.1

We first recall some properties of the function Θ defined in (1.12). In what follows, $N\{g\}$ stands for the total number of zeros of a function g that fall inside $\mathbb{D}$.

Lemma 2.1. *Let T, E, C and P be given by (1.4), (1.5) and (1.7), let P be invertible and let $\Theta(z)$ be defined as in (1.12). Then*

1. *$\Theta(t)$ is J-unitary at every point $t \in \mathbb{T}$:*

$$\Theta(t)^* J \Theta(t) = \Theta(t) J \Theta(t)^* = J := \begin{bmatrix} 1 & 0 \\ 0 & -1 \end{bmatrix}. \tag{2.1}$$

2. *For $z, \zeta \in \mathbb{D}$,*

$$\frac{J - \Theta(z) J \Theta(\zeta)^*}{1 - z\bar{\zeta}} = \begin{bmatrix} E^* \\ C^* \end{bmatrix} (I - zT^*)^{-1} P^{-1} (I - \bar{\zeta} T)^{-1} \begin{bmatrix} E & C \end{bmatrix} \tag{2.2}$$

and

$$\det \Theta(z) = \prod_{i=1}^{k} \left(\frac{(z - z_i)(1 - \bar{z}_i)}{(1 - z\bar{z}_i)(1 - z_i)} \right)^{n_i}. \tag{2.3}$$

3. *The function $(zI - T)^{-1} \begin{bmatrix} E & -C \end{bmatrix} \Theta(z)$ is analytic on $\mathbb{D}$.*
4. *The rational functions Θ_{11} and Θ_{22} do not vanish on the unit circle and have respectively $\mathrm{sq}_+(P)$ and $\mathrm{sq}_-(P)$ zeros in $\mathbb{D}$:*

$$N\{\Theta_{11}\} = \mathrm{sq}_+(P) \quad \text{and} \quad N\{\Theta_{22}\} = \mathrm{sq}_-(P). \tag{2.4}$$

5. *For every $z \in \mathbb{C}$,*

$$|\Theta_{21}(z)| + |\Theta_{22}(z)| > 0. \tag{2.5}$$

6. *For every $\mathcal{E} \in \mathcal{S}_{\widetilde{\kappa}}$, the function $f = \mathbf{T}_\Theta[\mathcal{E}] := \frac{\Theta_{11}\mathcal{E} + \Theta_{12}}{\Theta_{21}\mathcal{E} + \Theta_{22}}$ belongs to $\mathcal{S}_\kappa$ with $\kappa \le \widetilde{\kappa} + \mathrm{sq}_-(P)$.*

Proof. Identity (2.2) follows by a straightforward calculation (see, e.g., [3, Section 7.1]) based on the identity (1.6). A similar calculation gives

$$\begin{aligned}\frac{J-\Theta(\zeta)^*J\Theta(z)}{1-z\bar{\zeta}} &= \begin{bmatrix} E^* \\ -C^* \end{bmatrix}(I-T^*)^{-1}P^{-1}(I-\bar{\zeta}T)^{-1}P(1-zT^*)^{-1} \\ &\quad \times P^{-1}(I-T)^{-1}\begin{bmatrix} E & -C \end{bmatrix}. \end{aligned} \tag{2.6}$$

Identities (2.1) follow from (2.2) and (2.6), since Θ is rational and has no poles on $\mathbb{T}$. Equality (2.3) follows from (1.6) by the standard properties of determinants (see, e.g., [5, Lemma 2.2] for the proof). The third statement of the lemma is yet another consequence of identity (1.6) due to which we have

$$\begin{aligned}\begin{bmatrix} E & -C \end{bmatrix}\Theta(z) &= \begin{bmatrix} E & -C \end{bmatrix} + (z-1)\begin{bmatrix} E & -C \end{bmatrix}\begin{bmatrix} E^* \\ C^* \end{bmatrix}(I-zT^*)^{-1} \\ &\quad \times P^{-1}(I-T)^{-1}\begin{bmatrix} E & -C \end{bmatrix} \\ &= \left(I+(z-1)(P-TPT^*)(I-zT^*)^{-1}P^{-1}(I-T)^{-1}\right) \\ &\quad \times \begin{bmatrix} E & -C \end{bmatrix} \\ &= (zI-T)P(I-T^*)(I-zT^*)^{-1}P^{-1}(I-T)^{-1}\begin{bmatrix} E & -C \end{bmatrix}. \end{aligned}$$

Therefore,

$$(zI-T)^{-1}\begin{bmatrix} E & -C \end{bmatrix}\Theta(z) = P(I-T^*)(I-zT^*)^{-1}P^{-1}(I-T)^{-1}\begin{bmatrix} E & -C \end{bmatrix}$$

and the function on the right-hand side is analytic on $\mathbb{D}$.

Equalities (2.1) imply in particular

$$|\Theta_{11}(t)|^2-|\Theta_{21}(t)|^2=1, \quad |\Theta_{21}(t)|^2-|\Theta_{22}(t)|^2=-1 \quad (t\in\mathbb{T}) \tag{2.7}$$

and thus, Θ_{11} and Θ_{22} do not vanish on $\mathbb{T}$. For the proof of the second equality in (2.4), we refer to [3, Theorem 13.2.3] or to [8, Lemma 4]. This equality tells us that if T and E are defined as in (1.4) and M is an arbitrary vector in $\mathbb{C}^{|\mathbf{n}|}$ such that the unique solution R (which is necessarily Hermitian) of the Stein equation

$$R-T^*RT = EE^*-MM^* \tag{2.8}$$

is invertible, then the function

$$F_M(z) = 1-(z-1)M^*(I-zT^*)^{-1}R^{-1}(I-T)^{-1}M \tag{2.9}$$

has $\mathrm{sq}_-(R)$ zeros inside $\mathbb{D}$:

$$N\{F_M\} = \mathrm{sq}_-(R). \tag{2.10}$$

Let C be the vector associated with the problem $\mathbf{IP}_\kappa$ and decomposed as in (1.5). For an $\varepsilon>0$, define $C_\varepsilon := C+\varepsilon E$ and the matrix P_ε, a unique solution of the Stein equation

$$P_\varepsilon - TP_\varepsilon T^* = EE^*-C_\varepsilon C_\varepsilon^*. \tag{2.11}$$

Due to the structure (1.4) of E, the above perturbation changes only the top entries $f_{i,0}$ in each of the blocks C_i replacing them by $f_{i,0}+\varepsilon$. It is clear that there exists ε_0 so that for every $\varepsilon\in(0,\varepsilon_0)$,

$$\mathrm{sq}_\pm(P_\varepsilon)=\mathrm{sq}_\pm(P) \quad \text{and} \quad f_{i,0}+\varepsilon\neq 0 \quad \text{for} \quad i=1,\dots,k.$$

Now we let $\mathbf{C}_\varepsilon$ to be the block diagonal matrix with lower triangular Toeplitz diagonal blocks:

$$\mathbf{C}_\varepsilon = \begin{bmatrix} \mathbf{C}_{\varepsilon,1} & & 0 \\ & \ddots & \\ 0 & & \mathbf{C}_{\varepsilon,k} \end{bmatrix}, \quad \mathbf{C}_{\varepsilon,i} = \begin{bmatrix} f_{i,0}+\varepsilon & 0 & \dots & 0 \\ f_{i,1} & f_{i,0}+\varepsilon & \ddots & \vdots \\ \vdots & \ddots & \ddots & 0 \\ f_{i,n_i-1} & \dots & f_{i,1} & f_{i,0}+\varepsilon \end{bmatrix}.$$

It is obvious that $\mathbf{C}_\varepsilon$ is invertible and satisfies relations

$$\mathbf{C}_\varepsilon E = C_\varepsilon \quad \text{and} \quad \mathbf{C}_\varepsilon T = T\mathbf{C}_\varepsilon.$$

Multiplying both parts in (2.11) by $\mathbf{C}_\varepsilon^{-1}$ on the left and by its adjoint on the right and making use of the two last equalities, we get

$$\mathbf{C}_\varepsilon^{-1} P_\varepsilon \mathbf{C}_\varepsilon^{-*} - \mathbf{C}_\varepsilon^{-1} T P_\varepsilon \mathbf{C}_\varepsilon^{-*} T^* = \mathbf{C}_\varepsilon^{-1} E E^* \mathbf{C}_\varepsilon^{-*} - EE^*$$

which can be written in the form (2.8) upon letting

$$R = -\mathbf{C}_\varepsilon^{-1} P_\varepsilon \mathbf{C}_\varepsilon^{-*} \quad \text{and} \quad M = \mathbf{C}_\varepsilon^{-1} E.$$

For this setting, the formula (2.9) takes the form

$$\begin{aligned} F_\varepsilon(z) &= 1 + (z-1)E^* \mathbf{C}_\varepsilon^{-*}(I - zT^*)^{-1}(\mathbf{C}_\varepsilon^{-1} P_\varepsilon \mathbf{C}_\varepsilon^{-*})^{-1}(I-T)^{-1}\mathbf{C}_\varepsilon^{-1} E \\ &= 1 + (z-1)E^*(I - zT^*)^{-1} P_\varepsilon^{-1}(I-T)^{-1}E \end{aligned}$$

from which it follows that

$$\lim_{\varepsilon \to 0} F_\varepsilon(z) = 1 + (z-1)E^*(I - zT^*)^{-1} P^{-1}(I-T)^{-1}E = \Theta_{11}(z).$$

Due to (2.10) we have

$$N\{F_\varepsilon\} = \text{sq}_-(-\mathbf{C}_\varepsilon^{-1} P_\varepsilon \mathbf{C}_\varepsilon^{-*}) = \text{sq}_+(\mathbf{C}_\varepsilon^{-1} P_\varepsilon \mathbf{C}_\varepsilon^{-*}) = \text{sq}_+(P_\varepsilon) = \text{sq}_+(P).$$

Passing to the limit as $\varepsilon \to 0$ implies that Θ_{11} has $\text{sq}_+(P)$ zeros in the closed unit disk and since it does not have zeros on $\mathbb{T}$, the first equality in (2.4) follows.

To prove the fifth statement, note that if $\Theta_{21}(z) = \Theta_{22}(z) = 0$, then $\det \Theta(z) = 0$ and thus by formula (2.3), inequality (2.5) may fail only at $z \in \{z_1, \dots, z_d\}$. Let us show that it doesn't. Assuming that $\Theta_{21}(z_1) = \Theta_{22}(z_1) = 0$ we have by (1.12) and (1.6)

$$\begin{aligned} 0 &= \Theta_{21}(z_1)E^* + \Theta_{22}(z_1)C^* \\ &= C^* + (z_1 - 1)C^*(I - z_1 T^*)^{-1} P^{-1}(I-T)^{-1}(EE^* - CC^*) \\ &= C^* + (z_1 - 1)C^*(I - z_1 T^*)^{-1} P^{-1}(I-T)^{-1}(P - TPT^*) \\ &= C^*(I - z_1 T^*)^{-1} P^{-1}(z_1 I - T)(I-T)^{-1} P(I - T^*) \end{aligned}$$

which is equivalent to

$$0 = C^*(I - z_1 T^*)^{-1} P^{-1}(z_1 I - T).$$

Due to the Jordan structure (2.2) of T and since $z_1 \neq z_i$ for $i = 2, \dots, d$, it follows from the last equality that the row-vector $C^*(I - z_1T^*)^{-1}P^{-1}$ must be of the form

$$C^*(I - z_1T^*)^{-1}P^{-1} = \begin{bmatrix} \alpha & 0 & \dots & 0 \end{bmatrix}$$

and then we have

$$\begin{aligned} \begin{bmatrix} 0 & 0 \end{bmatrix} &= \begin{bmatrix} \Theta_{21}(z_1) & \Theta_{22}(z_1) \end{bmatrix} \\ &= \begin{bmatrix} 0 & 1 \end{bmatrix} + (z_1 - 1) \begin{bmatrix} \alpha & 0 & \dots & 0 \end{bmatrix} (I - T)^{-1} \begin{bmatrix} E & -C \end{bmatrix} \\ &= \begin{bmatrix} 0 & 1 \end{bmatrix} - \alpha \begin{bmatrix} 1 & -f_{1,0} \end{bmatrix} = \begin{bmatrix} -\alpha & 1 + \alpha f_{1,0} \end{bmatrix} \end{aligned}$$

which is a contradiction. Thus, inequality (2.5) holds for $z = z_1$ and similarly for $z_2, \dots, z_d$ which completes the proof of the fifth statement.

Finally, let us observe that the function $f = \mathbf{T}_\Theta[\mathcal{E}]$ is well defined, i.e., that the function $G_\mathcal{E} = \Theta_{21}\mathcal{E} + \Theta_{22}$ does not vanish identically for any generalized Schur function $\mathcal{E}$. Indeed, if $G_\mathcal{E} \equiv 0$, then the function $-\frac{\Theta_{22}}{\Theta_{21}} = \mathcal{E}$ belongs to a generalized Schur class which is impossible, since due to the second equality in (2.7), $\left|\frac{\Theta_{22}(t)}{\Theta_{21}(t)}\right| > 1$ at every $t \in \mathbb{T}$. Now the last statement in the lemma follows from the identity

$$G(z)K_f(z,\zeta)G(\zeta)^* = K_\mathcal{E}(z,\zeta) + \begin{bmatrix} \mathcal{E}(\zeta)^* & 1 \end{bmatrix} \frac{J - \Theta(\zeta)^*J\Theta(z)}{1 - z\bar{\zeta}} \begin{bmatrix} \mathcal{E}(z) \\ 1 \end{bmatrix}$$

since $\operatorname{sq}_-\left(\frac{J-\Theta(\zeta)^*J\Theta(z)}{1-z\bar{\zeta}}\right) \leq \operatorname{sq}_-(P)$, by (2.6). □

For notational convenience, we will often write f^* rather than $\overline{f}$.

Theorem 2.2. *Let P satisfy conditions (1.11), let Θ be given by (1.12) and let f be a solution of the problem $\mathbf{IP}_\kappa$. Then*

1. *The kernel*

$$\mathbf{K}_f(z,\zeta) = \begin{bmatrix} P & (I - \bar{\zeta}T)^{-1}(E - Cf(\zeta)^*) \\ (E^* - f(z)C^*)(I - zT^*)^{-1} & K_f(z,\zeta) \end{bmatrix} \tag{2.12}$$

has κ negative squares on $\rho(f)$.

2. *The function f is of the form (1.15) for some $\mathcal{E} \in \mathcal{S}_{\kappa - \operatorname{sq}_-(P)}$.*

Proof. Let $\Gamma \in \mathbb{D}$ be any contour enclosing the points $z_1, \dots, z_k$ and such that $\overline{\operatorname{Int}\Gamma} \subset \rho(f)$. Since $\operatorname{sq}_-(K_f) = \kappa$, approximation arguments show that the kernel

$$\widetilde{\mathbf{K}}_f(z,\zeta) = -\frac{1}{4\pi^2}\int_\Gamma\int_\Gamma \begin{bmatrix} (\xi - T)^{-1}E \\ (\xi - z)^{-1} \end{bmatrix} K_f(\xi,\omega) \begin{bmatrix} E^*(\bar{\omega} - T^*)^{-1} & (\bar{\omega} - \bar{\zeta})^{-1} \end{bmatrix} d\xi d\bar{\omega} \tag{2.13}$$

has at most κ negative squares. Since f is a solution of the problem $\mathbf{IP}_\kappa$, we have $P_\mathbf{n}(f;\mathbf{z}) = P$ and $M_\mathbf{n}(f;\mathbf{z}) = C$ which together with (1.9) and (1.10) lead us to

$$P = -\frac{1}{4\pi^2}\int_\Gamma\int_\Gamma (\xi - T)^{-1}EK_f(\xi,\omega)E^*(\bar{\omega} - T^*)^{-1}\, d\xi d\bar{\omega} \tag{2.14}$$

and

$$C=\frac{1}{2\pi i}\int_{\Gamma}(\xi-T)^{-1}Ef(\xi)d\xi.$$

Using the latter equality along with (1.2) gives

$$\begin{aligned}-\frac{1}{4\pi^2}\int_{\Gamma}\int_{\Gamma}(\xi-T)^{-1}EK_f(\xi,\omega)\frac{d\xi d\bar{\omega}}{\bar{\omega}-\bar{\zeta}} &= \frac{1}{2\pi i}\int_{\Gamma}(\xi-T)^{-1}E\frac{1-f(\xi)f(\zeta)^*}{1-\xi\bar{\zeta}}d\xi \\ &= (I-\bar{\zeta}T)^{-1}E-(I-\bar{\zeta}T)^{-1}Cf(\zeta)^*.\end{aligned}$$

Finally,

$$-\frac{1}{4\pi^2}\int_{\Gamma}\int_{\Gamma}\frac{K_f(\xi,\omega)\,d\xi d\bar{\omega}}{(\xi-z)(\bar{\omega}-\bar{\zeta})}=K_f(z,\zeta).$$

Substituting the two last equalities and (2.14) into (2.13) and comparing the resulting matrix with (2.12) we conclude that $\mathbf{K}_f(z,\zeta)=\widetilde{\mathbf{K}}_f(z,\zeta)$. Therefore, $\mathrm{sq}_-(\mathbf{K}_f)\leq\kappa$. On the other hand, it follows from (2.12) that $\mathrm{sq}_-(\mathbf{K}_f)\geq\mathrm{sq}_-(K_f)=\kappa$ which completes the proof of the first statement of the theorem. To prove the second, we first observe that the kernel

$$\mathbf{S}(z,\zeta)=K_f(z,\zeta)-(E^*-f(z)C^*)(I-zT^*)^{-1}P^{-1}(I-\bar{\zeta}T)^{-1}(E-Cf(\zeta)^*)$$

is the Schur complement of the block P in $\mathbf{K}_f(z,\zeta)$ and since $\mathrm{sq}_-(\mathbf{K}_f)=\kappa$ by the first part, it follows that

$$\mathrm{sq}_-(\mathbf{S})=\mathrm{sq}_-(\mathbf{K}_f)-\mathrm{sq}_-(P)=\kappa-\mathrm{sq}_-(P). \tag{2.15}$$

Making use of relation

$$K_f(z,\zeta)=\frac{\begin{bmatrix}1 & -f(z)\end{bmatrix}J\begin{bmatrix}1\\-f(\zeta)^*\end{bmatrix}}{1-z\bar{\zeta}}\quad\text{where}\quad J:=\begin{bmatrix}1&0\\0&-1\end{bmatrix},$$

we represent $\mathbf{S}$ in the form

$$\begin{aligned}\mathbf{S}(z,\zeta) &= \begin{bmatrix}1 & -f(z)\end{bmatrix}\left\{\frac{J}{1-z\bar{\zeta}}\right. \\ &\quad\left.-\begin{bmatrix}E^*\\C^*\end{bmatrix}(I-zT^*)^{-1}P^{-1}(I-\bar{\zeta}T)^{-1}\begin{bmatrix}E & C\end{bmatrix}\right\}\begin{bmatrix}1\\-f(\zeta)^*\end{bmatrix}\end{aligned} \tag{2.16}$$

which in turn, can be written as

$$\mathbf{S}(z,\zeta)=\begin{bmatrix}1 & -f(z)\end{bmatrix}\frac{\Theta(z)J\Theta(\zeta)^*}{1-z\bar{\zeta}}\begin{bmatrix}1\\-f(\zeta)^*\end{bmatrix},$$

due to (2.2). The last equality can be written in terms of the pair $\{u,v\}$ defined by

$$\begin{bmatrix}v(z) & -u(z)\end{bmatrix}=\begin{bmatrix}1 & -f(z)\end{bmatrix}\Theta(z) \tag{2.17}$$

as

$$\mathbf{S}(z,\zeta)=\begin{bmatrix}v(z) & -u(z)\end{bmatrix}\frac{J}{1-z\bar{\zeta}}\begin{bmatrix}v(\zeta)^*\\-u(\zeta)^*\end{bmatrix}=\frac{v(z)v(\zeta)^*-u(z)u(\zeta)^*}{1-z\bar{\zeta}}. \tag{2.18}$$

Let us show that $v(z)\not\equiv 0$. Indeed, by the first equality in (2.7), $\left|\frac{\Theta_{11}(t)}{\Theta_{21}(t)}\right|>1$ for every $t\in\mathbb{T}$. Thus, $\frac{\Theta_{11}(t)}{\Theta_{21}(t)}\notin\mathcal{S}_\kappa$. However, if $v=\Theta_{11}-f\Theta_{21}\equiv 0$ we have that $f=\frac{\Theta_{11}}{\Theta_{21}}\in\mathcal{S}_\kappa$ which is a contradiction. Thus, $v\not\equiv 0$ and the function $\mathcal{E}=\dfrac{u}{v}$ is meromorphic on $\mathbb{D}$. Equality (2.18) can be written in terms of this function as

$$\mathbf{S}(z,\zeta)=v(z)\frac{1-\mathcal{E}(z)\mathcal{E}(\zeta)^*}{1-z\bar{\zeta}}v(\zeta)^*=v(z)K_{\mathcal{E}}(z,\zeta)v(\zeta)^*$$

which together with (2.15) implies $\mathrm{sq}_-(K_{\mathcal{E}})=\kappa-\mathrm{sq}_-(P)$ so that $\mathcal{E}\in\mathcal{S}_{\kappa-\mathrm{sq}_-(P)}$. Finally, it follows from (2.17) that

$$f=\frac{\Theta_{11}u+\Theta_{12}v}{\Theta_{21}u+\Theta_{22}v}=\frac{\Theta_{11}\mathcal{E}+\Theta_{12}}{\Theta_{21}\mathcal{E}+\Theta_{22}}.$$

Thus, f is of the form (1.15) with an $\mathcal{E}\in\mathcal{S}_{\kappa-\mathrm{sq}_-(P)}$ which completes the proof of the theorem. □

Now we will take a closer look at the numerator and the denominator in the linear fractional formula (1.15). Let $\mathcal{E}$ be a fixed function from $\mathcal{S}_{\widetilde{\kappa}}$ with the coprime representation

$$\mathcal{E}(z)=\frac{S(z)}{B(z)},\quad S\in\mathcal{S},\ B\in\mathcal{B}_{\widetilde{\kappa}},\tag{2.19}$$

and let Θ be the rational matrix function defined as in (1.12). Let

$$U_{S,B}=\Theta_{11}S+\Theta_{12}B,\quad V_{S,B}(z)=\Theta_{21}S+\Theta_{22}B,\tag{2.20}$$

so that (1.15) takes the form

$$f(z)=\frac{U_{S,B}(z)}{V_{S,B}(z)}.\tag{2.21}$$

For the rest of the paper we assume that $V_{S,B}$ has zeros at z_i of respective multiplicities $m_i\geq 0$, i.e., that

$$V_{S,B}(z_i)=\cdots=V_{S,B}^{(m_i-1)}(z_i)=0\quad\text{and}\quad V_{S,B}^{(m_i)}(z_i)\neq 0\quad(i=1,\dots,k).\tag{2.22}$$

Since the case where $m_i=0$ is not excluded, the latter assumption is not restrictive.

Theorem 2.3. *Let P be invertible, let $S\in\mathcal{S}$, $B\in\mathcal{B}_{\widetilde{\kappa}}$, let Θ, $U_{S,B}$ and $V_{S,B}$ be given as in* (1.12) *and* (2.20). *Then*

1. $N\{V_{S,B}\}=\mathrm{sq}_-(P)+\widetilde{\kappa}$. *If in addition, S is a finite Blaschke product of degree m (i.e., if $S\in\mathcal{B}_m$), then $N\{U_{S,B}\}=\mathrm{sq}_+(P)+m$.*
2. *$U_{S,B}$ and $V_{S,B}$ can have a common zero at no point inside $\mathbb{D}$, but $z_1,\dots,z_d$.*
3. *$U_{S,B}$ and $V_{S,B}$ cannot have a common zero at z_j of multiplicity greater than n_j.*
4. *If $V_{S,B}$ has the zero of multiplicity $m_j>n_j$ at z_j, then $U_{S,B}$ has the zero of multiplicity n_j at z_j.*
5. *If $V_{S,B}$ has the zero of multiplicity $m_j\leq n_j$ at z_j, then $U_{S,B}$ has the zero of multiplicity at least m_j at z_j.*

Proof. By the second equality in (2.7), $|\Theta_{22}(t)| > |\Theta_{21}(t)|$ on $\mathbb{T}$ and since $S \in \mathcal{S}$ and B is unimodular on $\mathbb{T}$, it follows that

$$|\Theta_{22}(t)B(t)| > |\Theta_{21}(t)S(t)|$$

at almost every point $t \in \mathbb{T}$. Then, by Rouché's theorem, the functions $V_{S,B} = \Theta_{21}S + \Theta_{22}B$ and $\Theta_{22}B$ have the same number of zeros in the disk $\{z : |z| < r\}$ for every r close enough to 1. Since the rational function Θ_{22} and the finite Blaschke product B have finitely many zeros in $\mathbb{D}$, we let $r \to 1$ to conclude that

$$N\{V_{S,B}\} = N\{\Theta_{22}B\} = N\{\Theta_{22}\} \cdot N\{B\} = \mathrm{sq}_-(P) + \widetilde{\kappa},$$

where the last equality holds since $N\{\Theta_{22}\} = \mathrm{sq}_-(P)$ (see (2.4)) and since $N\{B\} = \widetilde{\kappa}$. Furthermore, $|\Theta_{11}(t)| > |\Theta_{12}(t)|$ on $\mathbb{T}$ by the first equality in (2.7) and if S is a finite Blaschke product, we have

$$|\Theta_{11}(t)S(t)| > |\Theta_{12}(t)B(t)|$$

at every $t \in \mathbb{T}$. Then we use the preceding arguments to conclude that

$$N\{U_{S,B}\} = N\{\Theta_{11}S\} = N\{\Theta_{11}\} \cdot N\{S\} = \mathrm{sq}_+(P) + m,$$

where the last equality holds since $N\{\Theta_{11}\} = \mathrm{sq}_+(P)$ (see (2.4)) and since $N\{S\} = m$. This completes the proof of the first statement.

To prove the second statement, we write (2.20) in the matrix form as

$$\begin{bmatrix} U_{S,B}(z) \\ V_{S,B}(z) \end{bmatrix} = \Theta(z) \begin{bmatrix} S(z) \\ B(z) \end{bmatrix} \tag{2.23}$$

and assuming that $U_{S,B}(w) = V_{S,B}(w) = 0$ at some point $w \in \mathbb{D}$, we get

$$\Theta(w) \begin{bmatrix} S(w) \\ B(w) \end{bmatrix} = 0$$

which implies, since $|S(w)| + |B(w)| > 0$, that $\det \Theta(w) = 0$. But by (2.3), $z_1, \ldots, z_k$ are the only zeros of $\det \Theta$, which completes the proof of the second statement.

Assuming that $U_{S,B}$ and $V_{S,B}$ have the common zero of order $m_j > n_j$ at z_j, we conclude by (2.23) that the vector-valued function $\Theta(z) \begin{bmatrix} S(z) \\ B(z) \end{bmatrix}$ has the zero of multiplicity $m_j > n_j$ at z_j. But then, $\det \Theta(z)$ has the zero of multiplicity $m_j > n_j$ at z_j, which contradicts to (2.3) and completes the proof of the third statement.

By statement (3) in Lemma 2.1, the function

$$\begin{aligned} Q(z) &:= (zI - T)^{-1} \begin{bmatrix} E & -C \end{bmatrix} \Theta(z) \begin{bmatrix} S(z) \\ B(z) \end{bmatrix} \\ &= (zI - T)^{-1} \begin{bmatrix} E & -C \end{bmatrix} \begin{bmatrix} U_{S,B}(z) \\ V_{S,B}(z) \end{bmatrix} \end{aligned}$$

(the second equality follows by (2.23)) is analytic on $\mathbb{D}$ and in particular, at $z_1,\dots,z_k$. The block structure (1.4), (1.5) of matrices T, C and E leads to the conformal block structure of Q:

$$Q(z)=\begin{bmatrix} Q_1(z)\\ \vdots \\ Q_k(z)\end{bmatrix}, \quad \text{where } Q_i(z)=(zI-J_{n_i}(z_i))^{-1}\left[E_{n_i}U_{S,B}(z)-C_iV_{S,B}(z)\right] \tag{2.24}$$

and to the conclusion that $Q_i(z)$ is analytic at z_i for $i=1,\dots,k$. It is readily seen from the definition of Q_i that the residue of Q_i at z_i is equal to

$$\operatorname{Res}_{z=z_i}Q_i(z)=M_{n_i}(U_{S,B};z_i)-\begin{bmatrix} f_{i,0} & 0 & \dots & 0\\ f_{i,1} & f_{i,0} & \ddots & \vdots\\ \vdots & \ddots & \ddots & 0\\ f_{i,n_i-1} & \dots & f_{i,1} & f_{i,0}\end{bmatrix} M_{n_i}(V_{S,B};z_i)$$

where $M_{n_i}(U_{S,B};z_i)$ and $M_{n_i}(V_{S,B};z_i)$ are defined in accordance to (1.8). Since Q_i is analytic at z_i and therefore, $\operatorname{Res}_{z=z_i}Q_i(z)=0$, the last displayed equality implies

$$\begin{bmatrix} f_{i,0} & 0 & \dots & 0\\ f_{i,1} & f_{i,0} & \ddots & \vdots\\ \vdots & \ddots & \ddots & 0\\ f_{i,n_i-1} & \dots & f_{i,1} & f_{i,0}\end{bmatrix}\begin{bmatrix} V_{S,B}(z_i)\\ \frac{V'_{S,B}(z_i)}{1!}\\ \vdots\\ \frac{V^{(n_i-1)}_{S,B}(z_i)}{(n_i-1)!}\end{bmatrix}=\begin{bmatrix} U_{S,B}(z_i)\\ \frac{U'_{S,B}(z_i)}{1!}\\ \vdots\\ \frac{U^{(n_i-1)}_{S,B}(z_i)}{(n_i-1)!}\end{bmatrix}.$$

Thus, if $m_i\le n_i$, then conditions (2.22) force

$$U^{(j)}_{S,B}(z_i)=0 \quad \text{for } j=0,\dots,m_i-1,$$

which means that $U_{S,B}$ has a zero at z_i of at least the same multiplicity as $V_{S,B}$ does. If $m_i>n_i$, then the same arguments show that $U_{S,B}$ has zero of multiplicity $\widetilde{m}_i\ge n_i$ at z_i. If $\widetilde{m}_j>n_i$, then z_i is a common zero of $U_{S,B}$ and $V_{S,B}$ of multiplicity greater than n_j, which is impossible, by Statement 3. Thus, $\widetilde{m}_i=n_i$, which completes the proof of the theorem. □

Proof of Theorem 1.1. By Theorem 2.2, every solution f of the problem $\mathbf{IP}_\kappa$ is of the form (1.15), which is equivalent to representation (1.13). Thus, to prove Theorem 1.1, it suffices to show that a function f of the form (1.13) is a solution of the problem $\mathbf{IP}_\kappa$ if and only if the parameters $S\in\mathcal{S}$ and $B\in\mathcal{B}_\kappa$ meet conditions (1.14). To this end, take f in the form (2.21) and represent the function Q_i from (2.24) as

$$Q_i(z)=(zI-J_{n_i}(z_i))^{-1}\left[E_{n_i}f(z)-C_i\right]V_{S,B}(z).$$

If conditions (1.14) are satisfied, i.e., if $V_{S,B}(z_i) \neq 0$ for $i = 1, \dots, k$, then f is analytic at $z_1, \dots, z_d$ and the residue of Q_i at z_i equals

$$0 = \operatorname{Res}_{z=z_i} Q_i(z) = \begin{bmatrix} r_{i,0} & 0 & \dots & 0 \\ r_{i,1} & r_{i,0} & \ddots & \vdots \\ \vdots & \ddots & \ddots & 0 \\ r_{i,n_i-1} & \dots & r_{i,1} & r_{i,0} \end{bmatrix} \begin{bmatrix} V_{S,B}(z_i) \\ \frac{V'_{S,B}(z_i)}{1!} \\ \vdots \\ \frac{V_{S,B}^{(n_i-1)}(z_i)}{(n_i-1)!} \end{bmatrix}, \tag{2.25}$$

where

$$r_{i,j} = f_{i,j} - \frac{f^{(j)}(z_i)}{j!} \quad (j = 0, \dots, n_i - 1). \tag{2.26}$$

Note that analyticity of f at z_i is required to establish the second equality in (2.25); the first holds in any event since Q_i is analytic at z_i. Since $V_{S,B}(z_i) \neq 0$, it follows from (2.25) that

$$r_{i,j} = 0 \quad (j = 0, \dots, n_i - 1;\ i = 1, \dots, k),$$

which is equivalent to (2.31), by (2.26). Furthermore, $V_{S,B}(z)$ has $\mathrm{sq}_-(P) + (\kappa - \mathrm{sq}_-(P)) = \kappa$ zeros inside $\mathbb{D}$ by Theorem 2.3 (part (1)) and none of them are cancelled by zeros of $U_{S,B}(z)$ by part (2) in the same theorem. Therefore, f has κ poles inside $\mathbb{D}$ and it is a generalized Schur function by Lemma 2.1 (part (4)). Therefore, f belongs to $\mathcal{S}_\kappa$ and since it satisfies conditions (2.31), it solves the problem $\mathbf{IP}_\kappa$.

Let us assume that at least one of the conditions (1.14) fails, i.e., that $V_{S,B}(z_i) = 0$ for some $i \in \{1, \dots, k\}$. Then $U_{S,B}(z_i) = 0$, by statement (5) in Theorem 2.3 and after zero cancellation, it turns out that f has $\kappa' < \kappa$ poles inside $\mathbb{D}$ and therefore, it does not belong to $\mathcal{S}_\kappa$. This is one reason why f is not a solution of the $\mathbf{P}_\kappa$ problem. Besides, a function $f \in \mathcal{S}_{\kappa'}$ cannot satisfy all the interpolation conditions in (2.31). If it had, then by virtue of Theorem 2.2 it would be of the form $f = \mathbf{T}_\Theta[\mathcal{E}']$ for some $\mathcal{E}' \in \mathcal{S}_{\kappa' - \mathrm{sq}_-(P)}$ and the same coefficient matrix Θ. Since the map $\mathcal{E} \to \mathbf{T}_\Theta[\mathcal{E}]$ is invertible, we would have $\mathcal{E} \equiv \mathcal{E}'$ which is impossible since the latter functions have different numbers of poles in $\mathbb{D}$. This completes the proof of Theorem 1.1. □

In conclusion we consider the functions f obtained via the formula (1.13) from the parameters $\{S, B\}$ which fail to satisfy all the conditions (1.14). We will be interested in two questions: how many negative squares f may lose and which interpolation conditions it still satisfies. In addition to the tuple $\mathbf{n} = (n_1, \dots, n_k) \in \mathbb{Z}_+^k$ associated with the problem $\mathbf{IP}_\kappa$ we consider another tuple $\mathbf{m} = (m_1, \dots, m_k) \in \mathbb{Z}_+^k$ and introduce

$$\mathcal{I}_+ := \{i \in \{1, \dots, k\} :\ n_i > m_i\}, \quad \mathcal{I}_- := \{i \in \{1, \dots, k\} :\ n_i < m_i\}, \tag{2.27}$$

$$\mathcal{I}_0 = \{i \in \{1, \dots, k\} :\ n_i = m_i\}, \quad \gamma_{\mathbf{m}} := \sum_{i=1}^{k} \min\{m_i,\ n_i\}. \tag{2.28}$$

Theorem 2.4. *Let f be of the form (1.13) with $S \in \mathcal{S}$ and $B \in \mathcal{B}_{\widetilde{\kappa}}$ having no common zeros on $\mathbb{D}$ and such that*

$$(\Theta_{21}S + \Theta_{22}B)^{(j)}(z_i) = 0 \quad (i = 1, \dots, k;\ j = 0, \dots, m_i - 1) \tag{2.29}$$

and

$$(\Theta_{21}S + \Theta_{22}B)^{(m_i)}(z_i) \neq 0 \quad (i = 1, \dots, k). \tag{2.30}$$

Then f belongs to the class $\mathcal{S}_{\widetilde{\kappa}+\mathrm{sq}_-(P)-\gamma_{\mathbf{m}}}$, where $\gamma_{\mathbf{m}}$ is given in (2.28). Furthermore, f has a pole of multiplicity $m_i - n_i$ at z_i if $i \in \mathcal{I}_-$, and satisfies interpolation conditions

$$f^{(j)}(z_i) = j!\, f_{i,j} \quad (i \in \mathcal{I}_+;\ j = 0, \dots, n_i - m_i - 1). \tag{2.31}$$

Proof. We take f in the form (2.21) with $U_{S,B}$ and $V_{S,B}$ defined as in (2.20). Conditions (2.29), (2.30) say that $V_{S,B}$ has zeros of order m_i at z_i for $j = 1, \dots, k$. If $m_i \leq n_i$, then $U_{S,B}$ has zero of order at least m_i at z_i (statement (5) in Theorem 2.3) and therefore f admits an analytic continuation to z_i. If $m_i > n_i$, then the same arguments show that $U_{S,B}$ has zero of multiplicity n_i at z_i (statement (4) in Theorem 2.3) and after cancellation f will have a pole of multiplicity $m_i - n_i$ at z_i. By statement (1) in Theorem 2.3, the total number of zeros of $V_{S,B}$ inside $\mathbb{D}$ is equal to $\widetilde{\kappa} + \mathrm{sq}_-(P)$. Therefore $\widetilde{\kappa} + \mathrm{sq}_-(P) - |\mathbf{m}|$ zeros fall into $\mathbb{D} \setminus \{z_1, \dots, z_k\}$ and cannot be cancelled by zeros of $U_{S,B}$ by statement (2) in Theorem 2.3. After all zero cancellations, the function $V_{S,B}$ will have $m_i - n_i$ zeros every z_i for $i \in \mathcal{I}_-$ and still $\widetilde{\kappa} + \mathrm{sq}_-(P) - |\mathbf{m}|$ zeros in $\mathbb{D} \setminus \{z_1, \dots, z_k\}$. Thus, the function $f = \frac{U_{S,B}}{V_{S,B}}$ will have

$$\widetilde{\kappa} + \mathrm{sq}_-(P) - |\mathbf{m}| + \sum_{i \in \mathcal{I}_-} (m_i - n_i) = \widetilde{\kappa} + \mathrm{sq}_-(P) - \gamma_{\mathbf{m}}$$

poles inside $\mathbb{D}$. By statement (4) in Lemma 2.1, f is a generalized Schur function and therefore, it belongs to $\mathcal{S}_{\widetilde{\kappa}+\mathrm{sq}_-(P)-\gamma_{\mathbf{m}}}$. Furthermore, if $i \in \mathcal{I}_+$, then f is analytic at z_i and therefore equality (2.25) holds. Since $V_{S,B}^{(m_i)}(z_i) \neq 0$ by (2.30), it follows from (2.25) that

$$r_{i,j} = 0 \quad (i \in \mathcal{I}_+;\ j = 0, \dots, n_i - m_i - 1),$$

which is equivalent to (2.31), by (2.26). □

3. The divisor-remainder version

The problem $\mathbf{IP}_\kappa$ can be formulated in the divisor-remainder form (3.3) as follows. Let H_κ^∞ be the set of all functions f of the form (1.1) where $s \in H^\infty$ and $b \in \mathcal{B}_\kappa$ may have common zeros. From this definition it follows that $\mathcal{S}_\kappa = (H_\kappa^\infty \setminus H_{\kappa-1}^\infty) \cap \mathcal{B}L^\infty$ where $\mathcal{B}L^\infty$ denotes the unit ball of $L^\infty(\mathbb{T})$. Let $\varphi \in H^\infty$ be any function satisfying interpolation conditions (1.3):

$$\varphi^{(j)}(z_i) = j!\, f_{i,j} \qquad (i = 1, \dots, k;\ j = 0, \dots, n_i - 1), \tag{3.1}$$

and let θ be a finite Blaschke product defined by

$$\theta(z) = \prod_{i=1}^{k} \left(\frac{z - z_i}{1 - z\bar{z}_i} \right)^{n_i}. \tag{3.2}$$

Proposition 3.1. *A function f is a solution of the problem $\mathbf{IP}_\kappa$ if and only if it belongs to $\mathcal{S}_\kappa$ and admits a representation*

$$f(z) = \varphi(z) + \theta(z)h(z) \quad \text{for some} \quad h \in H^\infty_\kappa. \tag{3.3}$$

Proof. If $f = s/b$ (where $s \in \mathcal{S}$ and $b \in \mathcal{B}_\kappa$) belongs to $\mathcal{S}_\kappa$ and satisfies conditions (1.3), then by (3.1), the function $s - \varphi b$ belongs to H^∞ and satisfies homogeneous conditions

$$(s - \varphi b)^{(j)}(z_i) = 0 \qquad (i = 1, \dots, k;\ j = 0, \dots, n_i - 1) \tag{3.4}$$

By the maximum modulus principle, the function $(s - \varphi b)/\theta$ belongs to H^∞ where θ is defined in (3.2). Since $b \in \mathcal{B}_\kappa$, the function $h := \dfrac{s - \varphi b}{\theta b}$ belongs to H^∞_κ and therefore f can be represented as in (3.3), since

$$f = \frac{s}{b} = \varphi + \theta \cdot \frac{s - \varphi b}{\theta b} = \varphi + \theta h.$$

Conversely, let $f \in \mathcal{S}_\kappa$ be of the form (3.3). Since f has κ poles and $\varphi \in H^\infty$, it follows that h has κ poles in $\mathbb{D}$ none of which are in $\{z_1, \dots, z_k\}$, the zero set of θ. Therefore, h is analytic at $z_1, \dots, z_k$. Therefore, the function $f - \varphi = \theta h$ satisfies the homogeneous conditions (3.4), so that f satisfies (1.3) due to (3.4). □

By Proposition 3.1, the solution set for the problem $\mathbf{IP}_\kappa$ is equal to $(\varphi + \theta H^\infty_\kappa) \cap \mathcal{S}_\kappa$. Thus if the Pick matrix P of the problem meets conditions (1.11), then the set $(\varphi + \theta H^\infty_\kappa) \cap \mathcal{S}_\kappa$ is not empty (and is parametrized as in Theorem 1.1) and therefore, a larger set

$$\Omega_\kappa(\varphi, \theta) := (\varphi + \theta H^\infty_\kappa) \cap \mathcal{B}L^\infty = (\varphi + \theta H^\infty_\kappa) \cap \left(\bigcup_{\alpha=0}^{\kappa} \mathcal{S}_\alpha \right) \tag{3.5}$$

is not empty. The second equality in (3.5) is easily verified: since $\varphi, \theta \in H^\infty$, it follows that $\varphi + \theta H^\infty_\kappa \subset H^\infty_\kappa$ and on the other hand, $H^\infty_\kappa \cap \mathcal{B}L^\infty = \cup_{\alpha=0}^{\kappa} \mathcal{S}_\alpha$. Clearly, the elements of $\Omega_\kappa(\varphi, \theta)$ are solutions of certain L^∞-norm constraint interpolation problem; the next theorem characterizes $\Omega_\kappa(\varphi, \theta)$ in terms of the kernel $\mathbf{K}_f$ defined in (2.12) as well as in terms of the linear fractional transformation $\mathbf{T}_\Theta$ defined in (1.15).

Theorem 3.2. *Let the Pick matrix P defined in (1.7) be invertible and let $\kappa \geq \operatorname{sq}_-(P)$. Let Θ be given by (1.12), let φ be an H^∞-function satisfying conditions (3.1), let θ be given by (3.2) and let f be a function meromorphic on $\mathbb{D}$. The following are equivalent:*

1. *f belongs $(\varphi + \theta H^\infty_\kappa) \cap \mathcal{B}L^\infty$.*
2. *$\operatorname{sq}_-(\mathbf{K}_f(z, \zeta)) \leq \kappa$ where the kernel $\mathbf{K}_f$ is defined in (2.12).*
3. *$f = \mathbf{T}_\Theta[\mathcal{E}]$ for some $\mathcal{E} \in H^\infty_{\kappa - \operatorname{sq}_-(P)} \cap \mathcal{B}L^\infty$.*

Proof. **(1)** $\Rightarrow$ **(2)**. Let us assume that f is of the form (3.3) and belongs to $\mathcal{S}_{\widetilde{\kappa}}$ for some $\widetilde{\kappa} \le \kappa$. Let us assume that the function h in representation (3.3) has poles of multiplicities m_i at z_i for $i = 1, \dots, k$ (the case where $m_i = 0$ is not excluded). Let $\mathcal{I}_{\pm}$, $\mathcal{I}_0$ and $\gamma_{\mathbf{m}}$ be defined as in (2.27), (2.28). Since $h \in H^{\infty}_{\kappa}$, it may have at most $\kappa - |\mathbf{m}|$ poles outside the set $\{z_1, \dots, z_k\}$. After cancellation of the poles of h with the zeros of θ, we obtain the following representation for f:

$$f(z) = \varphi(z) + \widetilde{\theta}(z)\widetilde{h}(z), \tag{3.6}$$

where

$$\widetilde{\theta}(z) = \prod_{i \in \mathcal{I}^+} \left(\frac{z - z_i}{1 - z\bar{z}_i}\right)^{n_i - m_i} \quad \text{and} \quad \widetilde{h}(z) = h(z) \cdot \prod_{i \in \mathcal{I}^-} \left(\frac{z - z_i}{1 - z\bar{z}_i}\right)^{m_i - n_i}. \tag{3.7}$$

It is clear that $\widetilde{h}$ has poles of multiplicities $m_i - n_i$ at z_i for every $i \in \mathcal{I}_-$ and at most $\kappa - |\mathbf{m}|$ poles in $\mathbb{D} \setminus \{z_1, \dots, z_k\}$. It follows from (3.6) that f has the same poles and since f belongs to $\mathcal{S}_{\widetilde{\kappa}}$, we get

$$\widetilde{\kappa} \le \sum_{i \in \mathcal{I}_-} (m_i - n_i) + \kappa - |\mathbf{m}| = \kappa - \gamma_{\mathbf{m}}. \tag{3.8}$$

On the other hand, $\widetilde{h}$ is analytic at every z_i for $i \in \mathcal{I}_+$ and therefore, f of the form (3.6) satisfies interpolation conditions

$$f^{(j)}(z_i) = \varphi^{(j)}(z_i) = j!\, f_{i,j} \qquad (i \in \mathcal{I}_+;\ j = 0, \dots, n_i - m_i - 1). \tag{3.9}$$

Therefore, f is a solution of the problem $\mathbf{IP}_{\widetilde{\kappa}}$ with interpolation conditions (3.9). The Pick matrix $\widetilde{P}$ of this interpolation problem is a principal submatrix of the Pick matrix P of the original $\mathbf{IP}_{\kappa}$, and for a suitable permutation matrix U, we have

$$UPU = \begin{bmatrix} P_1 & P_2^* \\ P_2 & \widetilde{P} \end{bmatrix}. \tag{3.10}$$

Furthermore, associating the matrices $\widetilde{T}$, $\widetilde{E}$ and $\widetilde{C}$ with the problem $\mathbf{IP}_{\widetilde{\kappa}}$ via formulas (1.4) and (1.7), it is easy to check the block decompositions

$$UTU = \begin{bmatrix} T_1 & 0 \\ T_2 & \widetilde{T} \end{bmatrix}, \quad UE = \begin{bmatrix} E_1 \\ \widetilde{E} \end{bmatrix}, \quad UE = \begin{bmatrix} C_1 \\ \widetilde{C} \end{bmatrix} \tag{3.11}$$

conformal with (3.10). By Theorem 2.2 applied to the problem $\mathbf{IP}_{\widetilde{\kappa}}$, the kernel

$$\widetilde{\mathbf{K}}_f(z,\zeta) = \begin{bmatrix} \widetilde{P} & (I - \bar{\zeta}\widetilde{T})^{-1}\left(\widetilde{E} - \widetilde{C}f(\zeta)^*\right) \\ \left(\widetilde{E}^* - f(z)\widetilde{C}^*\right)(I - z\widetilde{T}^*)^{-1} & K_f(z,\zeta) \end{bmatrix} \tag{3.12}$$

has $\widetilde{\kappa}$ negative squares on $\rho(f)$. Due to (3.10)–(3.12), the kernel $\mathbf{K}_f$ defined in (2.12) can be represented as

$$\begin{bmatrix} U & 0 \\ 0 & 1 \end{bmatrix} \mathbf{K}_f(z,\zeta) \begin{bmatrix} U^* & 0 \\ 0 & 1 \end{bmatrix} = \begin{bmatrix} P_1 & B(\zeta)^* \\ B(z) & \widetilde{\mathbf{K}}_f(z,\zeta) \end{bmatrix}, \tag{3.13}$$

where $B(z)$ is analytic on $\rho(f)$ (the explicit formula for B is not that important for now). Then the number of negative squares of the kernel on the right-hand side of (3.13) can be estimated as follows

$$\mathrm{sq}_-\left(\begin{bmatrix} P_1 & B(\zeta)^* \\ B(z) & \widetilde{\mathbf{K}}_f(z,\zeta) \end{bmatrix}\right) \leq \mathrm{sq}_-(\widetilde{\mathbf{K}}_f) + d = \widetilde{\kappa} + d, \tag{3.14}$$

where d is the size of the square matrix P_1 (see [7, Proposition 4.1] for the proof). The number d is equal to the difference between the sizes of the Pick matrices P and $\widetilde{P}$ or, which is the same, to the difference between the numbers of interpolation conditions in (1.3) and (3.9). Thus,

$$d = |\mathbf{n}| - \sum_{i \in \mathcal{I}_+} (n_i - m_i) = \sum_{i \in \mathcal{I}_- \cup \mathcal{I}_0} n_i + \sum_{i \in \mathcal{I}_+} m_i = \gamma_{\mathbf{m}} \tag{3.15}$$

Now we combine (3.8), (3.14) and (3.15) to conclude from (3.13) that

$$\mathrm{sq}_-(\mathbf{K}_f) = \mathrm{sq}_-\left(\begin{bmatrix} P_1 & B(\zeta)^* \\ B(z) & \widetilde{\mathbf{K}}_f(z,\zeta) \end{bmatrix}\right) \leq \widetilde{\kappa} + d = \widetilde{\kappa} + \gamma_{\mathbf{m}} \leq \kappa - \gamma_{\mathbf{m}} + \gamma_{\mathbf{m}} = \kappa,$$

which completes the proof of the implication $(\mathbf{1}) \Rightarrow (\mathbf{2})$.

$(\mathbf{2}) \Rightarrow (\mathbf{3})$. Let us assume that $\mathrm{sq}_-(\mathbf{K}_f) = \widetilde{\kappa} \leq \kappa$. Then by the arguments used in the proof of Theorem 2.2 to derive statement (2) from statement (1) we conclude that f is of the form $f = \mathbf{T}_\Theta[\mathcal{E}]$ for some $\mathcal{E} \in \mathcal{S}_{\widetilde{\kappa} - \mathrm{sq}_-(P)}$. Since $\mathcal{S}_{\widetilde{\kappa} - \mathrm{sq}_-(P)} \subset H^\infty_{\widetilde{\kappa} - \mathrm{sq}_-(P)} \cap \mathcal{B}L^\infty \subset H^\infty_{\kappa - \mathrm{sq}_-(P)} \cap \mathcal{B}L^\infty$, the proof is completed.

$(\mathbf{3}) \Rightarrow (\mathbf{1})$. Let f be of the form $f = \mathbf{T}_\Theta[\mathcal{E}]$ for some $\mathcal{E} \in \mathcal{S}_{\widetilde{\kappa}}$ where $\widetilde{\kappa} \leq \kappa - \mathrm{sq}_-(P)$. Then we equivalently can take f in the form (1.13) with $S \in \mathcal{S}$ and $B \in \mathcal{B}_{\widetilde{\kappa}}$ having no common zeros on $\mathbb{D}$. Let $m_1, \dots, m_k$ be the integers uniquely determined from conditions (2.29) and (2.30). Then we conclude by Theorem 2.4 that f satisfies interpolation conditions (2.31) (or (3.9) which is the same) and belongs to the class $\mathcal{S}_{\kappa_1}$ where

$$\kappa_1 = \widetilde{\kappa} + \mathrm{sq}_-(P) - \gamma_{\mathbf{m}}. \tag{3.16}$$

Thus, f solves the problem $\mathbf{IP}_{\kappa_1}$ with interpolation conditions (3.9) and therefore by virtue of Proposition 3.1, it admits a representation (3.6) with $\widetilde{\theta}(z)$ defined as in (3.7) and some $\widetilde{h} \in H^\infty_{\kappa_1}$. From (3.2) and (3.7) we observe that the ratio $\theta_1 := \theta/\widetilde{\theta}$ is a finite Blaschke product of degree

$$\deg \theta_1 = \sum_{i \in \mathcal{I}_+} m_i + \sum_{i \in \mathcal{I}_- \cup \mathcal{I}_0} n_i = \gamma_{\mathbf{m}}$$

and therefore the function $h := \widetilde{h}/\theta_1$ belongs to $H^\infty_{\kappa_2}$ where $\kappa_2 = \kappa_1 + \gamma_{\mathbf{m}}$ and due to (3.16), we have $\kappa_2 = \widetilde{\kappa} + \mathrm{sq}_-(P) \leq \kappa$, so that $h \in H^\infty_\kappa$. Now we get from (3.6)

$$f = \varphi + \widetilde{\theta} \cdot \widetilde{h} = \varphi + \theta \cdot \frac{\widetilde{h}}{\theta_1} = \varphi + \theta h.$$

Since $f \in \mathcal{S}_{\kappa'}$ and $h \in H^\infty_\kappa$, it follows that

$$f \in (\varphi + \theta H^\infty_\kappa) \cap \mathcal{S}_{\kappa'} \subset (\varphi + \theta H^\infty_\kappa) \cap \mathcal{B}L^\infty$$

which completes the proof of the theorem. □

As corollary, we obtain the following "if and only if" version of Theorem 2.2.

Theorem 3.3. *Let P satisfy conditions* (1.11), *let Θ be given by* (1.12) *and let $f \in \mathcal{S}_\kappa$. The following are equivalent:*

1. *f is a solution of the problem $\mathbf{IP}_\kappa$.*
2. $\operatorname{sq}_-(\mathbf{K}_f(z,\zeta)) = \kappa$ *where the kernel $\mathbf{K}_f$ is defined in* (2.12).
3. *$f = \mathbf{T}_\Theta[\mathcal{E}]$ for some $\mathcal{E} \in \mathcal{S}_{\kappa - \operatorname{sq}_-(P)}$.*

Proof. Implications (**1**) $\Rightarrow$ (**2**) $\Rightarrow$ (**3**) are proved in Theorem 2.2. If $f = \mathbf{T}_\Theta[\mathcal{E}]$ for some $\mathcal{E} \in \mathcal{S}_{\kappa - \operatorname{sq}_-(P)}$, then $f \in (\varphi + \theta H^\infty_\kappa) \cap \mathcal{B}L^\infty$, by Theorem 3.2, where ϕ and θ are the functions associated with the problem $\mathbf{IP}_\kappa$. By the assumption of the theorem, $f \in \mathcal{S}_\kappa$ and therefore

$$f \in (\varphi + \theta H^\infty_\kappa) \cap \mathcal{B}L^\infty \cap \mathcal{S}_\kappa = (\varphi + \theta H^\infty_\kappa) \cap \mathcal{S}_\kappa$$

and the latter set coincides with the solution set for the problem $\mathbf{IP}_\kappa$, by Proposition 3.1. This completes the proof of the implication (**3**) $\Rightarrow$ (**1**) and therefore, of the theorem. □

References

[1] V.M. Adamjan, D.Z. Arov and M.G. Kreĭn, *Analytic properties of the Schmidt pairs of a Hankel operator and the generalized Schur-Takagi problem*, Mat. Sb. **86** (1971), 34–75.

[2] J.A. Ball, *A non-Euclidean Lax-Beurling theorem with applications to matricial Nevanlinna-Pick interpolation*, Toeplitz centennial (Tel Aviv, 1981), pp. 67–84, Operator Theory: Adv. Appl., **OT 4**, Birkhäuser, 1982.

[3] J.A. Ball, I. Gohberg, and L. Rodman, *Interpolation of rational matrix functions*, OT45, Birkhäuser Verlag, 1990.

[4] J.A. Ball, J.W. Helton, *A Beurling-Lax theorem for the Lie group* U(m,n) *which contains most classical interpolation theory*, J. Operator Theory **9** (1983), no. 1, 107–142.

[5] V. Bolotnikov, *On Carathéodory–Fejér problem for generalized Schur functions*, Integral Equations Operator Theory, **50** (2004), no. 1, 9–41.

[6] V. Bolotnikov, *Nevanlinna–Pick meromorphic interpolation: The degenerate case and minimal norm solutions*, J. Math. Anal. Appl., **353** (2009), no. 2, 642–651.

[7] V. Bolotnikov and A. Kheifets, *Boundary Nevanlinna–Pick interpolation problems for generalized Schur functions*, Operator Theory: Advances and Applications **OT 165** (2006), 67–119.

[8] L.B. Golinskii. *A generalization of the matrix Nevanlinna–Pick problem*, Izv. Akad. Nauk Armyan. SSR Ser. Mat. 18 (1983), 187–205.

[9] M.G. Kreĭn and H. Langer, *Über die verallgemeinerten Resolventen und die charakteristische Funktion eines isometrischen Operators im Raume* Π_κ, Colloq. Math. Soc. János Bolyai **5** (1972), 353–399.

[10] M.G. Kreĭn and H. Langer, *Über einige Fortsetzungsprobleme, die eng mit der Theorie hermitescher Operatoren im Raume* Π_κ *zusammenhängen. I. Einige Funktionenklassen und ihre Darstellungen* Math. Nachr. **77** (1977), 187–236.

[11] H. Woracek, *Nevanlinna-Pick interpolation: the degenerated case*, Linear Algebra Appl. **252** (1997), 141–158.

Vladimir Bolotnikov
Department of Mathematics
The College of William and Mary
Williamsburg, VA 23187-8795, USA
e-mail: `vladi@math.wm.edu`

Operator Theory:
Advances and Applications, Vol. 195, 103–119

The Algebra of τ-measurable Operators

Jurie Conradie and Charlotte Crowther

Abstract. Properties of the measure topology on the algebra of τ-measurable operators affiliated to a semi-finite von Neumann algebra with a distinguished faithful semi-finite normal trace τ, in particular its local convexity and order continuity, are investigated.

Mathematics Subject Classification (2000). Primary 46E30; Secondary 46L52.

Keywords. von Neumann algebras, τ-measurable operators, local convexity, direct sum

The setting for classical integration theory is the algebra of all measurable functions on a measure space, with functions equal almost everywhere identified. In non-commutative integration theory the corresponding setting is an algebra of (possibly unbounded) measurable operators affiliated to a von Neumann algebra. A number of different notions of measurability for operators affiliated to a von Neumann algebra have been used (see, for example, [14]). For the case where the von Neumann algebra is semi-finite and equipped with a distinguished semi-finite faithful normal trace τ, Nelson ([12], see also [17]) has introduced the more restricted notion of τ-measurability. This yields an algebra of τ-measurable operators which is still large enough to accommodate non-commutative analogues of all the classical Banach function spaces ([6]). In the commutative case this algebra corresponds to the space of all measurable functions that are bounded except possibly on a set of finite measure. This is the largest space on which the topology of convergence in measure is a vector topology, and for every function in it we can define a (finite-valued) decreasing rearrangement. It is possible to define a topology on this algebra (the measure topology) which turns it into a topological $*$-algebra; in the commutative case the topology is the familiar topology of convergence in measure. The fact that there exists an analogue of the decreasing rearrangement for all τ-measurable operators makes it possible to use classical function theory in the study of these operators and has proved extremely useful in proving non-commutative versions of many classical results.

In this paper we look at some properties of the algebra of τ-measurable operators. We show that the structure of the algebra and its topology is heavily dependent on the nature of the projection lattice of the von Neumann algebra. In particular, the measure topology can be, but need not be locally convex. We also characterise the largest ideal in the algebra on which the topology is order continuous.

1. The algebra $\widetilde{\mathcal{M}}$

In what follows $\mathcal{M}$ will denote a semi-finite von Neumann algebra, a subalgebra of the algebra $B(H)$ of all bounded linear operators on a Hilbert space H. The lattice of projections in $\mathcal{M}$ is denoted by $\mathcal{M}^p$, and the identity of $\mathcal{M}$ by 1. A closed densely defined linear operator x in H with domain $D(x)$ is affiliated with $\mathcal{M}$ (written $x\,\eta\,\mathcal{M}$) if $u^*xu = x$ (with equality of domains) for all unitary operators u in the commutant $\mathcal{M}'$ of $\mathcal{M}$. Let $x = v|x|$ be the polar decomposition of x, with v a partial isometry and $|x| = (x^*x)^{\frac{1}{2}}$. If $\{e_t(|x|) : t \geq 0\}$ is the spectral family of $|x|$, then

$$|x| = \int_0^\infty t\, de_t(|x|)$$

and $x\,\eta\,\mathcal{M}$ if and only if $v \in \mathcal{M}$ and $e_t(|x|) \in \mathcal{M}$ for every $t > 0$. If B is a Borel subset of $[0,\infty)$, we denote by $\chi_B(|x|)$ the projection in $\mathcal{M}$ corresponding to the characteristic function χ_B via the Borel functional calculus for $|x|$. In particular, we have $e_t(|x|) = \chi_{[0,t]}(|x|)$ and $1 - e_t(|x|) = \chi_{(t,\infty)}(|x|)$.

If $x \in \widetilde{\mathcal{M}}$, we denote by $\mathbf{n}(x)$ the projection onto the kernel of x, and write $\mathbf{r}(x) = 1 - \mathbf{n}(x)$ for the right support of x.

We assume further that $\mathcal{M}$ is equipped with a distinguished faithful semi-finite normal trace τ. A closed densely defined operator x affiliated with $\mathcal{M}$ is called τ-measurable if and only if for every $\epsilon > 0$ there is a $p \in \mathcal{M}^p$ such that $p(H) \subset D(x)$ and $\tau(1-p) < \epsilon$. The set of all τ-measurable operators will be denoted by $\widetilde{\mathcal{M}}$. If $x\,\eta\,\mathcal{M}$, then it can be shown that $x \in \widetilde{\mathcal{M}}$ if and only if there is a $t \geq 0$ such that $\tau(\chi_{(t,\infty)}(|x|)) = \tau(1 - e_t(|x|)) < \infty$.

For $x \in \widetilde{\mathcal{M}}$, the decreasing rearrangement $\mu_.(x)$ of x is the real-valued decreasing right-continuous function defined on $[0,\infty)$ by

$$\mu_t(x) = \inf\{s \geq 0 : \tau(\chi_{(s,\infty)}) \leq t\}, \qquad t \geq 0.$$

The set $\widetilde{\mathcal{M}}$ becomes a $*$-algebra when sums and products are defined as the closures of respectively the algebraic sum and algebraic product. For a self-adjoint $x \in \widetilde{\mathcal{M}}$ we write $x \geq 0$ if $(x\xi,\xi) \geq 0$ for all ξ in the domain of x (where $(\cdot,\cdot)$ denotes the inner product on H), and we put $\widetilde{\mathcal{M}}^+ = \{x \in \widetilde{\mathcal{M}} : x \geq 0\}$. The cone $\widetilde{\mathcal{M}}^+$ defines a partial order on the self-adjoint elements of $\widetilde{\mathcal{M}}$. Further information on the order structure of $\widetilde{\mathcal{M}}$ can be found in [7], Section 1.

The trace τ extends naturally to $\widetilde{\mathcal{M}}^+$ so that the faithfulness, semi-finiteness and normality of τ are preserved (cf. [8], Theorem 3.5 or [7] Section 3).

The set of all $x \in \widetilde{\mathcal{M}}$ such that $\mu_t(x) \to 0$ as $t \to \infty$ is denoted by $\widetilde{\mathcal{M}}_0$. Equivalently, $\widetilde{\mathcal{M}}_0$ consists of all $x \in \widetilde{\mathcal{M}}$ such that $\tau(\chi_{(t,\infty)}(|x|)) < \infty$ for all $t > 0$. It is a two-sided self-adjoint ideal in $\mathcal{M}$. Clearly $p \in \mathcal{M}^p \cap \widetilde{\mathcal{M}}_0$ if and only if $\tau(p) < \infty$.

If $\mathcal{M}$ is a commutative von Neumann algebra, we may identify $\mathcal{M}$ with $L_\infty(X, \Sigma, \mu)$, where (X, Σ, μ) is a localizable measure space. In this case $\tau(f) = \int_X f \, d\mu$ defines a faithful normal trace on $L_\infty(X, \Sigma, \mu)$. Then $\widetilde{\mathcal{M}} = \widetilde{L}_\infty(X, \Sigma, \mu)$ is the space of all (equivalence classes of) complex-valued measurable functions on X which are bounded except on a set of finite measure.

It follows at once from the definition of $\widetilde{\mathcal{M}}$ that if $x \in \widetilde{\mathcal{M}}$, then there is a $p \in \mathcal{M}^p$ such that $x = xp + x(1-p)$, with $||xp|| < \infty$ and $\tau(1-p) < \infty$. Since xp is bounded and $xp \, \eta \, \mathcal{M}$, it follows that $xp \in \mathcal{M}$. From $x(1-p)(1-p) = x(1-p)$ it follows that $\mathbf{r}(x(1-p)) \leq 1 - p$. Hence we can write $x = x_1 + x_2$, with $x_1 \in \mathcal{M}, x_2 \in \widetilde{\mathcal{M}}$ and $\tau(\mathbf{r}(x_2)) < \infty$. Let us put

$$\widetilde{\mathcal{M}}_f = \{x \in \widetilde{\mathcal{M}} : \tau(\mathbf{r}(x)) < \infty\}.$$

Then $\widetilde{\mathcal{M}}_f$ is a linear subspace of $\widetilde{\mathcal{M}}$ and the above discussion shows that $\widetilde{\mathcal{M}}$ can be written as the sum (but not a direct sum) of $\mathcal{M}$ and $\widetilde{\mathcal{M}}_f$. It is clear that we always have $\widetilde{\mathcal{M}}_f \subseteq \widetilde{\mathcal{M}}_0$. This gives:

Proposition 1.1. $\widetilde{\mathcal{M}} = \mathcal{M} + \widetilde{\mathcal{M}}_f = \mathcal{M} + \widetilde{\mathcal{M}}_0$.

In the case $\tau(1) < \infty$, $\mathcal{M} \subseteq \widetilde{\mathcal{M}}_f$ and hence $\widetilde{\mathcal{M}} = \widetilde{\mathcal{M}}_f$; in fact, in this case $x \in \widetilde{\mathcal{M}}$ if and only if $x \, \eta \, \mathcal{M}$.

The structure of $\widetilde{\mathcal{M}}$ depends heavily on the nature of the lattice of projections of $\mathcal{M}$. Recall that a projection $p \in \mathcal{M}^p$ is atomic (or minimal) if it has no nonzero subprojections, and $\mathcal{M}$ is an atomic von Neumann algebra if every non-zero $p \in \mathcal{M}^p$ majorizes a non-zero atomic projection. In what follows we look more closely at $\widetilde{\mathcal{M}}$ in the case where $\mathcal{M}$ is an atomic algebra.

If $\inf\{\tau(p) : 0 \neq p \in \mathcal{M}^p\} > 0$, then it is easily checked that $\mathcal{M}$ is atomic.

Proposition 1.2 (**[15], Examples 2.2(3)**). *The following statements are equivalent:*

1. $\widetilde{\mathcal{M}} = \mathcal{M}$.
2. $\inf\{\tau(p) : 0 \neq p \in \mathcal{M}^p\} > 0$.

Now let $\mathcal{M}$ be atomic and suppose $\inf\{\tau(p) : 0 \neq p \in \mathcal{M}^p\} = 0$. We show that under certain conditions we can improve on Proposition 1.1 by writing $\widetilde{\mathcal{M}}$ as a *direct* sum. To do this we need to recall some basic results on reduced algebras. If $q \in \mathcal{M}^p$ and $x \in \mathcal{M}$, we write x_q for the restriction of qx to qH. The set $\mathcal{M}_q = \{x_q : x \in \mathcal{M}\}$ is called the reduction of $\mathcal{M}$ by q and is known to be a von Neumann algebra on qH. A faithful semi-finite normal trace τ_q can be defined on $\mathcal{M}_q$ by putting $\tau_q(x_q) = \tau(qxq)$. It is also possible to define reductions of $\widetilde{\mathcal{M}}$. If $x \in \widetilde{\mathcal{M}}$ and $q \in \mathcal{M}^p$, we denote by x_q the restriction of the closure of qx to qH and put $\widetilde{\mathcal{M}}_q = \{x_q : x \in \widetilde{\mathcal{M}}\}$. We could also start with the von Neumann algebra $\mathcal{M}_q$, equipped with the trace τ_q, and form the algebra $\widetilde{\mathcal{M}_q}$ of τ_q-measurable operators affiliated with the von Neumann algebra $\mathcal{M}_q$.

Proposition 1.3 ([15], Theorem 2.3). *If* $q \in \mathcal{M}^p$, $\widetilde{\mathcal{M}}_q = \widetilde{\mathcal{M}_q}$.

Proposition 1.4. *Let* $\mathcal{M}$ *be atomic, with* $\inf\{\tau(p) : 0 \neq p \in \mathcal{M}^p\} = 0$. *Suppose further that there is a* $K > 0$ *such that*

$$\sum_{\substack{\tau(p)<K \\ p \text{ atomic}}} \tau(p) < \infty.$$

Then there is a central projection $q \in \mathcal{M}^p$ *such that* $\tau(q) < \infty$ *and*

$$\widetilde{\mathcal{M}} = q\widetilde{\mathcal{M}}q \oplus (1-q)\widetilde{\mathcal{M}}(1-q) = \widetilde{\mathcal{M}_q} \oplus \mathcal{M}_{1-q}.$$

Proof. Put $\mathcal{P} = \{p \in \mathcal{M}^p : p \text{ atomic}, \tau(p) < K\}$ and put $q = \vee\{p : p \in \mathcal{P}\}$. We show that q is central. For this it will suffice to show that $u^*qu = q$ for every unitary u in $\mathcal{M}$. Suppose $p \in \mathcal{P}$. If u is unitary in $\mathcal{M}$, then $u^*pu \in \mathcal{P}$ since τ is unitarily invariant. Similarly, if u is unitary in $\mathcal{M}$ and $u^*pu \in \mathcal{P}$, then $p \in \mathcal{P}$. It follows that for every unitary u in $\mathcal{M}$ we have

$$q = \bigvee_{p\in\mathcal{P}} p = \bigvee_{p\in\mathcal{P}} u^*pu.$$

Clearly $q \geq p$ for all $p \in \mathcal{P}$ and hence we have that for all unitary $u \in \mathcal{M}$, $u^*qu \geq u^*pu$ for all $p \in \mathcal{P}$. Now suppose $e \in \mathcal{M}^p$ also satisfies $e \geq u^*pu$ for all $p \in \mathcal{P}$. Then $ueu^* \geq uu^*puu^* = p$ for all $p \in \mathcal{P}$. Hence $ueu^* \geq q$ and thus $e \geq u^*qu$, and so $u^*qu = \bigvee_{p\in\mathcal{P}} u^*pu$. Therefore

$$q = \bigvee_{p\in\mathcal{P}} p = \bigvee_{p\in\mathcal{P}} u^*pu = u^*qu.$$

This holds for all unitary $u \in \mathcal{M}$ and hence q is central. It follows from the assumption that $\tau(q) < \infty$.

We now have a direct sum decomposition

$$\widetilde{\mathcal{M}} = q\widetilde{\mathcal{M}}q \oplus (1-q)\widetilde{\mathcal{M}}(1-q)$$

since $q\widetilde{\mathcal{M}}(1-q)$ and $(1-q)\widetilde{\mathcal{M}}q$ are trivial since q is central. It follows from Proposition 1.3 that $q\widetilde{\mathcal{M}}q = \widetilde{\mathcal{M}_q}$ and $(1-q)\widetilde{\mathcal{M}}(1-q) = \widetilde{\mathcal{M}_{1-q}}$. The definition of q and the fact that $(\mathcal{M}_{1-q})^p = \{p \in \mathcal{M}^p : p \leq 1-q\}$ shows that $\tau_{1-q}(p) \geq K > 0$ for all non-zero $p \in (\mathcal{M}_{1-q})^p$, and therefore we can use Proposition 1.2 and Proposition 1.3 to deduce that $\widetilde{\mathcal{M}}_{1-q} = \mathcal{M}_{1-q}$. □

2. The measure topology

The algebra $\widetilde{\mathcal{M}}$ can be equipped in a natural way with a topology such that it becomes a topological $*$-algebra. We recall the basic definitions and results and refer the reader to [8], [12], [17] for further detail.

For $\epsilon > 0, \delta > 0$ let

$$N(\epsilon,\delta) = \{x \in \widetilde{\mathcal{M}} : \exists p \in \mathcal{M}^p, p(H) \subset D(x), ||xp|| < \epsilon, \tau(1-p) < \delta\}.$$

Then the family $\{N(\epsilon,\delta) : \epsilon > 0, \delta > 0\}$ forms a base for the neighbourhoods of 0 for a Hausdorff topology τ_m on $\widetilde{\mathcal{M}}$. We write $N(\epsilon)$ for $N(\epsilon,\epsilon)$ and note that $\{N(\epsilon) : \epsilon > 0\}$ also forms a base for the τ_m-neighbourhoods of 0. We shall refer to τ_m as the measure topology, for reasons that will become apparent soon. Equipped with this topology, $\widetilde{\mathcal{M}}$ becomes a complete metrizable topological $*$-algebra in which $\mathcal{M}$ is dense. The measure topology is induced by the F-norm (or pseudonorm) ρ on $\widetilde{\mathcal{M}}$ defined by

$$\rho(x) = \inf\{\epsilon > 0 : x \in N(\epsilon)\}.$$

It follows from the fact that $x \in N(\epsilon,\delta)$ if and only if $\mu_\delta(x) < \epsilon$ that $N(\epsilon,\delta) = \{x \in \widetilde{\mathcal{M}} : \mu_\delta(x) < \epsilon\}$ and hence that the F-norm ρ can also be written as $\rho(x) = \inf\{t \geq 0 : \mu_t(x) \leq t\}$. From this we get that

$$d(x,y) = \inf\{t \geq 0 : \mu_t(x-y) \leq t\} \qquad (x, y \in \widetilde{\mathcal{M}})$$

defines a metric on $\widetilde{\mathcal{M}}$ that induces the measure topology.

If $\mathcal{M}$ is commutative and we identify $\mathcal{M}$ with an L_∞ space and take the trace to be the integral, the measure topology is the familiar topology of convergence in measure, and $\widetilde{L}_\infty$ is the largest vector space on which it is a linear topology.

It is known that $\widetilde{L}_\infty$ can be regarded as an Orlicz space. We show that this result extends easily to the non-commutative setting. Let the function $\varphi : [0,\infty) \to [0,\infty)$ be defined by $\varphi(t) = \chi_{(1,\infty)}(t)$. For $x \,\eta\, \mathcal{M}$, $\varphi(|x|) = \chi_{(1,\infty)}(|x|)$. Hence

$$m_\varphi(x) \stackrel{def}{=} \tau(\varphi(|x|)) = \tau(\chi_{(1,\infty)}(|x|)).$$

Now $m_\varphi(\lambda x) = \tau(\chi_{(1,\infty)}(|\lambda x|)) = \tau(\chi_{(\lambda^{-1},\infty)}(|x|)) < \infty$ for some $\lambda > 0$ if and only if $\tau(\chi_{(t,\infty)}(|x|) < \infty$ for some $t > 0$. But this happens if and only if $x \in \widetilde{\mathcal{M}}$, so

$$\widetilde{\mathcal{M}} = \{x \,\eta\, \mathcal{M} : m_\varphi(\lambda x) < \infty \text{ for some } \lambda > 0\},$$

and hence $\widetilde{\mathcal{M}}$ may be regarded as a (non-commutative) Orlicz space. The corresponding Orlicz F-norm is given by

$$\begin{aligned} ||x||_\varphi &= \inf\{\lambda > 0 : m_\varphi(\lambda^{-1}x) \leq \lambda\} \\ &= \inf\{\lambda > 0 : \tau(\chi_{(\lambda,\infty)}(|x|) \leq \lambda\} \\ &= \rho(x). \end{aligned}$$

In some cases the measure topology can be identified with a familiar topology.

Theorem 2.1 ([15]). *The following are equivalent:*

1. $\inf\{\tau(p) : p \in \mathcal{M}^p,\ \tau(p) \neq 0\} > 0$.
2. $\widetilde{\mathcal{M}} = \mathcal{M}$.
3. *The measure topology is equal to the norm topology on* $\mathcal{M}$.

We have seen earlier that $\widetilde{\mathcal{M}} = \widetilde{\mathcal{M}}_f + \mathcal{M}$. We now show that $\widetilde{\mathcal{M}}$, equipped with the measure topology, may be regarded as the sum of a pseudonormed and a normed space. To this end we define, for $x \in \widetilde{\mathcal{M}}$,

$$\theta(x) = \inf\{\rho(y) + \|z\| : x = y + z, y \in \widetilde{\mathcal{M}}_f, z \in \mathcal{M}\}.$$

It is easily checked that θ is a pseudonorm on $\widetilde{\mathcal{M}}$.

Theorem 2.2. *Let θ denote the pseudonorm defined above and also the topology on $\widetilde{\mathcal{M}}$ induced by it. Then*

$$\left(\widetilde{\mathcal{M}}, \tau_m\right) = \left(\widetilde{\mathcal{M}}_f + \mathcal{M}, \theta\right)$$

Proof. Let $\epsilon > 0$ and put $M(\epsilon) = \{x \in \widetilde{\mathcal{M}} : \theta(x) < \epsilon\}$. Suppose $x \in N(\epsilon) = \{y \in \widetilde{\mathcal{M}} : \mu_\epsilon(y) < \epsilon\}$. Then $\tau(\chi_{(\epsilon,\infty)}(|x|)) \leq \epsilon$. Put $x_1 = x\chi_{(\epsilon,\infty)}(|x|)$ and $x_2 = x\chi_{[0,\epsilon]}(|x|)$. Then $x = x_1 + x_2$, $\tau(\mathbf{r}(x_1)) = \tau(\mathbf{r}(x\chi_{(\epsilon,\infty)}(|x|))) \leq \tau(\chi_{(\epsilon,\infty)}(|x|)) \leq \epsilon$ and $\|x_2\| = \|x\chi_{[0,\epsilon]}(|x|)\| \leq \epsilon$. Hence $x_1 \in \widetilde{\mathcal{M}}_f$ and $x_2 \in \mathcal{M}$. Since $\mu_\epsilon(x_1) \leq \mu_\epsilon(x) < \epsilon$, $\rho(x_1) \leq \epsilon$, and hence $\theta(x) \leq \rho(x_1) + \|x_2\| \leq 2\epsilon$. It follows that $N(\epsilon) \subseteq M(2\epsilon)$.

Conversely, suppose $\theta(x) < \epsilon$. Then we can find $x_1 \in \widetilde{\mathcal{M}}_f, x_2 \in \mathcal{M}$ such that $x = x_1 + x_2$ and $\rho(x_1) + \|x_2\| < \epsilon$. Then

$$\mu_{2\epsilon}(x) \leq \mu_\epsilon(x_1) + \mu_\epsilon(x_2) < \rho(x_1) + \|x_2\| < \epsilon,$$

since $\rho(x_1) < \epsilon$. Hence $x \in N(2\epsilon)$, and $M(\epsilon) \subseteq N(2\epsilon)$. □

If (X, Σ, μ) is a finite measure space, then $\widetilde{L_\infty}(X, \Sigma, \mu) = L_0(X, \Sigma, \mu)$, the space of all (equivalence classes of) almost everywhere finite measurable functions. It is known that the topology of convergence in measure on $L_0(X, \Sigma, \mu)$ is induced by the pseudonorm

$$\rho_0(f) = \int_X \frac{|f(x)|}{1 + |f(x)|}\, d\mu(x), \qquad f \in L_0(X, \Sigma, \mu).$$

We note that if (X, Σ, μ) is a not necessarily finite measure space, then we still have $\rho_0(f) < \infty$ for every $f \in L_0(X, \Sigma, \mu)$ such that $\mu(\{x \in X : f(x) \neq 0\}) < \infty$. This motivates the following definition.

Definition 2.3. For $x \in \widetilde{\mathcal{M}}_f$, put

$$\varrho(x) = \int_0^\infty \frac{\mu_t(x)}{1 + \mu_t(x)}\, dt.$$

To show that ϱ is a pseudonorm, we shall need the following "triangle inequality".

Lemma 2.4. *Let $x, y \in \widetilde{\mathcal{M}}$. Then there exist partial isometries $u, v \in \mathcal{M}$ such that*

$$|x+y| \leq u|x|u^* + v|y|v^*.$$

Proof. The case where x and y are bounded was considered in [1], Theorem 2.2. Using the arguments in the proofs of Lemma 4 and Corollary 5 of [11], the inequality can be extended to unbounded operators in $\widetilde{\mathcal{M}}$. □

Proposition 2.5. *ϱ is a pseudonorm on $\widetilde{\mathcal{M}}_f$.*

Proof. Clearly $\varrho(x) \geq 0$ for all $x \in \widetilde{\mathcal{M}}_f$. We show that $\varrho(x) < \infty$ for all $x \in \widetilde{\mathcal{M}}_f$. Define the function $f : [0,\infty) \to [0,\infty)$ by $f(t) = \frac{t}{1+t}$. Then f is continuous, concave, monotone increasing and $f(0) = 0$. Let $x \in \widetilde{\mathcal{M}}_f$. Since $\varrho(x) = \varrho(|x|)$ we may assume that $x \geq 0$. Then

$$\begin{aligned} \varrho(x) &= \int_0^\infty \frac{\mu_t(x)}{1+\mu_t(x)}\, dt \\ &= \int_0^\infty f\left(\mu_t(x)\right)\, dt \\ &= \int_0^\infty \mu_t\left(f(x)\right)\, dt \end{aligned}$$

where the last equality follows by [8] Theorem 2.5 (iv) since f is continuous and increasing. It also follows by [8] Theorem 2.5 (iv) that

$$\int_0^\infty \mu_t\left(f(x)\right)\, dt = \tau\left(f(x)\right).$$

Since $f(t) \leq 1$ for all $t \geq 0$ we have that $f(x) \leq e_{(0,\infty)}(x) = \mathbf{r}(x)$ and thus by the above that $\varrho(x) = \tau\left(f(x)\right) \leq \tau(\mathbf{r}(x)) < \infty$.

We verify the properties of a pseudonorm.

Let $x \in \widetilde{\mathcal{M}}_f$ and let λ be such that $|\lambda| \leq 1$. Then, since f is increasing,

$$\begin{aligned} \varrho(\lambda x) &= \int_0^\infty \frac{\mu_t(\lambda x)}{1+\mu_t(\lambda x)}\, dt \\ &= \int_0^\infty \frac{|\lambda|\mu_t(x)}{1+|\lambda|\mu_t(x)}\, dt \\ &\leq \int_0^\infty \frac{\mu_t(x)}{1+\mu_t(x)}\, dt \\ &= \varrho(x). \end{aligned}$$

Let $x \in \widetilde{\mathcal{M}}_f$. We have that

$$\begin{aligned} \lim_{n\to\infty} \varrho\left(\frac{1}{n}x\right) &= \lim_{n\to\infty} \int_0^\infty \frac{\mu_t\left(\frac{1}{n}x\right)}{1+\mu_t\left(\frac{1}{n}x\right)}\, dt \\ &= \lim_{n\to\infty} \int_0^\infty \frac{\mu_t(x)}{n+\mu_t(x)}\, dt \\ &= 0 \end{aligned}$$

by the Dominated Convergence Theorem. Hence $\lim_{n\to\infty} \varrho\left(\frac{1}{n}x\right) = 0$ for all $x \in \widetilde{\mathcal{M}}_f$.

Let $0 \le x,\ y \in \widetilde{\mathcal{M}}_f$.

Then

$$\begin{aligned} \varrho(x+y) &= \int_0^\infty \frac{\mu_t(x+y)}{1+\mu_t(x+y)}\, dt \\ &= \int_0^\infty f\left(\mu_t(x+y)\right)\, dt \\ &= \int_0^\infty \mu_t\left(f(x+y)\right)\, dt \\ &= \tau\left(f(x+y)\right) \end{aligned}$$

as before.

Since f is continuous, concave and $f(0) = 0$ we have by [3] Proposition 15 that

$$\tau\left(f(x+y)\right) \le \tau\left(f(x)\right) + \tau\left(f(y)\right).$$

Then it follows as before that

$$\begin{aligned} \tau\left(f(x)\right) + \tau\left(f(y)\right) &= \int_0^\infty \mu_t\left(f(x)\right)\, dt + \int_0^\infty \mu_t\left(f(y)\right)\, dt \\ &= \int_0^\infty f\left(\mu_t(x)\right)\, dt + \int_0^\infty f\left(\mu_t(y)\right)\, dt \\ &= \varrho(x) + \varrho(y). \end{aligned}$$

Hence

$$\varrho(x+y) \le \varrho(x) + \varrho(y)$$

for $0 < x, y \in \widetilde{\mathcal{M}}_f$.

Now let $x,\ y \in \widetilde{\mathcal{M}}_f$. By Lemma 2.4 there exist partial isometries $u,\ v \in \mathcal{M}$ such that

$$|x+y| \le u|x|u^* + v|y|v^*.$$

Hence

$$\mu_t(|x+y|) \le \mu_t\left(u|x|u^* + v|y|v^*\right)$$

and since $f(t) = \frac{t}{1+t}$ is increasing we have

$$\varrho(x+y) = \varrho(|x+y|) \le \varrho\left(u|x|u^* + v|y|v^*\right).$$

By the above we have

$$\varrho(x+y) \leq \varrho(u|x|u^*) + \varrho(v|y|v^*).$$

Now $\mu_t(u|x|u^*) \leq \mu_t(|x|) = \mu_t(x)$ and $\mu_t(v|y|v^*) \leq \mu_t(y)$ by [8] Theorem 2.5 and by using the monotonicity of f again we finally have that

$$\varrho(x+y) \leq \varrho(x) + \varrho(y)$$

for all $x, y \in \widetilde{\mathcal{M}}_f$. □

Definition 2.6. Let $x \in \widetilde{\mathcal{M}}$. Put

$$\vartheta(x) = \inf\{\varrho(y) + ||z|| : x = y + z, y \in \widetilde{\mathcal{M}}_f, z \in \mathcal{M}\}.$$

It is easy to check that ϑ is a pseudonorm on $\widetilde{\mathcal{M}}$.

Theorem 2.7. *The topology induced on $\widetilde{\mathcal{M}}$ by the pseudonorm ϑ coincides with the measure topology τ_m.*

Proof. Let $\gamma > 0$ be given, and choose $\epsilon = \frac{1}{2}\gamma$. If $x \in \widetilde{\mathcal{M}}(\epsilon)$ then $d_\epsilon(|x|) = \tau\left(e_{(\epsilon,\infty)}(|x|)\right) \leq \epsilon$. Put $y = xe_{(\epsilon,\infty)}(|x|)$ and $z = xe_{[0,\epsilon]}(|x|)$. Then $x = y + z$, $y \in \widetilde{\mathcal{M}}_f$ since $\tau\left(\mathbf{r}\left(xe_{(\epsilon,\infty)}(|x|)\right)\right) \leq \tau\left(e_{(\epsilon,\infty)}(|x|)\right) \leq \epsilon$, and $z \in \mathcal{M}$, with $||z|| \leq \epsilon$. Furthermore,

$$\begin{aligned} \varrho(y) &= \int_0^\infty \frac{\mu_t(y)}{1+\mu_t(y)}\, dt \\ &= \int_0^{d_\epsilon(|x|)} \frac{\mu_t(y)}{1+\mu_t(y)}\, dt \\ &\leq d_\epsilon(|x|) \leq \epsilon. \end{aligned}$$

Hence $\vartheta(x) \leq \varrho(y) + ||z|| \leq 2\epsilon = \gamma$.

Conversely, let $\epsilon, \delta > 0$ and choose $n \in \mathbb{N}$ such that $(n-1)\epsilon > 2$. Then choose $\gamma > 0$ such that $4n\gamma < \delta$ and $4\gamma < \epsilon$. Suppose $x \in \widetilde{\mathcal{M}}$ and $\vartheta(x) < \gamma$. Then we can find $y \in \widetilde{\mathcal{M}}_f$ and $z \in \mathcal{M}$ such that $\varrho(y) + ||z|| < \vartheta(x) + \gamma \leq 2\gamma$. Now

$$\begin{aligned} \varrho(y) < 2\gamma &\implies \int_0^{2n\gamma} \frac{\mu_t(y)}{1+\mu_t(y)}\, dt < 2\gamma \\ &\implies 2n\gamma \frac{\mu_{2n\gamma}(y)}{1+\mu_{2n\gamma}(y)} < 2\gamma \\ &\implies \mu_{2n\gamma}(y) < \frac{2\gamma}{2(n-1)\gamma} = \frac{1}{n-1} < \frac{1}{2}\epsilon \\ &\implies \mu_{\frac{1}{2}\delta}(y) \leq \mu_{2n\gamma}(x) < \frac{1}{2}\epsilon. \end{aligned}$$

Hence

$$\mu_\delta(x) = \mu_\delta(y+z) \leq \mu_{\frac{1}{2}\delta}(y) + \mu_{\frac{1}{2}\delta}(z) < \frac{1}{2}\epsilon + ||z|| \leq \epsilon.$$ □

We look more closely at the case where $\widetilde{\mathcal{M}}$ can be written as a direct sum. If $q \in \mathcal{M}^p$ is a central projection, then $\widetilde{\mathcal{M}}$ is the direct sum of $\widetilde{\mathcal{M}}_q$ and $\widetilde{\mathcal{M}}_{1-q}$. We may then regard $\widetilde{\mathcal{M}}_q$ as a subset of $\widetilde{\mathcal{M}}$ and as such it may be equipped with the topology induced by the measure topology τ_m of $\widetilde{\mathcal{M}}$. As noted before, a trace τ_q can be defined on $\widetilde{\mathcal{M}}_q$, and this gives rise to the measure topology $(\tau_q)_m$ on $\widetilde{\mathcal{M}}_q$.

Lemma 2.8. *If q is a central projection in $\mathcal{M}$, then the subspace topology induced by τ_m on $\widetilde{\mathcal{M}}_q$ equals the topology $(\tau_q)_m$.*

Proof. Let $(\mu_q)_t(x_q)$ denote the generalised singular functional of x_q determined by the trace τ_q on $\widetilde{\mathcal{M}}_q$. Then it was shown in [15] that $(\mu_q)_t(x_q) = \mu_t(qxq)$ for $t > 0$. The result now follows from the characterisation of basic neighbourhoods of 0 in the measure topology in terms of the generalised singular functional noted before. □

In the case where $\widetilde{\mathcal{M}}$ can be written as a direct sum (Proposition 1.4), the measure topology is a product topology.

Proposition 2.9. *Let $\mathcal{M}$ and q be as in Proposition 1.4. Then the measure topology on $\widetilde{\mathcal{M}}$ is the product of the measure topology on $\widetilde{\mathcal{M}}_q$ and the norm topology on $\mathcal{M}_{1-q}$.*

Proof. If (x_n) converges to 0 in $\widetilde{\mathcal{M}}$ with respect to τ_m, then it follows from continuity of multiplication and Lemma 2.8 that qx_nq and $(1-q)x_n(1-q)$ converge to 0 in the measure topologies of $\widetilde{\mathcal{M}}_q$ and $\widetilde{\mathcal{M}}_{1-q}$ respectively. But by Theorem 2.1 the measure topology on $\widetilde{\mathcal{M}}_{1-q} = \mathcal{M}_{1-q}$ equals the norm topology.

Conversely, suppose $y_n \to 0$ in the measure topology on $\widetilde{\mathcal{M}}_q$ and $z_n \to 0$ in the norm topology on $\mathcal{M}_{1-q}$, and let $x_n = y_n + z_n$ (where we regard $\widetilde{\mathcal{M}}_q$ and $\mathcal{M}_{1-q}$ as subspaces of $\widetilde{\mathcal{M}}$). It will suffice to show that $\mu_t(x_n) \to 0$ for all $t > 0$. But this follows easily from the inequality $\mu_t(x_n) \leq \mu_{\frac{t}{2}}(y_n) + \mu_{\frac{t}{2}}(z_n)$ and the fact that $\mu_{\frac{t}{2}}(z_n) \leq \lim_{s \to 0^+} \mu_s(z_n) = \|z_n\|$. □

3. Local convexity of the measure topology

In this section we investigate the local convexity of the measure topology on $\widetilde{\mathcal{M}}$ and indicate how this depends on the nature of the projection lattice of $\mathcal{M}$.

It follows at once from Theorem 2.1 that if $\inf\{\tau(p) : p \in \mathcal{M}^p,\ \tau(p) \neq 0\} > 0$, then the measure topology on $\widetilde{\mathcal{M}} = \mathcal{M}$ equals the norm topology and is thus locally convex.

Recall that a projection $p \in \mathcal{M}^p$ is called *nonatomic* if there is no atomic projection $q \in \mathcal{M}^p$ such that $0 < q \leq p$. The algebra $\mathcal{M}$ is called nonatomic if it contains no atomic projections, and atomic if it contains no nonatomic projections.

The convex hull of a set A will be denoted by $\mathrm{co}(A)$.

Theorem 3.1. *If $\mathcal{M}^p$ contains a nonatomic projection, then the measure topology on $\widetilde{\mathcal{M}}$ is not locally convex.*

Proof. Suppose $\mathcal{M}^p$ contains a nonatomic projection p; since $\mathcal{M}$ is semi-finite, there is no loss in generality in assuming that $0 < \tau(p) < \infty$. We show that if $0 < \delta < \frac{1}{2}\tau(p)$ then for all ϵ such that $0 < \epsilon < \delta$ we have $N(\epsilon) \subseteq \mathrm{co}(N(\epsilon)) \nsubseteq N(\delta)$. If such an ϵ is given, choose $n \in \mathbb{N}$ such that $\frac{2\delta}{n} < \epsilon$ and find $q \in \mathcal{M}^p$, $q \leq p$ such that $\tau(q) = 2\delta$. Now since q is nonatomic, we can subdivide it into n disjoint projections $q_1, \ldots, q_n$ such that $\tau(q_k) < \epsilon$ for all $k = 1, \ldots, n$, i.e., $\sum_{k=1}^{n} q_k = q$. For $k = 1, \ldots, n$ define $x_k = 2n\delta q_k$. Then $x_k \in N(\epsilon)$ for each $k = 1, \ldots, n$, since $d_\epsilon(x_k) = \tau(q_k) < \epsilon$. Define

$$x = \sum_{k=1}^{n} \frac{1}{n} x_k = 2\delta \bigvee_{k=1}^{n} q_k = 2\delta q.$$

Then $x \in \mathrm{co}(N(\epsilon))$ but $d_\delta(x) = \tau(q) = 2\delta > \delta$ so that $x \notin N(\delta)$. Hence $N(\epsilon) \subseteq \mathrm{co}(N(\epsilon)) \nsubseteq N(\delta)$. It follows that the measure topology is not locally convex. □

We are left with those atomic algebras $\mathcal{M}$ for which $\inf\{\tau(p) : p \in \mathcal{M}^p, \tau(p) \neq 0\} = 0$. We show that in the case where $\widetilde{\mathcal{M}}$ can be written as a direct sum (cf. Proposition 1.4), the measure topology is locally convex.

Theorem 3.2. *Let $\mathcal{M}$ be atomic, with $\inf\{\tau(p) : p \in \mathcal{M}^p,\ \tau(p) \neq 0\} = 0$ and suppose there exists a constant $K > 0$ such that*

$$\sum_{\substack{\tau(p)<K \\ p \text{ atomic}}} \tau(p) < \infty.$$

Then the measure topology on $\widetilde{\mathcal{M}}$ is locally convex.

Proof. We show that for all $\delta > 0$ there exists $0 < \epsilon < \delta$ such that $N(\epsilon) \subseteq \mathrm{co}(N(\epsilon)) \subseteq N(\delta)$. Suppose that $\delta > 0$ and let $K > 0$ be such that $\sum\{\tau(p) : \tau(p) < K, p \text{ atomic }\} < \infty$. There are at most countably many non-zero terms in this sum, say $\tau(p_1), \tau(p_2), \ldots$. We may assume that the terms $\tau(p_n)$ have been arranged in descending order, so that $\tau(p_n) \downarrow_n 0$. Since $\sum_{n=1}^{\infty} \tau(p_n) < \infty$, we can choose $n_0 \in \mathbb{N}$ such that $\sum_{n=n_0}^{\infty} \tau(p_n) < \delta$. If $\delta < K$ we can find an $\epsilon > 0$ such that $\epsilon < \delta < K$ and $m_0 > n_0$ such that $\tau(p_{m_0}) > \epsilon \geq \tau(p_{m_0+1})$. If $\delta \geq K$ we can find an $\epsilon > 0$ such that $\epsilon < K \leq \delta$ and $m_0 > n_0$ such that $\tau(p_{m_0}) > \epsilon \geq \tau(p_{m_0+1})$. Now let $x \in \mathrm{co}(N(\epsilon))$. Then there exists $\alpha_1, \ldots, \alpha_n \geq 0$, and $x_1, \ldots, x_n \in N(\epsilon)$ such that $\sum_{i=1}^{n} \alpha_i = 1$ and $x = \sum_{i=1}^{n} \alpha_i x_i$. For $i = 1, \ldots, n$ there exists $q_i \in \mathcal{M}^p$ such that $q_i H \subseteq D(x_i)$, $\|x_i q_i\| \leq \epsilon$ and $\tau(1 - q_i) \leq \epsilon$. Hence for $i = 1, \ldots, n$ we have that $1 - q_i \neq p_k$ for $k = 1, \ldots, m_0$ since $\tau(p_k) > \epsilon$ and also $1 - q_i \neq p$ for p atomic with $\tau(p) \geq K$ since $K > \epsilon$. Therefore $1 - q_i \leq \bigvee_{k=m_0+1}^{\infty} p_k$ and $\bigvee_{i=1}^{n}(1 - q_i) \leq \bigvee_{k=m_0+1}^{\infty} p_k$.

Put $q = \bigwedge_{i=1}^{n} q_i$. Then

$$q(H) = \bigcap_{i=1}^{n} q_i(H) \subseteq \bigcap_{i=1}^{n} D(x_i) = D(x) ,$$

and since $q \leq q_i$,

$$\|xq\| \leq \sum_{i=1}^{n} \alpha_i \|x_i q\| = \sum_{i=1}^{n} \alpha_i \|x_i q_i q\| \leq \sum_{i=1}^{n} \alpha_i \|x_i q_i\| \|q\| \leq \epsilon \sum_{i=1}^{n} \alpha_i = \epsilon < \delta.$$

Lastly,

$$\tau(1-q) == \tau\left(\vee_{i=1}^{n}(1-q_i)\right) \leq \tau\left(\vee_{k=m_0+1}^{\infty} p_k\right) \leq \sum_{k=m_0+1}^{\infty} \tau(p_k) < \delta,$$

since $m_0 > n_0$. Thus $x \in N(\delta)$ and therefore $N(\epsilon) \subseteq \mathrm{co}(N(\epsilon)) \subseteq N(\delta)$. □

It remains to consider the case where $\inf\{\tau(p) : p \in \mathcal{M}^p,\ \tau(p) \neq 0\} = 0$ and $\sum\{\tau(p) : \tau(p) < K, p \text{ atomic}\} = \infty$ for every $K > 0$. The following characterisation will be helpful:

Lemma 3.3. *Let $\mathcal{M}$ be atomic and $\inf\{\tau(p) : p \in \mathcal{M}^p,\ \tau(p) \neq 0\} = 0$. Then $\sum\{\tau(p) : \tau(p) < K, p \text{ atomic}\} = \infty$ for every $K > 0$ if and only if there is a sequence of atomic projections (p_n) in $\mathcal{M}^p$ with $\tau(p_n) \downarrow_n 0$ such that $\sum_{n=1}^{\infty} \tau(p_n) = \infty$.*

Proof. If the first condition is satisfied, then for every $n, k \in \mathbb{N}$ there exists a finite number of atomic projections each with trace less than $\frac{1}{n}$ such that the sum of their traces is larger than $\frac{1}{k}$. This enables us to choose a sequence (p_i) such that $\dfrac{1}{k+1} < \sum_{i=n_{k-1}}^{n_k} \tau(p_i) < \dfrac{1}{k}$ for all k, with $n_0 = 1$. Then $\tau(p_i) \downarrow_i 0$ and

$$\sum_{i=1}^{\infty} \tau(p_i) > 1 + \frac{1}{2} + \frac{1}{3} + \cdots = \infty.$$

Suppose now there exists such a sequence of atomic projections (p_n) and that $K > 0$. There exists $N \in \mathbb{N}$ such that $\tau(p_n) < K$ for all $n \geq N$ since $\tau(p_n) \downarrow_n 0$. Then

$$\sum_{\substack{\tau(p)<K \\ p \text{ atomic}}} \tau(p) \geq \sum_{n=N}^{\infty} \tau(p_n) = \infty.$$ □

If the sequence of atomic projections above can be chosen mutually orthogonal, the measure topology is not locally convex. The proof is similar to that of Theorem 3.1.

Theorem 3.4. *Let $\mathcal{M}$ be atomic and suppose there exists a sequence of mutually orthogonal atomic projections (p_n) in $\mathcal{M}^p$ with $\tau(p_n) \downarrow_n 0$ such that $\sum_{n=1}^{\infty} \tau(p_n) = \infty$. Then the measure topology on $\widetilde{\mathcal{M}}$ is not locally convex.*

Proof. Suppose $0 < \epsilon < \delta$. Choose $M \in \mathbb{N}$ such that $\tau(p_n) \leq \frac{\epsilon}{2}$ for all $n \geq M$ and $N \in \mathbb{N}$ such that $\sum_{n=M}^{M+N} \tau(p_n) > \delta$. For $k = M, \ldots, M+N$ define $x_k = (N+1)2\delta p_k$. Then $x_k \in N(\epsilon)$ for $k = M, \ldots, M+N$ since $\tau(\chi_{(\epsilon,\infty)}(x_k) = \tau(p_k) < \epsilon$. Hence

$$x = \sum_{k=M}^{M+N} \frac{1}{N+1} x_k = 2\delta \bigvee_{k=M}^{M+N} p_k \in co(N(\epsilon))$$

but $x \notin N(\delta)$ since

$$\tau(\chi_{(\delta,\infty}(x) = \tau(\bigvee_{k=M}^{M+N} p_k) = \sum_{k=M}^{M+N} \tau(p_k) > \delta$$

by orthogonality of the p_k's. Thus $co(N(\epsilon) \not\subseteq N(\delta)$, $N(\delta)$ does not contain a convex neighbourhood of 0 and so the measure topology is not locally convex. □

The following example shows that it is not always possible to find a sequence of orthogonal projections as above.

Example. Let $M_2(\mathbb{C})$ be the von Neumann algebra of 2×2 matrices over the complex field $\mathbb{C}$. There exists a unique, up to multiplication by a positive constant, faithful finite normal trace, τ, on the factor $M_2(\mathbb{C})$, see [10] Proposition 8.5.3. Put

$$\mathcal{M} = \oplus_{n=1}^{\infty} M_2(\mathbb{C})_n$$

where $M_2(\mathbb{C})_n$ is a copy of $M_2(\mathbb{C})$ with trace $\tau_n = \frac{1}{2^n}\tau$. The trace on $\mathcal{M}$ is $\tau_{\mathcal{M}} = \oplus_{n=1}^{\infty} \tau_n$. Note that $\mathcal{M}$ is atomic, that $\inf\{\tau_{\mathcal{M}}(p) : p \in \mathcal{M}^p,\ \tau_{\mathcal{M}}(p) \neq 0\} = 0$ and that for any $K > 0$

$$\sum_{\substack{\tau_{\mathcal{M}}(p)<K \\ p \text{ atomic}}} \tau_{\mathcal{M}}(p) = \infty.$$

Yet, the trace of any family of orthogonal projections is finite since the trace on $\mathcal{M}$ is finite.

It is easy to see that if $\mathcal{M}$ is atomic then $\mathcal{M}$ is commutative if and only if every atomic projection is central, and this is the case if and only if any two atomic projections are mutually orthogonal. This allows a complete characterization in the commutative case of the conditions under which the measure topology is locally convex.

Corollary 3.5. *Let $\mathcal{M}$ be commutative, Then the measure topology on $\widetilde{\mathcal{M}}$ is locally convex if and only if $\mathcal{M}$ is atomic and either* $\inf\{\tau(p) : p \in \mathcal{M}^p, \tau(p) \neq 0\} > 0$ *or* $\inf\{\tau(p) : p \in \mathcal{M}^p, \tau(p) \neq 0\} = 0$ *and there is a $K > 0$ such that* $\sum\{\tau(p) : p \in \mathcal{M}^p, p \text{ atomic }, \tau(p) < K\} < \infty$.

We finish this section by mentioning some closely related results. In [13] it was shown that the non-commutative L^p spaces, with $0 < p < 1$, have trivial duals (and therefore cannot be locally convex) if and only if $\mathcal{M}$ has no minimal projections. This result was generalised to the Haagerup L^p spaces associated with a not necessarily semi-finite von Neumann algebra $\mathcal{M}$ in [18]. A similar result was obtained in [4] for the non-commutative $L^{p\infty}$ spaces, with $0 < p < 1$.

More recently, it was shown in [2] that $\widetilde{\mathcal{M}}$, when equipped with the topology of local convergence in measure rather than the topology of convergence in measure, is locally convex if and only if $\mathcal{M}$ is atomic.

4. The ideal $\widetilde{\mathcal{M}}_0$

We recall that the ideal of all $x \in \widetilde{\mathcal{M}}$ such that $\mu_t(x) \to 0$ as $t \to \infty$ is denoted by $\widetilde{\mathcal{M}}_0$.

The topology of convergence in measure is "order continuous" on $\widetilde{\mathcal{M}}_0$:

Theorem 4.1 ([7] Lemma 3.5). *If $x \in \widetilde{\mathcal{M}}_0$ and $x \geq x_\alpha \downarrow_\alpha 0$ in $\widetilde{\mathcal{M}}$ then $\mu_t(x_\alpha) \downarrow_\alpha 0$ for all $t > 0$, and so $x_\alpha \to_\alpha 0$ in the measure topology.*

Definition 4.2. We define

$$\widetilde{\mathcal{M}}_a = \{x \in \widetilde{\mathcal{M}} : |x| \geq x_\alpha \downarrow 0 \Rightarrow \mu_t(x_\alpha) \to 0 \text{ for all } t > 0\}.$$

Thus $\widetilde{\mathcal{M}}_a$ is the largest subspace of $\widetilde{\mathcal{M}}$ on which the measure topology is "order continuous". Since $|y| \leq |x|$ implies that $\mu_t(y) \leq \mu_t(x)$ for all $t > 0$, it is easy to see that both $\widetilde{\mathcal{M}}_0$ and $\widetilde{\mathcal{M}}_a$ are *solid* in $\widetilde{\mathcal{M}}$ in the sense that if $x \in \widetilde{\mathcal{M}}_0$ (respectively $\widetilde{\mathcal{M}}_a$), $y \in \widetilde{\mathcal{M}}$ with $|y| \leq |x|$, then $y \in \widetilde{\mathcal{M}}_0$ (respectively $\widetilde{\mathcal{M}}_a$).

Proposition 4.3.

$$\widetilde{\mathcal{M}}_a = \widetilde{\mathcal{M}}_0.$$

Proof. By Theorem 4.1 we have that $\widetilde{\mathcal{M}}_0 \subseteq \widetilde{\mathcal{M}}_a$.
Conversely, suppose that $x \notin \widetilde{\mathcal{M}}_0$. Then there exists $c > 0$ such that $\lim_{t\to\infty} \mu_t(x) = c$ and hence $\tau\left(e_{(t,\infty)}(|x|)\right) = \infty$ for $0 < t \leq c$. Denote $e_{(c,\infty)}(|x|)$ by e. Let $\{p_\alpha\}_{\alpha\in A}$ be a family of projections maximal with respect to $p_\alpha \neq 0$ for all $\alpha \in A$, $\tau(p_\alpha) < \infty$ for all $\alpha \in A$ and $\sum_{\alpha\in F} p_\alpha \leq e$ for all finite $F \subset A$. Since $\mathcal{M}$ is semi-finite, $A \neq \emptyset$. Let $\mathcal{F} = \{F \subset A : F \text{ is finite}\}$ be ordered by inclusion. Then $\left(\sum_{\alpha\in F} p_\alpha\right)_{F\in\mathcal{F}}$ is an increasing net bounded above by e and hence its supremum, say p, exists in $\mathcal{M}$ and $p \leq e$. Note that $\tau\left(\sum_{\alpha\in F} p_\alpha\right) \leq \sum_{\alpha\in F} \tau(p_\alpha) < \infty$ for every finite $F \subset A$. Suppose $e - p > 0$. By faithfulness of the trace $\tau(e-p) > 0$ and then by semi-finiteness of $\mathcal{M}$ there exists $0 < q \in \mathcal{M}^p$ such that $q \leq e - p$ and $0 < \tau(q) < \infty$. This contradicts the maximality of the family $\{p_\alpha\}$, and hence we must have $e = p$. For $F \in \mathcal{F}$, let $q_F = e - \sum_{\alpha\in F} p_\alpha$. Then $q_F \downarrow 0$. Note that $\tau(q_F) = \tau(e) - \tau\left(\sum_{\alpha\in F} p_\alpha\right) = \infty$ for all $F \in \mathcal{F}$ since $\tau\left(\sum_{\alpha\in F} p_\alpha\right) < \infty$. Now $|x| \geq |x| e_{(c,\infty)}(|x|) = |x| e \geq ce \geq cq_F \downarrow 0$ but $\mu_t(cq_F) = c\mu_t(q_F) = c\chi_{(0,\tau(q_F))}(t) = c$ for all $t > 0$ since $\tau(q_F) = \infty$ for all $F \in \mathcal{F}$. Hence we have found a net $(cq_F)_{F\in\mathcal{F}}$ with $|x| \geq cq_F \downarrow_{F\in\mathcal{F}} 0$ but $\mu_t(cq_F) \not\to 0$ for all $t > 0$. Thus $x \notin \widetilde{\mathcal{M}}_a$ and $\widetilde{\mathcal{M}}_0 = \widetilde{\mathcal{M}}_a$. $\square$

If we assume that $\mathcal{M}$ is nonatomic, we can give another characterisation of the subspace $\widetilde{\mathcal{M}_0}$. This generalizes a result, for the commutative case, of Dilworth and Trautman ([5] Lemma 3.2).

Proposition 4.4. *Let $\mathcal{M}$ be nonatomic. Then*

$$\widetilde{\mathcal{M}_0} = \bigcap_{\epsilon,\delta>0} co(N(\epsilon,\delta)).$$

Proof. A typical neighbourhood of zero in $\widetilde{\mathcal{M}_0}$ is

$$N_0(\epsilon,\delta) = \{x \in \widetilde{\mathcal{M}_0} : \exists p \in \mathcal{M}^p, p(H) \subset D(x), ||xp|| < \epsilon, \tau(1-p) < \delta\},$$

where $\epsilon, \delta > 0$.

We first show that $\widetilde{\mathcal{M}_0} = \mathrm{co}(N_0(\epsilon,\delta))$ for any $\epsilon,\delta > 0$. Clearly $\mathrm{co}(N_0(\epsilon,\delta)) \subset \widetilde{\mathcal{M}_0}$. Conversely, let $x \in \widetilde{\mathcal{M}_0}$. Then $d_\epsilon(x) = \tau(e_{(\epsilon,\infty)}(|x|)) < \infty$ by [15] Theorem 2.1. Put $p = e_\epsilon(|x|)$ and notice that $\|xp\| \le \epsilon$ and $\tau(1-p) < \infty$. Since $\mathcal{M}$ is nonatomic, there exist disjoint $e_1, \ldots, e_n \in \mathcal{M}^p, n \in \mathbb{N}$ such that $1 - p = \bigvee_{i=1}^n e_i$ and with $\tau(e_i) \le \delta$ for all $i = 1, \ldots, n$. Note that $p(1-e_i) = p$ for all i since $p = 1 - \bigvee_{i=1}^n e_i = \bigwedge_{i=1}^n (1-e_i)$.

For $1 \le i \le n$, put $x_i = xp + nxe_i$. Then for each $1 \le i \le n$, $x_i \in N_0(\epsilon,\delta)$ since $1 - e_i \in \mathcal{M}^p$,

$$\|(xp + nxe_i)(1-e_i)\| = \|xp(1-e_i) + 0\| = \|xp\| \le \epsilon$$

and $\tau(1-(1-e_i)) = \tau(e_i) \le \delta$. Finally,

$$\begin{aligned} \sum_{i=1}^n \frac{1}{n} x_i &= \frac{1}{n}(xp + nxe_1 + xp + nxe_2 + \cdots + xp + nxe_n) \\ &= xp + x(1-p) \\ &= x. \end{aligned}$$

Thus $x \in \mathrm{co}(N_0(\epsilon,\delta))$ and so $\widetilde{\mathcal{M}_0} = \mathrm{co}(N_0(\epsilon,\delta))$ for any $\epsilon,\delta > 0$. Hence

$$\widetilde{\mathcal{M}_0} = \bigcap_{\epsilon,\delta>0} \mathrm{co}(N_0(\epsilon,\delta) \subseteq \bigcap_{\epsilon,\delta>0} \mathrm{co}(N(\epsilon,\delta)).$$

Conversely, we will show that $x \in \widetilde{\mathcal{M}} \backslash \widetilde{\mathcal{M}_0}$ implies that $x \notin \mathrm{co}(N(\epsilon,\delta))$ for some $\epsilon,\delta > 0$. Suppose that $x \in \widetilde{\mathcal{M}} \backslash \widetilde{\mathcal{M}_0}$. Then by [15] Theorem 2.1 there exists $c > 0$ such that $\tau\left(e_{(c,\infty)}(|x|)\right) = \infty$.

Let $0 < \epsilon < c$, let $\alpha_1, \ldots, \alpha_n, n \in \mathbb{N}$ be positive real numbers with $\sum_{i=1}^n \alpha_i = 1$ and let $x_1, \ldots, x_n \in N(\epsilon,\epsilon)$. Then for each $i = 1, \ldots, n$ we have that

$$\alpha_i x_i \in \alpha_i N(\epsilon,\epsilon) = N(\alpha_i\epsilon,\epsilon)$$

and

$$\sum_{i=1}^n \alpha_i x_i \in N(\alpha_1\epsilon,\epsilon) + \cdots + N(\alpha_n\epsilon,\epsilon) \subseteq N(\alpha_1\epsilon + \cdots + \alpha_n\epsilon, n\epsilon) = N(\epsilon, n\epsilon).$$

This implies that $\mu_{n\epsilon}\left(\sum_{i=1}^{n}\alpha_i x_i\right)\le\epsilon$. When we take

$$\delta=\mu_{n\epsilon}\left(\sum_{i=1}^{n}\alpha_i x_i\right)=\inf\left\{\theta\ge 0:\tau\left(e_{(\theta,\infty)}\left(|\sum_{i=1}^{n}\alpha_i x_i|\right)\right)\le n\epsilon\right\}$$

then, since $\delta\le\epsilon$, we have that

$$\tau\left(e_{(\epsilon,\infty)}\left(|\sum_{i=1}^{n}\alpha_i x_i|\right)\right)\le n\epsilon$$

and since $c>\epsilon$

$$\tau\left(e_{(c,\infty)}\left(|\sum_{i=1}^{n}\alpha_i x_i|\right)\right)\le n\epsilon,$$

but $\tau\left(e_{(c,\infty)}(|x|)\right)=\infty$ and it thus follows that x cannot be written as a linear combination of elements in $N(\epsilon)$, that is, $x\notin \mathrm{co}(N(\epsilon))$, which proves the result. □

References

[1] C.A. Akeman, J. Anderson and G.K. Pedersen, *Triangle inequalities in operator algebras.* Linear and Multilinear Algebra **11** (1982), 167–178.

[2] A.M. Bikchentaev, *Local convergence in measure on semi-finite von Neumann algebras II.* Mat. Zametki **82** (2007),703–707 (translation in Math. Notes **82** (2007), 703–707).

[3] L.G. Brown and H. Kosaki, *Jensen's inequality in semi-finite von Neumann algebras.* J. Operator Theory **23** (1990), 3–19.

[4] L.J. Ciach, *Dual of non-commutative L^{p1}-spaces with* $0<p\le 1$. Demonstratio Math. **30** (1997), 353–358.

[5] S.J. Dilworth and D.A. Trautman, *On two function spaces that are similar to* L_0. Proc. Amer. Math. Soc. **108** (1990), 451–456.

[6] P.G. Dodds, T.K. Dodds and B. de Pagter, *Non-commutative Banach function spaces.* Math. Proc. Camb. Phil. Soc. **201** (1989), 583–597.

[7] P.G. Dodds, T.K. Dodds and B. de Pagter, *Non-commutative Köthe duality.* Trans. Amer. Math. Soc. **339** (1993), 717–750.

[8] T. Fack and H. Kosaki, *Generalised s-numbers of τ-measurable operators.* Pacific J. Math. **123** (1986), 269–300.

[9] R.V. Kadison and J.R. Ringrose, *Fundamentals of the theory of operator algebras. Volume* 1. Academic Press, 1983.

[10] R.V. Kadison and J.R. Ringrose, *Fundamentals of the theory of operator algebras. Volume* 2. Academic Press, 1986.

[11] H. Kosaki, *On the continuity of the map* $\phi\to|\phi|$ *from the predual of a* W^**-algebra.* J. Funct. Anal. **59** (1984), 123–131.

[12] E. Nelson, *Notes on non-commutative integration.* J. Funct. Anal. **15** (1974), 103–116.

[13] K.-S. Saito, *Noncommutative L^p-spaces with* $0 < p < 1$. Math. Proc. Cambridge Philos. Soc. 89 (1981), 405–411.

[14] I.E. Segal, *A non-commutative extension of abstract integration.* Ann. of Math. **57** (1953), 401–457.

[15] A. Ströh and G. West, *τ-compact operators affiliated to a semi-finite von Neumann algebra.* Proc. Roy. Irish Acad. Sect A 93 (1993), 73–86.

[16] M. Takesaki, *Theory of operator algebras I.* Springer-Verlag, 1979.

[17] M. Terp, *L^p-spaces associated with von Neumann algebras.* Technical report, Institute of Mathematical Sciences, Copenhagen University, 1981.

[18] K. Watanabe, *Dual of noncommutative L^p-spaces with* $0 < p < 1$. Math. Proc. Cambridge Philos. Soc. 103 (1988), 503–509.

Jurie Conradie
Department of Mathematics and Applied Mathematics
University of Cape Town
Rondebosch 7701, South Africa
e-mail: `Jurie.Conradie@uct.ac.za`

Charlotte Crowther
Value Based Management
Group Capital Management
Nedbank, South Africa
e-mail: `CharlotteCr@Nedbank.co.za`

Operator Theory:
Advances and Applications, Vol. 195, 121–138

Holomorphic $\mathcal{N}_K$ and Bergman-type Spaces

A. El-Sayed Ahmed and M.A. Bahkit

Abstract. In this paper we introduce a new class of functions, called $\mathcal{N}_K$-type space of analytic functions by the help of a nondecreasing function $K : [0,\infty) \to [0,\infty)$. Further, under mild conditions on the weight function K we characterize lacunary series in $\mathcal{N}_K$ space. Finally, we study the boundedness and compactness of composition operators between $\mathcal{N}_K$ and Bergman spaces.

Mathematics Subject Classification (2000). Primary 47B33, 47B38 Secondary 46E15 and 30H05.

Keywords. $\mathcal{N}_K$-space, Composition operators, Bergman-type spaces, Hadamard gap series.

1. Introduction

Let $\Delta = \{z : |z| < 1\}$ be the unit disk in the complex plane $\mathbb{C}$, $\partial\Delta$ its boundary, H(Δ) be the class of all analytic functions on Δ, while $dA(z)$ denotes the Lebesgue measure on the plane, normalized so that $A(\Delta) = 1$. For each $a \in \Delta$, the Green's function with logarithmic singularity at $a \in \Delta$ is denoted by $g(z,a) = \log \frac{1}{|\varphi_a(z)|}$, where $\varphi_a(z) = \frac{a-z}{1-\bar{a}z}$ is the Möbius transformation of Δ. We will frequently use the following easily verified equality:

$$(1 - |\varphi_a(z)|^2) = \frac{(1-|a|^2)(1-|z|^2)}{|1-\overline{a}z|^2}.$$

For $p \in (0,\infty)$, the Bergman spaces L^p_a are defined by

$$L^p_a = \left\{f \in H(\Delta) : \|f\|_p = \int_\Delta |f(z)|^p dA(z) \ < \infty\right\}.$$

For $p \in (0,\infty)$ and $-1 < \alpha < \infty$, the Bergman-type spaces $\mathcal{A}^p_\alpha$ are defined by

$$\mathcal{A}^p_\alpha = \{f \in H(\Delta) : \|f\|_{\mathcal{A}^p_\alpha} = \sup_{z\in\Delta} |f(z)|^p(1-|z|^2)^\alpha \ < \infty\}.$$

$$\mathcal{A}_\alpha = \{f \in H(\Delta) : \|f\|_{\mathcal{A}_\alpha} = \sup_{z\in\Delta} |f(z)|(1-|z|^2)^\alpha \ < \infty\}.$$

For more information about several studied on Bergman-type spaces we refer to [7, 31]. In this paper, if $p = \alpha = 1$, we say that $f \in \mathcal{A}^{-1}$. If $p = 2$ and $\alpha = 1$, we say that $f \in \mathcal{A}^2$. Moreover, $f \in \mathcal{A}_0^{-1}$ if and only if

$$\lim_{|z|\to 1} |f(z)|(1-|z|^2) = 0.$$

If $f \in \mathcal{A}_\alpha$, where $\alpha \in (0,\infty)$ then the subharmonicity of $|f|^\alpha$ for $\delta < 1-|z|$, implies that

$$|f(z)|^\alpha = \frac{1}{\delta^{2n}} \int_{z+\delta\Delta} |f(z_1)|^\alpha dA(z_1) \leq \frac{1}{\delta^{2n}} \|f\|^\alpha_{\mathcal{A}_\alpha} .$$

Hence,

$$|f(z)| \leq (1-|z|)^{-2n/\alpha} \|f\|_{\mathcal{A}_\alpha}.$$

For $0 \leq p < \infty$, the spaces $\mathcal{Q}_p$ consist of all functions $f \in H(\Delta)$ such that

$$\|f\|^2_{\mathcal{Q}_p} = \sup_{a\in\Delta} \int_\Delta |f'(z)|^2 g^p(z,a) dA(z) < \infty \text{ (see, e.g., [3, 4, 10] and others).}$$

Given a function $K : [0,\infty) \to [0,\infty)$, the class of all functions $f \in H(\Delta)$ for which

$$\|f\|^2_{\mathcal{Q}_K} = \sup_{a\in\Delta} \int_\Delta |f'(z)|^2 K(g(z,a)) dA(z) < \infty, \text{ (see [26])}$$

is called $\mathcal{Q}_K$ space. Clearly, if $K(t) = t^p$, then $\mathcal{Q}_K = \mathcal{Q}_p$. The study of $\mathcal{Q}_K$ spaces has mainly been on understanding the relationship between the properties of K and the resulting spaces $\mathcal{Q}_K$. For more information about these spaces we refer to [10, 11, 18, 27, 28]. For an analytic function f on Δ and $0 < \alpha < \infty$, if

$$\|f\|_{\mathcal{B}^\alpha} = \sup_{z\in\Delta} (1-|z|^2)^\alpha |f'(z)| < \infty,$$

then f belongs to α-Bloch spaces $\mathcal{B}^\alpha$ (see [30]).

Palmberg in [21] introduced the $\mathcal{N}_p$-spaces, (with $p \in (0,\infty)$) consist of $f \in H(\Delta)$ such that

$$\|f\|^2_{\mathcal{N}_p} = \sup_{a\in\Delta} \int_\Delta |f(z)|^2 (1-|\varphi_a(z)|^2)^p dA(z) < \infty.$$

When $p = 1$, $\mathcal{N}_p$ coincides $\mathcal{N}_1$. The $\mathcal{N}_1$-space was introduced in [17].

Two quantities A_f and B_f, both depending on an analytic function f on Δ, are said to be equivalent, written as $A_f \approx B_f$, if there exists a finite positive constant C not depending on f such that for every analytic function f on Δ we have:

$$\frac{1}{C} B_f \leq A_f \leq C B_f.$$

If the quantities A_f and B_f, are equivalent, then in particular we have $A_f < \infty$ if and only if $B_f < \infty$.

In this paper, we introduce the new space $\mathcal{N}_K$ by the right continuous and nondecreasing function $K : [0,\infty) \to [0,\infty)$. The $\mathcal{N}_K$ space consists of $f \in H(\Delta)$ such that

$$\|f\|^2_{\mathcal{N}_K} = \sup_{a\in\Delta} \int_\Delta |f(z)|^2 K(g(z,a))dA(z) < \infty.$$

If

$$\lim_{|a|\to 1} \int_\Delta |f(z)|^2 K(g(z,a))dA(z) = 0,$$

then f is said to belong to $\mathcal{N}_{K,0}$. Clearly, if $K(t) = t^p$, then $\mathcal{N}_K = \mathcal{N}_p$; since $g(z,a) \approx (1 - |\varphi_a(z)|^2)$. For $K(t) = 1$ it gives the Bergman space $\mathcal{A}^2$. It is easy to check that $\|\cdot\|_{\mathcal{N}_K}$ is a complete semi-norm on $\mathcal{N}_K$. If $\mathcal{N}_K$ consists of just the constant functions, we say that it is trivial. We assume from now on that all $K : [0,\infty) \to [0,\infty)$ to appear in this paper are right-continuous and nondecreasing functions such that the integral

$$\int_0^{1/e} K(\log(1/\rho))\rho\, d\rho = \int_1^\infty K(t)e^{-2t}\, dt \quad \text{is convergent.}$$

From a change of variable we see that the coordinate function z belongs to $\mathcal{N}_K$ space if and only if

$$\sup_{a\in\Delta} \int_\Delta \frac{(1-|a|^2)^2}{|1-\bar{a}z|^4}\, K\left(\log\frac{1}{|z|}\right) dA(z) < \infty.$$

Simplifying the above integral in polar coordinates, we conclude that $\mathcal{N}_K$ space is nontrivial if and only if

$$\sup_{t\in(0,1)} \int_0^1 \frac{(1-t)^2}{(1-tr^2)^3}\, K\left(\log\frac{1}{r}\right) r\, dr < \infty. \tag{1.1}$$

Throughout this paper we always assume that condition (1.1) is satisfied, so that the $\mathcal{N}_K$ space we study is not trivial. An important tool in the study of $\mathcal{N}_K$ space is the auxiliary function ϕ_K defined by

$$\phi_K(s) = \sup_{0<t<1} \frac{K(st)}{K(t)}, \qquad 0 < s < \infty.$$

The following condition has played a crucial role in the study of $\mathcal{N}_K$ space:

$$\int_1^\infty \phi_K(s)\frac{ds}{s^2} < \infty. \tag{1.2}$$

The function theory of $\mathcal{N}_K$ obviously depends on the properties of K. Given two weight functions K_1 and K_2, we are going to write $K_1 \lesssim K_2$ if there exists a constant $C > 0$, independent of t, such that $K_1(t) \le CK_2(t)$ for all t. The notation $K_1 \lesssim K_2$ is used in a similar fashion. When $K_1 \lesssim K_2 \lesssim K_1$, we write $K_1 \approx K_2$.

Corollary 1.1. [28] *If K satisfies condition (1.2), then there exists a constant $C > 0$ such that $K(2t) \le C\,K(t)$ for all $0 \le 2t \le 1$.*

Lemma 1.2. [28] *If K satisfies condition* (1.2), *then the function*

$$K^*(t) = t\int_1^\infty \phi_K(s)\frac{ds}{s^2} \quad (\textit{where, } 0 < t < \infty),$$

has the following properties:

(A) K^* *is nondecreasing on* $(0, \infty)$.
(B) $K^*(t)/t$ *is nondecreasing on* $(0, \infty)$.
(C) $K^*(t) \geq K(t)$ *for all* $t \in (0, \infty)$.
(D) $K^* \lesssim K$ *on* $(0, 1]$.

If $K(t) = K(1)$ *for* $t \geq 1$, *then we also have*

(E) $K^*(t) = K^*(1) = K(1)$ *for* $t \geq 1$, *so* $K^* \approx K$ *on* $(0, \infty)$.

Theorem 1.3. [28] *If K satisfies condition* (1.2), *then for any* $\alpha \geq 1$ *and* $0 \leq \beta < 1$ *we have*

$$\int_0^1 r^{\alpha-1}\left(\log\frac{1}{r}\right)^{-\beta} K\left(\log\frac{1}{r}\right)dr \approx C(\beta)\left(\frac{1-\beta}{\alpha}\right)^{1-\beta} K\left(\frac{1-\beta}{\alpha}\right),$$

where $C(\beta)$ *is a constant depending only on* β.

Now, we give the following proposition.

Proposition 1.4. *If K satisfies condition* (1.2), *then we can find another non-negative weight function given by*

$$K^*(t) = t\int_1^\infty \phi_K(s)\frac{ds}{s^2} \quad (\textit{where, } 0 < t < \infty),$$

such that $\mathcal{N}_K = \mathcal{N}_K^*$ *and that the new function* K^* *has the following properties:*

(a) K^* *is nondecreasing on* $(0, \infty)$.
(b) K^* *satisfies condition* 1.1.
(c) K^* *satisfies condition* 1.2.
(d) $K^*(t)/t$ *is nondecreasing on* $(0, \infty)$.
(e) $K^*(2t) \approx K^*(t)$ *on* $(0, \infty)$.
(f) K^* *is concave on* $(0, \infty)$.
(g) K^* *is differentiable* (*up to any given order*) *on* $(0, \infty)$.
(h) $K^*(t) = K^*(1)$ *for all* $t \geq 1$.
(i) $K^* \lesssim K$ *on* $(0, 1]$.

Proof. The proof of this proposition can be obtained from Lemma 1.2, and also using some similar steps to that used in the proof of proposition 4 in [28], so it will be omitted.

Recall that a linear operator $T : X \to Y$ is said to be bounded if there exists a constant $M > 0$ such that $||T(f)||_Y \leq M||f||_X$ for all maps $f \in X$. Moreover, $T : X \to Y$ is said to be compact if it takes bounded sets in X to sets in Y which have compact closure. For Banach spaces X and Y of $H(\Delta)$, T is compact from X to Y if and only if for each bounded sequence $\{x_n\} \in X$, the sequence $\{Tx_n\} \in Y$ contains a subsequence converging to some limit in Y.

2. Holomorphic $\mathcal{N}_K$ space

In this section we prove some basic Banach space properties of $\mathcal{N}_K$-space.

Let $\Delta(a,r) = \{z : |\varphi_a(z)| < r\}$ be the pseudo-hyperbolic disk with radius r, where $0 < r < 1$. For a point $a \in \Delta$ and $0 < r < 1$, we can get that $\Delta(a,r)$ with pseudo-hyperbolic center a and pseudo-hyperbolic radius r, its Euclidean center and Euclidean radius are $\frac{(1-r^2)a}{1-r^2|a|^2}$ and $\frac{(1-|a|^2)r}{1-r^2|a|^2}$, respectively (see [23]).

Lemma 2.1. (see [29] Lemma 3) *For any fixed $r \in (0,1)$ and there exists a positive constant C such that*

$$|f(a)|^p(1-|a|^2)^p \le C \int_{\Delta(a,r)} |f(z)|^p dA(z)$$

for all $a \in \Delta$, all $p > 0$, and all analytic f in Δ.

Lemma 2.2. (*Test functions in $\mathcal{N}_K$*) *For $w \in \Delta$ we define*

$$h_w(z) = \frac{1-|w|^2}{(1-\overline{w}z)^2}.$$

Suppose that condition (1.1) *is satisfied. Then $h_w \in \mathcal{N}_K$ and* $\sup \|h_w\|_{\mathcal{N}_K} \le 1$.

Proof. Trivially $h_w \in H(\Delta)$ Then

$$\|h_w\|^2_{\mathcal{N}_K} = \sup_{w\in\Delta} \int_\Delta |\varphi'_w(z)|^2 K(g(z,a)\, dA(z) \le 1,$$

by a change of variables and since condition (1.1) is satisfied.

Proposition 2.3. *For each nondecreasing function $K : [0,\infty) \to [0,\infty)$. Then*

$$\mathcal{N}_K \subset \mathcal{A}^{-1}.$$

Proof. Suppose that $f \in \mathcal{N}_K$, and let C be a constant such that

$$\sup_{a\in\Delta} \int_\Delta |f(z)|^2 K(g(z,a))\, dA(z) = C < \infty.$$

By the fact K is nondecreasing, for all $r, 0 < r < 1$, we have

$$\begin{aligned} C &\ge \int_\Delta |f(z)|^2 K(g(z,a))\, dA(z) \\ &\ge \int_{\Delta(a,r)} |f(z)|^2 K\left(\log \frac{1}{|\varphi_a(z)|}\right) dA(z) \\ &\ge K(\log \frac{1}{r}) \int_{\Delta(a,r)} |f(z)|^2\, dA(z) \end{aligned}$$

By Lemma 2.1 for any fixed $r_0 \in (0,1)$, there exists a constant C' such that

$$\begin{aligned} |f(a)|^2(1-|a|)^2 &\le C' \int_{\Delta(a,r_0)} |f(z)|^2 dA(z) \\ &\le \frac{C\,C'}{K(\log 1/r_0)}. \end{aligned}$$

Since r_0 is fixed, then

$$\sup_{a\in\Delta} |f(a)|(1-|a|) \leq \sqrt{\frac{C\,C'}{K(\log 1/r_0)}}\,.$$

Thus $f \in \mathcal{A}^{-1}$ in Δ. Hence, $\mathcal{N}_K \subset \mathcal{A}^{-1}$.

Theorem 2.4. $\mathcal{N}_K = \mathcal{A}^{-1}$ *if and only if*

$$\int_0^1 (1-r^2)^{-2}\, K(\log(1/r)) r dr < \infty. \tag{2.1}$$

Proof. Let us first assume that (2.1) holds. Then $\mathcal{N}_K \subset \mathcal{A}^{-1}$. To prove that $\mathcal{A}^{-1} \subset \mathcal{N}_K$, we assume that $f \in \mathcal{A}^{-1}$ and observe that

$$\begin{aligned} \int_\Delta |f(z)|^2\, K(g(z,a))\, dA(z) &\leq \|f\|^2_{\mathcal{A}^{-1}} \int_\Delta (1-|z|^2)^{-2}\, K(g(z,a))\, dA(z) \\ &= \|f\|^2_{\mathcal{A}^{-1}} \int_\Delta (1-|w|^2)^{-2}\, K(\log(1/|w|))\, dA(w) \\ &= 2\pi\, \|f\|^2_{\mathcal{A}^{-1}} \int_0^1 (1-r^2)^{-2} K(\log(1/r)) r dr < \infty. \end{aligned}$$

Hence $f \in \mathcal{N}_K$ and we have proved that (2.1) is a sufficient condition that $\mathcal{N}_K = \mathcal{A}^{-1}$. Conversely, we assume that $\mathcal{N}_K = \mathcal{A}^{-1}$. To prove that (2.1) is a necessary condition, we study the Hadamard gap function $f(z) = \sum_{j=1}^{\infty} 2^j z^{2^j}$. Let

$$L(r) = \int_0^{2\pi} |f(re^{i\theta})|^2 d\theta = 2\pi \sum_{j=1}^{\infty} 2^{2j} z^{2^{j+1}}.$$

In the interval $[\frac{1}{2}, 1)$, we have $\log r \geq 2(r-1)$, and it follows that

$$r^{2^{j+1}} \geq \exp\{2^{j+2}(r-1)\}, \quad r \in [\frac{1}{2}, 1),$$

and thus that there is a positive constant C (details are given below) such that

$$L(r) \geq C\,(1-r)^{-2}, \quad r \in [\frac{3}{4}, 1).$$

Since $\mathcal{N}_K = \mathcal{A}^{-1}$, we have

$$\begin{aligned} \infty &> \int_\Delta |f(z)|^2\, K(\log(1/|z|))\, dA(z) \\ &= \int_0^1 K(\log(1/r)) \left(\int_0^{2\pi} |f(re^{i\theta})|^2 d\theta \right) r\, dr \\ &= \int_0^1 K(\log(1/r)) L(r) r\, dr \\ &\geq C \int_{\frac{3}{4}}^1 K(\log(1/r))(1-r)^{-2}\, r\, dr. \end{aligned}$$

It remains to prove the estimate of $L(r)$ used above. The general term in the series defining $L(r)$ can be estimated from below by $(1-r)^{-2}t^2\exp\{-4t\}$, where $t = 2^j(1-r)$. It is easy to see that

$$\sup_{t>0} t^2 e^{-4t} = \frac{1}{4e^2}$$

is assumed for $t = \frac{1}{2}$. For $r \in [\frac{3}{4}, 1)$, we find j so that $\frac{1}{2} \le 2^j(1-r) < 1$.

For this j, we have

$$2^{2j}(1-r)^2 \exp\{-4(2^j)(1-r)\} \ge e^{-4},$$

and

$$L(r) \ \ge \ 2\pi\, e^{-4}(1-r)^{-2} = \ C\,(1-r)^{-2}, \quad r \in [\frac{3}{4}, 1).$$

Hence (2.1) holds and the proof is complete.

Theorem 2.5. *Let* $K_1(1) > 0$ *and* $K_2(t) = \inf\{K_1(t), K_1(1)\}$. *Also, we suppose that* $\mathcal{N}_{K_2} = \mathcal{A}^{-1}$. *Then* $\mathcal{N}_{K_2} = \mathcal{N}_{K_1}$.

Proof. Since $K_2 \le K_1$ and K_2 is nondecreasing, it is clear that $\mathcal{N}_{K_1} \subset \mathcal{N}_{K_2}$. It remains to prove that $\mathcal{N}_{K_2} \subset \mathcal{N}_{K_1}$. We note that

$$\begin{aligned} g(z,a) &> 1, \quad z \in \Delta(a, 1/e), \\ g(z,a) &\le 1, \quad z \in \Delta \setminus \Delta(a, 1/e). \end{aligned}$$

Thus $K_1(g(z,a)) = K_2(g(z,a))$ in $\Delta \setminus \Delta(a, 1/e)$. It suffices to deal with integrals over $\Delta(a, 1/e)$. If $f \in \mathcal{N}_{K_2}$, then $f \in \mathcal{A}^{-1}$, hence

$$\begin{aligned} & \int_{\Delta(a,1/e)} |f(z)|^2\, K_1(g(z,a))dA(z) \\ \le \ & \|f\|^2_{\mathcal{A}^{-1}} \int_{\Delta(a,1/e)} (1-|z|^2)^{-2}\, K_1(g(z,a))dA(z) \\ = \ & \|f\|^2_{\mathcal{A}^{-1}} \int_{\Delta(0,1/e)} (1-|w|^2)^{-2}\, K_1(\log(1/|w|))dA(w) \\ = \ & 2\pi\, \|f\|^2_{\mathcal{A}^{-1}} \int_0^{1/e} (1-r^2)^{-2}\, K_1(\log(1/r))rdr \\ \le \ & \|f\|^2_{\mathcal{A}^{-1}}. \end{aligned}$$

The right-hand member gives a bound for the supremum over $a \in \Delta$ of the first term in this chain of inequalities. Hence $f \in \mathcal{N}_{K_1}$ and the theorem is proved.

Theorem 2.6. *Let* $K : [0,\infty) \to [0,\infty)$ *be nondecreasing function. Then*

(i) $\mathcal{N}_{K,0} \subset \mathcal{A}_0^{-1}$.
(ii) $\mathcal{N}_{K,0} = \mathcal{A}_0^{-1}$ *if and only if* (2.1) *holds.*

Proof. (i) Without loss of generality, we note that if $f \in \mathcal{N}_{K,0}$ and $K(1) > 0$, then

$$K(1) \int_{\Delta(a,1/e)} |f(z)|^2 dA(z) \leq \int_{\Delta} |f(z)|^2 K(g(z,a)) dA(z) \longrightarrow 0, \quad |a| \to 1.$$

By subharmonicity,

$$\begin{aligned} \pi(1/e)^2 \, K(1) \, (1-|a|^2)^2 |f(a)|^2 &\leq K(1) \int_{E(a)} |f(z)|^2 dA(z) \\ &\leq K(1) \int_{\Delta(a,1/e)} |f(z)|^2 dA(z) \\ &\leq \int_{\Delta} |f(z)|^2 K(g(z,a)) dA(z), \end{aligned}$$

where $E(a) = \{z \in \Delta, |z-a| < \frac{1}{e}(1-|a|)\}$. If $f \in \mathcal{N}_{K,0}$, we have

$$\lim_{|a|\to 1} (1-|a|^2)^2 |f(a)|^2 = 0.$$

(ii) We only need to prove that $\mathcal{A}_0^{-1} \subset \mathcal{N}_{K,0}$. Assume that

$$\mathcal{L} = \int_0^1 (1-r^2)^{-2} \, K(\log \frac{1}{r}) r dr < \infty.$$

For a given $\varepsilon > 0$ there exists an $r_1, 0 < r_1 < 1$, such that

$$\int_{r_1}^1 (1-r^2)^{-2} K(\log \frac{1}{r}) r dr < \varepsilon. \tag{2.2}$$

Thus,

$$\begin{aligned} &\int_{\Delta \setminus \Delta(a,r)} |f(z)|^2 K(g(z,a)) dA(z) \\ \leq\ & \|f\|^2_{\mathcal{A}^{-1}} \int_{\Delta \setminus \Delta(a,r)} (1-|z|^2)^{-2} \, K(g(z,a)) dA(z) \\ =\ & \|f\|^2_{\mathcal{A}^{-1}} \int_{r \leq |w| < 1} (1-|w|^2)^{-2} \, K(\log(1/|w|)) dA(w) \\ =\ & 2\pi \, \|f\|^2_{\mathcal{A}^{-1}} \int_{r_1}^1 (1-|r|^2)^{-2} \, K(\log \frac{1}{r}) r dr \\ <\ & 2\pi\varepsilon \|f\|^2_{\mathcal{A}^{-1}}. \end{aligned} \tag{2.3}$$

Since $f \in \mathcal{A}_0^{-1}$, $|f(\varphi_a(w))|(1-|\varphi_a(w)|^2) \longrightarrow 0$ uniformly for $|w| \leq r_1$ if $|a| \to 1$, where r_1 is fixed and $0 < r_1 < 1$. Then, by changing variable $z = \varphi_a(w)$, we see that

$$\begin{aligned}
&\lim_{|a|\to 1}\int_{\Delta(a,r_1)} |f(z)|^2 K(g(z,a))dA(z)\\
= &\lim_{|a|\to 1}\int_{|w|\le r_1} |f(\varphi_a(w))|^2(1-|\varphi_a(w)|^2)^2(1-|w|^2)^{-2}\,K\left(\log\frac{1}{|w|}\right)dA(w)\\
\le &\ \mathcal{L}\lim_{|a|\to 1}\sup_{|w|\le r_1} |f(\varphi_a(w))|^2(1-|\varphi_a(w)|^2)^2=0. \qquad (2.4)
\end{aligned}$$

By (2.3) and (2.4), it is easy to see that

$$\lim_{|a|\to 1}\int_{\Delta} |f(z)|^2 K(g(z,a))dA(z)=0.$$

Then, $f\in\mathcal{N}_{K,0}$ and hence (2.1) is a sufficient condition for $\mathcal{N}_{K,0}=\mathcal{A}_0^{-1}$ to hold. Conversely, suppose that (2.1) does not hold; that is

$$\int_0^1 K\left(\log\frac{1}{r}\right)\frac{r}{(1-r^2)^2}\,dr=\infty.$$

Thus we find a continuous strictly decreasing function $h:[0,1)\longrightarrow[0,\infty)$ tending to zero at 1 such that

$$\int_0^1 K\left(\log\frac{1}{r}\right)\frac{h(r)}{(1-r^2)^2}\,r\,dr=\infty. \qquad (2.5)$$

For $\frac{3}{4}\le r<1$ such that $\frac{1}{2}\le 2^k(1-r)<1$, we have $1-2^{-k}<r$. Now, we define $a_k=\big(h(1-2^{-k})\big)^{\frac{1}{2}}$ and consider the gap series

$$f_0(z)=\sum_{k=1}^{\infty} a_k\,2^{\frac{2k}{p}}\,z^{2^k},$$

If

$$L_0(r)=\int_0^{2\pi}|f_0(re^{i\theta})|^2d\theta=2\pi\sum_{k=1}^{\infty}a_k^2r^{2^{k+1}},$$

we argue as in the second half of the proof of Theorem 2.4 to deduce that

$$L_0(r)\ge 2\pi h(r)(1-r)^{-2}e^{-4}=C\,h(r)(1-r)^{-2};\quad \frac{3}{4}\le r<1.$$

Then, it follows that

$$\begin{aligned}
&\int_{\Delta}|f_0(z)|^2K(g(z,a))dA(z)\\
\ge\ &C\int_{\frac{3}{4}}^1 h(r)K\left(\log\frac{1}{r}\right)(1-r^2)^{-2}\,rdr=\infty.
\end{aligned}$$

This means that $f_0\in\mathcal{A}_0^{-1}\setminus\mathcal{N}_{K,0}$, which is a contradiction. Hence (2.1) holds. This completes the proof of our theorem.

3. Hadamard gaps in $\mathcal{N}_K$ space

Recall that the analytic function

$$f(z) = \sum_{k}^{\infty} a_k z^{n_k} \quad (\text{with } n_k \in \mathbb{N}\,;\ \text{for all } k \in \mathbb{N})$$

is said to belong to the Hadamard gap class (also known as lacunary series) if there exists a constant $c > 1$ such that $\frac{n_{k+1}}{n_k} \geq c$ for all $k \in \mathbb{N}$. It is well known that a lacunary series belongs to BMOA space if and only if it is in the Hardy space H^2; see [6]. It is also well known that a lacunary series is in the Bloch space if and only if its Taylor coefficients are bounded; see [31] for instance. In [19], Miao gave characterizations of Besov space by the help of a lacunary series. These characterizations extended to higher dimensions using several complex variables and quaternion sense (see [2, 15]). Lacunary series in Q_p spaces is characterized in [5, 12]. Very recently lacunary series in $\mathcal{N}_p$ spaces are also characterized in [21]. In this section, we obtain the following sufficient condition for functions to be in $\mathcal{N}_K$ classes in terms of Taylor coefficients using a lacunary series.

Theorem 3.1. *If K satisfies condition* (1.2) *and*

$$f(z) = \sum_{n=1}^{\infty} a_n z^{n-1},$$

then

$$\int_{\Delta} |f(z)|^2 \, K\Big(\log \frac{1}{|z|}\Big) dA(z) \approx \sum_{n=1}^{\infty} |a_n|^2 \, K\Big(\frac{1}{n}\Big).$$

Proof. Write

$$I(f) = \int_{\Delta} |f(z)|^2 \, K\Big(\log \frac{1}{|z|}\Big) dA(z).$$

Integrating in polar coordinates leads to

$$I(f) = \sum_{n=1}^{\infty} |a_n|^2 \int_0^1 r^{2n-1} \, K\left(\log \frac{1}{r}\right) dr.$$

We apply Theorem 1.3 with $\beta = 0$ and $\alpha = 2n$ to obtain

$$I(f) \approx \sum_{n=1}^{\infty} |a_n|^2 \, K\left(\frac{1}{2n}\right),$$

by Corollary 1.1. Since $K(2t) \leq cK(t)$, we then obtain

$$I(f) \approx \sum_{n=1}^{\infty} |a_n|^2 \, K\Big(\frac{1}{n}\Big).$$

Theorem 3.2. *If K satisfies condition* (1.2) *and*

$$f(z) = \sum_{j=1}^{\infty} b_j \, z^{n_j - 1},$$

is in the Hadamard gap class, then $f \in \mathcal{N}_K$ if and only if

$$\sum_{j=1}^{\infty} |b_j|^2 \; K\left(\frac{1}{n_j}\right) < \infty. \tag{3.1}$$

Proof. First assume that condition (3.1) holds. We write $z = re^{i\theta}$ in polar form and observe that

$$|f(z)| \leq \sum_{j=1}^{\infty} |b_j| \, r^{n_j - 1}.$$

Then,

$$\begin{aligned}
\|f\|^2_{\mathcal{N}_K} &= \sup_{a\in\Delta} \int_{\Delta} |f(z)|^2 \, K\big(g(z,a)\big) dA(z) \\
&\leq \sup_{a\in\Delta} \int_0^1 \left(\sum_{j=1}^{\infty} |b_j| \, r^{n_j-1}\right)^2 \left(\frac{1}{2\pi}\int_0^{2\pi} K\big(g(re^{i\theta},a)\big) d\theta\right) r dr \\
&\leq \sup_{a\in\Delta} \int_0^1 r^{-1} \left(\sum_{j=1}^{\infty} |b_j| \, r^{n_j}\right)^2 \left(\frac{1}{2\pi}\int_0^{2\pi} K\big(g(re^{i\theta},a)\big) d\theta\right) dr.
\end{aligned}$$

By Proposition 1.4, we may well assume that K is concave. Then

$$\frac{1}{2\pi}\int_0^{2\pi} K\big(g(re^{i\theta},a)\big) d\theta \leq K\left(\frac{1}{2\pi}\int_0^{2\pi} g(re^{i\theta},a) \, d\theta\right)$$

by Jensen's formula (see [13, 24]). We proceed to estimate the integral

$$I(a) = \frac{1}{2\pi}\int_0^{2\pi} g(re^{i\theta},a) \, d\theta = \frac{1}{2\pi}\int_0^{2\pi} \log\left|\frac{1-\overline{a}re^{i\theta}}{re^{i\theta}-a}\right| d\theta.$$

Hence,

$$I(a) \leq \begin{cases} \log\left(\frac{1}{|a|}\right), & 0 < r \leq |a|; \\ \log\left(\frac{1}{r}\right), & |a| < r < 1. \end{cases}$$

In particular,

$$I(a) = \frac{1}{2\pi}\int_0^{2\pi} g(re^{i\theta},a) \, d\theta \leq \log\left(\frac{1}{r}\right).$$

From this we have,

$$\|f\|^2_{\mathcal{N}_K} \leq \sup_{a\in\Delta} \int_0^1 r^{-1} \left[\sum_{j=1}^{\infty} |b_j| \, r^{n_j}\right]^2 K\left(\log\frac{1}{r}\right) dr. \tag{3.2}$$

Use the Cauchy-Schwarz inequality (see [13, 24]) to produce

$$\begin{aligned}\left[\sum_{j=1}^{\infty}|b_j|\,r^{n_j}\right]^2 &= \left[\sum_{n=0}^{\infty}\sum_{n_j\in I_n}|b_j|\,r^{n_j}\right]^2 \leq \left[\sum_{n=0}^{\infty}\sum_{n_j\in I_n}|b_j|\,r^{2^n}\right]^2\\ &\leq \left[\sum_{n=0}^{\infty}2^{\frac{n}{2}}r^{2^n}\right]\left[\sum_{n=0}^{\infty}2^{-\frac{n}{2}}r^{2^n}\Big(\sum_{n_j\in I_n}|b_j|\Big)^2\right]\end{aligned}$$

where $I_n = \{j : 2^n \leq j < 2^{n+1}, j \in \mathbb{N}\}$. To this end, we combine the elementary estimates (see [28] for more details):

$$\begin{aligned}\sum_{n=0}^{\infty}2^{\frac{n}{2}}r^{2^n} &\leq \sqrt{2}\sum_{n=0}^{\infty}\int_{2^n}^{2^{n+1}}t^{-\frac{1}{2}}r^{\frac{t}{2}}\,dt\\ &\leq \sqrt{2}\int_0^{\infty}t^{-\frac{1}{2}}r^{\frac{t}{2}}\,dt\\ &\leq 2\,\Gamma\big(\tfrac{1}{2}\big)\Big(\log\frac{1}{r}\Big)^{-\frac{1}{2}}.\end{aligned}$$

This very useful tool can now be applied to the calculation above to obtain

$$\|f\|^2_{\mathcal{N}_K} \leq 2\,\Gamma\big(\tfrac{1}{2}\big)\sum_{n=0}^{\infty}2^{-\frac{n}{2}}\left[\sum_{n_j\in I_n}|b_j|\right]^2\int_0^1 r^{2n-1}\Big(\log\frac{1}{r}\Big)^{-\frac{1}{2}}K\Big(\log\frac{1}{r}\Big)dr.$$

This together with (3.2) and Theorem 1.3 for $\alpha = 2n, \beta = \frac{1}{2}$, as well as Corollary 1.1 gives

$$\|f\|^2_{\mathcal{N}_K} \lesssim \sum_{n=0}^{\infty}\left[\sum_{n_j\in I_n}|b_j|\right]^2\frac{1}{2^n}K\Big(\frac{1}{2^n}\Big). \tag{3.3}$$

If $n_j \in I_n$, then $n_j < 2^{n+1}$. It follows from the monotonicity of K and Corollary 1.1 that

$$\frac{1}{n_j}K\big(\frac{1}{n_j}\big) \geq \frac{1}{2^{n+1}}K\big(\frac{1}{2^{n+1}}\big) \gtrsim \frac{1}{2^n}K\Big(\frac{1}{2^n}\Big).$$

Combining this with (3.3), we obtain

$$\|f\|^2_{\mathcal{N}_K} \lesssim \sum_{n=0}^{\infty}\left[\sum_{n_j\in I_n}|b_j|\sqrt{\frac{1}{n_j}K\Big(\frac{1}{n_j}\Big)}\ \right]^2 \tag{3.4}$$

Since f is in the Hadamard gap class, there exists a constant c such that $n_{j+1} \geq cn_j$ for all $j \in \mathbb{N}$. Hence, the Taylor series of $f(z)$ has at most $([\log_c 2]+1)$ terms $a_j z^{n_j}$

such that $n_j \in I_n$. By (3.2) and Hölder's inequality,

$$\begin{aligned}\|f\|^2_{\mathcal{N}_K} &\lesssim (\log_c 2+1)\sum_{n=0}^{\infty}\sum_{n_j\in I_n}|b_j|^2\,K\left(\frac{1}{n_j}\right)\\ &= (\log_c 2+1)\sum_{n=1}^{\infty}|b_j|^2\,K\left(\frac{1}{n_j}\right).\end{aligned}$$

This shows that condition (2.1) implies $f \in \mathcal{N}_K$.

Conversely, assume that

$$f(z)=\sum_{n=1}^{\infty} b_j\, z^{n_j-1}$$

is the Hadamard gap in $\mathcal{N}_K$. Then

$$\int_\Delta |f(z)|^2\,K\left(\log\frac{1}{|z|}\right)dA(z)=\int_\Delta |f(z)|^2\,K(g(z,0))dA(z)<\infty,$$

which, according to Theorem 3.1 implies condition (2.1). This completes the proof.

Theorem 3.3. *Let K satisfy condition* (1.2) *and let*

$$\int_0^1 r^{-1}\left[\sum_{j=1}^{\infty}|b_j|\,r^{n_j}\right]^2 K\left(\log\frac{1}{r}\right)dr\ <\infty.$$

If

$$f(z)=\sum_{j=1}^{\infty} b_j\, z^{n_j-1},$$

is in the Hadamard gap class, then $f\in\mathcal{N}_K$ if and only if $f\in\mathcal{N}_{K,0}$.

Proof. First assume that $f\in\mathcal{N}_K$ with Hadamard gap class. We must show that $I(a)\to 0$ as $|a|\to 1^-$, where

$$I(a)=\int_\Delta |f(z)|^2\,K(g(z,a))dA(z)\ ,\qquad a\in\Delta\,.$$

Since,

$$\int_0^1 r^{-1}\left[\sum_{j=1}^{\infty}|b_j|\,r^{n_j}\right]^2 K\left(\log\frac{1}{r}\right)dr\ <\infty,$$

for any given $\varepsilon>0$ there exists some $\delta\in(0,1)$ such that

$$2\int_\delta^1 r^{-1}\left[\sum_{j=1}^{\infty}|b_j|\,r^{n_j}\right]^2 K\left(\log\frac{1}{r}\right)dr\ <\ \varepsilon.$$

We may assume that

$$\lim_{|a|\to 1^-} K\left(\log\frac{1}{|a|}\right)=0.$$

Otherwise, $\mathcal{N}_K$ coincides with the Bergman space $\mathcal{A}^2_\infty$, and the desired result is obvious. We write $I(a) = I_1(a) + I_2(a)$, where

$$I_1(a) = \int_{|z|<\delta} |f(z)|^2 \, K\big(g(z,a)\big) dA(z),$$

and

$$I_2(a) = \int_{\delta\leq|z|<1} |f(z)|^2 \, K\big(g(z,a)\big) dA(z).$$

By arguments used in the proof of Theorem 3.2, we have

$$I_1(a) \leq 2 \; K\left(\log\frac{1}{|a|}\right)\int_0^\delta r^{-1}\left[\sum_{j=1}^\infty |b_j| \, r^{n_j}\right]^2 dr,$$

whenever $\delta < |a| < 1$, because in this case

$$\frac{1}{2\pi}\int_0^{2\pi} g(re^{i\theta},a)\, d\theta = \log\left(\frac{1}{|a|}\right).$$

In particular, $I_1(a) \to 0$ as $|a| \to 1^-$. Similarly, we have

$$I_2(a) \leq 2 \int_0^\delta r^{-1}\left[\sum_{j=1}^\infty |b_j| \, r^{n_j}\right]^2 K\left(\log\frac{1}{r}\right) dr \leq \varepsilon.$$

It follows that

$$\lim_{|a|\to 1^-} \sup I(a) \leq \varepsilon.$$

Since ε is arbitrary, we conclude that $I(a) \to 0$ as $|a| \to 1^-$. So $f \in \mathcal{N}_{K,0}$ and the proof is complete.

Carefully checking the proof of (3.2) and (3.3), we also obtain the following sufficient condition for a function to be in $\mathcal{N}_{K,0}$(and hence in $\mathcal{N}_K$) in terms of Taylor coefficients.

Theorem 3.4. *If K satisfies condition* (1.2) *and*

$$f(z) = \sum_{j=1}^\infty b_j \, z^{n_j-1},$$

satisfies the condition

$$\sum_{n=1}^\infty\left[\sum_{j\in I_n} |b_j|\right]^2 \frac{1}{2^{n+1}} \, K\left(\frac{1}{n}\right) < \infty,$$

then $f \in \mathcal{N}_{K,0}$, where $I_n = \{j : 2^n \leq j < 2^{n+1}, j \in \mathbb{N}\}$.

4. Composition operators on $\mathcal{N}_K$ classes

There have been several attempts to study compactness and boundedness of composition operators in many function spaces (see, e.g., [1, 8, 14, 16, 18, 20, 25] and others). For comprehensive treatments of composition operators we refer to [9] and [22]. In this section we study composition operators acting on $\mathcal{N}_K$ space.

In this section we let $\phi \in H(\Delta)$ to denote a non-constant function satisfying $\phi(\Delta) \subseteq \Delta$. The composition operator $C_\phi : H(\Delta) \to H(\Delta)$ is defined by $C_\phi = f \circ \phi$. First, in the following result, we describe boundedness for our $\mathcal{N}_K$ classes.

Theorem 4.1. *Let $K : [0, \infty) \to [0, \infty)$ be a nondecreasing function and $\alpha \in (0, \infty)$. Then $C_\phi : \mathcal{N}_K \to \mathcal{A}_\alpha$ is bounded if and only if*

$$\sup_{z\in\Delta} \frac{(1-|z|^2)^\alpha}{1-|\phi(z)|^2} < \infty. \tag{4.1}$$

Proof. First assume that condition (4.1) holds. Then

$$\begin{aligned}
\|C_\phi f\|_{\mathcal{A}_\alpha} &= \sup_{z\in\Delta} |f(\phi(z))|(1-|z|^2)^\alpha \\
&\le \sup_{z\in\Delta} \frac{(1-|z|^2)^\alpha}{1-|\phi(z)|^2} \sup_{z\in\Delta} |f(\phi(z))|(1-|\phi(z)|^2) \\
&\le \|f\|_{\mathcal{A}^{-1}} \sup_{z\in\Delta} \frac{(1-|z|^2)^\alpha}{1-|\phi(z)|^2} \\
&\le \lambda \|f\|_{\mathcal{N}_K},
\end{aligned}$$

where λ is a positive constant. Conversely, assume that $C_\phi : \mathcal{N}_K \to \mathcal{A}_\alpha$ is bounded, then $\|C_\phi f\|_{\mathcal{A}_\alpha} \le \|f\|_{\mathcal{N}_K}$. Fix a point $z_0 \in \Delta$, and let h_w be the test function in Lemma 2.2 with $w = \phi(z_0)$. Then,

$$1 \ge ||h_w||_{\mathcal{N}_K} \ge \lambda_1 ||h_w \circ (\phi)||_{\mathcal{A}_\alpha} \ge \frac{1-|w|^2}{|1-\overline{w}\phi(z_0)|^2}(1-|z_0|^2)^\alpha = \frac{(1-|z_0|^2)^\alpha}{1-|\phi(z_0)|^2},$$

where λ_1 is a positive constant.

Theorem 4.2. *$C_\phi : \mathcal{N}_K \to \mathcal{A}_\alpha$ is compact if and only if*

$$\lim_{r\to 1} \sup_{|\phi(z)|>r} \frac{(1-|z|^2)^\alpha}{1-|\phi(z)|^2} = 0. \tag{4.2}$$

Proof. First assume that $C_\phi : \mathcal{N}_K \to \mathcal{A}_\alpha$ is compact and suppose that there exists $\varepsilon_0 > 0$ a sequence $\{z_n\} \subset \Delta$ such that

$$\frac{(1-|z_n|^2)^\alpha}{1-|\phi(z_n)|^2} \ge \varepsilon_0 \quad \text{whenever} \quad |\phi(z_n)| > 1 - \frac{1}{n}.$$

Clearly, we can assume that $w_n = \phi(z_n)$ tends to $w_0 \in \partial\Delta$ as $n \to \infty$. Let $h_{w_n} = \frac{(1-|w_n|^2)}{(1-\overline{w_n}z)^2}$ be the function in Lemma 2.2. Then $h_{w_n} \to h_{w_0}$ with respect to the compact-open topology. Define $f_n = h_{w_n} - h_{w_0}$. Then $\|f_n\|_{\mathcal{N}_K} \le 1$ (see Lemma

2.2) and $f_n \to 0$ uniformly on compact subsets of Δ. Thus, $f_n \circ \phi \to 0$ in $\mathcal{A}_\alpha$ by assumption. But, for n big enough,

$$\begin{aligned} \|C_\phi f_n\|_{\mathcal{A}_\alpha} &\geq |h_{w_n}(\phi(z_n)) - h_{w_0}(\phi(z_n))|(1-|z_n|^2)^\alpha \\ &\geq \underbrace{\frac{(1-|z_n|^2)^\alpha}{1-|\phi(z_n)|^2}}_{\geq \varepsilon_0} \underbrace{\left|1 - \frac{(1-|w_n|^2)(1-|w_0|^2)}{|1-\overline{w_0}\, w_n|}\right|}_{=1}, \end{aligned}$$

which is a contradiction. Conversely, assume that for all $\varepsilon > 0$ there exists $\delta \in (0,1)$ such that $\frac{(1-|z|^2)^\alpha}{1-|\phi(z)|^2} < \varepsilon$ whenever $|\phi(z)| > \delta$. Let $\{f_n\}$ be a bounded sequence in $\mathcal{N}_K$ norm which converges to zero on compact subsets of Δ. Clearly, we may assume that $|\phi(z)| > \delta$. Then

$$\begin{aligned} \|C_\phi f_n\|_{\mathcal{A}_\alpha} &= \sup_{z\in\Delta} |f_n(\phi(z))|(1-|z|^2)^\alpha \\ &= \sup_{z\in\Delta} \frac{(1-|z|^2)^\alpha}{1-|\phi(z)|^2} |f_n(\phi(z))|(1-|\phi(z)|^2). \end{aligned}$$

It is not hard to show that $\|.\|_{\mathcal{A}^{-1}} \lesssim \|.\|_{\mathcal{N}_K}$. Thus, we obtain that

$$\|C_\phi f_n\|_{\mathcal{A}_\alpha} \leq \varepsilon \, \|f_n\|_{\mathcal{A}^{-1}} \leq \varepsilon \, \|f_n\|_{\mathcal{N}_K} \leq \varepsilon.$$

It follows that C_ϕ is a compact operator. This completes the proof of the theorem.

Acknowledgement

The authors would like to thank the referees for their valuable comments and interesting suggestions on this paper.

References

[1] A. El-Sayed Ahmed and M.A. Bakhit, *Composition operators on some holomorphic Banach function spaces*, Mathematica Scandinavica 104 (2), (2009), 275–295.

[2] A. Avetsiyan, *Hardy-Bloch type spaces and lacunary series in the polydisk*, Glasg. Math. J. 49 No. 2 (2007), 345–356.

[3] R. Aulaskari, *Complex function spaces.* Proceedings of the summer school, Mekrijärvi, Finland, August 30–September 3, 1999. Report Series. Department of Mathematics, University of Joensuu. 4. Joensuu: Joensuu University, ISSN 1455-805X (2001).

[4] R. Aulaskari and P. Lappan, *Criteria for an analytic function to be Bloch and a harmonic or meromorphic function to be normal*, Complex Analysis and its Applications (eds. Y. Chung-Chun et al.), Pitman Research Notes in Mathematics 305, Longman (1994), 136–146.

[5] R. Aulakari, D. Stegenga and J. Xiao, *Some subclasses of BMOA and their characterization in terms of Carleson measures*, Rocky Mountain J. Math. 26 (1996), 485–506.

[6] A. Bernstein, *Analytic functions of Bounded mean oscillation II*, In aspects of contemporary complex Analysis, Academic Press London (1980), 3–36.

[7] A. Borichev, H. Hedenmalm and K. Zhu, *Bergman spaces and related topics in Complex Analysis*, American Mathematical Society, 2006.

[8] B.S. Bourdon, J.A. Cima and A.L. Matheson, *Compact composition operators on BMOA*, Trans. Amer. Math. Soc. 351 (1999), 2183–2169.

[9] C.C. Cowen and B.D. MacCluer, *Composition operators on spaces of analytic functions*, CRC Press, (1995).

[10] M. Essén and H. Wulan, *On analytic and meromorphic functions and spaces of $\mathcal{Q}_K$*, Illinois J. Math. 46 (2002), 1233–1258.

[11] M. Essén, H. Wulan and J. Xiao, *Function-theoretic aspects of Möbius invariant Q_K spaces*, J. Funct. Anal. 230 (2006), 78–115.

[12] P. Galanopoulos, *On $\mathcal{B}_{\log}$ to $\mathcal{Q}^p_{\log}$ pullbacks*, J. Math. Anal. Appl. Vol. 337(2008), 712–725.

[13] J. Garnett, *Bounded analytic functions*, Academic Press, New York, (1981).

[14] P. Ghatage, J. Yan, and D. Zheng, *Composition operators with closed range on the Bloch space*, Proc. Amer. Math. Soc. Vo. 129 N. 7 (2000), 2039–2044.

[15] K. Gürlebeck and A. El-Sayed Ahmed, *On series expansions of hyperholomorphic* $\mathbf{B}^{\mathbf{q}}$ *functions*, In Tao Qian et al. (Eds.), Trends in Mathematics Advances in Analysis and Geometry, (Basel/Switzerland: Birkhäuser Verlag Publisher) (2004), 113–129.

[16] L. Jiang and Y. He, *Composition operators from $\mathcal{B}^\alpha$ to $F(p,q,s)$*, Acta Math. Sci. Ser B 32, (2003), 252–260

[17] M. Lindström and N. Palmberg, *Spectra of composition operators on BMOA*, Integr. Equ. Oper. Theory, 53 (2005), 75–86.

[18] S. Li and H. Wulan, *Composition operators on $\mathcal{Q}_K$ spaces*, J. Math. Anal. Appl. 327(2007), 948–958.

[19] J. Miao, *A property of Analytic functions with Hadamard gaps*, Bull. Austral. Math. Soc. 45(1992), 105–112.

[20] K. Madigan and A. Matheson, *Compact composition operators on the Bloch space*, Trans. Amer. Math. Soc. 347 (1995), 2679–2687.

[21] N. Palmberg, *Composition operators acting on $\mathcal{N}_p$-spaces*, Bull. Belg. Math. Soc. Simon Stevin, 14 (2007), 545–554.

[22] J.H. Shapiro, *Composition operators and classical function theory*, Springer, (1993).

[23] K. Stroethoff, *Besov-type characterisations for the Bloch space*, Bull. Austral. Math. Soc. 39 (1989), 405–420.

[24] W. Rudin, *Real and Complex Analysis, McGraw-Hill, New York*, 1987.

[25] W. Smith, *Composition operators between Bergman and Hardy spaces*, Trans. Amer. Math. Soc. 348(1996), 2331–2348.

[26] H. Wulan and P. Wu, *Characterizations of $\mathcal{Q}_T$ spaces*, J. Math. Anal. Appl. 254(2001), 484–497.

[27] H. Wulan and J. Zhou, *The higher order derivatives of Q_K type spaces*, J. Math. Anal. Appl. 332, No. 2(2007), 1216–1228.

[28] H. Wulan and K. Zhu, *Lacunary spaces in $\mathcal{Q}_K$ spaces*, Studia Math. 178 (2007), 217–230.

[29] H. Wulan and K. Zhu, *Lipschitz type characterizations for Bergman spaces*, to appear.

[30] R. Zhao, *On α-Bloch functions and $VMOA$*, Acta. Math. Sci. Vol. 3(1996), 349–360.

[31] K. Zhu, *Operator theory in function spaces*, Marcel Dekker, New York, 1990.

A. El-Sayed Ahmed and M.A. Bahkit
Sohag University
Faculty of Science
Mathematics Department
82524 Sohag – Egypt
e-mail: ahsayed80@hotmail.com
mabakhit2008@yahoo.com

Operator Theory:
Advances and Applications, Vol. 195, 139–166

A Time-variant Norm Constrained Interpolation Problem Arising from Relaxed Commutant Lifting

A.E. Frazho, S. ter Horst and M.A. Kaashoek

Abstract. A time-variant analogue of an interpolation problem equivalent to the relaxed commutant lifting problem is introduced and studied. In a somewhat less general form the problem already appears in the analysis of the set of all solutions to the three chain completion problem. The interpolants are upper triangular operator matrices of which the columns induce contractive operators. The set of all solutions of the problem is described explicitly. The results presented are time-variant analogues of the main theorems in [23].

Mathematics Subject Classification (2000). Primary 47A57; Secondary 47A20, 30E05.

Keywords. Norm constrained interpolation, contractions, upper triangular operator matrices, positive real operator matrices, harmonic majorants, parametrization of all solutions.

0. Introduction

Time-variant versions of metric constrained interpolation problems and time-varying linear system theory have been intensively studied since the early 1990's; see the papers [1, 2, 7, 8, 13] and the books [14, 25, 20, 15] for a general overview and additional references. The connection with commutant lifting theory was made in [18], where a time-varying analogue of the commutant lifting theorem, known as the three chain completion theorem, was proved. An early version of a time-variant commutant lifting theorem appeared in Ball-Gohberg [6], which was later extended to the setting of nest algebras in [30] (see also [12]); the connection with the three chain theorem is explained in [5]. One of the recent developments in commutant lifting theory is the introduction of a relaxation of the commutant lifting setting in [21]. In the present paper we consider a time-variant norm constrained abstract interpolation problem, which in the time-invariant case is equivalent to the relaxed commutant lifting problem [23].

To state the interpolation problem considered in this paper we need some notation. Throughout $\mathcal{U}_k$ and $\mathcal{Y}_k$ are Hilbert spaces with k being an arbitrary integer, and the symbols $\mathbf{U}$ and $\mathbf{Y}$ stand for the Hilbert direct sums $\oplus_{k\in\mathbb{Z}}\mathcal{U}_k$ and $\oplus_{k\in\mathbb{Z}}\mathcal{Y}_k$, respectively. We shall consider operator matrices $H = [H_{j,\,k}]_{j,\,k\in\mathbb{Z}}$ of which the $(j,\,k)$th entry $H_{j,\,k}$ is an operator from $\mathcal{U}_k$ into $\mathcal{Y}_j$. The set of all such operator matrices will be denoted by $\mathbb{M}(\mathbf{U},\mathbf{Y})$. By $\mathbb{UM}(\mathbf{U},\mathbf{Y})$ we denote the subset of $\mathbb{M}(\mathbf{U},\mathbf{Y})$ consisting of all $H = [H_{j,\,k}]_{j,\,k\in\mathbb{Z}}$ that are *upper triangular*, that is, $H_{j,\,k} = 0$ for each $k < j$.

In the present paper we are particularly interested in those $H = [H_{j,\,k}]_{j,\,k\in\mathbb{Z}}$ in $\mathbb{UM}(\mathbf{U},\mathbf{Y})$ that have the additional property

$$\sum_{j=-\infty}^{k} \|H_{j,\,k}u_k\|^2 \leq c_H\|u_k\|^2, \quad u_k \in \mathcal{U}_k \quad (k \in \mathbb{Z}), \tag{0.1}$$

where c_H is some constant depending on H only. The set of all such operator matrices is denoted by $\mathbb{UM}^2(\mathbf{U},\mathbf{Y})$. We say that H belongs to $\mathbb{UM}^2_{\text{ball}}(\mathbf{U},\mathbf{Y})$ whenever the constant c_H can be taken equal to one. Thus an upper triangular operator matrix H belongs to $\mathbb{UM}^2_{\text{ball}}(\mathbf{U},\mathbf{Y})$ if and only if for each $k \in \mathbb{Z}$ the kth column of H induces a contractive operator from $\mathcal{U}_k$ into $\mathbf{Y} = \oplus_{k\in\mathbb{Z}}\mathcal{Y}_k$. The following is the main problem treated in this paper.

Problem 0.1. *Assume that for each $k \in \mathbb{Z}$ we have given a subspace $\mathcal{F}_k$ of $\mathcal{U}_k$ and a contraction*

$$\omega_k = \begin{bmatrix}\omega_{k,\,1}\\ \omega_{k,\,2}\end{bmatrix} : \mathcal{F}_k \to \begin{bmatrix}\mathcal{Y}_k\\ \mathcal{U}_{k-1}\end{bmatrix}. \tag{0.2}$$

Given this data, find all $H = [H_{j,\,k}]_{j,\,k\in\mathbb{Z}}$ in $\mathbb{UM}^2_{\text{ball}}(\mathbf{U},\mathbf{Y})$ such that for each $k \in \mathbb{Z}$ the following interpolation conditions hold:

$$H_{k,\,k}|_{\mathcal{F}_k} = \omega_{k,\,1}, \quad H_{j,\,k}|_{\mathcal{F}_k} = H_{j,\,k-1}\omega_{k,\,2} \quad (j,k \in \mathbb{Z}, j < k). \tag{0.3}$$

In the time-invariant case, the spaces $\mathcal{F}_k = \mathcal{F}$, $\mathcal{U}_k = \mathcal{U}$, and $\mathcal{Y}_k = \mathcal{Y}$ and the contraction $\omega_k = \omega$ do not depend on k, and the operators $H_{j,k}$ depend only on the difference $j - k$. In this setting, the above problem reduces to the function theory problem considered in the first paragraph of [23]. To see this, note that in this time-invariant setting the operator matrix H can be identified with the $\mathcal{L}(\mathcal{U},\mathcal{Y})$-valued function F_H, analytic on the open unit disc $\mathbb{D}$, given by

$$F_H(\lambda) = \sum_{\nu=0}^{\infty} \lambda^\nu H_{-\nu}.$$

Moreover in this case the interpolation condition and the norm constraint in Problem 0.1 can be restated as

$$\omega_1 + \lambda F_H(\lambda)\omega_2 = F_H(\lambda)|_{\mathcal{F}} \quad (\lambda \in \mathbb{D}) \quad \text{and} \quad \sum_{\nu=0}^{\infty} \|H_{-\nu}u\|^2 \leq \|u\|^2 \quad (u \in \mathcal{U}).$$

For a particular choice of the contractions ω_k, Problem 0.1 appears in a natural way in the analysis of the set of all solutions to the three chain completion problem [18], [19]. Indeed, see Section 4 in [19] or Section XIV.3 in [20], where one can find Problem 0.1 with ω_k being an isometry for each $k \in \mathbb{Z}$.

To state our first main result some additional notation is needed. We use the symbol $\mathbb{UM}^\infty(\mathbf{U},\mathbf{Y})$ to denote the set of all double infinite upper triangular operator matrices H that induce bounded linear operators from the Hilbert space $\mathbf{U} = \oplus_{k\in\mathbb{Z}}\mathcal{U}_k$ into the Hilbert space $\mathbf{Y} = \oplus_{k\in\mathbb{Z}}\mathcal{Y}_k$. If this induced operator is a contraction, then we say that H belongs to $\mathbb{UM}^\infty_{\text{ball}}(\mathbf{U},\mathbf{Y})$. In particular, $\mathbb{UM}^\infty_{\text{ball}}(\mathbf{U},\mathbf{Y}) \subset \mathbb{UM}^2_{\text{ball}}(\mathbf{U},\mathbf{Y})$. We write $\mathbb{UM}^\infty_0(\mathbf{U},\mathbf{Y})$ and $\mathbb{UM}^\infty_{\text{ball},0}(\mathbf{U},\mathbf{Y})$ for the sets of all strictly upper triangular operator matrices in $\mathbb{UM}^\infty(\mathbf{U},\mathbf{Y})$ and $\mathbb{UM}^\infty_{\text{ball}}(\mathbf{U},\mathbf{Y})$, respectively. Finally, when $\mathcal{U}_k = \mathcal{Y}_k$ for each $k \in \mathbb{Z}$, and hence $\mathbf{U} = \mathbf{Y}$, we shall always replace the argument $(\mathbf{U},\mathbf{Y})$ by $(\mathbf{U})$. Thus $\mathbb{M}(\mathbf{U})$ stands for $\mathbb{M}(\mathbf{U},\mathbf{U})$, and $\mathbb{UM}(\mathbf{U})$ stands for $\mathbb{M}(\mathbf{U},\mathbf{U})$, etc. We are now ready to state the first main result.

Theorem 0.1. *For each $k \in \mathbb{Z}$ let ω_k be the contraction given by* (0.2). *Choose*

$$Z^{(1)} = \left[(Z^{(1)}_{j,k})\right]_{j,k\in\mathbb{Z}} \in \mathbb{UM}^\infty(\mathbf{U},\mathbf{Y}),\ Z^{(2)} = \left[Z^{(2)}_{j,k}\right]_{j,k\in\mathbb{Z}} \in \mathbb{UM}^\infty_0(\mathbf{U}), \tag{0.4}$$

such that

$$\|Z^{(1)}u\|^2 + \|Z^{(2)}u\|^2 \le \|u\|^2 \quad (u \in \mathbf{U}) \tag{0.5}$$

and

$$Z^{(1)}_{j,k}|_{\mathcal{F}_k} = \begin{cases} \omega_{k,1}, & \text{if } j = k, \\ 0, & \text{if } j \neq k, \end{cases} \qquad Z^{(2)}_{j,k}|_{\mathcal{F}_k} = \begin{cases} \omega_{k,2}, & \text{if } j = k-1, \\ 0, & \text{if } j \neq k-1. \end{cases} \tag{0.6}$$

Then

$$H = Z^{(1)}(I_{\mathbf{U}} - Z^{(2)})^{-1} \tag{0.7}$$

is a well-defined upper triangular operator matrix, H belongs to $\mathbb{UM}^2_{\text{ball}}(\mathbf{U},\mathbf{Y})$, and H is a solution of Problem 0.1.

Let us explain why formula (0.7) makes sense. In general, for arbitrary infinite operator matrices the usual matrix product is not defined. The situation is different for upper triangular matrices. For instance, for $A \in \mathbb{UM}(\mathbf{U},\mathbf{Y})$ and $B \in \mathbb{UM}(\mathbf{U})$ the matrix product AB is well defined and AB belongs to $\mathbb{UM}(\mathbf{U},\mathbf{Y})$. Furthermore, with the usual matrix product $\mathbb{UM}(\mathbf{U})$ is an algebra with the identity matrix $I_{\mathbf{U}}$ as a unit and an operator matrix $M = [M_{j,k}]_{j,k\in\mathbb{Z}} \in \mathbb{UM}(\mathbf{U})$ is invertible in $\mathbb{UM}(\mathbf{U})$ if and only if for each $j \in \mathbb{Z}$ the jth diagonal entry $M_{j,j}$ is invertible as an operator on $\mathcal{U}_j$ (see Subsection 1.2 below for further details). From these remarks is clear that the operator $I_{\mathbf{U}} - Z^{(2)}$ in (0.7) is invertible in $\mathbb{UM}(\mathbf{U})$ and that the product in (0.7) is well defined.

One can always find $Z^{(1)}$ and $Z^{(2)}$ satisfying the conditions (0.4), (0.5) and (0.6) in Theorem 0.1. For instance one can take

$$Z^{(1)}_{j,k} = \delta_{j,k}\omega_{k,1}\Pi_{\mathcal{F}_k} \quad \text{and} \quad Z^{(2)}_{j,k} = \delta_{j,k-1}\omega_{k,2}\Pi_{\mathcal{F}_k}. \tag{0.8}$$

Here $\delta_{j,k}$ is the Kronecker delta and the map $\Pi_{\mathcal{F}_j}$ is the orthogonal projection of $\mathcal{U}_j$ onto $\mathcal{F}_j$. The solution of Problem 0.1 corresponding to this choice of $Z^{(1)}$ and $Z^{(2)}$ is given by $\tilde{H} = [\tilde{H}_{j,k}]_{j,k\in\mathbb{Z}}$ with

$$\tilde{H}_{j,k} = \begin{cases} 0, & j > k, \\ \omega_{k,1}\Pi_{\mathcal{F}_k}, & j = k, \\ (\omega_{j,1}\Pi_{\mathcal{F}_j})(\omega_{j+1,2}\Pi_{\mathcal{F}_{j+1}})\cdots(\omega_{k-1,2}\Pi_{\mathcal{F}_{k-1}})(\omega_{k,2}\Pi_{\mathcal{F}_k}), & j < k. \end{cases}$$

Thus Problem 0.1 is always solvable.

Our second main result shows, in particular, that the method of Theorem 0.1 gives all solutions to Problem 0.1, that is, given a solution H to Problem 0.1, there exists a pair of operator matrices $(Z^{(1)}, Z^{(2)})$ satisfying (0.4), (0.5) and (0.6) such that H is given by the formula (0.7). In general, such a pair $(Z^{(1)}, Z^{(2)})$ is not uniquely determined by H. This phenomenon already appears in the time-invariant case and can be illustrated by simple examples. For instance, assume all spaces $\mathcal{F}_k$ consist of the zero element only. In that case $H = 0$ is in $\mathbb{UM}^2_{\text{ball}}(\mathbf{U}, \mathbf{Y})$ and is a solution, while (0.7) holds with $Z^{(1)} = 0$ and with any $Z^{(2)}$ from $\mathbb{UM}^\infty_0(\mathbf{U})$.

Given a solution H to Problem 0.1, we shall describe the set of all pairs $(Z^{(1)}, Z^{(2)})$ satisfying (0.4), (0.5) and (0.6) stated in the above theorem such that H is given by (0.7). The precise result is given by Theorem 4.1 in Section 4. Here we only describe some of the main ingredients entering into the proof and present an abbreviated version of this theorem.

Let H be a solution to Problem 0.1, and let H_k be the kth column of H. Recall that H_k defines a contraction from $\mathcal{U}_k$ into $\mathbf{Y}$. Let $D_{H_k} = (I_{\mathcal{U}_k} - (H_k)^* H_k)^{1/2}$ denote the corresponding defect operator, and let $\mathcal{D}_{H_k}$ be the corresponding defect space, i.e., $\mathcal{D}_{H_k}$ is the closure of the range of D_{H_k} in $\mathcal{U}_k$. Since (0.3) is satisfied, for each $f \in \mathcal{F}_k$ we have

$$\begin{aligned} \|D_{H_k} f\|^2 &= \|f\|^2 - \|H_k f\|^2 = \|f\|^2 - \|\omega_{k,1} f\|^2 - \|H_{k-1}\omega_{k,2} f\|^2 \\ &= \|f\|^2 - \|\omega_{k,1} f\|^2 - \|\omega_{k,2} f\|^2 + \|\omega_{k,2} f\|^2 - \|H_{k-1}\omega_{k,2} f\|^2 \\ &= \|D_{\omega_k} f\|^2 + \|D_{H_{k-1}}\omega_{k,2} f\|^2. \end{aligned}$$

Hence we can define a contraction ω_{H_k} by

$$\omega_{H_k} : \mathcal{F}_{H_k} := \overline{D_{H_k}\mathcal{F}_k} \to \mathcal{D}_{H_{k-1}}, \quad \omega_{H_k} D_{H_k}|_{\mathcal{F}_k} = D_{H_{k-1}}\omega_{k,2}. \tag{0.9}$$

Now put $\mathbf{D}_H = \oplus_{k\in\mathbb{Z}}\mathcal{D}_{H_k}$, and let $\mathbf{C}_{H,\omega}$ be the set of all operator matrices $C = [C_{j,k}]_{j,k\in\mathbb{Z}}$ in $\mathbb{UM}^\infty_{\text{ball},0}(\mathbf{D}_H)$ such that

$$C_{j,k}|_{\mathcal{F}_{H_k}} = \begin{cases} \omega_{H_k}, & \text{if } j = k-1, \\ 0, & \text{if } j \neq k-1. \end{cases} \tag{0.10}$$

Thus

$$\mathbf{C}_{H,\omega} = \{C \in \mathbb{UM}^\infty_{\text{ball},0}(\mathbf{D}_H) \mid \text{the } (j,k)\text{th entry } C_{j,k} \text{ of } C \text{ satisfies (0.10) for each } j,k \in \mathbb{Z}\}. \tag{0.11}$$

We can now state the abbreviated version of our second main result.

Theorem 0.2. *Let* H *be a solution to* Problem 0.1. *Then there exists a pair of operator matrices* $(Z^{(1)}, Z^{(2)})$ *satisfying* (0.4), (0.5) *and* (0.6) *such that* H *is given by* (0.7). *Furthermore, the set of all such pairs* $(Z^{(1)}, Z^{(2)})$ *is in one-to-one correspondence with the set* $\mathbf{C}_{H,\omega}$.

The full version of the above theorem (see Theorem 4.1 below) will also present necessary and sufficient conditions guaranteeing that the set $\mathbf{C}_{H,\omega}$ consists of a single element only.

Let H be a solution to Problem 0.1. In the analysis of the set of all pairs $(Z^{(1)}, Z^{(2)})$ satisfying (0.4)–(0.7) the following problem enters in a natural way.

Problem 0.2. *Given* $H \in \mathbb{UM}^2_{\rm ball}(\mathbf{U}, \mathbf{Y})$, *describe the set of operator matrices* F *in* $\mathbb{UM}(\mathbf{U})$ *satisfying*

$$F + F^* \geq H^*H + I_{\mathbf{U}} \quad \text{and} \quad F_{j,j} = I_{\mathcal{U}_j} \quad (j \in \mathbb{Z}). \tag{0.12}$$

Note that the matrix product H^*H is well defined because each column of H induces a contractive operator (see Subsection 1.4 for further details). The inequality sign in (0.12) means that the operator matrix $\frac{1}{2}(F + F^*) - H^*H - I_{\mathbf{U}}$ is non-negative (see Subsection 1.3 for the definition of this notion and further details). The fact that (0.12) appears in the analysis, follows from the observation that $F = (I_{\mathbf{U}} - Z^{(2)})^{-1}$ satisfies (0.12) whenever the pair $(Z^{(1)}, Z^{(2)})$ satisfies the conditions (0.4)–(0.7). This connection will be made more precise in Theorem 2.1. The solution to Problem 0.2 will be obtained as a corollary to Theorem 3.2.

For the time-invariant case Theorems 0.1 and Theorem 0.2 can be found in [23]. By using the reduction techniques developed in Chapter X of [20] (also [17]) Problem 0.1 can be transformed into a problem of the type considered in [23]. This transformation together with techniques from [20] can be used to present an alternative way to prove our main results. We shall not develop this approach in the present paper. Theorems 0.1 and Theorem 0.2 can be used to solve a time variant analogue of the relaxed commutant lifting problem. We will describe this connection in the final section of the paper.

This paper consists of five sections not counting this introduction. The first section has a preliminary character. We introduce some additional notation and recall a number of elementary facts about operator matrices that will be used in the proofs. In Section 2 we outline a general approach to deal with Problem 0.1 and prove Theorem 0.1. Section 3 is divided into three subsections. In this section a time-variant analogue of the Cayley transform is used to relate operator matrices from $\mathbb{UM}^\infty_{\rm ball,0}(\mathbf{U}, \mathbf{Y})$ to positive real operator matrices from $\mathbb{UM}(\mathbf{U})$. We apply this result to solve Problem 0.2 and to parameterize the set of all its solutions. Theorem 0.2 is proved in Section 4; this section also presents the full version of Theorem 0.2 and its proof. Here we also discuss the problem of finding necessary and sufficient conditions for the existence of a unique solution to Problem 0.1. In the final section an example involving finite operator matrices will be presented and we discuss the connection with a time-variant analogue of the relaxed commutant lifting problem.

1. Preliminaries

In this section we bring together a number of elementary facts about operator matrices that will be used in the sequel. In what follows we assume the reader to be familiar with the notations introduced in the previous section.

1.1. The set $\mathbb{M}(\mathbf{U},\mathbf{Y})$ and bounded operators

The set $\mathbb{M}(\mathbf{U},\mathbf{Y})$ is a linear space with respect to the usual operation of matrix addition. Given $M = [M_{j,k}]_{j,k\in\mathbb{Z}}$ in $\mathbb{M}(\mathbf{U},\mathbf{Y})$ and $j \leq k$ we write $\Delta_{j,k}(M)$ for the $\{j,k\}$-*finite section* of M, that is,

$$\Delta_{j,k}(M) = \begin{bmatrix} M_{j,j} & \cdots & M_{j,k} \\ \vdots & & \vdots \\ M_{k,j} & \cdots & M_{k,k} \end{bmatrix}. \tag{1.1}$$

Note that $\Delta_{j,k}(M)$ defines a bounded linear operator from $\mathcal{U}_j \oplus \cdots \oplus \mathcal{U}_k$ into $\mathcal{Y}_j \oplus \cdots \oplus \mathcal{Y}_k$.

In general, an operator matrix $M \in \mathbb{M}(\mathbf{U},\mathbf{Y})$ does not induce in a canonical way a bounded operator from $\mathbf{U} = \oplus_{k\in\mathbb{Z}} \mathcal{U}_k$ into the space $\mathbf{Y} = \oplus_{k\in\mathbb{Z}} \mathcal{Y}_k$. In order for this to happen it is necessary and sufficient that

$$\sup_{j\leq k} \|\Delta_{j,k}(M)\| < \infty. \tag{1.2}$$

Furthermore, if (1.2) is satisfied, then the quantity in the left-hand side of (1.2) is equal to the norm of $M = [M_{j,k}]_{j,k\in\mathbb{Z}}$ as an operator from $\mathbf{U}$ into $\mathbf{Y}$.

1.2. Invertibility in the algebra $\mathbb{UM}(\mathbf{U})$

Let $\mathbf{X} = \oplus_{k\in\mathbb{Z}} \mathcal{X}_k$, $\mathbf{U} = \oplus_{k\in\mathbb{Z}} \mathcal{U}_k$, and $\mathbf{Y} = \oplus_{k\in\mathbb{Z}} \mathcal{Y}_k$ be Hilbert space direct sums. If $B \in \mathbb{UM}(\mathbf{X},\mathbf{U})$ and $A \in \mathbb{UM}(\mathbf{U},\mathbf{Y})$, then the (block) matrix product AB is well defined and $AB \in \mathbb{UM}(\mathbf{X},\mathbf{Y})$. Moreover, for $C \in \mathbb{UM}(\mathbf{X},\mathbf{Y})$ we have

$$AB = C \quad \Longleftrightarrow \quad \Delta_{j,k}(A)\Delta_{j,k}(B) = \Delta_{j,k}(C) \ (j \leq k). \tag{1.3}$$

In particular, the set $\mathbb{UM}(\mathbf{U})$ is closed under the usual multiplication of matrices. In fact, from (1.3) we see that $\mathbb{UM}(\mathbf{U})$ is an algebra with the identity matrix $I_{\mathbf{U}}$ as a unit. From (1.3) it also follows that the operator matrix $M = [M_{i,j}]_{i,j\in\mathbb{Z}} \in \mathbb{UM}(\mathbf{U})$ is invertible in $\mathbb{UM}(\mathbf{U})$ if and only if for each $j \in \mathbb{Z}$ the jth diagonal entry $M_{j,j}$ is an invertible operator on $\mathcal{U}_j$. In that case, we have

$$\Delta_{j,k}(M^{-1}) = \Delta_{j,k}(M)^{-1} \quad (j,k \in \mathbb{Z}, j \leq k). \tag{1.4}$$

In particular, the (j,j)th entry of M^{-1} is equal to $M_{k,k}^{-1}$.

The above-mentioned properties of $\mathbb{UM}(\mathbf{U})$ also follow from the fact that an operator matrix from $\mathbb{UM}(\mathbf{U})$ can be identified in the usual way with a linear transformation on the linear space $\mathbf{U}^+$. By definition, the space $\mathbf{U}^+$ consists of all double infinite one column matrices $\mathbf{u} = [u_j]_{j\in\mathbb{Z}}$, with $u_j \in \mathcal{U}_j$ for each $j \in \mathbb{Z}$, such that $u_\nu = 0$ for $\nu > \ell$, for some ℓ depending on $\mathbf{u}$.

1.3. Hermitian and non-negative operator matrices

An operator matrix $M \in \mathbb{M}(\mathbf{U})$, $M = [M_{j,k}]_{j,k\in\mathbb{Z}}$, is said to be *Hermitian* if $M^* = M$, where M^* is the operator matrix $M^* = [(M_{k,j})^*]_{j,k\in\mathbb{Z}}$. The *real part* of $M \in \mathbb{M}(\mathbf{U})$ is the operator matrix $\operatorname{Re} M$ given by

$$\operatorname{Re} M = \frac{1}{2}(M + M^*). \tag{1.5}$$

Obviously, $\operatorname{Re} M$ is Hermitian. We call $M \in \mathbb{M}(\mathbf{U})$ *non-negative* if for each $j, k \in \mathbb{Z}$, $j \leq k$, the finite section $\Delta_{j,k}(M)$ induces a non-negative operator on the Hilbert space direct sum $\mathcal{U}_j \oplus \cdots \oplus \mathcal{U}_k$. In that case M is Hermitian. For operator matrices M and N in $\mathbb{M}(\mathbf{U})$ we say that M is *greater than or equal to* N, and write $M \geq N$, if the operator matrix $M - N$ is non-negative. Hence $M \geq 0$ means that M is non-negative. Finally, an operator matrix $M \in \mathbb{UM}(\mathbf{U})$ is said to be *positive real* whenever $\operatorname{Re} M$ is non-negative. Positive real operator matrices $M \in \mathbb{UM}(\mathbf{U})$ that induce bounded operators on $\mathbf{U} = \oplus_{k\in\mathbb{Z}}\mathcal{U}_k$ (i.e., $M \in \mathbb{UM}^\infty(\mathbf{U})$) have been extensively studied in [3, 4].

1.4. The operator matrix H^*H

Let $H \in \mathbb{UM}^2_{\text{ball}}(\mathbf{U}, \mathbf{Y})$. Recall that for each $k \in \mathbb{Z}$ the kth column H_k of H induces a contractive operator, also denoted by H_k, from $\mathcal{U}_k$ into $\mathbf{Y} = \oplus_{k\in\mathbb{Z}}\mathcal{Y}_k$. It follows that for each j and k in $\mathbb{Z}$ the product $(H_j)^*H_k$ is a well-defined contraction from $\mathcal{U}_k$ into $\mathcal{Y}_j$. We define H^*H to be the operator matrix in $\mathbb{M}(\mathbf{U})$ given by

$$H^*H = \left[\ (H_j)^*H_k\ \right]_{j,k\in\mathbb{Z}}. \tag{1.6}$$

Clearly, $H^*H \in \mathbb{M}(\mathbf{U})$ is hermitian. In fact, since for each $j, k \in \mathbb{Z}$, $j \leq k$,

$$\Delta_{j,k}(H^*H) = \begin{bmatrix} (H_j)^* \\ \vdots \\ (H_k)^* \end{bmatrix} \begin{bmatrix} H_j & \cdots & H_k \end{bmatrix},$$

the operator matrix H^*H is non-negative. Note that H^*H is the real part of the operator matrix $V \in \mathbb{UM}(\mathbf{U})$ be given by

$$V = \left[\ V_{j,k}\ \right]_{j,k\in\mathbb{Z}}, \quad V_{j,k} = \begin{cases} 2(H_j)^*H_k, & \text{for } j < k, \\ (H_j)^*H_j, & \text{for } j = k, \\ 0, & \text{for } j > k. \end{cases} \tag{1.7}$$

2. The set $\mathbb{UM}^2_{\text{ball}}(\mathbf{U}, \mathbf{Y})$ and the proof of Theorem 0.1

The main result of this section (Theorem 2.1 below) shows how elements in $\mathbb{UM}^2_{\text{ball}}(\mathbf{U}, \mathbf{Y})$ can be constructed from a pair of operators $Z^{(1)} \in \mathbb{UM}^\infty(\mathbf{U}, \mathbf{Y})$ and $Z^{(2)} \in \mathbb{UM}^\infty_0(\mathbf{U})$ satisfying an additional norm constraint. This result together with Proposition 2.2 allows us to prove Theorem 0.1. Theorem 2.1 also shows the relevance of Problem 0.2 in the analysis of Problem 0.1.

Theorem 2.1. *Assume we have given operator matrices*

$$Z^{(1)} \in \mathbb{UM}^{\infty}(\mathbf{U},\mathbf{Y}), \quad Z^{(2)} \in \mathbb{UM}_0^{\infty}(\mathbf{U}) \tag{2.1}$$

satisfying the norm constraint

$$\|Z^{(1)}u\|^2 + \|Z^{(2)}u\|^2 \le \|u\|^2 \quad (u \in \mathbf{U}). \tag{2.2}$$

Then

$$H = Z^{(1)}(I_{\mathbf{U}} - Z^{(2)})^{-1} \quad \text{and} \quad F = (I_{\mathbf{U}} - Z^{(2)})^{-1} \tag{2.3}$$

are well-defined operator matrices,

$$H \in \mathbb{UM}^2_{\text{ball}}(\mathbf{U},\mathbf{Y}), \quad F \in \mathbb{UM}(\mathbf{U}) \tag{2.4}$$

and

$$F + F^* \ge H^*H + I_{\mathbf{U}}, \quad F_{j,j} = I_{\mathcal{U}_j} \quad (j \in \mathbb{Z}). \tag{2.5}$$

Conversely, if we have given H and F as in (2.4) *such that* (2.5) *holds, then F is an invertible element in $\mathbb{UM}(\mathbf{U})$, the operator matrices*

$$Z^{(1)} = HF^{-1} \quad \text{and} \quad Z^{(2)} = I_{\mathbf{U}} - F^{-1} \tag{2.6}$$

are well defined and these operator matrices satisfy (2.1) *and* (2.2). *Moreover, the map $(Z^{(1)}, Z^{(2)}) \mapsto (H,F)$ defined by* (2.3) *is a one-to-one map from the set of all pairs $(Z^{(1)}, Z^{(2)})$ satisfying* (2.1) *and* (2.2) *onto the set of all pairs (H,F) satisfying* (2.4) *and* (2.5). *The inverse of this map is given by the map $(H,F) \mapsto (Z^{(1)}, Z^{(2)})$ defined by* (2.6).

Proof. We split the proof into four parts. In the first two parts $Z^{(1)}$ and $Z^{(2)}$ are given and satisfy (2.1) and (2.2), and we prove that H and F in (2.3) are well defined and satisfy (2.4) and (2.5). In the third part we prove the reverse statement. In the final part we show that the maps $(Z^{(1)}, Z^{(2)}) \mapsto (H,F)$ and $(H,F) \mapsto (Z^{(1)}, Z^{(2)})$ in Theorem 2.1 are each others inverses.

Part 1. Assume that $Z^{(1)} \in \mathbb{UM}^{\infty}(\mathbf{U},\mathbf{Y})$ and $Z^{(2)} \in \mathbb{UM}_0^{\infty}(\mathbf{U})$ satisfy (2.2), and let H be given by the first part of (2.3). Our aim is to prove that $H \in \mathbb{UM}^2_{\text{ball}}(\mathbf{U},\mathbf{Y})$.

Fix $k, j \in \mathbb{Z}, j < k$, and $u_k \in \mathcal{U}_k$. Define

$$v = \begin{bmatrix} v_j \\ \vdots \\ v_k \end{bmatrix} = \Delta_{j,k}\left((I_{\mathbf{U}} - Z^{(2)})^{-1}\right)u, \quad \text{where } u = \begin{bmatrix} 0 \\ \vdots \\ 0 \\ u_k \end{bmatrix} \in \oplus_{n=j}^{k} \mathcal{U}_n.$$

Note that $v \in \oplus_{n=j}^{k} \mathcal{U}_n$. Since $Z^{(2)}$ is strictly upper triangular, we have $v_k = u_k$. Observe that

$$\begin{aligned} \Delta_{j,k}(Z^{(2)})\Delta_{j,k}\left((I_{\mathbf{U}} - Z^{(2)})^{-1}\right) &= \Delta_{j,k}\left(Z^{(2)}(I_{\mathbf{U}} - Z^{(2)})^{-1}\right) \\ &= \Delta_{j,k}\left((I_{\mathbf{U}} - Z^{(2)})^{-1} - I_{\mathbf{U}}\right) \\ &= \Delta_{j,k}\left((I_{\mathbf{U}} - Z^{(2)})^{-1}\right) - I_{\oplus_{n=j}^{k}\mathcal{U}_n} \end{aligned}$$

and

$$\Delta_{j,k}(Z^{(1)})\Delta_{j,k}\left((I_{\mathbf{U}}-Z^{(2)})^{-1}\right)=\Delta_{j,k}\left(Z^{(1)}(I_{\mathbf{U}}-Z^{(2)})^{-1}\right)=\Delta_{j,k}(H).$$

Hence

$$\Delta_{j,k}(Z^{(2)})v=v-u=:\alpha \quad \text{and} \quad \Delta_{j,k}(Z^{(1)})v=\Delta_{j,k}(H)u=:\beta,$$

where, using $v_k=u_k$,

$$\alpha=\begin{bmatrix} v_j \\ \vdots \\ v_{k-1} \\ 0 \end{bmatrix} \quad \text{and} \quad \beta=\begin{bmatrix} H_{j,k}u_k \\ \vdots \\ H_{k-1,k}u_k \\ H_{k,k}u_k \end{bmatrix}.$$

The assumption (2.2) implies that

$$\begin{bmatrix} \Delta_{j,k}(Z^{(1)}) \\ \Delta_{j,k}(Z^{(2)}) \end{bmatrix}$$

is a contraction. Thus

$$\sum_{\nu=j}^{k-1}\|v_\nu\|^2+\sum_{\nu=j}^{k}\|H_{\nu,k}u_k\|^2=\|\alpha\|^2+\|\beta\|^2\le\|v\|^2=\sum_{\nu=j}^{k}\|v_\nu\|^2.$$

By comparing the first term in the left-hand side of the above inequality with the term in the right side and using $v_k=u_k$, we see that $\sum_{\nu=j}^{k}\|H_{\nu,k}u_k\|^2$ is less than or equal to $\|u_k\|^2$. This holds for each $j\le k$ and $u_k\in\mathcal{U}_k$, and therefore the operator defined by the kth column of H is a contraction. Since $k\in\mathbb{Z}$ is arbitrary, $H\in\mathbb{UM}^2_{\mathrm{ball}}(\mathbf{U},\mathbf{Y})$.

Part 2. Under the same assumptions as in Part 1, let H and F be given by (2.3). Our aim is to prove that (2.5) holds. From the definition of F the right-hand side of (2.5) is clear. Thus we have to prove the inequality in the left-hand side of (2.5).

First assume that there exists an $N\ge 0$ such that $\mathcal{U}_k=\mathcal{Y}_k=\{0\}$ for $|k|>N$. In that case $\mathbb{M}(\mathbf{U})=\mathcal{L}(\mathbf{U})$, $\mathbb{UM}(\mathbf{U})=\mathbb{UM}^\infty(\mathbf{U})$ and $\mathbb{UM}(\mathbf{U},\mathbf{Y})=\mathbb{UM}^\infty(\mathbf{U},\mathbf{Y})$. In particular, H and F are bounded operators. We have

$$\begin{aligned} I-Z^{(2)*}Z^{(2)} &= I-(I-I+Z^{(2)*})(I-I+Z^{(2)}) \\ &= I-(I-F^{-*})(I-F^{-1}) \\ &= F^{-*}+F^{-1}-F^{-*}F^{-1}. \end{aligned}$$

Thus

$$\begin{aligned} H^*H &= F^*Z^{(1)*}Z^{(1)}F\le F^*(I-Z^{(2)*}Z^{(2)})F \\ &= F^*(F^{-*}+F^{-1}-F^{-*}F^{-1})F=F+F^*-I. \end{aligned}$$

So (2.5) holds in this case.

Now fix $j\le k$. For $N\ge|j|,|k|$ set

$$\mathbf{U}_N=\oplus_{i=-N}^{N}\mathcal{U}_i,\quad \mathbf{Y}_N=\oplus_{i=-N}^{N}\mathcal{Y}_i,$$

and notice that

$$\Delta_{-N,N}(F) = (I_{\mathbf{U}_N} - \Delta_{-N,N}(Z^{(2)}))^{-1} \quad \text{and}$$
$$\Delta_{-N,N}(H) = \Delta_{-N,N}(Z^{(1)})(I_{\mathbf{U}_N} - \Delta_{-N,N}(Z^{(2)}))^{-1}.$$

Next apply the result of the second paragraph of this part to

$$\Delta_{-N,N}(Z^{(1)}) \in \mathbb{UM}(\mathbf{U}_N, \mathbf{Y}_N) \quad \text{and} \quad \Delta_{-N,N}(Z^{(2)}) \in \mathbb{UM}_0(\mathbf{U}_N).$$

Using that $\Delta_{j,k}(\Delta_{-N,N}(M)) = \Delta_{j,k}(M)$ for each $M \in \mathbb{UM}(\mathbf{U})$, it follows that

$$\Delta_{j,k}(F) + \Delta_{j,k}(F^*) \geq \Delta_{j,k}(H^* Q_N H) + I, \tag{2.7}$$

where $Q_N \in \mathcal{L}(\mathbf{Y})$ is the orthogonal projection on $\mathbf{Y}_N$. Since

$$\Delta_{j,k}(H^* Q_N H) = \begin{bmatrix} (H_j)^* \\ \vdots \\ (H_k)^* \end{bmatrix} Q_n \begin{bmatrix} H_j & \cdots & H_k \end{bmatrix},$$

$\begin{bmatrix} H_j & \cdots & H_k \end{bmatrix}$ is a bounded linear operator, $Q_N \to I_{\mathbf{Y}}$ as $N \to \infty$ with convergence in the strong operator topology, and (2.7) holds for each $N \geq |j|, |k|$, it follows that

$$\Delta_{j,k}(F) + \Delta_{j,k}(F^*) \geq \Delta_{j,k}(H^* H) + I. \tag{2.8}$$

Thus (2.5) holds.

Part 3. In this part we assume that $H \in \mathbb{UM}^2_{\text{ball}}(\mathbf{U}, \mathbf{Y})$, $F \in \mathbb{UM}(\mathbf{U})$, and that condition (2.5) is fulfilled. We show that the operator matrices $Z^{(1)}$ and $Z^{(2)}$ defined by (2.6) are in $\mathbb{UM}^\infty(\mathbf{U}, \mathbf{Y})$ and $\mathbb{UM}^\infty_0(\mathbf{U})$, respectively, and that (2.2) is satisfied.

The second part of (2.5) implies that F is invertible in $\mathbb{UM}(\mathbf{U})$. Thus the operator matrices $Z^{(1)}$ and $Z^{(2)}$ in (2.6) are well defined. Moreover, for each $j \in \mathbb{Z}$ the jth diagonal entry of F^{-1} is the identity operator $I_{\mathcal{U}_j}$, and thus the matrix $Z^{(2)}$ is strictly upper triangular. It remains to show that $Z^{(1)}$ and $Z^{(2)}$ satisfy (2.2), since this automatically implies that $Z^{(1)} \in \mathbb{UM}^\infty(\mathbf{U}, \mathbf{Y})$ and $Z^{(2)} \in \mathbb{UM}^\infty_0(\mathbf{U})$.

First note that it suffices to show that

$$\Delta_{j,k}(Z^{(1)})^* \Delta_{j,k}(Z^{(1)}) + \Delta_{j,k}(Z^{(2)})^* \Delta_{j,k}(Z^{(2)}) \leq \Delta_{j,k}(I_{\mathbf{U}}) \quad (j \leq k). \tag{2.9}$$

Indeed, if the above inequalities have been established, then we can use the results reviewed in the second part of Subsection 1.1 to derive (2.2). Fix $j \leq k$, and write Δ in place of $\Delta_{j,k}$. From $Z^{(2)} = I_{\mathbf{U}} - F^{-1}$, we see that $\Delta(Z^{(2)}) = \Delta(I_{\mathbf{U}}) - \Delta(F^{-1})$. Now use formula (1.4) to see that $\Delta(F^{-1}) = \Delta(F)^{-1}$. But then

$$\begin{aligned} \Delta(Z^{(2)})^* \Delta(Z^{(2)}) &= \left(\Delta(I_{\mathbf{U}}) - \Delta(F)^{-*}\right)\left(\Delta(I_{\mathbf{U}}) - \Delta(F)^{-1}\right) \\ &= \Delta(I_{\mathbf{U}}) - \Delta(F)^{-1} - \Delta(F)^{-*} + \Delta(F)^{-*}\Delta(F)^{-1} \\ &= \Delta(F)^{-*}(\Delta(F)^* \Delta(F) - \Delta(F)^* - \Delta(F) + \Delta(I_{\mathbf{U}}))\Delta(F)^{-1}. \end{aligned}$$

Thus

$$\Delta(F)^*\left(\Delta(I_{\mathbf{U}}) - \Delta(Z^{(2)})^*\Delta(Z^{(2)})\right)\Delta(F)$$
$$= \Delta(F)^* + \Delta(F) - \Delta(I_{\mathbf{U}}) \geq \Delta(H^*H).$$

Here we used that the first part of (2.5) implies (2.8).

Next we consider the equality $H = Z^{(1)}F$. Again using (1.3), with $\Delta = \Delta_{j,k}$, we have $\Delta(H) = \Delta(Z^{(1)})\Delta(F)$, and thus

$$\Delta(F)^*\Delta(Z^{(1)})^*\Delta(Z^{(1)})\Delta(F) = \Delta(H)^*\Delta(H).$$

By combining this with the result of the previous paragraph we see that

$$\Delta(F)^*(\Delta(I_{\mathbf{U}}) - \Delta(Z^{(1)})^*\Delta(Z^{(1)}) - \Delta(Z^{(2)})^*\Delta(Z^{(2)}))\Delta(F)$$
$$\geq \Delta(H^*H) - \Delta(H)^*\Delta(H) \geq 0.$$

To see that the last inequality holds, let P denote the projection from $\mathbf{Y}$ onto $\mathcal{Y}_j \oplus \cdots \oplus \mathcal{Y}_k$ and observe that

$$\begin{aligned}\Delta(H^*H) &= \begin{bmatrix}(H_j)^*H_j & \cdots & (H_j)^*H_k\\ \vdots & & \vdots\\ (H_k)^*H_j & \cdots & (H_k)^*H_k\end{bmatrix} = \begin{bmatrix}(H_j)^*\\ \vdots\\ (H_k)^*\end{bmatrix}\begin{bmatrix}H_j & \cdots & H_k\end{bmatrix}\\ &\geq \begin{bmatrix}(H_j)^*\\ \vdots\\ (H_k)^*\end{bmatrix} P\begin{bmatrix}H_j & \cdots & H_k\end{bmatrix} = \Delta(H)^*\Delta(H).\end{aligned}$$

Since $\Delta(F)$ is invertible, we obtain (2.9).

Part 4. In case $Z^{(1)}$ and $Z^{(2)}$ satisfy the assumptions of Parts 1 and 2 and H and F are given by (2.3), it is clear that

$$HF^{-1} = Z^{(1)} \quad \text{and} \quad I - F^{-1} = Z^{(2)}.$$

If H and F satisfy the assumptions of Parts 3 and $Z^{(1)}$ and $Z^{(2)}$ are given by (2.6), then

$$(I - Z^{(2)})^{-1} = (I_{\mathbf{U}} - I_{\mathbf{U}} + F^{-1})^{-1} = F \quad \text{and} \quad Z^{(1)}(I_{\mathbf{U}} - Z^{(2)})^{-1} = HF^{-1}F = H.$$

Thus the maps $(Z^{(1)}, Z^{(2)}) \mapsto (H, F)$ and $(H, F) \mapsto (Z^{(1)}, Z^{(2)})$ in Theorem 2.1 are each others inverses. □

The next proposition is an addition to Theorem 2.1. It takes into account the interpolation condition on H in (0.3) and on the pair $(Z^{(1)}, Z^{(2)})$ in (0.6).

Proposition 2.2. *Let $Z^{(1)}$ and $Z^{(2)}$ be as in* (2.1) *and* (2.2), *and define H and F by* (2.3). *Then the pair $(Z^{(1)}, Z^{(2)})$ satisfies the interpolation conditions* (0.6) *if and only if H satisfies the interpolation condition* (0.3) *and*

$$F_{j,k}|_{\mathcal{F}_k} = F_{j,k-1}\omega_{k,2} \quad (j, k \in \mathbb{Z}, j < k). \tag{2.10}$$

Proof. Let $Z^{(1)}$ and $Z^{(2)}$ satisfy (2.1) and (2.2), and let H and F be defined by (2.3).

We begin with a general remark about the interpolation conditions (0.3), (0.6), and (2.10). Recall that for each $k \in \mathbb{Z}$ the space $\mathcal{F}_k$ is a subspace of $\mathcal{U}_k$. In what follows τ_k is the canonical embedding of $\mathcal{F}_k$ into $\mathcal{U}_k$. Furthermore, $\mathbf{F}$ will denote the Hilbert direct sum $\oplus_{k\in\mathbb{Z}} \mathcal{F}_k$. Now let

$$E = [E_{j,k}]_{j,k\in\mathbb{Z}}, \quad \Omega^{(1)} = \left[\Omega^{(1)}_{j,k}\right]_{j,k\in\mathbb{Z}}, \quad \Omega^{(2)} = \left[\Omega^{(2)}_{j,k}\right]_{j,k\in\mathbb{Z}}, \tag{2.11}$$

be the operator matrices defined by

$$E_{j,k} = \begin{cases} \tau_k, & \text{if } j = k, \\ 0, & \text{if } j \neq k, \end{cases} \tag{2.12}$$

and

$$\Omega^{(1)}_{j,k} = \begin{cases} \omega_{k,1}, & \text{if } j = k, \\ 0, & \text{if } j \neq k, \end{cases} \qquad \Omega^{(2)}_{j,k} = \begin{cases} \omega_{k,2}, & \text{if } j = k-1, \\ 0, & \text{if } j \neq k-1. \end{cases} \tag{2.13}$$

Observe that both E and $\Omega^{(1)}$ are diagonal operator matrices, $E \in \mathbb{UM}^\infty_{\text{ball}}(\mathbf{F}, \mathbf{U})$ and $\Omega^{(1)} \in \mathbb{UM}^\infty_{\text{ball}}(\mathbf{F}, \mathbf{Y})$, while $\Omega^{(2)} \in \mathbb{UM}^\infty_{\text{ball},0}(\mathbf{F}, \mathbf{U})$ is a shifted diagonal operator, that is, all the entries of $\Omega^{(2)}$ are zero except those in the first diagonal above the main diagonal. Using these operator matrices we can restate the interpolation conditions. In fact, we have

$$(0.3) \iff HE = \Omega^{(1)} + H\Omega^{(2)}, \tag{2.14}$$

$$(0.6) \iff Z^{(1)}E = \Omega^{(1)} \quad \text{and} \quad Z^{(2)}E = \Omega^{(2)}, \tag{2.15}$$

$$(2.10) \iff FE - E = F\Omega^{(2)}. \tag{2.16}$$

Now assume that the pair $(Z^{(1)}, Z^{(2)})$ satisfies the interpolation condition (0.6). Since H and F are given by (2.3), we have

$$H = Z^{(1)}(I_{\mathbf{U}} - Z^{(2)})^{-1} = Z^{(1)} + Z^{(1)}(I_{\mathbf{U}} - Z^{(2)})^{-1}Z^{(2)} = Z^{(1)} + HZ^{(2)},$$

$$F = (I_{\mathbf{U}} - Z^{(2)})^{-1} = I_{\mathbf{U}} + (I_{\mathbf{U}} - Z^{(2)})^{-1}Z^{(2)} = I_{\mathbf{U}} + FZ^{(2)}.$$

Hence, using (2.15), we see that

$$HE = Z^{(1)}E + HZ^{(2)}E = \Omega^{(1)} + H\Omega^{(2)}, \quad FE = E + FZ^{(2)}E = E + F\Omega^{(2)}.$$

But then we can use (2.14) to conclude that H satisfies (0.3), and we can use (2.16) to conclude that F satisfies (2.10).

Next assume that H and F satisfy the interpolation conditions (0.3) and (2.10), respectively. From (2.3) we see that

$$Z^{(1)} = HF^{-1} \quad \text{and} \quad Z^{(2)} = I - F^{-1}.$$

Note that (2.10) and (2.16) imply that $F^{-1}E = E - \Omega^{(2)}$. Hence

$$Z^{(1)}E = HF^{-1}E = HE - H\Omega^{(2)} = \Omega^{(1)} \quad \text{by (2.14)},$$
$$Z^{(2)}E = E - F^{-1}E = \Omega^{(2)}.$$

But then we can use the equivalence in (2.15) to conclude that the pair $(Z^{(1)}, Z^{(2)})$ satisfies the interpolation conditions (0.6) as desired. □

Proof of Theorem 0.1. Let $Z^{(1)}$ and $Z^{(2)}$ be a pair of operator matrices satisfying (0.4), (0.5) and (0.6). Define H by formula (0.7). Theorem 2.1 tells us that H is well defined and $H \in \mathbb{UM}^2_{\text{ball}}(\mathbf{U}, \mathbf{Y})$; see formula (2.4). Since $(Z^{(1)}, Z^{(2)})$ satisfies the interpolation conditions (0.6), we see from Proposition 2.2 that H satisfies (0.3). Thus H is a solution to Problem 0.1. □

3. Majorants of H^*H and the solution to Problem 0.2

In this section we solve Problem 0.2 and give a parametrization of the set of all its solutions. The first subsection, which has a preliminary character, deals with a time-variant version of the Cayley transform. The main result (Theorem 3.2 below) is presented in the second subsection, which is then used in the final subsection to solve Problem 0.2.

3.1. The Cayley transform

Let $C \in \mathbb{UM}^\infty_{\text{ball},0}(\mathbf{U})$. Then we know from Subsection 1.2 that $I_{\mathbf{U}} - C$ is invertible in the algebra $\mathbb{UM}(\mathbf{U})$. It follows that K given by

$$K = (I_{\mathbf{U}} + C)(I_{\mathbf{U}} - C)^{-1} \tag{3.1}$$

is a well-defined element of $\mathbb{UM}(\mathbf{U})$. We shall refer to K as the *Cayley transform* of C. The following proposition shows that K is positive real (see [3], page 94, for a related but somewhat less general result).

Proposition 3.1. *The map $C \mapsto K$ defined by* (3.1) *establishes a one-to-one correspondence between the set of all operator matrices C in $\mathbb{UM}^\infty_{\text{ball},0}(\mathbf{U})$ and the positive real $K \in \mathbb{UM}(\mathbf{U})$ satisfying $K_{j,j} = I_{\mathcal{U}_j}$ for all $j \in \mathbb{Z}$.*

Proof. Assume that K is defined by (3.1) for a $C \in \mathbb{UM}^\infty_{\text{ball},0}(\mathbf{D})$. Then

$$K = I_{\mathbf{U}} + 2C\,(I_{\mathbf{U}} - C)^{-1}.$$

Since $(I_{\mathbf{U}} - C)^{-1}$ is upper triangular and C is strictly upper triangular, it follows that $K_{j,j} = I_{\mathcal{U}_j}$ for each $j \in \mathbb{Z}$. Given an operator matrix M let $\Delta(M) = \Delta_{j,k}(M)$ denote the finite section of M for $j \le k$. We have to prove (see Subsection 1.3) that $\Delta(\operatorname{Re} K)$ is non-negative. To do this note that

$$\begin{aligned} 2\Delta(\operatorname{Re} K) &= (I - \Delta(C^*))^{-1}(I + \Delta(C^*)) + (I + \Delta(C))(I - \Delta(C))^{-1} \\ &= (I - \Delta(C^*))^{-1}\{(I + \Delta(C^*))(I - \Delta(C)) \\ &\qquad + (I - \Delta(C^*))(I + \Delta(C))\}\,(I - \Delta(C))^{-1} \\ &= 2(I - \Delta(C^*))^{-1}\{I - \Delta(C^*)\Delta(C)\}\,(I - \Delta(C))^{-1}. \end{aligned}$$

In other words,

$$\Delta(\operatorname{Re} K) = (I - \Delta(C^*))^{-1}\{I - \Delta(C)^*\Delta(C)\}\,(I - \Delta(C))^{-1}. \tag{3.2}$$

Here $I = I_{\mathcal{U}_j \oplus \cdots \oplus \mathcal{U}_k}$. Because C is a contraction, $\Delta(C)$ is also a contraction. Hence all finite sections $\Delta(\operatorname{Re} K)$ of $\operatorname{Re} K$ are non-negative. Therefore K is positive real.

Conversely, for a positive real matrix K in $\mathbb{UM}(\mathbf{U})$ satisfying $\operatorname{diag} K_{j,j} = I_{\mathcal{U}_j}$ for $j \in \mathbb{Z}$, consider the operator matrix C defined by

$$C = (K - I_{\mathbf{U}})(K + I_{\mathbf{U}})^{-1}. \tag{3.3}$$

We know that $K + I_{\mathbf{U}} \in \mathbb{UM}(\mathbf{U})$ and for each $j \in \mathbb{Z}$ the jth diagonal element of $K + I_{\mathbf{U}}$ is $2I_{\mathcal{U}_j}$. Hence $K + I$ is invertible in $\mathbb{UM}(\mathbf{U})$ (see Subsection 1.2). Moreover, $K - I$ is in $\mathbb{UM}_0(\mathbf{U})$. Thus C in (3.3) is a well-defined operator matrix in $\mathbb{UM}_0(\mathbf{U})$. We claim that C is in $\mathbb{UM}^{\infty}_{\text{ball},0}(\mathbf{U})$. To see this let Δ and I be as in the previous paragraph. Using (1.3) and (1.4) we have

$$\Delta(C) = (\Delta(K) - I)(\Delta(K) + I)^{-1}.$$

Thus

$$\begin{aligned} I - \Delta(C)^*\Delta(C) &= I - (\Delta(K)^* + I)^{-1}(\Delta(K)^* - I)(\Delta(K) - I)(\Delta(K) + I)^{-1} \\ &= (\Delta(K)^* + I)^{-1}\{(\Delta(K)^* + I)(\Delta(K) + I) \\ &\qquad -(\Delta(K)^* - I)(\Delta(K) - I)\}(\Delta(K) + I)^{-1} \\ &= 2(\Delta(K)^* + I)^{-1}(\Delta(K)^* + \Delta(K))(\Delta(K) + I)^{-1} \\ &= 4(\Delta(K)^* + I)^{-1}\Delta(\operatorname{Re} K)(\Delta(K) + I)^{-1} \geq 0. \end{aligned}$$

Hence any finite section of C is a contraction. Therefore C is a contraction, and thus in $\mathbb{UM}^{\infty}_{\text{ball},0}(\mathbf{U})$.

Finally, one easily verifies that the maps $C \mapsto K$ given by (3.1) and $K \mapsto C$ given by (3.3) are each others inverses. Hence the map $C \mapsto K$ defined by (3.1) has the desired properties. □

If K in $\mathbb{UM}(\mathbf{U})$ is positive real with $K_{j,j} = I_{\mathcal{U}_j}$ for $j \in \mathbb{Z}$, then C defined by (3.3) will be called the *inverse Cayley transform* of K.

3.2. Time-variant harmonic majorants of H^*H

Let $H \in \mathbb{UM}^2_{\text{ball}}(\mathbf{U}, \mathbf{Y})$ be given, and consider the operator matrix H^*H (see Subsection 1.4). In the present subsection we describe the operator matrices $W \in \mathbb{UM}(\mathbf{U})$ satisfying

$$\operatorname{Re} W \geq H^*H \quad \text{and} \quad W_{j,j} = I_{\mathcal{U}_j} \quad (j \in \mathbb{Z}). \tag{3.4}$$

This description will be used in the next subsection to give the solution to Problem 0.2.

When $W \in \mathbb{UM}(\mathbf{U})$ satisfies the first identity in (3.4) we call $\operatorname{Re} W$ a *time-variant harmonic majorant* of H^*H. In that case, since H^*H is non-negative, W is automatically positive real. Time-variant harmonic majorants of H^*H do exist. In fact (see Subsection 1.4) the operator matrix V defined by (1.7) belongs to $\mathbb{UM}(\mathbf{U})$ and $\operatorname{Re} V = H^*H$. Thus H^*H is its own time-variant harmonic majorant.

To describe all $W \in \mathbb{UM}(\mathbf{U})$ satisfying (3.4) recall that $\mathbf{D}_H$ is the Hilbert space direct sum $\oplus_{k\in\mathbb{Z}}\mathcal{D}_{H_k}$, where H_k is the kth column of H and $\mathcal{D}_{H_k}$ is the corresponding defect space. The latter space is well defined because H_k defines a contraction from $\mathcal{U}_k$ in to $\mathbf{Y}$. We define ∇_H and Π_H to be the diagonal operator matrices in $\mathbb{UM}(\mathbf{D}_H)$ and $\mathbb{UM}(\mathbf{U}, \mathbf{D}_H)$, respectively, given by

$$(\nabla_H)_{j,k} = \begin{cases} D_{H_k} & \text{for } j = k, \\ 0 & \text{for } j \neq k. \end{cases} \quad \text{and} \quad (\Pi_H)_{j,k} = \begin{cases} \Pi_{H_k} & \text{for } j = k, \\ 0 & \text{for } j \neq k. \end{cases} \tag{3.5}$$

In the definition of ∇_H we view D_{H_k} as an operator on $\mathcal{D}_{H_k}$, and in the definition of Π_H the operator Π_{H_k} is the orthogonal projection of $\mathcal{U}_k$ onto $\mathcal{D}_{H_k}$. We can now state the main result of this section.

Theorem 3.2. *Let $H \in \mathbb{UM}^2_{\mathrm{ball}}(\mathbf{U}, \mathbf{Y})$, and let ∇_H and Π_H be the diagonal operator matrices given by* (3.5). *Then all operator matrices $W \in \mathbb{UM}(\mathbf{U})$ satisfying* (3.4) *are determined by*

$$W = V + \Pi_H^* \nabla_H (I + C)(I - C)^{-1} \nabla_H \Pi_H, \tag{3.6}$$

where V in $\mathbb{UM}(\mathbf{U})$ is given by (1.7), *and C is an arbitrary operator matrix in $\mathbb{UM}^\infty_{\mathrm{ball},0}(\mathbf{D}_H)$. Moreover, W and C in* (3.4) *determine each other uniquely.*

Proof. Let $C \in \mathbb{UM}^\infty_{\mathrm{ball},0}(\mathbf{D}_H)$. Then W in (3.6) is equal to

$$W = V + \Pi_H^* \nabla_H^* K \nabla_H \Pi_H,$$

where K is the Cayley transform of C. Using $\mathrm{Re}\, K$ is non-negative this yields

$$\mathrm{Re}\, W = \mathrm{Re}\, V + \Pi_H^* \nabla_H^* \mathrm{Re}\, K \nabla_H \Pi_H \geq \mathrm{Re}\, V = H^* H,$$

and we have

$$W_{j,j} = V_{j,j} + D_{H_j} K_{j,j} D_{H_j} = H_j^* H_j + D_{H_j}^2 = I_{\mathcal{U}_j} \quad (j \in \mathbb{Z}).$$

So W satisfies (3.4).

To prove the converse implication, assume that $W \in \mathbb{UM}(\mathbf{U})$ satisfies (3.4). Since $\mathrm{Re}\, V = H^* H$, the first part of (3.4) implies that the real part of the operator matrix $\Lambda = W - V \in \mathbb{UM}(\mathbf{U})$ is non-negative. The second part of (3.4) gives $\Lambda_{j,j} = I_{\mathcal{U}_j} - H_j^* H_j = D_{H_j}^2$ for all $j \in \mathbb{Z}$. The fact that $\mathrm{Re}\, \Lambda$ is non-negative implies that for any $j < k$ the finite section $\Delta_{j,k}(\mathrm{Re}\, \Lambda)$ is a non-negative operator on $\oplus_{i=j}^k \mathcal{U}_i$. In particular, the two by two operator matrix

$$\begin{bmatrix} 2\Lambda_{j,j} & \Lambda_{j,k} \\ \Lambda_{j,k}^* & 2\Lambda_{k,k} \end{bmatrix} = \begin{bmatrix} 2D_{H_j}^2 & \Lambda_{j,k} \\ \Lambda_{j,k}^* & 2D_{H_k^2} \end{bmatrix}$$

is a non-negative operator on the Hilbert direct sum $\mathcal{U}_j \oplus \mathcal{U}_k$. Recall that an arbitrary operator matrix

$$\begin{bmatrix} A_1 & B^* \\ B & A_2 \end{bmatrix} \text{ acting on the Hilbert space direct sum } \begin{bmatrix} \mathcal{E}_1 \\ \mathcal{E}_2 \end{bmatrix}$$

is a non-negative operator if and only if A_1 and A_2 are non-negative and $B = A_2^{1/2} \Phi A_1^{1/2}$, where Φ is a contraction from $\overline{A_1 \mathcal{E}_1}$ into $\overline{A_2 \mathcal{E}_2}$. Moreover, in this case, B and Φ uniquely determine each other; see Theorem XVI.1.1 in [16] for further details. Thus there exists a unique operator $K_{j,k}$ from $\mathcal{D}_{H_j}$ into $\mathcal{D}_{H_k}$ such that

$\Lambda_{k,j} = D_{H_k} K_{k,j} D_{H_j}$. Now set $K_{j,j} = I_{\mathcal{D}_{H_j}}$ and $K_{j,k} = 0$ for $j > k$, and put $K = [\ K_{j,k}\ ]_{j,k\in\mathbb{Z}}$. Then

$$K \in \mathbb{UM}(\mathbf{D}_H) \quad \text{and} \quad \Lambda = \Pi_H^* \nabla_H^* K \nabla_H \Pi_H.$$

Since the range of ∇_H is a dense set in $\mathbf{D}_H$ and Λ is positive real, it follows that K is positive real and that Λ and K determine each other uniquely. Let C be the inverse Cayley transform of K. Then $C \in \mathbb{UM}^{\infty}_{\text{ball},0}(\mathbf{D}_H)$ and W is given by (3.6). It also follows from the above that C and W determine each other uniquely. □

Corollary 3.3. *There is only one operator matrix $W \in \mathbb{UM}(\mathbf{U})$ satisfying* (3.4) *if and only if for each $k \in \mathbb{Z}$ the kth column H_k of H defines an isometry from $\mathcal{U}_k$ into $\mathbf{Y}$. In this case $W = V$, where V is given by* (1.7)*, is the only operator matrix in $\mathbb{UM}(\mathbf{U})$ satisfying* (3.4).

Proof. The set $\mathbb{UM}^{\infty}_{\text{ball},0}(\mathbf{D}_H)$ consists of just one element if and only if $\mathbf{D}_H = \{0\}$, i.e., $\mathcal{D}_{H_j} = \{0\}$ for all j. The latter condition is equivalent to H_j being an isometry for each $j \in \mathbb{Z}$. □

3.3. All solutions to Problem 0.2

We now describe the solution to Problem 0.2. Fix a $H \in \mathbb{UM}^2_{\text{ball}}(\mathbf{U},\mathbf{Y})$. Define $N \in \mathbb{UM}(\mathbf{U})$ by

$$N = [\ N_{j,k}\ ]_{j,k\in\mathbb{Z}}, \quad N_{j,k} = \begin{cases} (H_j)^* H_k, & \text{for } j \le k, \\ 0, & \text{for } j > k, \end{cases} \tag{3.7}$$

Here, as before, H_k is the kth column of H.

Theorem 3.4. *Let $H \in \mathbb{UM}^2_{\text{ball}}(\mathbf{U},\mathbf{Y})$. Define $N \in \mathbb{UM}(\mathbf{U})$ by* (3.7)*, and let the matrices $\nabla_H \in \mathbb{UM}(\mathbf{D}_H)$ and $\Pi_H \in \mathbb{UM}(\mathbf{D}_H,\mathbf{U})$ be the diagonal operator matrices given by* (3.5). *For each $C \in \mathbb{UM}^{\infty}_{\text{ball},0}(\mathbf{D}_H)$ put*

$$F = N + \Pi_H^* \nabla_H (I_{\mathbf{D}_H} - C)^{-1} \nabla_H \Pi_H. \tag{3.8}$$

Then $F \in \mathbb{UM}(\mathbf{U})$, and the map $C \mapsto F$ defined by (3.8) *is a one-to-one map from $\mathbb{UM}^{\infty}_{\text{ball},0}(\mathbf{D}_H)$ onto the set of all $F \in \mathbb{UM}(\mathbf{U})$ that satisfy* (0.12). *In particular, $F \in \mathbb{UM}(\mathbf{U})$ satisfying* (0.12) *exist. Finally, there exists a unique $F \in \mathbb{UM}(\mathbf{U})$ satisfying* (0.12) *if and only if for each $k \in \mathbb{Z}$ the kth column H_k of H defines an isometry from $\mathcal{U}_k$ into $\mathbf{Y}$. In this case $F = N$ is the only $F \in \mathbb{UM}(\mathbf{U})$ satisfying* (0.12).

Proof. Assume W and F belong to $\mathbb{UM}(\mathbf{U})$ and determine each other uniquely via

$$W = 2F - I_{\mathbf{U}} \quad \text{and} \quad F = \frac{1}{2}(W + I). \tag{3.9}$$

Then $W_{j,j} = I_{\mathcal{U}_j}$ if and only if $F_{j,j} = I_{\mathcal{U}_j}$ for each $j \in \mathbb{Z}$. Moreover, $2\mathrm{Re}\, F = \mathrm{Re}\, W + I$. Hence F satisfies (2.5) (where $F + F^* = 2\mathrm{Re}\, F$) if and only if W satisfies (3.4).

To complete the proof it remains to show that the map $C \mapsto W$ given by (3.6) composed with the map $W \mapsto F$ in (3.9) gives the map $C \mapsto F$ in (3.8). Let $C \in \mathbb{UM}^{\infty}_{\mathrm{ball},0}(\mathbf{D}_H)$, define W by (3.6) and F by (3.9). Notice that V in (1.7) and N in (3.7) are related via $V = 2N + \Pi_H^* \nabla_H^2 \Pi_H - I$. Thus

$$\begin{aligned} W &= V + \Pi_H^* \nabla_H (I + C)(I - C)^{-1} \nabla_H \Pi_H \\ &= 2N + \Pi_H^* \nabla_H^2 \Pi_H + \Pi_H^* \nabla_H (I + C)(I - C)^{-1} \nabla_H \Pi_H - I \\ &= 2N + \Pi_H^* \nabla_H (I + C + I - C)(I - C)^{-1} \nabla_H \Pi_H - I \\ &= 2N + 2\Pi_H^* \nabla_H (I - C)^{-1} \nabla_H \Pi_H - I. \end{aligned}$$

Hence

$$\begin{aligned} F &= \frac{1}{2}(W + I) = \frac{1}{2}(2N + 2\Pi_H^* \nabla_H (I - C)^{-1} \nabla_H \Pi_H) \\ &= N + \Pi_H^* \nabla_H (I - C)^{-1} \nabla_H \Pi_H. \end{aligned}$$

So F is given by (3.8). □

A state space example. Consider the state space system $\{A, B, E, D\}$, where A is an operator on $\mathcal{X}$ whose spectrum is contained in the open unit disc. The input space $\mathbf{U} = \oplus_{j=1}^{n} \mathcal{U}_j$ and the output space $\mathbf{Y} = \oplus_{j=1}^{n} \mathcal{Y}_j$. Furthermore, B is an operator mapping $\mathbf{U}$ into $\mathcal{X}$ and E is an operator mapping $\mathcal{X}$ into $\mathbf{Y}$, while $D \in \mathbb{UM}(\mathbf{U}, \mathbf{Y})$ is a finite upper triangular operator matrix mapping $\mathbf{U} = \oplus_{j=1}^{n} \mathcal{U}_j$ into $\mathbf{Y} = \oplus_{j=1}^{n} \mathcal{Y}_j$.

Now, let H be the operator matrix (consisting of N doubly infinite columns) of which the kth column $H_k = \mathrm{col}\{H_{j,k}\}_{j \in \mathbb{Z}}$, $k = 1, \dots, n$, is given by

$$H_{j,k} = \begin{cases} EA^{-j-1}B_k & \text{when } j < 0, \\ D_{j,k} & \text{when } j = 1, \dots, n, \\ 0 & \text{when } j > n. \end{cases}$$

Here $D_{j,k}$ is the (j, k)th entry of the $n \times n$ operator matrix D, and B_k is the restriction of B to the kth component $\mathcal{U}_k$ of $\mathbf{U} = \oplus_{j=1}^{n} \mathcal{U}_j$. For this H we consider the finite operator matrix version of Problem 0.2, that is, we seek all $F \in \mathbb{UM}(\mathbf{U})$ satisfying

$$F + F^* \geq H^* H + I_{\mathbf{U}} \quad \text{and} \quad F_{j,j} = I_{\mathcal{U}_j} \quad (j = 1, \dots, n). \tag{3.10}$$

By setting $\mathcal{U}_j = \{0\}$ for $j < 0$ and $j > n$, we can identify $\mathbf{U}$ and $\oplus_{j \in \mathbb{Z}} \mathcal{U}_j$, and we may view H as an operator matrix in $\mathbb{M}(\mathbf{U}, \mathbf{E})$, where $\mathbf{E} = \oplus_{j \in \mathbb{Z}} \mathcal{E}_j$ with

$$\mathcal{E}_j = \begin{cases} \mathbf{Y} & \text{when } j < 0 \text{ or } j > n, \\ \mathcal{Y}_j & \text{when } j = 1, \dots, n. \end{cases}$$

The fact that D is upper triangular implies that $H \in \mathbb{UM}(\mathbf{U}, \mathbf{E})$. Hence the results obtained above apply.

Since A has its spectrum in the open unit disc, we know that the Lyapunov equation $P = A^*PA + E^*E$ has a unique solution which is given by

$$P = \sum_{j=0}^{\infty} A^{*j} E^* E A^j.$$

Using this, we obtain $H^*H = D^*D + B^*PB$. It follows that for each $k = 1, \ldots, n$ the kth column H_k of H induces a contraction from $\mathbf{U}$ into $\mathbf{E}$ if and only if $P_{\mathcal{U}_k}(D^*D + B^*PB)|_{\mathcal{U}_k}$ is a contraction, where $P_{\mathcal{U}_k}$ is the orthogonal projection onto $\mathcal{U}_k$. In other words, H is in $\mathbb{UM}^2_{\text{ball}}(\mathbf{U}, \mathbf{E})$ if and only if the block diagonal entries of $D^*D + B^*PB$ are contractions.

Now assume that H is in $\mathbb{UM}^2_{\text{ball}}(\mathbf{U}, \mathbf{E})$. Let N be the upper triangular part of $D^*D + B^*PB$; see (3.7). In this setting, D_{H_k} equals the positive square root of the operator $I - P_{\mathcal{U}_k}(D^*D + B^*PB)|_{\mathcal{U}_k}$, and ∇_H is the diagonal operator formed by $\{D_{H_k}\}_{k=1}^n$ on $\oplus_{k=1}^n \mathcal{D}_{H_k}$; see (3.5). Then the set of all operators $F \in \mathbb{UM}(\mathbf{U})$ satisfying (3.10) is determined by (3.8). Finally, $H = Z^{(1)}(I_{\mathbf{U}} - Z^{(2)})^{-1}$, where $Z^{(1)} = HF^{-1}$ and $Z^{(2)} = I - F^{-1}$. Moreover, the operator matrix $\begin{bmatrix} Z^{(1)*} & Z^{(2)*} \end{bmatrix}$ induces a contraction.

4. The full version of Theorem 0.2 and its proof

The following theorem is the full version of Theorem 0.2.

Theorem 4.1. *Let H be a solution to* Problem 0.1, *and let $\mathbf{C}_{H,\omega}$ be the set of all operator matrices C in $\mathbb{UM}^\infty_{\text{ball},0}(\mathbf{D}_H)$ defined by* (0.11). *Fix $C \in \mathbf{C}_{H,\omega}$, and put*

$$F = N + \Pi_H^* \nabla_H (I_{\mathbf{D}_H} - C)^{-1} \nabla_H \Pi_H. \tag{4.1}$$

Here $N \in \mathbb{UM}(\mathbf{U})$ is defined by (3.7), *and $\nabla_H \in \mathbb{UM}(\mathbf{D}_H)$ and $\Pi_H \in \mathbb{UM}(\mathbf{D}_H, \mathbf{U})$ are the diagonal operator matrices given by* (3.5). *Then $F \in \mathbb{UM}(\mathbf{U})$, F satisfies* (0.12), *and*

$$F_{j,k}|_{\mathcal{F}_k} = F_{j,k-1}\omega_{k,2} \quad (j, k \in \mathbb{Z}, j < k). \tag{4.2}$$

Put

$$Z^{(1)} = HF^{-1} \quad \text{and} \quad Z^{(2)} = I_{\mathbf{U}} - F^{-1}. \tag{4.3}$$

Then the pair of operator matrices $(Z^{(1)}, Z^{(2)})$ satisfies (0.4), (0.5), (0.6), *and H is given by* (0.7). *Furthermore, the map $C \mapsto (Z^{(1)}, Z^{(2)})$ is a one-to one map from the set $\mathbf{C}_{H,\omega}$ onto the set of all pairs $(Z^{(1)}, Z^{(2)})$ satisfying* (0.4), (0.5), (0.6), *and such that H is given by* (0.7). *In particular, there exists a unique pair of operator matrices $(Z^{(1)}, Z^{(2)})$ satisfying* (0.4), (0.5) *and* (0.6) *such that H is given by* (0.7) *if and only if there exists a $k \in \mathbb{Z} \cup \{\pm\infty\}$ such that for each $j \in \mathbb{Z}$ we have $\mathcal{F}_{H_j} = \mathcal{D}_{H_j}$ if $j > k$ and ω_{H_j} is a co-isometry if $j \leq k$.*

Regarding the criterion in the final sentence of the above theorem, note that in the two extreme cases $k = \infty$ and $k = -\infty$ the condition reduces to 'ω_{H_j} is a co-isometry for all $j \in \mathbb{Z}$' if $k = \infty$ and '$\mathcal{F}_{H_j} = \mathcal{D}_{H_j}$ for all $j \in \mathbb{Z}$' if $k = -\infty$.

Let H be a solution to Problem 0.1. Recall that, in particular, this implies that H belongs to $\mathbb{UM}^2_{\rm ball}(\mathbf{U},\mathbf{Y})$. Hence Theorem 3.4 applies to H. It will be convenient first to prove the following proposition and lemma.

Proposition 4.2. *Let H be a solution to Problem 0.1. Fix C in $\mathbb{UM}^\infty_{\rm ball,0}(\mathbf{D}_H)$, and let $F=[F_{j,\,k}]_{j,k\in\mathbb{Z}}$ in $\mathbb{UM}(\mathbf{U})$ by defined by (4.1). Then F satisfies (4.2) if and only if C belongs to the set $\mathbf{C}_{H,\,\omega}$ defined by (0.11).*

Proof. Let H be a solution to Problem 0.1. We first show that

$$C\in\mathbf{C}_{H,\,\omega}\quad\Longleftrightarrow\quad C\nabla_H\Pi_H E=\nabla_H\Pi_H\Omega^{(2)}. \tag{4.4}$$

Here E and $\Omega^{(2)}$ are the operator matrices in $\mathbb{UM}(\mathbf{F},\mathbf{U})$ defined by (2.12) and (2.13). To prove (4.4) we use (0.9) and (0.10). From these formulas it follows that the operator matrix $C=[C_{j,\,k}]_{j,k\in\mathbb{Z}}$ in $\mathbb{UM}^\infty_{\rm ball,0}(\mathbf{D}_H)$ belongs to $\mathbf{C}_{H,\,\omega}$ if and only if for each $j<k$ in $\mathbb{Z}$ and each $f_k\in\mathcal{F}_k$ we have

$$C_{j,\,k}D_{H_k}f_k=\begin{cases} D_{H_{k-1}}\omega_{k,\,2}f_k & \text{for } j=k-1,\\ 0, & \text{for } j\neq k-1,\end{cases}\qquad f_k\in\mathcal{F}_k\quad (j<k\in\mathbb{Z}). \tag{4.5}$$

In the language of operator matrices (4.5) is equivalent to the right-hand side of (4.4). Thus our claim follows.

By assumption H belongs to $\mathbb{UM}^2_{\rm ball}(\mathbf{U},\mathbf{Y})$. Hence Theorem 3.4 applies to H. In particular, since (3.8), and (4.1) are the same identities, F is well defined and belongs to $\mathbb{UM}(\mathbf{U})$. From (3.8) it follows that F is also given by the following formula:

$$F=I_{\mathbf{U}}+\tilde{N}+\Pi_H^*\nabla_H(I-C)^{-1}C\nabla_H\Pi_H, \tag{4.6}$$

where $\tilde{N}=[\tilde{N}_{i,\,j}]\in\mathbb{UM}(\mathbf{U})$ is the strictly upper triangular operator matrix given by

$$\tilde{N}_{i,\,j}=\begin{cases}(H_i)^*H_j, & \text{for } i<j,\\ 0, & \text{for } i\geq j.\end{cases} \tag{4.7}$$

We claim that

$$\tilde{N}E=N\Omega^{(2)}. \tag{4.8}$$

Note that both $\tilde{N}E$ and $N\Omega^{(2)}$ are strictly upper triangular operator matrices in $\mathbb{UM}(\mathbf{F},\mathbf{U})$. Furthermore, for $j<k$ the (j,k)th entry of $\tilde{N}E$ is equal to $(H_j)^*H_k\tau_k$, where τ_k is the canonical embedding of $\mathcal{F}_k$ into $\mathcal{U}_k$. Now observe, using the second part of (0.3), that for $j<k$ we have

$$\begin{aligned}(H_j)^*H_k\tau_kf_k &= \sum_{\nu=-\infty}^{j}H^*_{\nu,\,j}H_{\nu,\,k}f_k=\sum_{\nu=-\infty}^{j}H^*_{\nu,\,j}H_{\nu,\,k-1}\omega_{k,\,2}f_k\\ &= (H_j)^*H_{k-1}\omega_{k,\,2}f_k\quad (f_k\in\mathcal{F}_k).\end{aligned}$$

But $(H_j)^*H_{k-1}\omega_{k,\,2}$ is precisely the (j,k)th entry of $N\Omega^{(2)}$. Thus (4.8) is proved.

Next we use the two representations for F given by (4.1) and (4.6). We multiply (4.1) and (4.6) from the right by $\Omega^{(2)}$ and E, respectively, and subtract the resulting identities. Then, using (4.8), we obtain

$$F\Omega^{(2)} - FE + E = \Pi_H^* \nabla_H (I_{\mathbf{D}_H} - C)^{-1} (\nabla_H \Pi_H \Omega^{(2)} - C \nabla_H \Pi_H E).$$

Notice that $\nabla_H (I_{\mathbf{D}_H} - C)^{-1}$ acts as a one-to-one linear transformations on the linear space $\mathbf{D}_H^+$ (see the final paragraph of Subsection 1.2 for the definition of this space). Since Π_H^* is also one-to-one on vectors from $\mathbf{D}_H^+$, it follows that

$$F\Omega^{(2)} - FE + E = 0 \iff \nabla_H \Pi_H \Omega^{(2)} - C \nabla_H \Pi_H E = 0. \tag{4.9}$$

By comparing the left-hand side of (4.9) with (2.16) and the right-hand side of (4.9) with (4.4), we see that F satisfies (4.2) if and only if C belongs to $\mathbf{C}_{H,\omega}$. □

Lemma 4.3. *Let $\mathbf{V} = \oplus_{k\in\mathbb{Z}} \mathcal{V}_k$ and $\mathbf{W} = \oplus_{k\in\mathbb{Z}} \mathcal{W}_k$, and let $\mathcal{X}_k$ be a subspace of $\mathcal{V}_k$ and Λ_k a contraction from $\mathcal{X}_k$ into $\mathcal{W}_{k-1}$ for each $k \in \mathbb{Z}$. Then the set*

$$\mathbf{C}_\Lambda = \Big\{ C \in \mathbb{UM}^\infty_{\mathrm{ball},0}(\mathbf{V},\mathbf{W}) \mid C = \left[\ C_{j,k}\ \right]_{j,k\in\mathbb{Z}} \text{ and }$$
$$C_{j,k}|_{\mathcal{X}_k} = \begin{cases} \Lambda_k, & \text{if } j = k-1, \\ 0, & \text{if } j \neq k-1, \end{cases} \text{ for each } j,k \in \mathbb{Z} \Big\}$$

is a singleton if and only if there exists a $k \in \mathbb{Z} \cup \{\pm\infty\}$ such that for each $j \in \mathbb{Z}$

$$\mathcal{X}_j = \mathcal{V}_j \text{ if } j > k \text{ and } \Lambda_j \text{ is a co-isometry if } j \leq k. \tag{4.10}$$

Proof. Let $C = \left[\ C_{i,j}\ \right]_{i,j\in\mathbb{Z}} \in \mathbf{C}_\Lambda$. Fix a $k \in \mathbb{Z}$. Observe that $\mathcal{X}_k = \mathcal{V}_k$ implies that the kth column of C is completely determined by

$$C_{i,k} = \begin{cases} \Lambda_k, & \text{if } i = k-1, \\ 0, & \text{if } i \neq k-1. \end{cases}$$

Moreover, if Λ_k is a co-isometry, then we can use Corollary XXVII.5.3 from [24] to show that the $(k-1)$th row of C is completely determined by

$$C_{k-1,j} = \begin{cases} \Lambda_k \Pi_{\mathcal{X}_k}, & \text{if } j = k, \\ 0, & \text{if } j \neq k. \end{cases}$$

Here $\Pi_{\mathcal{X}_k}$ denotes the orthogonal projection from $\mathcal{V}_k$ onto $\mathcal{X}_k$. From these two observations and the fact that C is strictly upper triangular it follows that the existence of a $k \in \mathbb{Z} \cup \{\pm\infty\}$ such that (4.10) holds for all $j \in \mathbb{Z}$ is sufficient for $\mathbf{C}_\Lambda$ to be a singleton.

To see that the condition is also necessary, assume that no $k \in \mathbb{Z} \cup \{\pm\infty\}$ exists such that (4.10) holds for all $j \in \mathbb{Z}$, i.e., there exists a $k \in \mathbb{Z}$ with $\mathcal{X}_k \neq \mathcal{V}_k$ and such that Λ_{k-1} in not a co-isometry. In that case, set $\mathcal{Z}_k = \mathcal{V}_k \ominus \mathcal{X}_k$ and let $D_{\Lambda_{k-1}^*}$ and $\mathcal{D}_{\Lambda_{k-1}^*}$ denote the defect operator and defect space of Λ_{k-1}^*, respectively. Then both $\mathcal{Z}_k$ and $\mathcal{D}_{\Lambda_{k-1}^*}$ are not equal to $\{0\}$, and thus there exists a non-zero contraction N from $\mathcal{Z}_k$ into $\mathcal{D}_{\Lambda_{k-1}^*}$. Now define $C = \left[\ C_{i,j}\ \right]_{i,j\in\mathbb{Z}} \in \mathbb{UM}^\infty_{\mathrm{ball},0}(\mathbf{V},\mathbf{W})$ by setting $C_{i,j} = 0$ in case $i \neq j-1$ and $(i,j) \neq (k-2,k)$, $C_{j-1,j} = \Lambda_j \Pi_{\mathcal{X}_j}$ for each $j \in \mathbb{Z}$ and $C_{k-2,k} = D_{\Lambda_{k-1}^*} N \Pi_{\mathcal{Z}_k}$. One easily sees that C is in $\mathbf{C}_\Lambda$. Moreover, this is not the only element of $\mathbf{C}_\Lambda$, because $\mathbf{C}_\Lambda$ always contains the operator matrix from $\mathbb{UM}^\infty_{\mathrm{ball},0}(\mathbf{V},\mathbf{W})$ that has zeros in all entries accept for the first upper diagonal on which $\Lambda_k \Pi_{\mathcal{X}_k}$ is the entry in the $(k-1, k)$th position. □

Proof of Theorem 4.1. Let H be a solution to Problem 0.1. Let F be given by (4.1) with C from $\mathbf{C}_{H,\omega}$. By assumption H satisfies (0.3). Proposition 4.2 tells us that F satisfies (2.10). Thus we can apply Proposition 2.2 to show that the pair $(Z^{(1)}, Z^{(2)})$ satisfies the interpolation condition (0.6). To see that the map $C \mapsto (Z^{(1)}, Z^{(2)})$ is a one-to one map from the set $\mathbf{C}_{H,\omega}$ onto the set of all pairs $(Z^{(1)}, Z^{(2)})$ satisfying (0.4), (0.5), (0.6) and such that H is given by (0.7) it remains to apply Theorems 2.1 and 3.4.

In order to prove the claim in the final part of Theorem 4.1 note that there exists a unique pair of operator matrices $(Z^{(1)}, Z^{(2)})$ satisfying (0.4), (0.5) and (0.6) such that H is given by (0.7) if and only if the set $\mathbf{C}_{H,\omega}$ is a singleton. The result then follows from Lemma 4.3. □

Recall that for a given solution H to Problem 0.1 there may exist many pairs $(Z^{(1)}, Z^{(2)})$ satisfying (0.4)-(0.6) such that H is given by (0.7). The following proposition provides a necessary and sufficient condition for the existence of a unique pair of operator matrices $(Z^{(1)}, Z^{(2)})$ satisfying (0.4)–(0.6), and thus a sufficient condition for the existence of a unique solution to Problem 0.1. However, the condition is in general not necessary. This phenomenon already occurs in the time-invariant case; see [28, 27]. The problem to give necessary and sufficient conditions for the existence of a unique solution to Problem 0.1 remains open.

Proposition 4.4. *There exists a unique pair of operator matrices $(Z^{(1)}, Z^{(2)})$ satisfying* (0.4), (0.5) *and* (0.6) *if and only if there exists a $k \in \mathbb{Z} \cup \{\pm\infty\}$ such that for each $j \in \mathbb{Z}$ we have $\mathcal{F}_j = \mathcal{U}_j$ if $j > k$ and ω_j is a co-isometry if $j \leq k$. In particular, in that case there exists a unique solution to Problem* 0.1.

Proof. Let $(Z^{(1)}, Z^{(2)})$ be a pair of operator matrices as in (0.4).

Set $\mathbf{Z} = \oplus_{i \in \mathbb{Z}} (\mathcal{Y}_{i+1} \oplus \mathcal{U}_i)$, and define

$$\widetilde{Z} = \begin{bmatrix} \widetilde{Z}_{i,j} \end{bmatrix}_{i,j \in \mathbb{Z}} \in \mathbb{UM}_0^\infty(\mathbf{U}, \mathbf{Z}), \text{ where } \widetilde{Z}_{i,j} = \begin{bmatrix} Z^{(1)}_{i+1,j} \\ Z^{(2)}_{i,j} \end{bmatrix} : \mathcal{U}_j \to \begin{bmatrix} \mathcal{Y}_{i+1} \\ \mathcal{U}_i \end{bmatrix}.$$

Observe that (0.5) is equivalent to $\widetilde{Z} \in \mathbb{UM}^\infty_{\mathrm{ball},0}(\mathbf{U}, \mathbf{Z})$, while (0.6) corresponds to

$$\widetilde{Z}_{j,k}|_{\mathcal{F}_k} = \begin{cases} \omega_k, & \text{if } j = k-1, \\ 0, & \text{if } j \neq k-1. \end{cases} \tag{4.11}$$

Thus the pair $(Z^{(1)}, Z^{(2)})$ satisfies (0.5) and (0.6) if and only if $\widetilde{Z}$ is an element of

$$\mathbf{C}_\omega = \{\widetilde{Z} \in \mathbb{UM}^\infty_{\mathrm{ball},0}(\mathbf{U}, \mathbf{Z}) \mid \text{the } (i,j)\text{th entry } \widetilde{Z}_{i,j} \text{ of } \widetilde{Z} \text{ satisfies (4.11) for each } i, j \in \mathbb{Z}\}.$$

The first statement now follows from Lemma 4.3. The last statement of Proposition 4.4 follows immediately from the first part. □

5. An example involving finite operator matrices and time-variant relaxed commutant lifting

In this section we present some examples of how our results can be applied. In each case the contractions ω_k, $k \in \mathbb{Z}$, are not given beforehand but are constructed from the given data. The first subsection deals with a 4×4 operator matrix problem. In the second subsection we introduce a time-variant analogue of the relaxed commutant lifting problem, and show how this time-variant problem can be solved by using Theorem 0.1.

5.1. An example involving finite operator matrices

When in Problem 0.1 all spaces $\mathcal{U}_k$ and $\mathcal{Y}_j$ are set to zero, with the exception of a finite numbers of k's and j's, finite operator matrix problems appear. We illustrate this with an example.

Consider the problem of finding all 4×4 operator matrices

$$A = \begin{bmatrix} A_{1,1} & A_{1,2} & A_{1,3} & A_{1,4} \\ A_{2,1} & A_{2,2} & A_{2,3} & A_{2,4} \\ A_{3,1} & A_{3,2} & A_{3,3} & A_{3,4} \\ A_{4,1} & A_{4,2} & A_{4,3} & A_{4,4} \end{bmatrix} : \begin{bmatrix} \mathcal{X}_1 \\ \mathcal{X}_2 \\ \mathcal{X}_3 \\ \mathcal{X}_4 \end{bmatrix} \to \begin{bmatrix} \mathcal{Y}_1 \\ \mathcal{Y}_2 \\ \mathcal{Y}_3 \\ \mathcal{Y}_4 \end{bmatrix} \tag{5.1}$$

such that

$$\| \begin{bmatrix} A_{1,k} & A_{1,k+1} \\ A_{2,k} & A_{2,k+1} \\ A_{3,k} & A_{3,k+1} \\ A_{4,k} & A_{4,k+1} \end{bmatrix} \| \leq 1, \quad k = 1, 2, 3. \tag{5.2}$$

The problem is always solvable (the zero matrix is a solution). Moreover all solutions can be obtained by repeatedly applying Parrott's lemma [29] (cf., Section IV.1 in [16] or Section XXVII.5 in [24]). Here we show that a more direct description of the set of all solutions can be obtained from Theorem 0.1. To do this we set $\mathcal{Y}_j = \{0\}$ for $j \neq 1, 2, 3, 4$ and put

$$\begin{aligned}
&\mathcal{U}_k = \{0\}, \quad \mathcal{F}_k = \{0\} \quad (k \neq 4, 5, 6), \\
&\mathcal{U}_4 = \mathcal{X}_1 \oplus \mathcal{X}_2, \quad \mathcal{F}_4 = \{0\}, \\
&\mathcal{U}_5 = \mathcal{X}_2 \oplus \mathcal{X}_3, \quad \mathcal{F}_5 = \mathcal{X}_2 \oplus \{0\}, \quad \sigma_5(x_2 \oplus 0) = 0 \oplus x_2 \quad (x_2 \in \mathcal{X}_2), \\
&\mathcal{U}_6 = \mathcal{X}_3 \oplus \mathcal{X}_4, \quad \mathcal{F}_6 = \mathcal{X}_3 \oplus \{0\}, \quad \sigma_6(x_3 \oplus 0) = 0 \oplus x_3 \quad (x_2 \in \mathcal{X}_3).
\end{aligned}$$

To formulate this as an upper triangular operator matrix problem, consider

$$\begin{aligned}
H_{1,4} &= \begin{bmatrix} A_{1,1} & A_{1,2} \end{bmatrix}, & H_{1,5} &= \begin{bmatrix} A_{1,2} & A_{1,3} \end{bmatrix}, & H_{1,6} &= \begin{bmatrix} A_{1,3} & A_{1,4} \end{bmatrix} \\
H_{2,4} &= \begin{bmatrix} A_{2,1} & A_{1,2} \end{bmatrix}, & H_{2,5} &= \begin{bmatrix} A_{2,2} & A_{2,3} \end{bmatrix}, & H_{2,6} &= \begin{bmatrix} A_{2,3} & A_{2,4} \end{bmatrix} \\
H_{3,4} &= \begin{bmatrix} A_{3,1} & A_{3,2} \end{bmatrix}, & H_{3,5} &= \begin{bmatrix} A_{3,2} & A_{3,3} \end{bmatrix}, & H_{3,6} &= \begin{bmatrix} A_{3,3} & A_{3,4} \end{bmatrix} \\
H_{4,4} &= \begin{bmatrix} A_{1,1} & A_{1,2} \end{bmatrix}, & H_{4,5} &= \begin{bmatrix} A_{1,2} & A_{1,3} \end{bmatrix}, & H_{4,6} &= \begin{bmatrix} A_{1,3} & A_{1,4} \end{bmatrix}
\end{aligned}$$

Then (5.1) and (5.2) are satisfied if and only if

$$\left\| \begin{bmatrix} H_{1,k} \\ H_{2,k} \\ H_{3,k} \\ H_{4,k} \end{bmatrix} \right\| \leq 1 \quad (k = 4, 5, 6),$$

$$\begin{bmatrix} H_{1,5} \\ H_{2,5} \\ H_{3,5} \\ H_{4,5} \end{bmatrix} |_{\mathcal{F}_5} = \begin{bmatrix} H_{1,4} \\ H_{2,4} \\ H_{3,4} \\ H_{4,4} \end{bmatrix} \sigma_5, \quad \text{and} \quad \begin{bmatrix} H_{1,6} \\ H_{2,6} \\ H_{3,6} \\ H_{4,6} \end{bmatrix} |_{\mathcal{F}_5} = \begin{bmatrix} H_{1,5} \\ H_{2,5} \\ H_{3,5} \\ H_{4,5} \end{bmatrix} \sigma_6.$$

Now consider our main problem with $\mathcal{Y}_j$, $\mathcal{U}_k = \{0\}$ and $\mathcal{F}_k$ as above. Furthermore, put

$$\omega_{5,1} = 0, \quad \omega_{6,1} = 0, \quad \text{and} \quad \omega_{5,2} = \sigma_5 \quad \omega_{6,2} = \sigma_6.$$

Since $\mathcal{F}_k = \{0\}$ for $k \neq 5, 6$, we don't have to consider the operators $\omega_{k,1}$ and $\omega_{k,1}$ for $k \neq 5, 6$. Note that

$$\begin{bmatrix} \omega_{5,1} \\ \omega_{6,1} \end{bmatrix} \quad \text{and} \quad \begin{bmatrix} \omega_{5,2} \\ \omega_{6,2} \end{bmatrix} \quad \text{are both contractions.}$$

It is now clear how the problem regarding the operator matrix A in (5.1) can be solved via Theorem 0.1.

5.2. A time-variant relaxed commutant lifting problem

In this subsection we present a time-varying version of the relaxed commutant lifting problem from [21] and explain the connection with Problem 0.1. We plan to come back to this time-variant relaxed commutant lifting problem in more detail in a future publication, where we will also discuss the relation with the time-invariant version, the three chain completion problem [18, 19] and its weighted versions [9].

A data set for the time-variant relaxed commutant lifting problem is a set $\Lambda = \{A_n, T'_n, U'_n, R_n, Q_n \mid n \in \mathbb{Z}\}$ consisting of Hilbert space operators with for each $k \in \mathbb{Z}$

$$T'_k : \mathcal{H}'_{k-1} \to \mathcal{H}'_k, \quad A_k : \mathcal{H}_k \to \mathcal{H}'_k, \quad R_k : \mathcal{H}_{0,k} \to \mathcal{H}_k, \quad Q_k : \mathcal{H}_{0,k-1} \to \mathcal{H}_k,$$

and such that T'_k and A_k are contractions and

$$T'_k A_{k-1} R_{k-1} = A_k Q_k, \quad R^*_{k-1} R_{k-1} \leq Q^*_k Q_k.$$

The operator U'_k is completely determined by the set of operators $\{T'_n \mid n \in \mathbb{Z}\}$ and can be seen as a time-varying analogue of the Sz.-Nagy-Schäffer isometric lifting of T'_k; cf., [10]. To define U'_k we set

$$\mathbf{D}'_n = \oplus_{i \leq n} \mathcal{D}_{T'_n} \quad \text{and} \quad \mathcal{K}'_n = \mathcal{H}_n \oplus \mathbf{D}'_n \quad (n \in \mathbb{Z}).$$

Then U'_k is the isometric operator mapping $\mathcal{K}'_{k-1}$ into $\mathcal{K}'_k$ given by

$$U'_k = \begin{bmatrix} T'_k & 0 \\ E_{\mathcal{D}_{T'_n}} D_{T'_n} & E_{\mathbf{D}'_{k-1}} \end{bmatrix} : \begin{bmatrix} \mathcal{H}'_{k-1} \\ \mathbf{D}'_{k-1} \end{bmatrix} \to \begin{bmatrix} \mathcal{H}'_k \\ \mathbf{D}'_k \end{bmatrix}. \tag{5.3}$$

Here $E_{\mathcal{D}_{T'_k}}$ and $E_{\mathbf{D}'_{k-1}}$ are the canonical embeddings of $\mathcal{D}_{T'_k}$ and $\mathbf{D}'_{k-1}$ into $\mathbf{D}'_k$, respectively. We then consider the following problem.

Problem 5.1. *Given the data set* $\Lambda = \{A_n, T'_n, U'_n, R_n, Q_n \mid n \in \mathbb{Z}\}$, *describe the sets of operators* $\{B_n \mid n \in \mathbb{Z}\}$ *with the property that for each* $k \in \mathbb{Z}$ *the operator* B_k *is a contraction from* $\mathcal{H}_k$ *into* $\mathcal{K}'_k$ *satisfying*

$$\Pi_{\mathcal{H}'_k} B_k = A_k \quad \textit{and} \quad U'_k B_{k-1} R_{k-1} = B_k Q_k. \tag{5.4}$$

Here $\Pi_{\mathcal{H}'_k}$ *is the orthogonal projection from* $\mathcal{K}'_k$ *onto* $\mathcal{H}'_k$.

After some translation and reduction steps it follows that the special case of Problem 5.1 with $\mathcal{H}_{0,k} = \mathcal{H}_k$ and $R_k = I_{\mathcal{H}_k}$ for each $k \in \mathbb{Z}$ is just the nonstationary commutant lifting problem considered in [10] (see also Section 3.5 in [11]).

With the data set Λ we associate a set of contractions $\{\omega_n \mid n \in \mathbb{Z}\}$ of the form (0.2). For each $k \in \mathbb{Z}$, the contraction ω_k is defined on the subspace $\mathcal{F}_k = \overline{D_{A_k} Q_k \mathcal{H}_{0,k-1}}$ of $\mathcal{D}_{A_k}$ and is given by

$$\omega_k : \mathcal{F}_k \to \begin{bmatrix} \mathcal{D}_{T'_k} \\ \mathcal{D}_{A_{k-1}} \end{bmatrix}, \quad \omega_k D_{A_k} Q_k = \begin{bmatrix} D_{T'_k} A_{k-1} R_{k-1} \\ D_{A_{k-1}} R_{k-1} \end{bmatrix}.$$

We refer to ω_k as the *kth underlying contraction of the data set* Λ. To see that ω_k is in fact a contraction observe that for each $h \in \mathcal{H}_{0,k-1}$ we have

$$\begin{aligned} \|D_{A_k} Q_k h\|^2 &= \|Q_k h\|^2 - \|A_k Q_k h\|^2 \leq \|R_{k-1} h\|^2 - \|T'_k A_{k-1} R_{k-1} h\|^2 \\ &= \|R_{k-1} h\|^2 - \|A_{k-1} R_{k-1} h\|^2 + \|A_{k-1} R_{k-1} h\|^2 - \|T'_k A_{k-1} R_{k-1} h\|^2 \\ &= \|D_{A_{k-1}} R_{k-1} h\|^2 + \|D_{T'_k} A_{k-1} R_{k-1} h\|^2. \end{aligned}$$

It follows directly from this computation that ω_k is contractive, and moreover, that ω_k is an isometry if and only if $R^*_{k-1} R_{k-1} = Q^*_k Q_k$.

From a variation on Parrott's lemma [29] (cf., Section IV.1 in [16] or Section XXVII.5 in [24]) we obtain that an operator B_k from $\mathcal{H}_k$ into $\mathcal{K}'_k$ is a contraction that satisfies the first requirement of (5.4) if and only B_k is of the form

$$B_k = \begin{bmatrix} A_k \\ \Gamma_k D_{A_k} \end{bmatrix}, \quad \text{with } \Gamma_k : \mathcal{D}_{A_k} \to \mathbf{D}'_k \text{ a contraction.} \tag{5.5}$$

Moreover, the contraction Γ_k is uniquely determined by B_k. In this case, writing out $U'_k B_{k-1} R_{k-1}$ and $B_k Q_k$, it follows that the second requirement in (5.4) holds if and only if Γ_k and Γ_{k-1} satisfy

$$\Gamma_k|_{\mathcal{F}_k} = E_{\mathcal{D}_{T'_k}} \omega_{k,1} + E_{\mathbf{D}'_{k-1}} \Gamma_{k-1} \omega_{k,2}. \tag{5.6}$$

Here $\omega_{k,1}$ is the first component of ω_k mapping $\mathcal{F}_k$ into $\mathcal{D}_{T'_k}$ and $\omega_{k,2}$ is the second component of ω_k mapping $\mathcal{F}_k$ into $\mathcal{D}_{A_{k-1}}$. Thus, alternatively, we seek the sets of operators $\{\Gamma_n \mid n \in \mathbb{Z}\}$ such that for each $k \in \mathbb{Z}$ the operator Γ_k is a contraction from $\mathcal{D}_{A_k}$ into $\mathbf{D}'_k$ and (5.6) is satisfied.

Now set $\mathbf{D} = \oplus_{k\in\mathbb{Z}}\mathcal{D}_{A_k}$ and $\mathbf{D}' = \oplus_{k\in\mathbb{Z}}\mathcal{D}_{T'_n}$. With a set of contractions $\{B_n : \mathcal{H}'_n \to \mathcal{K}'_n \mid n \in \mathbb{Z}\}$ satisfying the first condition of (5.4) we associate an operator matrix $H \in \mathbb{UM}^2_{\text{ball}}(\mathbf{D}, \mathbf{D}')$ in the following way.

Procedure 5.1. *Let $\{B_n : \mathcal{H}'_n \to \mathcal{K}'_n \mid n \in \mathbb{Z}\}$ be a set of contractions such that the first condition of* (5.4) *holds for each $k \in \mathbb{Z}$, and let Γ_k be the contraction from $\mathcal{D}_{A_k}$ into $\mathbf{D}'_k$ determined by* (5.5). *For each $k \in \mathbb{Z}$ we view $\mathbf{D}'_k$ as a subspace of $\mathbf{D}'$ and write $\Pi_{\mathbf{D}'_k}$ for the orthogonal projection from $\mathbf{D}'$ onto $\mathbf{D}'_k$. Define H_k to be the contraction from $\mathcal{D}_{A_k}$ into $\mathbf{D}'$ given by $H_k = \Pi^*_{\mathbf{D}'_k}\Gamma_k$ and $H \in \mathbb{UM}^2_{\text{ball}}(\mathbf{D}, \mathbf{D}')$ by*

$$H = \begin{bmatrix} \cdots & H_{-1} & H_0 & H_1 & \cdots \end{bmatrix}.$$

One can reverse this procedure in order to obtain a set of contractions $\{B_n \mid n \in \mathbb{Z}\}$ that satisfy (5.4) for each $k \in \mathbb{Z}$ from a operator matrix in $\mathbb{UM}^2_{\text{ball}}(\mathbf{D}, \mathbf{D}')$.

Procedure 5.2. *Let $H \in \mathbb{UM}^2_{\text{ball}}(\mathbf{D}, \mathbf{D}')$. For each $k \in \mathbb{Z}$ let H_k be the kth column of H, i.e., $H_k = H|_{\mathcal{D}_{A_k}}$, set $\Gamma_k = \Pi_{\mathbf{D}'_k} H_k$ and define B_k by* (5.5).

It is straightforward that Γ_k defined in Procedure 5.2 is a contraction, and thus that B_k is a contraction satisfying the first requirement of (5.4), and moreover, these procedures are each others inverse. Furthermore, the condition (5.6) on the contractions $\{\Gamma_n \mid n \in \mathbb{Z}\}$ translates to the columns in the operator matrix H obtained by Procedure 5.1 in the form

$$H_k|_{\mathcal{F}_k} = \tau'_k\omega_{k,1} + H_{k-1}\omega_{k,2}, \tag{5.7}$$

where τ'_k is the canonical embedding of $\mathcal{D}_{T'_n}$ into $\mathbf{D}'$, or, equivalently, in the form (0.3) with $\omega_{k,1}$ and $\omega_{k,2}$ as defined in the present subsection, $\mathcal{U}_k = \mathcal{D}_{A_k}$ and $\mathcal{Y}_k = \mathcal{D}_{T'_k}$ for each $k \in \mathbb{Z}$. The converse is also true. For an operator matrix $H \in \mathbb{UM}^2_{\text{ball}}(\mathbf{D}, \mathbf{D}')$ such that the kth column H_k of H satisfies (5.7), the contraction Γ_k obtained by Procedure 5.2 satisfies (5.6). In conclusion, we have the following result.

Theorem 5.1. *Let $\Lambda = \{A_n, T'_n, U'_n, R_n, Q_n \mid n \in \mathbb{Z}\}$ be a data set as described above. For any solution H of Problem* 0.1 *with ω_k equal to the kth underlying contraction of Λ for each $k \in \mathbb{Z}$, the set $\{B_n \mid n \in \mathbb{Z}\}$ obtained from Procedure* 5.2 *is a solution of Problem* 5.1. *Conversely, for any solution $\{B_n \mid n \in \mathbb{Z}\}$ of Problem* 5.1, *the operator matrix H obtained from Procedure* 5.1 *is a solution of Problem* 0.1 *in case ω_k in* (0.2) *is the kth underlying contraction of Λ for each $k \in \mathbb{Z}$.*

From Theorem 5.1 in combination with Theorems 0.1 and 0.2 it is clear how all solutions of Problem 5.1 can be described. Furthermore, in combination with the full version of the second main result, Theorem 4.1, a time-variant analogue of Theorem 1.2 in [22] (see also Theorem 1.2 in [26]) is obtained.

Not only can Problem 5.1 be seen as a special case of Problem 0.1, the converse is also true, as explained in the following theorem.

Theorem 5.2. *For each $k \in \mathbb{Z}$ let ω_k be a contraction of the form* (0.2) *with $\mathcal{F}_k$ a subspace of $\mathcal{U}_k$. Set*

$$A_k = \begin{bmatrix} I_{\mathcal{Y}_{k+1}} & 0 \end{bmatrix} : \begin{bmatrix} \mathcal{Y}_{k+1} \\ \mathcal{U}_k \end{bmatrix} \to \mathcal{Y}_{k+1}, \quad T'_k = 0 : \mathcal{Y}_k \to \mathcal{Y}_{k+1},$$

$$R_k = \begin{bmatrix} \omega_{1,k+1} \\ \omega_{2,k+1} \end{bmatrix} : \mathcal{F}_{k+1} \to \begin{bmatrix} \mathcal{Y}_{k+1} \\ \mathcal{Y}_k \end{bmatrix}, \quad Q_k = \begin{bmatrix} 0 \\ \Pi^*_{\mathcal{F}_k} \end{bmatrix} : \mathcal{F}_k \to \begin{bmatrix} \mathcal{Y}_{k+1} \\ \mathcal{Y}_k \end{bmatrix},$$

and define U'_k by (5.3) *for each $k \in \mathbb{Z}$. Here $\Pi_{\mathcal{F}_k}$ denotes the orthogonal projection from $\mathcal{U}_k$ onto $\mathcal{F}_k$. Then $\Lambda = \{A_n, T'_n, U'_n, R_n, Q_n \mid n \in \mathbb{Z}\}$ is a data set for a time-variant relaxed commutant lifting problem, and ω_k is the kth underlying contraction of Λ for each $k \in \mathbb{Z}$.*

Proof. Fix $k \in \mathbb{Z}$. Clearly, T'_k and A_k are contraction. Moreover, $T'_k A_{k-1} R_{k-1}$ and $A_n Q_n$ are both equal to the zero operator from $\mathcal{F}_k$ into $\mathcal{Y}_{k+1}$, and, since $R_{k-1} = \omega_k$ is a contraction, we have $R^*_{k-1} R_{k-1} \leq I_{\mathcal{F}_k} = Q^*_k Q_k$. It then follows that Λ is a data set for a time-variant relaxed commutant lifting problem. Next, observe that

$$D_{A_k} Q_k = \Pi_{\mathcal{F}_k}, \quad D_{T'_k} A_{k-1} R_{k-1} = \omega_{1,k} \quad \text{and} \quad D_{A_{k-1}} R_{k-1} = \omega_{2,k}.$$

This implies that the kth underlying contraction of Λ is equal to ω_k. □

References

[1] D. Alpay and P. Dewilde, Time-varying signal approximation and estimation, in: *Signal Processing, Scattering and Operator Theory, and Numerical Methods* (*Amsterdam,* 1989), Progr. Systems Control Theory **5**, Birkhäuser, Boston, MA, 1990, pp. 1–22.

[2] D. Alpay, P. Dewilde, and H. Dym, Lossless inverse scattering and reproducing kernels for upper triangular operators, in: *Extension and interpolation of linear operators and matrix functions*, OT **47**, Birkhäuser Verlag, Basel, 1990, pp. 61–135.

[3] D. Alpay, A. Dijksma and Y. Peretz, Nonstationary analogs of the Herglotz representation theorem: the discrete case. *J. Funct. Anal.* **166** (1999), 85–129.

[4] D. Alpay, A. Dijksma and Y. Peretz, Nonstationary analogs of the Herglotz representation theorem: realizations centered at an arbitrary point, in *Recent advances in operator theory* (*Groningen,* 1998), OT **124**, Birkhäuser, Basel, 2001, pp. 63–87.

[5] J.A. Ball, Commutant lifting and interpolation: the time-varying case, *Integral Equations Operator Theory* **25** (1996), 377–405.

[6] J.A. Ball and I. Gohberg, A commutant lifting theorem for triangular matrices with diverse applications, *Integral Equations Operator Theory* **8** (1985), 205–267.

[7] J.A. Ball, I. Gohberg, and M.A. Kaashoek, Nevanlinna-Pick interpolation for time-varying input-output maps: the discrete case, in: *Time-Variant Systems and Interpolation*, OT **56**, Birkhäuser Verlag, Basel, 1992, pp. 1–51.

[8] J.A. Ball, I. Gohberg, and M.A. Kaashoek, H_∞-control and interpolation for time-varying systems, in: *Systems and networks: mathematical theory and applications, Vol. I* (*Regensburg,* 1993), Math. Res., **77**, Akademie-Verlag, Berlin, 1994, pp. 33–48.

[9] A. Biswas, C. Foias, and A.E. Frazho, Weighted variants of the three chain completion theorem, in: *Recent Advances in Operator Theory and Related Topics (Szeged, 1999)*, **OT 127**, Birkhäuser Verlag, Basel, 2001, pp. 127–144.

[10] T. Constantinescu, Some aspects of nonstationarity. I, *Acta Sci. Math. (Szeged)* **54** (1990), 379–389.

[11] T. Constantinescu, *Schur parameters, factorization and dilation problems*, OT **82**, Birkhäuser Verlag, Basel, 1996.

[12] K.R. Davidson, *Nest algebras. Triangular forms for operator algebras on Hilbert space*, Pitman Research Notes in Mathematics Series, **191**, Longman Scientific & Technical, Harlow, New York, 1988.

[13] P. Dewilde, Time-varying (matrix) interpolation, deadbeat control and the Löwner case, in: Proceedings of MTNS 2008 (Blacksburg, 2008).

[14] P. Dewilde, M.A. Kaashoek and M. Verhagen (Editors), *Challenges of a Generalized System Theory*, KNAW Wetensch. Verh. Afd. Natuurk. Eerste Reeks, **40**, North-Holland, Amsterdam 1993.

[15] P. Dewilde and A.-J. van der Veen, *Time-varying systems and computations*, Kluwer Academic Publishers, Boston, MA, 1998.

[16] C. Foias and A.E. Frazho, *The Commutant Lifting Approach to Interpolation Problems*, OT **44**, Birkhäuser Verlag, Basel, 1990.

[17] C. Foias, A.E. Frazho, I. Gohberg, and M.A. Kaashoek, Discrete time-variant interpolation as classical interpolation with an operator argument, *Integral Equations and Operator Theory* **26** (1996), 371–403.

[18] C. Foias, A.E. Frazho, I. Gohberg, and M.A. Kaashoek, A time-variant version of the commutant lifting theorem and nonstationary interpolation problems, *Integral Equations Operator Theory* **28** (1997), 158–190.

[19] C. Foias, A.E. Frazho, I. Gohberg, and M.A. Kaashoek, Parameterization of all solutions of the three chains completion problem, *Integral Equations Operator Theory* **29** (1997), 455–490.

[20] C. Foias, A.E. Frazho, I. Gohberg, and M.A. Kaashoek, *Metric constrained interpolation, commutant lifting and systems*, Birkhäuser Verlag, Basel, 1998.

[21] C. Foias, A.E. Frazho, and M.A. Kaashoek, Relaxation of metric constrained interpolation and a new lifting theorem, *Integral Equations and Operator Theory*, **42** (2002), 253–310.

[22] A.E. Frazho, S. ter Horst, and M.A. Kaashoek, All solutions to the relaxed commutant lifting problem, *Acta Sci. Math. (Szeged)* **72** (2006), 299–318.

[23] A.E. Frazho, S. ter Horst, and M.A. Kaashoek, Relaxed commutant lifting: an equivalent version and a new application, in: *Recent Advances in Operator Theory (Seoul, South Korea,* 2006), pp. 157–168, OT **187**, Birkhäuser Verlag, Basel, 2008.

[24] I. Gohberg, S. Goldberg and M.A. Kaashoek, *Classes of Linear Operators*, Volume II, OT **49**, Birkhäuser Verlag, Basel, 1993.

[25] A. Halanay and V. Ionescu, *Time-varying discrete linear systems*, OT **68**, Birkhäuser Verlag, Basel, 1994.

[26] S. ter Horst, *Relaxed commutant lifting and Nehari interpolation*, Ph.D. Thesis, Vrije Universiteit, Amsterdam, 2007, available online: `www.darenet.nl`.

[27] S. ter Horst, Relaxed commutant lifting: existence of a unique solution, Proc. Amer. Math. Soc. **137** (2009), 2697–2707.

[28] W.S. Li and D. Timotin, The relaxed intertwining lifting in the coupling approach, *Integral Equations and Operator theory* **54** (2006), pp. 97–111.

[29] S. Parrott, On a quotient norm and the Sz.-Nagy-Foias lifting theorem, *J. Funct. Anal.* **30** (1978), 311–328.

[30] V.I. Paulsen and S.C. Power, Lifting theorems for nest algebras, *J. Operator Theory* **20** (1988), 311–327.

A.E. Frazho
Department of Aeronautics and Astronautics
Purdue University
West Lafayette, IN 47907, USA
e-mail: frazho@ecn.purdue.edu

S. ter Horst
Department of Mathematics
Utrecht University
Budapestlaan 6
NL-3584 CD Utrecht, The Netherlands
e-mail: s.terhorst@uu.nl

M.A. Kaashoek
Afdeling Wiskunde, Faculteit der Exacte Wetenschappen
Vrije Universiteit
De Boelelaan 1081a
NL-1081 HV Amsterdam, The Netherlands
e-mail: ma.kaashoek@few.vu.nl

Operator Theory:
Advances and Applications, Vol. 195, 167–172

Nowhere Density of the Set of Non-norm-attaining Functionals

Francisco J. García-Pacheco

Abstract. In this short note we prove that every real Banach space with separable dual can be equivalently renormed so that the set of non-norm-attaining functionals on it is nowhere dense. This solves a question proposed by Enflo.

Mathematics Subject Classification (2000). Primary 46B03; Secondary 47A07.

Keywords. Non-norm-attaining functional, nowhere dense

1. Introduction

It is already proven that every (real or complex) Banach space can be equivalently renormed so that the set of norm-attaining functionals has non-empty interior (see for instance [2], [7], or [1]). In 2005 Enflo proposed the following question: *Is it possible to renorm a big class of Banach spaces to make the set of non-norm-attaining functionals be nowhere dense?* Precisely, in [1] we briefly considered the equivalent problem of renorming Banach spaces to make the set of norm-attaining functionals have dense interior. This is the problem that we take care of in this paper. Our result assures that if a real Banach space has separable dual then it can be done.

In much of this work we will make use of concepts and notation from the Geometry of the Banach spaces. For example, the concept of smooth face will be essential: Recall that a subset C of the unit sphere S_X of a Banach space X is said to be a face of the unit ball B_X of X if C is convex and verifies the extremal condition, in other words, if $x, y \in \mathsf{S}_X$ and there exists $t \in (0,1)$ such that $tx + (1-t)\,y \in C$, then $x, y \in C$; C is also said to be an smooth face of B_X if C is a face of B_X and has non-empty interior relative to the unit sphere S_X. Also, bear in mind that $\mathsf{NA}\,(X)$ denotes the set of norm-attaining functionals on a Banach space X. And, if $x \in X$ then $\widehat{x}$ will denote the element x seen in the bidual X^{**} of X for every non-zero x. Finally, for a smooth Banach space X, we will denote by $\mathsf{J}_X : X \longrightarrow X^*$ the dual mapping of X, that is, for every $x \in X$ we have that $\mathsf{J}_X(x)$ is the only functional in X^* verifying that $\|\mathsf{J}_X(x)\| = \|x\|$ and $\mathsf{J}_X(x)(x) = \|x\|^2$. We refer the reader to [4], [2], and [1] for a wider perspective of these concepts and further notation.

To finish the introduction we will briefly explain the basic notation used in this manuscript. Given a Banach space X, we have that B_X, U_X, and S_X denote respectively the closed unit ball, the open unit ball, and the unit sphere of X. When we talk about balls centered at a point $x \in X$ of radius $r \geq 0$ we will add (x, r) to the above expressions. If M is any subset of X, then $\operatorname{int}(M), \operatorname{cl}(M)$, and $\operatorname{bd}(M)$ will stand respectively for the interior of M, the closure of M, and the boundary of M.

2. Banach spaces with infinitely many smooth faces in their unit ball

The reason why we only work with real Banach spaces is simple: No complex space has smooth faces in its unit sphere (see [5, Theorem 2.1]). On the other hand, there are real Banach spaces whose unit ball can have many smooth faces. In [6, Lemma 2.7] it is shown that if a real Banach space is separable and has dimension greater than 1 then its unit ball has faces that are not smooth faces.

Theorem 2.1. *Let X be a real Banach space. If the union of the smooth faces of B_{X^*} is dense in S_{X^*} then $\mathsf{S}_{X^*} \setminus \mathsf{NA}(X)$ is nowhere dense in S_{X^*}.*

Proof. By virtue of [1, Theorem 3.10] we have that

$$\operatorname{cl}(\mathsf{S}_{X^*} \setminus \mathsf{NA}(X)) \subseteq \bigcup \{\operatorname{bd}_{\mathsf{S}_{X^*}}(C) : C \text{ is a face of } \mathsf{B}_{X^*}\}.$$

By hypothesis we have that

$$\operatorname{cl}\left(\bigcup \{\operatorname{int}_{\mathsf{S}_{X^*}}(C) : C \text{ is a face of } \mathsf{B}_{X^*}\}\right) = \mathsf{S}_{X^*}.$$

Now, observe that for any two faces A and B of B_{X^*} we have that

$$\operatorname{int}_{\mathsf{S}_{X^*}}(A) \cap \operatorname{bd}_{\mathsf{S}_{X^*}}(B) = \varnothing.$$

Indeed, otherwise the extremal condition verified by B implies that $A \subseteq B$; now A is a maximal face of B_{X^*} since it has non-empty interior relative to S_{X^*}, which means that $A = B$; and it is a well-known topological fact that the interior and the boundary of any set are always disjoint. Next,

$$\begin{aligned} \varnothing &= \operatorname{int}_{\mathsf{S}_{X^*}}\left(\bigcup \{\operatorname{bd}_{\mathsf{S}_{X^*}}(C) : C \text{ is a face of } \mathsf{B}_{X^*}\}\right) \\ &\supseteq \operatorname{int}_{\mathsf{S}_{X^*}}(\operatorname{cl}(\mathsf{S}_{X^*} \setminus \mathsf{NA}(X))). \end{aligned}$$

□

Lemma 2.2. *Let X be a real Banach space. Let $(\delta_k)_{k\in\mathbb{N}} \subset [1/2, 1)$ and $(y_k^*)_{k\in\mathbb{N}} \subset \mathsf{S}_{X^*} \cap \mathsf{NA}(X)$ be two sequences such that $(\delta_k)_{k\in\mathbb{N}}$ is strictly increasing and convergent to 1. Then, the set*

$$\mathcal{B} := \mathsf{B}_X \cap \bigcap_{k=1}^{\infty} (y_k^*)^{-1}([-\delta_k, \delta_k])$$

is the unit ball of an equivalent norm on X such that:

1. $\mathrm{int}(\mathcal{B}) = \{x \in \mathsf{U}_X : |y_k^*(x)| < \delta_k$ *for each* $k \in \mathbb{N}\}$.
2. *The set* $(y_1^*)^{-1}(\delta_1) \cap \mathcal{B}$ *is an smooth face of* $\mathcal{B}$.

Proof. In the first place, $\mathcal{B}$ is closed, convex, balanced, bounded, and has non-empty interior (since $\mathsf{B}_X(0, 1/2) \subseteq \mathcal{B}$). Thus, $\mathcal{B}$ is indeed the unit ball of a new equivalent norm on X. Let us see now that $\mathcal{B}$ verifies the desired properties.

1. On the one hand, it is clear that
$$\mathrm{int}(\mathcal{B}) \subseteq \{x \in \mathsf{U}_X : |y_k^*(x)| < \delta_k \text{ for each } k \in \mathbb{N}\}.$$
On the other hand, let then $x \in \mathsf{U}_X$ with $|y_k^*(x)| < \delta_k$ for each $k \in \mathbb{N}$. Since $\|x\| < 1$ there exists $k_0 \in \mathbb{N}$ such that $\|x\| < \delta_{k_0}$. Now, take $r > 0$ small enough so that $\mathsf{B}_X(x, r) \subseteq \mathsf{B}_X$, $\|x\| + r \leq \delta_{k_0}$, and $|y_k^*(x)| + r \leq \delta_k$ for $1 \leq k \leq k_0$. With these many conditions on r we have that $\mathsf{B}_X(x, r) \subseteq \mathcal{B}$.
2. Since $y_1^* \in \mathsf{S}_{X^*} \cap \mathsf{NA}(X^*)$ there exists $x \in \mathsf{S}_X$ such that $y_1^*(x) = 1$. Take $0 < r < \delta_2 - \delta_1$ (note that $r < 2\delta_1$). We will show that
$$\mathsf{B}_X(\delta_1, r) \cap \mathrm{bd}(\mathcal{B}) \subseteq (y_1^*)^{-1}(\delta_1) \cap \mathcal{B}.$$
Pick then $u \in \mathsf{B}_X(\delta_1 x, r) \cap \mathrm{bd}(\mathcal{B})$ and assume that $y_1^*(u) < \delta_1$. Since $r < 2\delta_1$ we have that $y_1^*(u) > -\delta_1$. Therefore, we can find $s > 0$ such that $\mathsf{B}_X(u, s) \subseteq (y_1^*)^{-1}([-\delta_1, \delta_1])$ and $\delta_1 + r + s \leq \delta_2$. With this choice of s we also have that $\mathsf{B}_X(u, s) \subseteq \mathsf{B}_X(0, \delta_2)$. As a consequence, we deduce that $\mathsf{B}_X(u, s) \subseteq \mathcal{B}$ and $u \in \mathrm{int}(\mathcal{B})$, which is a contradiction. □

Theorem 2.3. *Let X be a real Banach space. Assume that X is separable. Then, X can be equivalently renormed in such a way that the union of the smooth faces of B_X is dense in S_X.*

Proof. In the first place, X is weakly compactly generated and thus it admits an equivalent renorming that is locally uniformly rotund (see [8]). Therefore, we can assume without loss of generality that X is already locally uniformly rotund. Our purpose is to inductively construct two sequences $(\delta_k)_{k\in\mathbb{N}} \subset [1/2, 1)$ and $(y_k^*)_{k\in\mathbb{N}} \subset \mathsf{S}_{X^*} \cap \mathsf{NA}(X)$ verifying that $(\delta_k)_{k\in\mathbb{N}}$ is strictly increasing and convergent to 1. Let $(x_n)_{n\in\mathbb{N}} \subset \mathsf{S}_X$ be any dense sequence and. Since X is locally uniformly rotund we have that every point in its unit sphere is a strongly exposed point of the unit ball. Therefore, for every $n \in \mathbb{N}$, let x_n^* denote the functional of norm 1 that strongly exposes x_n on B_X. We will now proceed inductively.

1. For $k = 1$ we take $N_1 = \mathbb{N}$, $n_1 = 1$, $y_1^* := x_1^*$, and $\delta_1 = 1/2$.

2. For $k=2$ we take $N_2 := \{n \in N_1 : |y_1^*(x_n)| < \delta_1\}$, $n_2 := \min N_2$, and $y_2^* := x_{n_2}^*$. There exists $1/2+1/4 \leq \delta_2 < 1$ such that $|y_2^*(x_n)| < \delta_2$ for $n \in \mathbb{N} \setminus N_2$.
3. For $k = l+1$ we take $N_{l+1} := \{n \in N_l : |y_l^*(x_n)| < \delta_l\}$, $n_{l+1} := \min N_{l+1}$, and $y_{l+1}^* := x_{n_{l+1}}^*$. There exists $1/2+1/4+\cdots+1/2^{l+1} \leq \delta_{l+1} < 1$ such that $\left|y_{l+1}^*(x_n)\right| < \delta_{l+1}$ for $n \in \mathbb{N} \setminus N_{l+1}$.

Note that for every $m \in \mathbb{N}$ there exists a unique $k_m \in \mathbb{N}$ such that $\left|y_{k_m}^*(x_m)\right| \geq \delta_{k_m}$. According to Lemma 2.2, we can consider the equivalent norm on X whose unit ball is given by

$$\mathcal{B} := \mathsf{B}_X \cap \bigcap_{k=1}^{\infty} (y_k^*)^{-1}([-\delta_k, \delta_k]).$$

Let us finally see that this new norm verifies the desired properties.

1. For every $k \in \mathbb{N}$ we have that $(y_k^*)^{-1}(\delta_k) \cap \mathcal{B}$ is a smooth face of $\mathcal{B}$. Indeed, by Lemma 2.2 we have that $(y_1^*)^{-1}(\delta_1) \cap \mathcal{B}$ is an smooth face of $\mathcal{B}$. For $k > 1$ we only need to inductively apply Lemma 2.2 to

$$\left(\mathsf{B}_X \cap \bigcap_{i=1}^{k-1} (y_i^*)^{-1}([-\delta_i, \delta_i])\right) \cap \bigcap_{i=k}^{\infty} (y_i^*)^{-1}([-\delta_i, \delta_i]).$$

2. The set

$$\bigcup_{k=1}^{\infty} (y_k^*)^{-1}(\{-\delta_k, \delta_k\}) \cap \mathcal{B}$$

is dense in $\operatorname{bd}(\mathcal{B})$. Indeed, let $x \in \operatorname{bd}(\mathcal{B})$ and assume that $|y_k^*(x)| < \delta_k$ for each $k \in \mathbb{N}$. By Lemma 2.2 we have that $\|x\| = 1$. Taking into account that $(x_n)_{n\in\mathbb{N}}$ is dense in S_X, we deduce that there exists a sequence $(m_j)_{j\in\mathbb{N}} \subset \mathbb{N}$ such that $(x_{m_j})_{j\in\mathbb{N}}$ converges to x. For every $j \in \mathbb{N}$ let us denote x_{m_j} by y_j. As we noted above, for every $j \in \mathbb{N}$ there exists a unique $k_j \in \mathbb{N}$ such that $\left|y_{k_j}^*(y_j)\right| \geq \delta_{k_j}$. Suppose that the sequence $(k_j)_{j\in\mathbb{N}}$ is finite. Then, there exist $i \in \mathbb{N}$ and a subsequence $(y_{j_l})_{l\in\mathbb{N}} \subseteq (y_j)_{j\in\mathbb{N}}$ such that $\left|y_{k_i}^*(y_{j_l})\right| \geq \delta_{k_i}$ for all $l \in \mathbb{N}$. Since $(y_{j_l})_{l\in\mathbb{N}}$ converges to x we deduce that $\left|y_{k_i}^*(x)\right| \geq \delta_{k_i}$. However, the previous inequality contradicts the fact that $|y_k^*(x)| < \delta_k$ for each $k \in \mathbb{N}$. As a consequence, the sequence $(k_j)_{j\in\mathbb{N}}$ is infinite, and by extracting a subsequence if necessary, we can assume that $(k_j)_{j\in\mathbb{N}}$ is strictly increasing. Finally, we have that

$$\bigcup_{k=1}^{\infty} (y_k^*)^{-1}(\{-\delta_k, \delta_k\}) \cap \mathcal{B} \supset \left(\frac{\delta_{k_j}}{\left|y_{k_j}^*(y_j)\right|} y_j\right)_{j\in\mathbb{N}} \longrightarrow x \text{ as } j \to \infty.$$

□

3. Nowhere density of the set of non-norm-attaining functionals

In this section we intend to show that the new equivalent norm constructed in the proof of Theorem 2.3 can actually be chosen to be a dual norm on dual spaces. This will obviously give us the key to obtain the result.

Theorem 3.1. *Let X be a real Banach space. Assume that X^* is separable. Then, X can be equivalently renormed in such a way that $\mathsf{S}_{X^*} \setminus \mathsf{NA}(X)$ is nowhere dense in S_{X^*}.*

Proof. In the first place, X admits an equivalent Fréchet differentiable renorming (see [3]), therefore we can assume that X is Fréchet differentiable. Let $(x_n)_{n\in\mathbb{N}} \subset \mathsf{S}_X$ be a dense sequence. For every $n \in \mathbb{N}$ we will let x_n^* denote the unique functional in S_{X^*} that supports B_X at x_n. Observe that $x_n^* = \mathsf{J}_X(x_n)$ for every $n \in \mathbb{N}$. Recall (see [3]) that X is Fréchet differentiable if and only if every $x \in \mathsf{S}_X$ is a strongly smooth point of B_X, that is, if $(f_n)_{n\in\mathbb{N}} \subset \mathsf{S}_{X^*}$ and $(f_n(x))_{n\in\mathbb{N}}$ converges to 1, then $(f_n)_{n\in\mathbb{N}}$ is convergent. Recall also (see [3]) that a point $x \in \mathsf{S}_X$ is said to be a strongly exposed point of B_X if there exists $f \in \mathsf{S}_{X^*}$ such that if $(y_n)_{n\in\mathbb{N}} \subset \mathsf{S}_X$ and $(f(y_n))_{n\in\mathbb{N}}$ converges to 1 then $(y_n)_{n\in\mathbb{N}}$ converges to x. Since x_n is a strongly smooth point of B_X for every $n \in \mathbb{N}$, we have that x_n^* is a strongly exposed point of B_{X^*} and $\widehat{x_n}$ strongly exposes x_n^* on B_{X^*} for every $n \in \mathbb{N}$. Furthermore, since the dual mapping $\mathsf{J}_X : X \longrightarrow X^*$ is continuous, we deduce that $(x_n^*)_{n\in\mathbb{N}}$ is dense in S_{X^*}. By virtue of the proof of Theorem 2.3 there exist two sequences $(\delta_k)_{k\in\mathbb{N}} \subset [1/2, 1)$ and $(n_k)_{k\in\mathbb{N}} \subset \mathbb{N}$ such that

$$\mathcal{B} := \mathsf{B}_{X^*} \cap \bigcap_{k=1}^{\infty} \left(\widehat{x_{n_k}}\right)^{-1}([-\delta_k, \delta_k])$$

is the unit ball of an equivalent norm on X^* for which the union of the smooth faces is dense in the unit sphere. Finally, the set $\mathcal{B}$ is ω^*-closed, thus the norm induced by $\mathcal{B}$ is a dual norm on X^* (see [3]). And, in accordance to Theorem 2.1 we have that $\mathsf{S}_{\mathcal{B}} \setminus \mathsf{NA}(X_{\mathcal{B}})$ is nowhere dense in $\mathsf{S}_{\mathcal{B}}$. □

Remark 3.2. Observe that for separable spaces we can obtain the following result: *every separable real Banach space can be equivalently renormed so that the set of non-norm-attaining functionals of norm 1 is ω^*-nowhere dense in the unit sphere of the dual.* This fact follows from the given construction and from the fact that Theorem 2.1 remains true if we substitute *dense* by *ω^*-dense* and *nowhere dense* by *ω^*-nowhere dense.*

4. Related open questions

There are two natural questions related to the results in this manuscript.

Question 4.1. *Is Theorem 2.3 true in full generality? That is, can every real Banach space be equivalently renormed so that the union of the smooth faces is dense in the unit sphere?*

Question 4.2. *Is Theorem* 3.1 *true in full generality? That is, can every (real or complex) Banach space be equivalently renormed so that the set of non-norm-attaining functionals is nowhere dense?*

Acknowledgment

Many thanks to the organizers of the conference IWOTA 2007 for the nice environment.

References

[1] A. Aizpuru, M.D. Acosta, R.M. Aron, and F.J. García-Pacheco, *Functionals that do not attain their norm.* Bull. Belg. Math. Soc. Simon Stevin **14** (2007), 407–418.

[2] M.D. Acosta and M.R. Galán, *New characterizations of the reflexivity in terms of the set of norm-attaining functionals.* Canad. Math. Bull. **41** (1998), 279–289.

[3] R. Deville, G. Godefroy and V. Zizler, *Smoothness and renormings in Banach spaces.* Pitman Monographs and Surveys in Pure and Applied Mathematics **64**, New York, 1993.

[4] J. Diestel, *Geometry of Banach spaces-selected topics.* Lecture Notes in Mathematics **485**, Springer-Verlag, Berlin-New York, 1975.

[5] F.J. García-Pacheco and A. Miralles, *Real renormings on complex Banach spaces.* Chin. Ann. Math. Ser. B **29** (2008), 239–246.

[6] F.J. García-Pacheco, *The minimum-norm problem for translations.* Preprint.

[7] M. Jiménez-Sevilla and J.P. Moreno, *A note on norm-attaining functionals.* Proc. Amer. Math. Soc. **126** (1998), 1989–1997.

[8] W. Johnson and J. Lindenstrauss, *Handbook of the geometry of Banach spaces* **II**. North-Holland, Amsterdam, 2003.

Francisco J. García-Pacheco
Department of Mathematics
Mailstop 3368
Texas A&M University
College Station, TX 77843, USA
e-mail: `fgarcia@math.tamu.edu`

Operator Theory:
Advances and Applications, Vol. 195, 173–184

A Stochastic Double Product in Non-Fock Quantum Stochastic Calculus

R.L. Hudson

Abstract. Generalising the previous Fock case, we show that in an extremal universally invariant representation of the canonical commutation relations, a second quantised double product of infinitesimal rotations is a stochastic double product in the corresponding non-Fock quantum stochastic calculus.

Mathematics Subject Classification (2000). 81S25.

Keywords. Extremal universally invariant, quantum stochastic analysis, stochastic double product

1. Introduction

In earlier works [1],[2] we constructed a family of unitary operators formally described as double stochastic product integrals of the form $\coprod(1+\lambda(dA^{\dagger}\otimes dA-dA\otimes dA^{\dagger}))$, where $A^{\dagger}$ and A are the standard creation and annihilation processes of Fock space quantum stochastic calculus [7],[9] and λ is a fixed real parameter. These consist of the second quantisations, living in the double Fock space $\mathcal{F}(L^2(\mathbb{R}^+))\otimes\mathcal{F}(L^2(\mathbb{R}^+))=\mathcal{F}(L^2(\mathbb{R}^+)\oplus L^2(\mathbb{R}^+))$, of double products of infinitesimal rotations, living in $L^2(\mathbb{R}^+)\oplus L^2(\mathbb{R}^+)$, for which explicit formulas related to the time-orthogonal unitary dilation [4] can be found. Similar double products have been applied in explicit constructions of quantisations of Lie bialgebras [3] and, in their related causal forms $\prod_{0<x<y<t}(1+\lambda(dA^{\dagger}(s)dA(t)-dA(s)dA^{\dagger}(t)))$ living in the single Fock space $\mathcal{F}(L^2(\mathbb{R}^+))$ will be used in subsequent work to generate rotational quantum Cameron-Martin-Girsanov theorems and associated quantum versions of the Black-Scholes model in mathematical finance.

The purpose of the present work is to describe similar constructions in which the creation and annihilation processes $A^{\dagger}$ and A are replaced by those of the quantum stochastic calculus of [6] based on non-Fock space extremal universally invariant representation of the canonical commutation relations (CCR) [11]. This calculus is better adapted to some applications, for example to quantum Black-Scholes models, than the Fock space calculus, because of the availability of a simple and general martingale representation theorem precisely generalising the classical Kunita-Watanabe martingale representation theorem for square-integrable martingales [8], which may be compared with the Fock space quantum martingale representation theorem of [10] for whose validity heavy restrictions on the martingale are required. This superiority may, in turn, be attributed to the fact that, in the non-Fock case, the vacuum vector is both cyclic and separating [11] for the von Neumann algebra $\mathcal{N}$ generated by the creation and annihilation processes, allowing replacement of operator-valued processes by the vector-valued processes got by their actions on the vacuum, unlike the Fock case, where the vacuum is cyclic but not separating.

It turns out that, to obtain unitarity in the non-Fock case, it is necessary to include an additional time term in the generator of the double product integral. When this is done, the natural unitary second quantisation of the direct sum of the double product of rotations in $L^2(\mathbb{R}^+)\oplus L^2(\mathbb{R}^+)$ and its dual factorises into factors belonging to $\mathcal{N}\otimes\mathcal{N}$ and to its commutant by an argument based on Shale's theorem [12] about the unitary implentability in the Fock case of linear canonical transformations sufficiently close to the identity. A similar argument was used in [5] to construct an analog, for the non-Fock second quantisation, of the second quantised time-orthogonal unitary dilation of [4] and an associated Feynman-Kac formula.

In Section 2 we describe in detail the realisation we shall use, living in the double Fock space $\mathcal{F}(L^2(\mathbb{R}^+))\otimes\ \mathcal{F}(L^2(\mathbb{R}^+))^- = \mathcal{F}(L^2(\mathbb{R}^+)\oplus L^2(\mathbb{R}^+)^-)$, where $\overline{\mathfrak{h}}$ denotes the dual of the Hilbert space $\mathfrak{h}$, of the extremal universally invariant representation of the CCR of variance $\sigma^2 > 1$, and review the corresponding quantum stochastic calculus. In Section 3 we investigate conditions for unitarity of double product integrals and find the necessary time correction. Finally in Section 4 we give the argument based on Shale's theorem to show that our unitary double product arises from factorising natural second quantisations of the direct sums of the processes in $L^2(\mathbb{R}^+)\oplus L^2(\mathbb{R}^+)$ of [1] and their duals.

2. Non-Fock second quantisation[6]

Let $\mathcal{H}$ be the *Fock space* $\mathcal{F}(\mathfrak{h})$ over a Hilbert space $\mathfrak{h}$, which we regard as the closed linear span of the *exponential vectors* $e(f)$, $f\in\mathfrak{h}$ satisfying $\langle e(f),e(g)\rangle = \exp\langle f,g\rangle$, $f,g\in\mathfrak{h}$. Since $\overline{\exp\langle f,g\rangle} = \exp\overline{\langle f,g\rangle}$ we can identify the dual space $\overline{\mathcal{H}}$ of $\mathcal{H}$ with the Fock space over $\overline{\mathfrak{h}}$ by setting $e(\bar{f}) = \overline{e(f)}$ for each $f\in\mathfrak{h}$. The *Fock representation of the CCR* over $\mathfrak{h}$ is the family of unitary operators $W(f)$, $f\in\mathfrak{h}$ defined by their actions on exponential vectors

$$W(f)e(g) = \exp\left\{-\frac{1}{2}\|f\|^2 - \langle f,g\rangle\right\}e(g+f);$$

they do indeed satisfy the Weyl form of the CCR

$$W(f)W(g) = \exp\{-i\,\mathrm{Im}\,\langle f,g\rangle\}\,W(f+g) \tag{1}$$

and the *Fock vacuum vector* $e(0)$ is cyclic for them.

Let a fixed real number $\sigma > 1$ be given once and for all, called the *variance*, and let α and β be real numbers such that $\sigma^2 = \alpha^2 + \beta^2$ and $\alpha^2 - \beta^2 = 1$. The extremal universally invariant (in the sense of [11]) representation of the CCR over $\mathfrak{h}$ of variance σ^2 is the family of operators on the Hilbert space $\mathcal{K} = \mathcal{H}\otimes\overline{\mathcal{H}}$ defined in terms of the Fock representation W by

$$\mathfrak{h} \ni f \mapsto W_\sigma(f) = W(\alpha f) \otimes \overline{W(\beta f)}. \tag{2}$$

The von Neumann algebra $\mathcal{N}$ generated by the operators $W_\sigma(f)$, $f \in \mathfrak{h}$ is a type III factor whose commutant $\mathcal{N}'$ is generated by the representation of the CCR over $\overline{\mathfrak{h}}$

$$\mathfrak{h} \ni f \mapsto W_\sigma(f) = W(\beta f) \otimes \overline{W(\alpha f)}. \tag{3}$$

The *vacuum vector* $\Omega = e(0) \otimes \overline{e(0)}$ is cyclic and separating.

The representations (2) and (3) were called *universally invariant* by Segal [11] for the following reason. Let R be a unitary operator in the Hilbert space $\mathfrak{h}$. For an arbitrary representation $\mathfrak{h} \ni f \mapsto W(f)$ of the CCR (1), the map $W^R : \mathfrak{h} \ni f \mapsto W(Rf)$ is another representation of the CCR with the same enveloping von Neumann algebra. In the case of the representations W_σ and W'_σ, W_σ^R is unitarily equivalent to W_σ and $(W'_\sigma)^R$ to W'_σ under the same unitary operator $\Gamma(R \oplus \bar{R})$, the second quantisation of the unitary operator $R \oplus \bar{R}$ which acts on arbitrary exponential vectors $e(f,\bar{g}) \in \mathcal{H}\otimes\overline{\mathcal{H}}$ by $\Gamma(R \oplus \bar{R})e(f,\bar{g}) = e(Rf, \bar{R}\bar{g})$. Conjugation by $\Gamma(R \oplus \bar{R})$ thus gives automorphisms of both $\mathcal{N}$ and $\mathcal{N}'$ even though $\Gamma(R \oplus \bar{R})$ does not belong to either algebra. Prior to Segal's work some physicists had believed that the Fock representation was characterised by its universal invariance in this sense.

It will be important to us below to know when these automorphisms are inner, in other words when the unitary operator $\Gamma(R \oplus \bar{R})$ can be factorised as a product

$$\Gamma(R \oplus \bar{R}) = \Gamma_\sigma(R)\Gamma'_\sigma(R) \tag{4}$$

of unitary operators $\Gamma_\sigma(R) \in \mathcal{N}$ and $\Gamma'_\sigma(R) \in \mathcal{N}'$. In [5] it was shown, using Shale's theorem on unitary implementability of Bogolubov transformations in the Fock representation [12], that such factorisation occurs if and only if R differs from the identity by a Hilbert-Schmidt operator on $\mathfrak{h}$.

Now let $\mathfrak{h}$ be the Hilbert space $L^2(\mathbb{R}_+)$. For each $t \in \mathbb{R}_+$ we identify $\mathcal{K}$ with the tensor product $\mathcal{K}_t\otimes\mathcal{K}^t$ where $\mathcal{K}_t = \mathcal{H}_t\otimes\overline{\mathcal{H}_t}$ and $\mathcal{K}^t = \mathcal{H}^t\otimes\overline{\mathcal{H}^t}$ using the canonical splitting at time t, $\mathcal{H} = \mathcal{H}_t\otimes\mathcal{H}^t$, into the Fock spaces over $\mathfrak{h}_t = L^2([0,t]$ and its orthogonal complement in $L^2(\mathbb{R}_+)$ in which each exponential vector splits as $e(f) = e(f|_{[0,t]}) \otimes e(f|_{]t,\infty[})$, and the isomorphism from

$$\mathcal{K} = \mathcal{H}\otimes\overline{\mathcal{H}} = \mathcal{H}_t\otimes\mathcal{H}^t \otimes \overline{\mathcal{H}_t\otimes\mathcal{H}^t} = \mathcal{H}_t\otimes\mathcal{H}^t \otimes \overline{\mathcal{H}_t}\otimes\overline{\mathcal{H}^t}$$

to

$$\mathcal{K}_t \otimes \mathcal{K}^t = \mathcal{H}_t \otimes \overline{\mathcal{H}_t} \otimes \mathcal{H}^t \otimes \overline{\mathcal{H}^t}$$

which appropriately permutes components of product vectors in $\mathcal{H}_t \otimes \mathcal{H}^t \otimes \overline{\mathcal{H}_t} \otimes \overline{\mathcal{H}^t}$ to give product vectors in $\mathcal{H}_t \otimes \overline{\mathcal{H}_t} \otimes \mathcal{H}^t \otimes \overline{\mathcal{H}^t}$. Ω is a product vector $\Omega_t \otimes \Omega^t$ for the splitting $\mathcal{K} = \mathcal{K}_t \otimes \mathcal{K}^t$. The sub-von Neumann algebra $\mathcal{N}_t$ of $\mathcal{N}$ generated by the $W_\sigma(f)$ with f of the form $f = fI_{[0,t]}$ where $I_{[0,t]}$ is the indicator function of $[0,t]$, and the similarly defined subalgebra $\mathcal{N}'_t$ of $\mathcal{N}'$ are of the form $\mathcal{N}(t) \otimes 1_{\mathcal{K}^t}$, $\mathcal{N}'(t) \otimes 1_{\mathcal{K}^t}$ where $\mathcal{N}(t)$ and $\mathcal{N}'(t)$ are the mutually commutant von Neumann algebras generated by the extremal universally invariant representations of the CCR over $\mathfrak{h}_t$ and $\overline{\mathfrak{h}_t}$ respectively.

In the extremal universally invariant quantum stochastic calculus of [6], adapted processes are pairs $(E, E^\dagger)$ of families of mutually adjoint algebraic tensor product operators operators $E(t) = E_t \otimes 1_{\mathcal{K}^t}, E^\dagger(t) = E_t^\dagger \otimes 1_{\mathcal{K}^t}$ defined on the domain $\mathcal{E}'$ spanned by the vectors $W'_\sigma(\bar{f})\Omega$, $f \in \mathfrak{h}$, where the E_t, and $E_t^\dagger$ are weakly affiliated to $\mathcal{N}_t$, in the sense that for arbitrary $f \in \mathfrak{h}_t$

$$\left\langle W'_\sigma(\bar{f})\Omega_t, E_t\Omega_t \right\rangle = \left\langle E_t^\dagger \Omega_t, W'_\sigma(\bar{f})^* \Omega_t \right\rangle.$$

Equivalently, such pairs are characterised by pairs

$$(\psi_E, \psi_E^\dagger) = (\psi_E, \psi_{E^\dagger}) = (E(.)\Omega, E^\dagger(.)\Omega)$$

of vector-valued functions on $\mathbb{R}_+$ such that for all $t \in \mathbb{R}_+$ $\psi_E(t) = \psi_t \otimes \Omega^t, \psi_E^\dagger(t) = \psi_t^\dagger \otimes \Omega^t \in \mathcal{K}_t \otimes \Omega^t$ and for all $f \in \mathfrak{h}$

$$\left\langle W'_\sigma(\bar{f})\Omega, \psi_E(t) \right\rangle = \left\langle \psi_E^\dagger(t), W'_\sigma(\bar{f})^* \Omega_t \right\rangle. \tag{5}$$

The component creation and annihilation processes $A^\dagger_\sigma$ and A_σ are defined in terms of the corresponding Fock space processes $A^\dagger$ and A [4], [9], whose actions on exponential vectors are

$$A^\dagger(t)e(f) = \left.\frac{d}{dz}e(f + zI_{[0,t]})\right|_{z=0}, \quad A(t)e(f) = \left\langle I_{[0,t]}, f \right\rangle e(f) = \int_0^t f\, e(f),$$

by

$$A^\dagger_\sigma(t) = \alpha A^\dagger(t) \otimes 1 + \beta 1 \otimes \bar{A}(t), \quad A_\sigma(t) = \alpha A(t) \otimes 1 + \beta 1 \otimes \bar{A}^\dagger(t).$$

Equivalently

$$\psi_{A^\dagger}(t) = \alpha(A^\dagger(t)e(0)) \otimes \overline{e(0)}, \quad \psi_A(t) = \beta e(0) \otimes \overline{A^\dagger(t)e(0)}.$$

The processes $A^\dagger_\sigma$ and A_σ satisfy the Heisenberg commutation relations

$$\left[A_\sigma(s), A^\dagger_\sigma(t)\right] = s \wedge t, \quad \left[A_\sigma(s), A_\sigma(t)\right] = \left[A^\dagger_\sigma(s), A^\dagger_\sigma(t)\right] = 0 \tag{6}$$

in the weak sense; thus for example the first relation of (6) is more precisely that for arbitrary $f, g, f', g' \in \mathfrak{h}$,

$$\left\langle A^\dagger_\sigma(s)e(f) \otimes e(\bar{g}), A^\dagger_\sigma(t)e(f') \otimes e(\bar{g}') \right\rangle - \left\langle A_\sigma(t)e(f) \otimes e(\bar{g}), A_\sigma(s)e(f') \otimes e(\bar{g}') \right\rangle$$
$$= s \wedge t \left\langle e(f) \otimes e(\bar{g}), e(f') \otimes e(\bar{g}') \right\rangle.$$

The quantum stochastic calculus, in which $A^\dagger{}_\sigma$ and A_σ, together with the time process $T(t) = T^\dagger(t) = t1$ are the basic integrator processes, is technically simpler than the usual Fock space based calculus of [7] because mutually adjoint pairs of operator-valued processes are replaced consistently by pairs of vector-valued processes $(\psi, \psi^\dagger)$ satisfying (5). The stochastic integral process pair $(M, M^\dagger)$, with

$$M = \int (E dA^\dagger{}_\sigma + F dA_\sigma),\ M^\dagger = \int (E^\dagger dA_\sigma + F^\dagger dA) \tag{7}$$

where $(E, E^\dagger)$ and $(F, F^\dagger)$ are adapted processes, can be conveniently defined in terms of the corresponding vector process $(\psi_M, \psi_M^\dagger = \psi_{M^\dagger})$ as follows. For elementary processes, of form $E = E(t_0)\chi_{[t_0,t_1]}$,

$$\psi_{\int EdK}(t) = \psi_E(t_0)\left(\psi_K(t \wedge t_1) - \psi_K(t \wedge t_0)\right)$$

where the product of vectors is defined as $\psi_1 \otimes \psi_2$ where $\psi_E(t_0) = \psi_1 \otimes \Omega^{t_0}$ and $\psi_K(t \wedge t_1) - \psi_K(t \wedge t_0) = \Omega_{t_0} \otimes \psi_2$. Extension to simple processes is by additivity and to arbitrary adapted processes $(E, E^\dagger)$ for which $\|\psi_E\|$ is locally square-integrable by approximation by simple processes using the isometry conditions

$$\|\psi_M(t)\|^2 = \int_{0<s<t} \left(\alpha^2 \|\psi_E(s)\|^2 + \beta^2 \|\psi_F(s)\|^2\right) ds \tag{8}$$

for M given by (7). No such isometry condition is available in the Fock quantum stochastic calculus; instead families of seminorm estimates must be used to extend integration from simple processes.

For locally square-integrable f the Weyl process $W_\sigma^f(t) = W_\sigma(f\chi_{[0,t]})$ satisfies the stochastic differential equation

$$dW_\sigma^f = W_\sigma^f\left(f dA^\dagger{}_\sigma - \bar{f} dA_\sigma - \frac{\sigma^2}{2}|f|^2 dT\right),\ W_\sigma^f(0) = 1. \tag{9}$$

The quantum Ito multiplication table in this calculus is

	$dA^\dagger{}_\sigma$	dA_σ	dT
$dA^\dagger{}_\sigma$	0	$\beta^2 dT$	0
dA_σ	$\alpha^2 dT$	0	0
dT	0	0	0

. (10)

The *Ito algebra* $\mathcal{I}_\sigma = \mathbb{C}\langle dA^\dagger{}_\sigma, dA_\sigma, dT\rangle$ is the complex associative algebra with this multiplication table, equipped with the involution $\dagger$ under which $dA^\dagger{}_\sigma$ and dA_σ are mutually adjoint and dT is self-adjoint.

3. Unitary double product integrals

Let dr be an element of the tensor product involuted algebra $\mathcal{I}_\sigma \otimes \mathcal{I}_\sigma$. Let $]a,b]$ and $]s,t]$ be subintervals of $\mathbb{R}_+$. The double product integral $\overrightarrow{\overleftarrow{\prod}}_{]a,b]\times]s,t]} (1 + dr)$ is an operator in the Hilbert space $\mathcal{K} \otimes \mathcal{K}$ affiliated to $\mathcal{N} \otimes \mathcal{N}$, defined in two different ways, the first of which is as follows. We begin by solving the stochastic differential equation

$$(\mathrm{id}_{\mathcal{I}_\sigma} \otimes d)Q(t)^{1,2} = dr^{1,3}(Q(t)^{1,2} + 1^2),\ Q(s) = 0, \tag{11}$$

for a process Q belonging to the tensor product $\mathcal{I}_\sigma \otimes \mathcal{P}_s$ where $\mathcal{P}_s$ denotes the space of processes affiliated to $\mathcal{N}$ starting at time s; here the superscripts $1, 2, 3$ refer to places in the tensor product $\mathcal{I}_\sigma \otimes \mathcal{P}_s \otimes \mathcal{I}_\sigma$ and the nilpotence of the algebra $\mathcal{I}_\sigma$ ensures that the solution is well defined as the sum

$$Q(t) = \int_{s<x<t} dr(.,x) + \int_{s<x_1<x_2<t} dr(.,x_2)dr(.,x_1). \tag{12}$$

Writing $Q(t)$ in the form $dA_\sigma^\dagger \otimes Q_1(t) + dA_\sigma \otimes Q_2(t) + dT \otimes Q_3(t)$ we can use it to formulate and solve a second stochastic differential equation

$$(d \otimes \mathrm{id}_{\mathcal{P}_s})P(b,t)^{1,3} = P(b,t)^{1,3}Q^{2,3}(t),\ P(a,t) = 1 \tag{13}$$

for a process P belonging to $\mathcal{P}_a \otimes \mathcal{P}_s$ where now the superscripts $1, 2, 3$ refer to places in the tensor product $\mathcal{P}_a \otimes \mathcal{I}_\sigma \otimes \mathcal{P}_s$.

Alternatively we may first solve the differential equation

$$(d \otimes \mathrm{id}_{\mathcal{I}_\sigma})X(b)^{1,3} = (X(b)^{1,3} + 1^3)dr^{1,2},\ X(a) = 0. \tag{14}$$

for the process X in $\mathcal{P}_a \otimes \mathcal{I}_\sigma$ as

$$X(b) = \int_{a<x<b} dr(x.,.) + \int_{a<x_1<x_2<b} dr(x_1,.)dr(x_2,.) \tag{15}$$

and then the equation

$$(\mathrm{id}_{\mathcal{P}_a} \otimes d)Y(b,t)^{1,2} = X(b)^{1,3}Y(b,t)^{1,2},\ Y(a,t) = 1 \tag{16}$$

for $Y \in \mathcal{P}_a \otimes \mathcal{P}_s$. Then $Y(b,t) = P(b,t)$, as may be seen by comparing their explicit expressions as sums of double iterated integrals over $]a,b]\times]s,t]$ obtained by the iterative solution of the second quantum stochastic differential equation (13), (16) in each case [3]. Their common value is defined to be $\overrightarrow{\overleftarrow{\prod}}_{]a,b]\times]s,t]}(1+dr)$. From the first definition we see that $\overrightarrow{\overleftarrow{\prod}}_{]a,b]\times]s,t]}(1+dr)$ is a forward evolution in a, b; for $a < b < c$

$$\overrightarrow{\overleftarrow{\prod}}_{]a,b]\times]s,t]}(1+dr) \overrightarrow{\overleftarrow{\prod}}_{]b,c]\times]s,t]}(1+dr) = \overrightarrow{\overleftarrow{\prod}}_{]a,c]\times]s,t]}(1+dr),$$

and is covariant under shifts in a, b; for arbitrary $p \in r$,

$$\left(\Gamma(S_p \oplus \mathrm{id}_\mathfrak{h}) \otimes \overline{\Gamma(S_p \oplus \mathrm{id}_\mathfrak{h})}\right)^* \overrightarrow{\overleftarrow{\prod}}_{]a,b]\times]s,t]}(1+dr) \left(\Gamma(S_p \oplus \mathrm{id}_\mathfrak{h}) \otimes \overline{\Gamma(S_p \oplus \mathrm{id}_\mathfrak{h})}\right)$$

$$= \overrightarrow{\overleftarrow{\prod}}_{]a+p,b+p]\times]s,t]}(1+dr)$$

where S_p is the isometric forward shift in $\mathfrak{h}$. Similarly, from the second definition, we see that $\overrightarrow{\overleftarrow{\prod}}_{]a,b]\times]s,t]} (1+dr)$ is a backward evolution in s,t, covariant under shifts in s,t.

The double product $\overleftarrow{\overrightarrow{\prod}}_{]a,b]\times]s,t]} (1+dr)$ is defined analogously by reversing the directions of the stochastic differential equations. If it exists then the adjoint of $\overrightarrow{\overleftarrow{\prod}}_{]a,b]\times]s,t]} (1+dr)$ also exists and is given by $\overleftarrow{\overrightarrow{\prod}}_{]a,b]\times]s,t]} (1+dr^\dagger)$ where $dr^\dagger$ is the image of dr under the product involution in $\mathcal{I}_\sigma \otimes \mathcal{I}_\sigma$ under which $(dL\otimes dM)^\dagger = dL^\dagger \otimes dM^\dagger$. The multiplicative inverse of $\overrightarrow{\overleftarrow{\prod}}_{]a,b]\times]s,t]} (1+dr)$ is $\overleftarrow{\overrightarrow{\prod}}_{]a,b]\times]s,t]} (1+dr')$ [3] where dr' is the quasiinverse of dr in $\mathcal{I}_\sigma \otimes \mathcal{I}_\sigma$, so that

$$dr + dr' + dr\ dr' = dr' + dr + dr'\ dr = 0;$$

again the nilpotence of the algebra $\mathcal{I}_\sigma$ ensures the existence of dr'. Thus $\overrightarrow{\overleftarrow{\prod}}_{]a,b]\times]s,t]} (1+dr)$ is unitary if and only if

$$dr + dr^\dagger + dr\ dr^\dagger = dr^\dagger + dr + dr^\dagger\ dr = 0.$$

In this calculus the double products $\overleftarrow{\overrightarrow{\prod}}_{[a,b]\times[s,d]} (1+\lambda(dA^\dagger{}_\sigma \otimes dA_\sigma - dA_\sigma \otimes dA^\dagger{}_\sigma))$ are not unitary, indeed for the generator $dr = \lambda(dA^\dagger{}_\sigma \otimes dA_\sigma - dA_\sigma \otimes dA^\dagger{}_\sigma) = -dr^\dagger$ we find using (10) that

$$dr + dr^\dagger + dr\ dr^\dagger = dr^\dagger + dr + dr^\dagger\ dr = -dr^2 = 2\lambda^2\alpha^2\beta^2 dT \otimes dT.$$

Unitarity is restored by replacing dr by

$$dr_\sigma = \lambda(dA^\dagger{}_\sigma \otimes dA_\sigma - dA_\sigma \otimes dA^\dagger{}_\sigma) - \lambda^2\alpha^2\beta^2 dT \otimes dT. \tag{17}$$

4. Explicit construction

In [1] a family of matrix operators was constructed consisting of limits of discrete double products of rotations in two-dimensional subspaces of $L^2(\mathbb{R}^+) \oplus L^2(\mathbb{R}^+)$,given by

$$W^{b,s}_{a,t} = \begin{bmatrix} \mathrm{id}_{L^2(\mathbb{R}^+)} + A^{b,s}_{a,t} & B^{b,s}_{a,t} \\ C^{b,s}_{a,t} & \mathrm{id}_{L^2(\mathbb{R}^+)} + D^{b,s}_{a,t} \end{bmatrix} \tag{18}$$

where $A^{b,s}_{a,t}$, $B^{b,s}_{a,t}$, $C^{b,s}_{a,t}$, $D^{b,s}_{a,t}$ are the integral operators on $L^2(\mathbb{R}^+)$ whose kernels take values at $(x,y) \in (\mathbb{R}^+)^2$

$$A^{b,s}_{a,t}(x,y) = <^b_a (x,y) \sum_{n=0}^{\infty} \frac{(y-x)^n}{n!} \frac{(-\lambda^2(t-s))^{n+1}}{(n+1)!},$$

$$\mathcal{B}^{b,s}_{a,t}(x,y) = I_{]a,b]}(x)I_{]s,t]}(y)\sum_{n=0}^{\infty}\frac{(t-y)^n}{n!}\frac{(-\lambda^2(b-x))^n}{n!},$$

$$\mathcal{C}^{b,s}_{a,t}(x,y) = I_{]s,t]}(x)I_{]a,b]}(y)\sum_{n=0}^{\infty}\frac{(x-s)^n}{n!}\frac{(-\lambda^2(y-a))^n}{n!},$$

$$\mathcal{D}^{b,s}_{a,t}(x,y) =>^t_s (x,y)\sum_{n=0}^{\infty}\frac{(x-y)^n}{n!}\frac{(-\lambda^2(b-a))^{n+1}}{(n+1)!}$$

where $<^b_a (x,y) = 1$ if $a < x < y \leq b$ and 0 otherwise and similarly, $>^t_s (x,y) = 1$ if $t \geq x > y > s$ and 0 otherwise. It was shown that the operators $W^{b,s}_{a,t}$ are unitary and satisfy the bievolution and covariance properties

$$W^{b,s}_{a,t}W^{c,s}_{b,t} = W^{c,s}_{a,t},\ W^{b,s}_{a,t}W^{b,r}_{a,s} = W^{b,r}_{a,t},\ a<b<c,\ r<s<t, \tag{19}$$

$$(S_p \oplus S_q)^* W^{b+p,s+q}_{a+p,t+q}(S_p \oplus S_q) = W^{b,s}_{a,t},\ (R^b_a \oplus R^t_s)^* W^{b,s}_{a,t}(R^b_a \oplus R^t_s) = \left(W^{b,s}_{a,t}\right)^*$$

where S_p and R^b_a denote the forward shift and time reversal operators on $L^2(\mathbb{R}^+)$

$$S_p f(x) = \begin{cases} 0 \text{ if } x \leq p \\ f(x-p) \text{ if } x > p \end{cases},\ R^b_a f(x) = \begin{cases} f(x) \text{ if } x \notin]a,b] \\ f(a+b-x) \text{ if } x \in]a,b] \end{cases}.$$

In [2] it was shown that the second quantisation $\Gamma(W^{b,s}_{a,t})$ of $W^{b,s}_{a,t}$ in the Fock space $\mathcal{F}(L^2(\mathbb{R}^+)) \otimes \mathcal{F}(L^2(\mathbb{R}^+)) = \mathcal{F}(L^2(\mathbb{R}^+) \oplus L^2(\mathbb{R}^+))$, that is, the unitary operator which acts on exponential vectors as $\Gamma(W^{b,s}_{a,t})e(f,g) = (W^{b,s}_{a,t}(f,g))$, can be identified as the Fock double product $\overrightarrow{\overleftarrow{\prod}}_{]a,b]\times]s,t]}(1 + \lambda(dA^\dagger \otimes dA - dA \otimes dA^\dagger))$ in so far as the Fock versions of (13) and (16) are satisfied.

The following theorem was proved in [5].

Theorem 4.1. *Let W be a unitary operator on a Hilbert space $\mathfrak{h}$ and let $\Gamma(W)$, $\overline{\Gamma(W)} = \Gamma(\overline{W})$ be the corresponding Fock space second quantisations in the Fock spaces $\mathcal{H}$ and $\overline{\mathcal{H}}$. In order that the unitary operator $\Gamma(W \oplus \overline{W}) = \Gamma(W) \otimes \overline{\Gamma(W)}$ on $\mathcal{H} \otimes \overline{\mathcal{H}}$ admit a factorisation into a product of unitary operators of the form*

$$\Gamma(W \oplus \overline{W}\ \ _\sigma) = \Gamma\ _\sigma(W)\Gamma'\ _\sigma(\overline{W})$$

where $\Gamma_\sigma(W)$ and $\Gamma_\sigma(\overline{W})$ belong to the von Neumann algebra $\mathcal{N}$ and to its commutant $\mathcal{N}'$ respectively, it is necessary and sufficient that the operator $W - \mathrm{id}_{\mathfrak{h}}$ be Hilbert-Schmidt.

By (18) the operators $W^{b,s}_{a,t}$ differ from the identity by the Hilbert-Schmidt operator $\Delta^{b,s}_{a,t} = \begin{bmatrix} A^{b,s}_{a,t} & B^{b,s}_{a,t} \\ C^{b,s}_{a,t} & D^{b,s}_{a,t} \end{bmatrix}$ on $L^2(\mathbb{R}^+) \oplus L^2(\mathbb{R}^+)$, so $\Gamma(W^{b,s}_{a,t} \oplus \overline{W}^{b,s}_{a,t})$ factorises into a product of unitary operators

$$\Gamma(W^{b,s}_{a,t} \oplus \overline{W}^{b,s}_{a,t}) = \Gamma\ _\sigma(W^{b,s}_{a,t})\Gamma'\ _\sigma(\overline{W}^{b,s}_{a,t})$$

with $\Gamma_\sigma(W^{b,s}_{a,t}) \in \mathcal{N}\otimes\mathcal{N}$ and $\Gamma'(\overline{W}^{b,s}_{a,t}) \in (\mathcal{N}\otimes\mathcal{N})' = \mathcal{N}'\otimes\mathcal{N}'$. Since $\mathcal{N}\otimes\mathcal{N}$ is a factor the factorisation is nonunique only to the extent that $\Gamma_\sigma(W^{b,s}_{a,t})$ can be replaced by a scalar multiple $\omega^{b,s}_{a,t}\Gamma_\sigma(W^{b,s}_{a,t})$ and $\Gamma'_\sigma(\overline{W}^{b,s}_{a,t})$ by $\bar\omega^{b,s}_{a,t}\Gamma'_\sigma(\overline{W}^{b,s}_{a,t})$ where $\omega^{b,s}_{a,t}$ is a complex number of unit modulus. Since $\Delta^{b,s}_{a,t}$ depends continuously on the parameters $a, b, s, t \in \mathbb{R}_+$ in Hilbert-Schmidt norm, by a theorem of Wigner [13] the representatives $\Gamma_\sigma(W^{b,s}_{a,t})$ and $\Gamma'_\sigma(\overline{W}^{b,s}_{a,t})$ can be chosen to depend strongly continuously on the parameters. The following theorem can then be proved exactly in analogy with the corresponding uniqueness theorem for the time-orthogonal unitary dilation [4], Theorem 6.1 in [5].

Theorem 4.2. *There is a unique system of representatives $\Gamma_\sigma(W^{b,s}_{a,t})$ and $\Gamma'_\sigma(\overline{W}^{b,s}_{a,t})$ satisfying the following evolution and covariance properties for arbitrary $a < b < c$, $r < s < t$ and $p, q \in \mathbb{R}_+$:*

$$\Gamma_\sigma(W^{b,s}_{a,t})\Gamma_\sigma(W^{c,s}_{b,t}) = \Gamma_\sigma(W^{c,s}_{a,t}),\ \Gamma_\sigma(W^{b,s}_{a,t})\Gamma_\sigma(W^{b,r}_{a,s}) = \Gamma_\sigma(W^{b,r}_{a,t}),$$

$$\begin{aligned}
&\left(\Gamma(S_p\oplus S_q)\otimes\overline{\Gamma(S_p\oplus S_q)}\right)^*\Gamma_\sigma(W^{b+p,s+q}_{a+p,t+q})\left(\Gamma(S_p\oplus S_q)\otimes\overline{\Gamma(S_p\oplus S_q)}\right)\\
=\ &\Gamma_\sigma(W^{b,s}_{a,t}),\\
&\left(\Gamma(R^b_a\oplus R^t_s)\otimes\overline{\Gamma(R^b_a\oplus R^t_s)}\right)^*\Gamma_\sigma(W^{b,s}_{a,t})\left(\Gamma(R^b_a\oplus R^t_s)\otimes\overline{\Gamma(R^b_a\oplus R^t_s)}\right)\\
=\ &\left(\Gamma_\sigma(W^{b,s}_{a,t})\right)^{\cdot *}
\end{aligned}$$

together with corresponding properties for the representatives $\Gamma'_\sigma(\overline{W}^{b,s}_{a,t})$.

Henceforth $\Gamma_\sigma(W^{b,s}_{a,t})$ will denote this representative. We can now state our main theorem.

Theorem 4.3. $\Gamma_\sigma(W^{b,s}_{a,t}) = \overset{\rightarrow\leftarrow}{\prod\limits_{]a,b]\times]s,d]}}(1+\lambda(dA^\dagger_\sigma\otimes dA_\sigma - dA_\sigma\otimes dA^\dagger_\sigma)-\lambda^2\alpha^2\beta^2 dT\otimes dT)$ *in the sense that it satisfies the quantum stochastic differential equations* (13) *and* (16) *with $dr = dr_\sigma$ given by* (17).

Proof. We convert double products in the extremal universally invariant quantum stochastic calculus to double products in Fock calculus, for which we know [2] that the analogs of (13) and (16) hold, as follows. Since

$$\begin{aligned}
A^\dagger_\sigma &= \alpha A^\dagger\otimes 1 + \beta 1\otimes\bar{A},\ A_\sigma = \alpha A\otimes 1 + \beta 1\otimes\bar{A}^\dagger\\
A'^\dagger_\sigma &= \alpha 1\otimes\bar{A}^\dagger + \beta A\otimes 1,\ A'_\sigma = \alpha 1\otimes\bar{A} + \beta A^\dagger\otimes 1,
\end{aligned}$$

$$\begin{aligned}
dr_\sigma &= \lambda(dA^\dagger_\sigma\otimes dA_\sigma - dA_\sigma\otimes dA^\dagger_\sigma)\\
&= \lambda(\alpha dA^\dagger + \beta d\bar{A})\otimes(\alpha dA + \beta d\bar{A}^\dagger) - \lambda(\alpha dA + \beta d\bar{A}^\dagger)\otimes(\alpha dA^\dagger + \beta d\bar{A})\\
&\quad -\lambda^2\alpha^2\beta^2 dT\otimes dT.
\end{aligned}$$

Similarly

$$\begin{aligned} dr'_{\sigma} &= \lambda(dA'^{\dagger}_{\sigma}\otimes dA'_{\sigma} - dA'_{\sigma}\otimes dA'^{\dagger}_{\sigma}) \\ &= \lambda(\alpha d\bar{A}^{\dagger}+\beta dA)\otimes(\alpha d\bar{A}+\beta dA^{\dagger}) - \lambda(\alpha d\bar{A}+\beta dA^{\dagger})\otimes(\alpha d\bar{A}^{\dagger}+\beta dA) \\ &\quad -\lambda^2\alpha^2\beta^2 dT\otimes dT. \end{aligned}$$

Since

$$dr_{\sigma} dr'_{\sigma} = 2\lambda^2\alpha\beta dT\otimes dT = dr'_{\sigma} dr_{\sigma}$$

we can use the product rule (which holds when as here the generators commute)

$$\begin{aligned} &\overrightarrow{\overleftarrow{\prod}}_{]a,b]\times]s,d]}(1+dr_{\sigma})\overrightarrow{\overleftarrow{\prod}}_{]a,b]\times]s,d]}(1+dr'_{\sigma}) \\ &= \overrightarrow{\overleftarrow{\prod}}_{]a,b]\times]s,d]}(1+dr_{\sigma}+dr'_{\sigma}+dr_{\sigma}dr'_{\sigma}) \\ &= \overrightarrow{\overleftarrow{\prod}}_{]a,b]\times]s,d]}(1+dr_{\sigma}+dr'_{\sigma}+.2\lambda^2\alpha\beta dT\otimes dT) \end{aligned}$$

In terms of the Fock calculus this becomes

$$\begin{aligned} &\overrightarrow{\overleftarrow{\prod}}_{]a,b]\times]s,d]}(1+\lambda(\alpha dA^{\dagger}+\beta d\bar{A})\otimes(\alpha dA+\beta d\bar{A}^{\dagger}) \\ &\quad -\lambda(\alpha dA+\beta d\bar{A}^{\dagger})\otimes(\alpha dA^{\dagger}+\beta d\bar{A}) \\ &\quad +\lambda(\alpha d\bar{A}^{\dagger}+\beta dA)\otimes(\alpha d\bar{A}+\beta dA^{\dagger}) - \lambda(\alpha d\bar{A}+\beta dA^{\dagger})\otimes(\alpha d\bar{A}^{\dagger}+\beta dA)) \\ &= \overrightarrow{\overleftarrow{\prod}}_{]a,b]\times]s,d]}(1+\lambda(\alpha^2-\beta^2)(dA^{\dagger}\otimes dA - dA\otimes dA^{\dagger} + d\bar{A}^{\dagger}\otimes d\bar{A} - d\bar{A}\otimes d\bar{A}^{\dagger})) \\ &= \overrightarrow{\overleftarrow{\prod}}_{]a,b]\times]s,d]}(1+\lambda(dA^{\dagger}\otimes dA - dA\otimes dA^{\dagger} + d\bar{A}^{\dagger}\otimes d\bar{A} - d\bar{A}\otimes d\bar{A}^{\dagger})) \\ &= \overrightarrow{\overleftarrow{\prod}}_{]a,b]\times]s,d]}(1+\lambda(dA^{\dagger}\otimes dA - dA\otimes dA^{\dagger})) \\ &\quad \overrightarrow{\overleftarrow{\prod}}_{]a,b]\times]s,d]}(1+\lambda(d\bar{A}^{\dagger}\otimes d\bar{A} - d\bar{A}\otimes d\bar{A}^{\dagger})) \end{aligned}$$

since the generators $\lambda(dA^{\dagger}\otimes dA - dA\otimes dA^{\dagger})$ and $\lambda(d\bar{A}^{\dagger}\otimes d\bar{A} - d\bar{A}\otimes d\bar{A}^{\dagger})$ also commute and their product vanishes. But we know from [2] that the double product

$$\overleftarrow{\overrightarrow{\prod}}_{]a,b]\times]s,d]}(1+\lambda(dA^{\dagger}\otimes dA - dA\otimes dA^{\dagger}))$$

is indeed given by the Fock second quantisation $\Gamma(W_{a,t}^{b,s})\otimes\bar{1}$ in the sense that the latter satisfies the stochastic differential equations which are the Fock analogs of

(13) and (16) with $dr = \lambda(dA^\dagger \otimes dA - dA \otimes dA^\dagger)$ and it follows by dualisation that

$$\overset{\rightarrow\leftarrow}{\prod_{]a,b]\times]s,d]}} (1 + \lambda(d\bar{A}^\dagger \otimes d\bar{A} - d\bar{A} \otimes d\bar{A}^\dagger))$$

is likewise given by $1 \otimes \Gamma(\bar{W}_{a,t}^{b,s})$ in the same sense. Hence

$$\begin{aligned}
&\overset{\rightarrow\leftarrow}{\prod_{]a,b]\times]s,d]}} (1 + dr\ {}_\sigma) \overset{\rightarrow\leftarrow}{\prod_{]a,b]\times]s,d]}} (1 + dr'\ {}_\sigma) \\
&= \overset{\rightarrow\leftarrow}{\prod_{]a,b]\times]s,d]}} (1 + \lambda(dA^\dagger \otimes dA - dA \otimes dA^\dagger)) \\
&\quad \overset{\rightarrow\leftarrow}{\prod_{]a,b]\times]s,d]}} (1 + \lambda(d\bar{A}^\dagger \otimes d\bar{A} - d\bar{A} \otimes d\bar{A}^\dagger)) \\
&= (\Gamma(W_{a,t}^{b,s}) \otimes \bar{1})(1 \otimes \Gamma(\bar{W}_{a,t}^{b,s})) \\
&= \Gamma(W_{a,t}^{b,s}) \otimes \Gamma(\bar{W}_{a,t}^{b,s}) \\
&= \Gamma\ {}_\sigma(W_{a,t}^{b,s})\Gamma'{}_\sigma(W_{a,t}^{b,s}).
\end{aligned}$$

Since $\overset{\rightarrow\leftarrow}{\prod_{]a,b]\times]s,d]}} (1 + dr\ {}_\sigma)$ and $\overset{\rightarrow\leftarrow}{\prod_{]a,b]\times]s,d]}} (1 + dr'\ {}_\sigma)$ satisfy the evolution and covariance conditions of Theorem 2 they constitute the unique such factorisation of $\Gamma(W) \otimes \overline{\Gamma(W)}$. □

References

[1] R.L. Hudson, *A double dilation constructed from a double product of rotations*, Markov Processes and Related Fields **13**, J.T. Lewis Memorial Volume, 169–190 (2007).

[2] R.L. Hudson, *An analytic double product integral*, pp. 241–250, in *Quantum Probability and Infinite-dimensional Analysis XX* Proceedings, Levico, 2005, eds. L. Accardi, W. Freudenberg and M. Schürmann, World Scientific (2007).

[3] R.L. Hudson, *Itô calculus and quantisation of Lie bialgebras*, Annales de l'Institue H. Poincaré-PR **41** (P.-A. Meyer memorial volume), 375–390 (2005).

[4] R.L. Hudson, P.D.F. Ion and K.R. Parthasarathy, *Time-orthogonal unitary dilations and non-commutative Feynman-Kac formulae*, Communications in Mathematical Physics **83**, 261–280 (1982).

[5] R.L. Hudson, P.D.F. Ion and K.R. Parthasarathy, *Time-orthogonal unitary dilations and non-commutative Feynman-Kac formulae II*, Publications of Research Institute for Mathematical Sciences, Kyoto, **20**, 607–633 (1984).

[6] R.L. Hudson and J.M. Lindsay, *A non-commutative martingale representation theorem for non-Fock quantum Brownian motion*, Journal of Functional Analysis **61**, 202–221 (1985).

[7] R.L. Hudson and K.R. Parthasarathy, *Quantum Ito's formula and stochastic evolutions*, Communications in Mathematical Physics **93**, 301–323 (1984).

[8] H. Kunita and S. Watanabe, *On square integrable martingales*, Nagoya Math. Jour. **30**, 201–245 (1967).

[9] K.R. Parthasarathy, *An introduction to quantum stochastic calculus*, Birkhäuser (1992).

[10] K.R. Parthasarathy and K.B. Sinha, *Stochastic integral representation of bounded quantum martingales in Fock space*, Jour. Funct. Anal. **67**, 126–151 (1986).

[11] I.E. Segal, *Mathematical characterisation of the physical vacuum*, Illinois Journal of Mathematics **6**, 500–523 (1962).

[12] D. Shale, *Linear symmetries of free Boson fields*, Trans. Amer. Math. Soc. **103**, 149–167 (1962).

[13] E. Wigner, *On unitary representations of the inhomogeneous Lorentz group*, Ann. of Math. **40**, 149–204 (1939).

R.L. Hudson
Department of Mathematics,
University of Loughborough,
Loughborough,
Leicestershire LE11 3TU, Great Britain
e-mail: `R.Hudson@lboro.ac.uk`

Operator Theory:
Advances and Applications, Vol. 195, 185–212

Singular Integral Operators on Variable Lebesgue Spaces with Radial Oscillating Weights

Alexei Yu. Karlovich

To the memory of Igor Borisovich Simonenko (1935–2008)

Abstract. We prove a Fredholm criterion for operators in the Banach algebra of singular integral operators with matrix piecewise continuous coefficients acting on a variable Lebesgue space with a radial oscillating weight over a logarithmic Carleson curve. The local spectra of these operators are massive and have a shape of spiralic horns depending on the value of the variable exponent, the spirality indices of the curve, and the Matuszewska-Orlicz indices of the weight at each point. These results extend (partially) the results of A. Böttcher, Yu. Karlovich, and V. Rabinovich for standard Lebesgue spaces to the case of variable Lebesgue spaces.

Mathematics Subject Classification (2000). Primary 47B35; Secondary 45E05, 46E30, 47A68.

Keywords. Variable Lebesgue space, Carleson curve, variable exponent, radial oscillating weight, Matuszewska-Orlicz indices, submultiplicative function

1. Introduction

Let X be a Banach space and $\mathcal{B}(X)$ be the Banach algebra of all bounded linear operators on X. An operator $A \in \mathcal{B}(X)$ is said to be n-normal (resp. d-normal) if its image $\operatorname{Im} A$ is closed in X and the defect number $n(A; X) := \dim \operatorname{Ker} A$ (resp. $d(A; X) := \dim \operatorname{Ker} A^*$) is finite. An operator A is said to be semi-Fredholm on X if it is n-normal or d-normal. Finally, A is said to be Fredholm if it is simultaneously n-normal and d-normal. Let N be a positive integer. We denote by X_N the direct sum of N copies of X with the norm

$$\|f\| = \|(f_1, \dots, f_N)\| := (\|f_1\|^2 + \dots + \|f_N\|^2)^{1/2}.$$

The author is partially supported by the grant FCT/FEDER/POCTI/MAT/59972/2004.

Let Γ be a Jordan curve, that is, a curve that is homeomorphic to a circle. We suppose that Γ is rectifiable. We equip Γ with Lebesgue length measure $|d\tau|$ and the counter-clockwise orientation. The *Cauchy singular integral* of $f \in L^1(\Gamma)$ is defined by

$$(Sf)(t) := \lim_{R\to 0} \frac{1}{\pi i} \int_{\Gamma\setminus\Gamma(t,R)} \frac{f(\tau)}{\tau - t} d\tau \quad (t \in \Gamma),$$

where $\Gamma(t,R) := \{\tau \in \Gamma : |\tau - t| < R\}$ for $R > 0$. David [11] (see also [4, Theorem 4.17]) proved that the Cauchy singular integral generates the bounded operator S on the Lebesgue space $L^p(\Gamma)$, $1 < p < \infty$, if and only if Γ is a *Carleson (Ahlfors-David regular) curve*, that is,

$$\sup_{t\in\Gamma} \sup_{R>0} \frac{|\Gamma(t,R)|}{R} < \infty,$$

where $|\Omega|$ denotes the measure of a measurable set $\Omega \subset \Gamma$. We can write $\tau - t = |\tau - t| e^{i\arg(\tau - t)}$ for $\tau \in \Gamma \setminus \{t\}$, and the argument can be chosen so that it is continuous on $\Gamma \setminus \{t\}$. It is known [4, Theorem 1.10] that for an arbitrary Carleson curve the estimate

$$\arg(\tau - t) = O(-\log|\tau - t|) \quad (\tau \to t)$$

holds for every $t \in \Gamma$. One says that a Carleson curve Γ satisfies the *logarithmic whirl condition* at $t \in \Gamma$ if

$$\arg(\tau - t) = -\delta(t)\log|\tau - t| + O(1) \quad (\tau \to t) \tag{1.1}$$

with some $\delta(t) \in \mathbb{R}$. Notice that all piecewise smooth curves satisfy this condition at each point and, moreover, $\delta(t) \equiv 0$. For more information along these lines, see [3], [4, Chap. 1], [5].

A measurable function $w : \Gamma \to [0, \infty]$ is referred to as a *weight function* or simply a *weight* if $0 < w(\tau) < \infty$ for almost all $\tau \in \Gamma$. Suppose $p : \Gamma \to (1, \infty)$ is a continuous function. Denote by $L^{p(\cdot)}(\Gamma, w)$ the set of all measurable complex-valued functions f on Γ such that

$$\int_\Gamma |f(\tau)w(\tau)/\lambda|^{p(\tau)} |d\tau| < \infty$$

for some $\lambda = \lambda(f) > 0$. This set becomes a Banach space when equipped with the Luxemburg-Nakano norm

$$\|f\|_{p(\cdot),w} := \inf\left\{\lambda > 0 : \int_\Gamma |f(\tau)w(\tau)/\lambda|^{p(\tau)} |d\tau| \le 1\right\}.$$

If p is constant, then $L^{p(\cdot)}(\Gamma, w)$ is nothing but the weighted Lebesgue space. Therefore, it is natural to refer to $L^{p(\cdot)}(\Gamma, w)$ as a *weighted generalized Lebesgue space with variable exponent* or simply as a *weighted variable Lebesgue space*. This is a special case of Musielak-Orlicz spaces [40] (see also [31]). Nakano [41] considered these spaces (without weights) as examples of so-called modular spaces, and sometimes the spaces $L^{p(\cdot)}(\Gamma, w)$ are referred to as weighted Nakano spaces. Since Γ is compact, we have

$$1 < \min_{t\in\Gamma} p(t), \quad \max_{t\in\Gamma} p(t) < \infty.$$

Therefore, if $w \in L^{p(\cdot)}(\Gamma)$ and $1/w \in L^{q(\cdot)}(\Gamma)$, where

$$q(t) := p(t)/(p(t) - 1)$$

is the conjugate exponent for p, then $L^{p(\cdot)}(\Gamma, w)$ is reflexive and its Banach dual can be identified with $L^{q(\cdot)}(\Gamma, 1/w)$ (see, e.g., [40, Section 13], [31, Corollary 2.7] and also [21, Section 2]).

Following [29, Section 2.3], denote by W the class of all continuous functions $\varrho : [0, |\Gamma|] \to [0, \infty)$ such that $\varrho(0) = 0$, $\varrho(x) > 0$ if $0 < x \le |\Gamma|$, and ϱ is almost increasing, that is, there is a universal constant $C > 0$ such that $\varrho(x) \le C\varrho(y)$ whenever $x \le y$. Further, let $\mathbb{W}$ be the set of all functions $\varrho : [0, |\Gamma|] \to [0, \infty]$ such that $x^\alpha \varrho(x) \in W$ and $x^\beta/\varrho(x) \in W$ for some $\alpha, \beta \in \mathbb{R}$. Clearly, the functions $\varrho(x) = x^\gamma$ belong to $\mathbb{W}$ for all $\gamma \in \mathbb{R}$. For $\varrho \in \mathbb{W}$, put

$$\Phi^0_\varrho(x) := \limsup_{y\to 0} \frac{\varrho(xy)}{\varrho(y)}, \quad x \in (0, \infty).$$

Since $\varrho \in \mathbb{W}$, one can show (see Subsection 2.2) that the limits

$$m(\varrho) := \lim_{x\to 0} \frac{\log \Phi^0_\varrho(x)}{\log x}, \quad M(\varrho) := \lim_{x\to\infty} \frac{\log \Phi^0_\varrho(x)}{\log x} \tag{1.2}$$

exist and $-\infty < m(\varrho) \le M(\varrho) < +\infty$. These numbers were defined under some extra assumptions on ϱ by Matuszewska and Orlicz [38, 39] (see also [36] and [37, Chapter 11]). We refer to $m(\varrho)$ (resp. $M(\varrho)$) as the *lower* (resp. *upper*) *Matuszewska-Orlicz index* of ϱ. For $\varrho(x) = x^\gamma$ one has $m(\varrho) = M(\varrho) = \gamma$. Examples of functions $\varrho \in \mathbb{W}$ with $m(\varrho) < M(\varrho)$ can be found, for instance, in [1], [37, p. 93], [46, Section 2].

Fix pairwise distinct points $t_1, \dots, t_n \in \Gamma$ and functions $w_1, \dots, w_n \in \mathbb{W}$. Consider the following weight

$$w(t) := \prod_{k=1}^{n} w_k(|t - t_k|), \quad t \in \Gamma. \tag{1.3}$$

Each function $w_k(|t - t_k|)$ is a radial oscillating weight. This is a natural generalization of so-called Khvedelidze weights $w(t) = \prod_{k=1}^n |t - t_k|^{\lambda_k}$, where $\lambda_k \in \mathbb{R}$ (see, e.g., [4, Section 2.2], [14, Section 1.4], [26], [28]).

Theorem 1.1. *Suppose Γ is a Carleson Jordan curve and $p : \Gamma \to (1, \infty)$ is a continuous function satisfying*

$$|p(\tau) - p(t)| \le -A_\Gamma/\log|\tau - t| \quad \text{whenever} \quad |\tau - t| \le 1/2, \tag{1.4}$$

where A_Γ is a positive constant depending only on Γ. Let $w_1, \dots, w_n \in \mathbb{W}$ and the weight w be given by (1.3). The Cauchy singular integral operator S is bounded on

$L^{p(\cdot)}(\Gamma, w)$ *if and only if*

$$0 < 1/p(t_k) + m(w_k), \quad 1/p(t_k) + M(w_k) < 1 \quad \textit{for all} \quad k \in \{1, \dots, n\}. \tag{1.5}$$

For $w(t) = \prod_{k=1}^{n} |t - t_k|^{\lambda_k}$, (1.5) reads as $0 < 1/p(t_k) + \lambda_k < 1$ for all $k \in \{1, \dots, n\}$ and for these weights Theorem 1.1 is obtained in [28]. The sufficiency portion of Theorem 1.1 in the form stated above is obtained by V. Kokilashvili, N. Samko, and S. Samko [29, Theorem 4.3]. The necessity portion is a new result. It will be proved in Section 4.1.

We define by $PC(\Gamma)$ as the set of all $a \in L^\infty(\Gamma)$ for which the one-sided limits

$$a(t \pm 0) := \lim_{\tau \to t \pm 0} a(\tau)$$

exist and finite at each point $t \in \Gamma$; here $\tau \to t - 0$ means that τ approaches t following the orientation of Γ, while $\tau \to t + 0$ means that τ goes to t in the opposite direction. Functions in $PC(\Gamma)$ are called *piecewise continuous* functions.

The operator S is defined on $L_N^{p(\cdot)}(\Gamma, w)$ elementwise. We let stand $PC_{N\times N}(\Gamma)$ for the algebra of all $N \times N$ matrix functions with entries in $PC(\Gamma)$. Writing the elements of $L_N^{p(\cdot)}(\Gamma, w)$ as columns, we can define the multiplication operator aI for $a \in PC_{N\times N}(\Gamma)$ as multiplication by the matrix function a. Let $\operatorname{alg}(S, PC, L_N^{p(\cdot)}(\Gamma, w))$ denote the smallest closed subalgebra of $\mathcal{B}(L_N^{p(\cdot)}(\Gamma, w))$ containing the operator S and the set $\{aI : a \in PC_{N\times N}(\Gamma)\}$.

For the case of constant $p \in (1, \infty)$, Khvedelidze weights w, and piecewise Lyapunov curves Γ, the algebra $\operatorname{alg}(S, PC, L_N^p(\Gamma, w))$ was well understood in 1970–1980's (see, e.g., [13, 14, 33, 45, 50]). In the earlier 1990's, Spitkovsky [51] discovered that local spectra of singular integral operators on Lebesgue spaces with Muckenhoupt weights on smooth curves have a shape of horns bounded by two circular arcs depending on the so-called indices of powerlikeness of the weight. In the middle of 1990's, Böttcher, Yu. Karlovich, and Rabinovich further observed that these horns metamorphose to spiralic horns bounded by logarithmic double spirals [3, 6, 42] if one passes from nice curves to Carleson curves satisfying (1.1). These spiralic horns become even more interesting creatures (so-called "general leaves") in case of arbitrary Carleson curves [4, 5]. The shape of a general leaf depends on the spirality indices of the curve and the indices of powerlikeness of the weight. The Fredholm theory for the algebra $\operatorname{alg}(S, PC, L_N^p(\Gamma, w))$ under the most general conditions on the curve Γ and the weight w is constructed by Böttcher and Yu. Karlovich and is presented in the monograph [4] (although, we advise to start the study of this theory from the nice survey [5]).

Partially, the results of [4] are extended to Orlicz spaces [17], and further to rearrangement-invariant spaces [18], and rearrangement-invariant spaces with Muckenhoupt weights [20]. The results of these papers resemble those of [4] with one important difference: the number $1/p$ is replaced by the Boyd indices α_X, β_X of the rearrangement-invariant space. Since the Boyd indices may be different, one more factor (a general space) leads to massive local spectra (for standard Lebesgue spaces these factors were general weights and general curves). Local spectra of singular integral operators can be massive also on weighted Hölder spaces on Carleson curves [43].

The study of singular integral operators (SIOs) with discontinuous coefficients on weighted variable Lebesgue spaces on sufficiently nice curves was started in [21, 23, 27, 30]. A Fredholm criterion for an arbitrary operator in $\mathrm{alg}(S, PC, L_N^{p(\cdot)}(\Gamma, w))$ for Carleson curves satisfying (1.1) and Khvedelidze weights is proved in [22]. Local spectra in that paper have a shape of logarithmic double spirals and cannot be massive. Under the same assumptions on the curve and the weight, it is proved in [24] that every semi-Fredholm operator in the algebra $\mathrm{alg}(S, PC, L_N^{p(\cdot)}(\Gamma, w))$ is Fredholm.

In this paper we extend the results of [22, 23, 24] to the case of Carleson curves satisfying (1.1) and weights of the form (1.3) satisfying the conditions of Theorem 1.1. Local spectra of SIOs corresponding to the points $t \in \Gamma \setminus \{t_1, \dots, t_n\}$ are logarithmic double spirals depending on $\delta(t)$ and $1/p(t)$. These spirals blow up to spiralic horns at $t = t_k$ if $k \in \{1, \dots, n\}$ and $m(w_k) < M(w_k)$. These spiralic horns are bounded by two logarithmic double spirals depending on $\delta(t_k)$ and on $1/p(t_k)+m(w_k)$ and $1/p(t_k)+M(w_k)$, respectively. Up to our knowledge, this paper is the first work, where massive local spectra of singular integral operators appear in the setting of weighted variable Lebesgue spaces. These results resemble those of [3, 6] for weighted standard Lebesgue spaces, although the weights considered in [3, 6] are more general than in the present paper.

The paper is organized as follows. In Section 2 we collect all necessary information on indices of submultiplicative functions associated with curves and weights. We prove that the indices of powerlikeness (see [4, Chap. 3]) of the weight (1.3) at t_k coincide with the Matuszewska-Orlicz indices of w_k. Section 3 contains standard results on singular integral operators with L^∞ coefficients: the necessary condition for Fredholmness, the local principle of Simonenko type, and the theorem on a Wiener-Hopf factorization. In Section 4, we prove a Fredholm criterion for a singular integral operator $aP + Q$, where $a \in PC(\Gamma)$ and $P := (I + S)/2$, $Q := (I - S)/2$. In a sense, the main result of Section 4 is the heart of the paper. It follows from a more general necessary condition for the Fredholmness of $aP+Q$ (see [21, Theorem 8.1]) and a sufficient condition for the Fredholmness of $aP+Q$, whose proof is based on the local principle of Simonenko type and the Wiener-Hopf factorization of a local representative of a. In fact, [21, Therem 8.1] together with the results of Section 2 imply the necessity of the conditions (1.5) for the boundedness of the operator S on $L^{p(\cdot)}(\Gamma, w)$. Section 5 contains the Allan-Douglas local principle and the two projections theorem. These results are our main tools in the construction of a symbols calculus for the algebra $\mathrm{alg}(S, PC, L_N^{p(\cdot)}(\Gamma, w))$ on the basis of the main result of Section 4. In Section 6, following the well-known scheme (see, e.g., [3], [4, Chap. 8] and also [17, 20, 22, 23]), we prove a Fredholm criterion for an arbitrary operator $A \in \mathrm{alg}(S, PC, L_N^{p(\cdot)}(\Gamma, w))$. Finally, we collect some remarks on index formulas and semi-Fredholm operators in the algebra $\mathrm{alg}(S, PC, L_N^{p(\cdot)}(\Gamma, w))$ and discuss open problems.

2. Submultiplicative functions and their indices

2.1. Submultiplicative functions

Following [4, Section 1.4], we say a function $\Phi : (0,\infty) \to (0,\infty]$ is *regular* if it is bounded in an open neighborhood of 1. A function $\Phi : (0,\infty) \to (0,\infty]$ is said to be *submultiplicative* if

$$\Phi(xy) \le \Phi(x)\Phi(y) \quad \text{for all} \quad x, y \in (0,\infty).$$

It is easy to show that if Φ is regular and submultiplicative, then Φ is bounded away from zero in some open neighborhood of 1. Moreover, in this case $\Phi(x)$ is finite for all $x \in (0,\infty)$. Given a regular and submultiplicative function $\Phi : (0,\infty) \to (0,\infty)$, one defines

$$\alpha(\Phi) := \sup_{x\in(0,1)} \frac{\log \Phi(x)}{\log x}, \quad \beta(\Phi) := \inf_{x\in(1,\infty)} \frac{\log \Phi(x)}{\log x}.$$

Clearly, $-\infty < \alpha(\Phi)$ and $\beta(\Phi) < \infty$.

Theorem 2.1 (see [4], Theorem 1.13 or [32], Chap. 2, Theorem 1.3). *If a function $\Phi : (0,\infty) \to (0,\infty)$ is regular and submultiplicative, then*

$$\alpha(\Phi) = \lim_{x\to 0} \frac{\log \Phi(x)}{\log x}, \quad \beta(\Phi) = \lim_{x\to\infty} \frac{\log \Phi(x)}{\log x}$$

and $-\infty < \alpha(\Phi) \le \beta(\Phi) < +\infty$.

The quantities $\alpha(\Phi)$ and $\beta(\Phi)$ are called the *lower* and *upper indices of the regular and submultiplicative function* Φ, respectively.

2.2. Matuszewska-Orlicz indices

Lemma 2.2. *Let $A > 0$ and $\varrho : (0, A] \to (0,\infty)$ be a continuous function. Then for every $B \in (0, A]$ the function*

$$\Phi_{\varrho,B}(x) := \begin{cases} \displaystyle\sup_{0<y\le B} \frac{\varrho(xy)}{\varrho(y)} & \text{if } x \in (0,1], \\ \displaystyle\sup_{0<y\le B} \frac{\varrho(y)}{\varrho(x^{-1}y)} & \text{if } x \in (1,\infty) \end{cases} \tag{2.1}$$

is submultiplicative.

This statement is proved similarly to [4, Lemma 1.15].

Lemma 2.3. *Let $A > 0$ and $\varrho : (0, A] \to (0,\infty)$ be a continuous function. If for some $B \in (0, A]$ the function $\Phi_{\varrho,B}$ given by (2.1) is regular, then*

$$\Phi^0_\varrho(x) := \limsup_{y\to 0} \frac{\varrho(xy)}{\varrho(y)}$$

is regular and submultiplicative and both functions have the same lower and upper indices

$$\alpha(\Phi_{\varrho,B}) = \alpha(\Phi^0_\varrho) =: m(\varrho), \quad \beta(\Phi_{\varrho,B}) = \beta(\Phi^0_\varrho) =: M(\varrho).$$

This statement is proved by analogy with [9, Lemma 2(a)] (see also [2, Theorem 8.18] and [4, Lemma 1.16]).

Lemma 2.4. *If $\varrho \in \mathbb{W}$, then for every $B \in (0, |\Gamma|]$ the function $\Phi_{\varrho,B}$ is regular.*

Proof. Since $\varrho \in \mathbb{W}$, there exist $a, b \in \mathbb{R}$ such that the functions $x^a \varrho(x)$ and $x^b/\varrho(x)$ are almost increasing on $(0, |\Gamma|]$, that is, there exist positive constants c_a, c_b such that $x^a \varrho(x) \le c_a y^a \varrho(y)$ and $x^b/\varrho(x) \le c_b y^b/\varrho(y)$ whenever $x \le y$ and $x, y \in (0, |\Gamma|]$. Suppose $x \in (0, 1]$ and $y \in (0, |\Gamma|]$. Then

$$\frac{\varrho(xy)}{\varrho(y)} = \frac{(xy)^a \varrho(xy)}{(xy)^a \varrho(y)} \le \frac{c_a y^a \varrho(y)}{(xy)^a \varrho(y)} = \frac{c_a}{x^a}.$$

Hence

$$\Phi_{\varrho,B}(x) \le \Phi_{\varrho,|\Gamma|}(x) \le c_a x^{-a}, \quad x \in (0, 1].$$

Similarly,

$$\Phi_{\varrho,B}(x) \le \Phi_{\varrho,|\Gamma|}(x) \le c_b x^b, \quad x \in (1, \infty).$$

Thus, $\Phi_{\varrho,B}$ is regular for every $B \in (0, |\Gamma|]$. □

From the above results and Theorem 2.1 we conclude that if $\varrho \in \mathbb{W}$, then its Matuszewska-Orlicz indices are well defined by (1.2).

2.3. Spirality indices of Carleson curves

Fix $t \in \Gamma$ and put

$$d_t := \max_{\tau \in \Gamma} |\tau - t|.$$

Suppose $\psi : \Gamma \setminus \{t\} \to (0, \infty)$ is a continuous function and consider

$$F_{\psi,t}(R_1, R_2) := \max_{\tau \in \Gamma, |\tau - t| = R_1} \psi(\tau) \Big/ \min_{\tau \in \Gamma, |\tau - t| = R_2} \psi(\tau), \quad R_1, R_2 \subset (0, d_t].$$

By [4, Lemma 1.15], the function

$$(W_t \psi)(x) := \begin{cases} \sup\limits_{0 < R \le d_t} F_{\psi,t}(xR, R) & \text{if} \quad x \in (0, 1], \\ \sup\limits_{0 < R \le d_t} F_{\psi,t}(R, x^{-1}R) & \text{if} \quad x \in (1, \infty) \end{cases}$$

is submultiplicative. For $t \in \Gamma$, we have,

$$\tau - t = |\tau - t| e^{i \arg(\tau - t)}, \quad \tau \in \Gamma \setminus \{t\},$$

and the argument $\arg(\tau - t)$ may be chosen to be continuous on $\Gamma \setminus \{t\}$. Consider

$$\eta_t(\tau) := e^{-\arg(\tau - t)}.$$

Lemma 2.5 (see [4], Theorem 1.18). *If Γ is a Carleson Jordan curve, then for every $t \in \Gamma$ the function $W_t \eta_t$ is regular and submultiplicative. If, in addition, it satisfies (1.1) at some point $t \in \Gamma$, then*

$$\alpha(W_t \eta_t) = \beta(W_t \eta_t) = \delta(t).$$

The numbers $\alpha(W_t\eta_t)$ and $\beta(W_t\eta_t)$ are called the *lower* and *upper spirality indices* of Γ at t, respectively. If Γ is a piecewise smooth curve, then $\alpha(W_t\eta_t) = \beta(W_t\eta_t) = 0$ for all $t \in \Gamma$ (non-spiral curves). Carleson curves satisfying (1.1) behave like perturbed logarithmic spirals. Examples of Carleson curves which have arbitrary prescribed and distinct spirality indices are given in [4, Section 1.6].

2.4. Indices of powerlikeness of continuous nonvanishing weights

Let ψ be a weight on Γ such that $\log\psi \in L^1(\Gamma(t,R))$ for every $R \in (0, d_t]$. Put

$$H_{\psi,t}(R_1, R_2) := \frac{\exp\left(\dfrac{1}{|\Gamma(t,R_1)|}\displaystyle\int_{\Gamma(t,R_1)} \log\psi(\tau)|d\tau|\right)}{\exp\left(\dfrac{1}{|\Gamma(t,R_2)|}\displaystyle\int_{\Gamma(t,R_2)} \log\psi(\tau)|d\tau|\right)}, \quad R_1, R_2 \in (0, d_t].$$

Consider the function

$$(V_t^0\psi)(x) := \limsup_{R\to 0} H_{\psi,t}(xR, R), \quad x \in (0,\infty).$$

Lemma 2.6. *Let Γ be a Carleson Jordan curve. Suppose $\psi : \Gamma \to [0,\infty]$ is a weight which is continuous and nonvanishing on $\Gamma \setminus \{t\}$. If $W_t\psi$ is regular, then $V_t^0\psi$ is regular and submultiplicative and*

$$\alpha(W_t\psi) = \alpha(V_t^0\psi), \quad \beta(W_t\psi) = \beta(V_t^0\psi).$$

This statement follows from [4, Theorem 3.3(c) and Lemma 3.16].

The numbers $\alpha(V_t^0\psi)$ and $\beta(V_t^0\psi)$ are called the *lower* and *upper indices of powerlikeness* of ψ at $t \in \Gamma$, respectively. This terminology can be explained by the simple fact that for the power weight $\psi(\tau) := |\tau - t|^\lambda$ its indices of powerlikeness coincide and are equal to λ. Examples of Muckenhoupt weights with distinct indices of powerlikeness are given in [4, Proposition 3.25 and Examples 3.26–3.27].

2.5. Indices of powerlikeness of radial oscillating weights

Lemma 2.7. *Let Γ be a Carleson Jordan curve and $\varrho \in \mathbb{W}$. Suppose*

$$\psi_t(\tau) := \varrho(|\tau - t|) \text{ for } \tau \in \Gamma \setminus \{t\}.$$

Then the functions $W_t\psi_t$ and $V_t^0\psi_t$ are regular and submultiplicative and

$$m(\varrho) = \alpha(W_t\psi_t) = \alpha(V_t^0\psi_t), \quad M(\varrho) = \beta(W_t\psi_t) = \beta(V_t^0\psi_t).$$

Proof. By Lemmas 2.2–2.4, the functions Φ_{ϱ,d_t} and Φ_ϱ^0 are regular and submultiplicative and

$$\alpha(\Phi_{\varrho,d_t}) = \alpha(\Phi_\varrho^0) =: m(\varrho), \quad \beta(\Phi_{\varrho,d_t}) = \beta(\Phi_\varrho^0) =: M(\varrho). \tag{2.2}$$

If $x \in (0,1]$ and $0 < R \le d_t$, then

$$F_{\psi_t,y}(xR,R) = \frac{\displaystyle\max_{\tau\in\Gamma,|\tau-t|=xR} \varrho(|\tau - t|)}{\displaystyle\min_{\tau\in\Gamma,|\tau-t|=R} \varrho(|\tau - t|)} = \frac{\varrho(xR)}{\varrho(R)}; \tag{2.3}$$

if $x \in [1,\infty)$ and $0 < R \le d_t$, then

$$F_{\psi_t,t}(R,x^{-1}R) = \frac{\max\limits_{\tau\in\Gamma,|\tau-t|=R} \varrho(|\tau-t|)}{\min\limits_{\tau\in\Gamma,|\tau-t|=x^{-1}R} \varrho(|\tau-t|)} = \frac{\varrho(R)}{\varrho(x^{-1}R)}. \tag{2.4}$$

From (2.3) and (2.4) it follows that

$$(W_t\psi_t)(x) = \Phi_{\varrho,d_t}(x), \quad x \in (0,\infty).$$

Hence $W_t\psi_t$ is regular because Φ_{ϱ,d_t} is so and

$$\alpha(W_t\psi_t) = \alpha(\Phi_{\varrho,d_t}), \quad \beta(W_t\psi_t) = \beta(\Phi_{\varrho,d_t}). \tag{2.5}$$

Combining (2.2), (2.5) and Lemma 2.6, we arrive at the desired statement. □

Theorem 2.8. *Suppose Γ is a Carleson Jordan curve. If $w_1,\dots,w_n \in \mathbb{W}$ and $w(\tau) = \prod_{k=1}^n w_k(|\tau-t_k|)$, then for every $t \in \Gamma$ the function $V_t^0 w$ is regular and submultiplicative and*

$$\alpha(V_{t_k}^0 w) = m(w_k), \quad \beta(V_{t_k}^0 w) = M(w_k) \quad \textit{for} \quad k \in \{1,\dots,n\},$$
$$\alpha(V_t^0 w) = 0, \qquad \beta(V_t^0 w) = 0 \qquad \textit{for} \quad t \in \Gamma\setminus\{t_1,\dots,t_n\}.$$

Proof. If $t \notin \{t_1,\dots,t_n\}$, then there exists $D_t > 0$ such that the portion $\Gamma(t,D_t)$ does not contain any point $t_1,\dots,t_n$. Since the weight w may vanish or go to infinity only at $t_1,\dots,t_n$ and $w_k(|\tau-t_k|)$ are continuous on $\Gamma\setminus\{t_k\}$, we can conclude that there exist constants $c_t, C_t \in (0,\infty)$ such that $c_t \le w(\tau) \le C_t$ for all $\tau \in \Gamma(t,D_t)$. Assume that $x \in (0,\infty)$ and $xR, R \in (0,D_t)$. Then

$$e^{c_t-C_t} \le H_{w,t}(xR,R) \le e^{C_t-c_t}.$$

Hence

$$e^{c_t-C_t} \le (V_t^0 w)(x) \le e^{C_t-c_t}, \quad x \in (0,\infty),$$

which implies that $V_t^0 w$ is regular. By [4, Theorem 3.3(c)], the function $V_t^0 w$ is submultiplicative. It is easy to see that $\alpha(V_t^0 w) = \beta(V_t^0 w) = 0$.

If $k \in \{1,\dots,n\}$, then there exists $D_k > 0$ such that the portion $\Gamma(t_k,D_k)$ does not contain any point of $\{t_1,\dots,t_n\}\setminus\{t_k\}$. As before, there exists positive constants c_k, C_k such that

$$c_k \le \frac{w(\tau)}{w_k(|\tau-t_k|)} \le C_k \quad \text{for all} \quad \tau \in \Gamma(t_k,D_k).$$

If $x \in (0,\infty)$ and $xR, R \in (0,D_k)$, then taking into account that

$$H_{w,t_k}(xR,R) = H_{w_k(|\tau-t_k|),t_k}(xR,R)\cdot H_{w/w_k(|\tau-t_k|),t_k}(xR,R),$$

we get

$$e^{c_k-C_k} H_{w_k(|\tau-t_k|),t_k}(xR,R) \le H_{w,t_k}(xR,R) \le e^{C_k-c_k} H_{w_k(|\tau-t_k|),t_k}(xR,R).$$

Therefore, for all $x \in (0,\infty)$,

$$e^{c_k-C_k}\big(V_{t_k}^0 w_k(|\tau-t_k|)\big)(x) \le (V_{t_k}^0 w)(x) \le e^{C_k-c_k}\big(V_{t_k}^0 w_k(|\tau-t_k|)\big)(x). \tag{2.6}$$

By Lemma 2.7, the function $V^0_{t_k} w_k(|\tau - t_k|)$ is regular and

$$m(w_k) = \alpha\big(V^0_{t_k} w_k(|\tau - t_k|)\big), \quad M(w_k) = \beta\big(V^0_{t_k} w_k(|\tau - t_k|)\big). \tag{2.7}$$

Since the function $V^0_{t_k} w_k(|\tau - t_k|)$ is regular, from (2.6) it follows that $V^0_{t_k} w$ is also regular and

$$\alpha\big(V^0_{t_k} w_k(|\tau - t_k|)\big) = \alpha(V^0_{t_k} w), \quad \beta\big(V^0_{t_k} w_k(|\tau - t_k|)\big) = \beta(V^0_{t_k} w). \tag{2.8}$$

Combining (2.7) and (2.8), we finally arrive at the equalities $\alpha(V^0_{t_k} w) = m(w_k)$ and $\beta(V^0_{t_k} w) = M(w_k)$. □

2.6. Indicator functions

Let Γ be a rectifiable Jordan curve and $p : \Gamma \to (1, \infty)$ be a continuous function. We say that a weight $w : \Gamma \to [0, \infty]$ belongs to $A_{p(\cdot)}(\Gamma)$ if

$$\sup_{t\in\Gamma} \sup_{R>0} \frac{1}{R} \|w\chi_{\Gamma(t,R)}\|_{p(\cdot)} \|w^{-1}\chi_{\Gamma(t,R)}\|_{q(\cdot)} < \infty.$$

If $p = const \in (1, \infty)$, then this class coincides with the well-known Muckenhoupt class. From the Hölder inequality for $L^{p(\cdot)}(\Gamma)$ (see, e.g., [31, Theorem 2.1]) it follows that if $w \in A_{p(\cdot)}(\Gamma)$, then Γ is a Carleson curve.

Theorem 2.9. *Let Γ be a rectifiable Jordan curve and let $p : \Gamma \to (1, \infty)$ be a continuous function. If $w : \Gamma \to [0, \infty]$ is an arbitrary weight such that the operator S is bounded on $L^{p(\cdot)}(\Gamma, w)$, then $w \in A_{p(\cdot)}(\Gamma)$.*

This statement can be proved by analogy with [17, Theorem 4.3], [18, Theorem 3.2] (see also [4, Theorem 4.8]). If $p = const \in (1, \infty)$, then $w \in A_p(\Gamma)$ is also sufficient for the boundedness of S on the weighted Lebesgue space $L^p(\Gamma, w)$ (see, e.g., [4, Theorem 4.15]).

Lemma 2.10. *Let Γ be a rectifiable Jordan curve, let $p : \Gamma \to (1, \infty)$ be a continuous function satisfying (1.4), and let $w : \Gamma \to [0, \infty]$ be an arbitrary weight. Suppose $w \in A_{p(\cdot)}(\Gamma)$. Then*

(a) *for every $t \in \Gamma$ and $x \in \mathbb{R}$, the function $V^0_t(\eta^x_t w)$ is regular and submultiplicative;*
(b) *for every $t \in \Gamma$,*

$$0 \le 1/p(t) + \alpha(V^0_t w), \quad 1/p(t) + \beta(V^0_t w) \le 1.$$

Part (a) follows from [21, Lemmas 4.5, 5.8, and 5.13]. Part (b) is the consequence of [21, Lemma 4.9 and Theorem 5.9].

Under the conditions of Lemma 2.10, the *indicator functions* of the pair (Γ, w) at every $t \in \Gamma$ are well defined by

$$\alpha_t(x) := \alpha\big(V^0_t(\eta^x_t w)\big), \quad \beta_t(x) := \beta\big(V^0_t(\eta^x_t w)\big)$$

because of Theorem 2.1.

Lemma 2.11. *Let Γ be a Carleson Jordan curve satisfying the logarithmic whirl condition* (1.1) *at every $t \in \Gamma$, let $p : \Gamma \to (1, \infty)$ be a continuous function satisfying* (1.4), *and let $w : \Gamma \to [0, \infty]$ be an arbitrary weight. If $w \in A_{p(\cdot)}(\Gamma)$, then for every $t \in \Gamma$ and $x \in \mathbb{R}$,*

$$\alpha_t(x) = \alpha(V_t^0 w) + \delta(t)x, \quad \beta_t(x) = \beta(V_t^0 w) + \delta(t)x.$$

This result follows from Lemma 2.5 and [21, Lemma 5.8, Corollary 5.16(b)].

3. Singular integral operators with L^∞ coefficients

3.1. General necessary condition for Fredholmness

In this section we will suppose that Γ is a Carleson Jordan curve, $p : \Gamma \to (1, \infty)$ is a continuous function, and $w : \Gamma \to [0, \infty]$ is an arbitrary weight (not necessarily of the form (1.3)) such that S is bounded on $L^{p(\cdot)}(\Gamma, w)$. Under these assumptions,

$$P := (I + S)/2, \quad Q := (I - S)/2$$

are bounded projections on $L^{p(\cdot)}(\Gamma, w)$ (see [21, Lemma 6.4]). The operators of the form $aP + Q$, where $a \in L^\infty(\Gamma)$, are called *singular integral operators* (SIOs).

Theorem 3.1. *Suppose $a \in L^\infty(\Gamma)$. If $aP + Q$ is Fredholm on $L^{p(\cdot)}(\Gamma, w)$, then $a^{-1} \in L^\infty(\Gamma)$.*

This result follows from [21, Theorem 6.11].

3.2. The local principle of Simonenko type

Two functions $a, b \in L^\infty(\Gamma)$ are said to be locally equivalent at a point $t \in \Gamma$ if

$$\inf \{ \|(a - b)c\|_\infty \ : \ c \in C(\Gamma), \ c(t) = 1 \} = 0.$$

Theorem 3.2. *Suppose $a \in L^\infty(\Gamma)$ and for each $t \in \Gamma$ there exists a function $a_t \in L^\infty(\Gamma)$ which is locally equivalent to a at t. If the operators $a_t P + Q$ are Fredholm on $L^{p(\cdot)}(\Gamma, w)$ for all $t \in \Gamma$, then $aP + Q$ is Fredholm on $L^{p(\cdot)}(\Gamma, w)$.*

For weighted Lebesgue spaces this theorem is known as Simonenko's local principle [48]. It follows from [21, Theorem 6.13].

3.3. Wiener-Hopf factorization

The curve Γ divides the complex plane $\mathbb{C}$ into the bounded simply connected domain D^+ and the unbounded domain D^-. Without loss of generality we assume that $0 \in D^+$. We say that a function $a \in L^\infty(\Gamma)$ admits a *Wiener-Hopf factorization on $L^{p(\cdot)}(\Gamma, w)$* if $1/a \in L^\infty(\Gamma)$ and a can be written in the form

$$a(t) = a_-(t) t^\kappa a_+(t) \quad \text{a.e. on } \Gamma, \tag{3.1}$$

where $\kappa \in \mathbb{Z}$, and the factors $a_\pm$ enjoy the following properties:

$$\begin{aligned} &\text{(i)} \quad a_- \in QL^{p(\cdot)}(\Gamma, w) \dot{+} \mathbb{C}, && 1/a_- \in QL^{q(\cdot)}(\Gamma, 1/w) \dot{+} \mathbb{C}, \\ &\phantom{\text{(i)}} \quad a_+ \in PL^{q(\cdot)}(\Gamma, 1/w), && 1/a_+ \in PL^{p(\cdot)}(\Gamma, w), \\ &\text{(ii)} \quad \text{the operator } (1/a_+)Sa_+I && \text{is bounded on } L^{p(\cdot)}(\Gamma, w). \end{aligned}$$

One can prove that the number κ is uniquely determined.

Theorem 3.3. *A function $a \in L^\infty(\Gamma)$ admits a Wiener-Hopf factorization* (3.1) *on $L^{p(\cdot)}(\Gamma, w)$ if and only if the operator $aP+Q$ is Fredholm on $L^{p(\cdot)}(\Gamma, w)$. If $aP+Q$ is Fredholm, then its index is equal to $-\kappa$.*

This theorem goes back to Simonenko [47, 49]. For more about this topic we refer to [4, Section 6.12], [8, Section 5.5], [14, Section 8.3] and also to [10, 35] in the case of weighted Lebesgue spaces. Theorem 3.3 follows from [21, Theorem 6.14].

4. Singular integral operators with PC coefficients

4.1. Necessary condition for Fredholmness in case of arbitrary weights

Theorem 4.1 (see [21], Theorem 8.1). *Let Γ be a Carleson Jordan curve and let $p : \Gamma \to (1, \infty)$ be a continuous function satisfying* (1.4). *Suppose $w : \Gamma \to [0, \infty]$ is an arbitrary weight such that the operator S is bounded on $L^{p(\cdot)}(\Gamma, w)$. If the operator $aP + Q$, where $a \in PC(\Gamma)$, is Fredholm on $L^{p(\cdot)}(\Gamma, w)$, then $a(t \pm 0) \neq 0$ and*

$$-\frac{1}{2\pi}\arg\frac{a(t-0)}{a(t+0)} + \frac{1}{p(t)} + \theta\alpha_t\left(\frac{1}{2\pi}\log\left|\frac{a(t-0)}{a(t+0)}\right|\right) + (1-\theta)\beta_t\left(\frac{1}{2\pi}\log\left|\frac{a(t-0)}{a(t+0)}\right|\right)$$

is not an integer number for all $t \in \Gamma$ and all $\theta \in [0, 1]$.

Notice that from Theorem 2.9 and Lemma 2.10 it follows that the indicator functions α_t and β_t are well defined for all $t \in \Gamma$.

For standard Lebesgue spaces with Muckenhoupt weights the converse to Theorem 4.1 is also true (see [4, Proposition 7.3]).

Corollary 4.2. *Let Γ be a Carleson Jordan curve and let $p : \Gamma \to (1, \infty)$ be a continuous function satisfying* (1.4). *If $w : \Gamma \to [0, \infty]$ is an arbitrary weight such that the operator S is bounded on $L^{p(\cdot)}(\Gamma, w)$, then*

$$0 < 1/p(t) + \alpha(V_t^0 w), \quad 1/p(t) + \beta(V_t^0 w) < 1 \tag{4.1}$$

for every $t \in \Gamma$.

Proof. From Theorem 2.9 and Lemma 2.10(b) it follows that

$$0 \le 1/p(t) + \alpha(V_t^0 w), \quad 1/p(t) + \beta(V_t^0 w) \le 1. \tag{4.2}$$

On the other hand, if we take $a = 1$, then $aP + Q = I$ is obviously invertible on $L^{p(\cdot)}(\Gamma, w)$. By Theorem 4.1,

$$1/p(t) + \theta\alpha(V_t^0 w) + (1-\theta)\beta(V_t^0 w) \notin \mathbb{Z} \tag{4.3}$$

for all $t \in \Gamma$ and all $\theta \in [0,1]$. Combining (4.2) and (4.3) with $\theta = 0$ and $\theta = 1$, we arrive at (4.1). □

Corollary 4.3. *Let Γ be a Carleson Jordan curve and $p : \Gamma \to (1,\infty)$ be a continuous function satisfying* (1.4)*. Suppose a weight w is given by* (1.3)*, where $w_1,\dots,w_n \in \mathbb{W}$. If the Cauchy singular integral operator S is bounded on the space $L^{p(\cdot)}(\Gamma, w)$, then the condition* (1.5) *is fulfilled.*

This result follows from Corollary 4.2 and Theorem 2.8. It gives the necessity portion of Theorem 1.1.

4.2. Wiener-Hopf factorization of local representatives

Fix $t \in \Gamma$. For a function $a \in PC(\Gamma)$ such that $a^{-1} \in L^\infty(\Gamma)$, we construct a "canonical" function $g_{t,\gamma}$ which is locally equivalent to a at the point $t \in \Gamma$. The interior and the exterior of the unit circle can be conformally mapped onto D^+ and D^- of Γ, respectively, so that the point 1 is mapped to t, and the points $0 \in D^+$ and $\infty \in D^-$ remain fixed. Let Λ_0 and Λ_∞ denote the images of $[0,1]$ and $[1,\infty)\cup\{\infty\}$ under this map. The curve $\Lambda_0 \cup \Lambda_\infty$ joins 0 to ∞ and meets Γ at exactly one point, namely t. Let $\arg z$ be a continuous branch of argument in $\mathbb{C}\setminus(\Lambda_0\cup\Lambda_\infty)$. For $\gamma \in \mathbb{C}$, define the function $z^\gamma := |z|^\gamma e^{i\gamma \arg z}$, where $z \in \mathbb{C}\setminus(\Lambda_0\cup\Lambda_\infty)$. Clearly, z^γ is an analytic function in $\mathbb{C}\setminus(\Lambda_0\cup\Lambda_\infty)$. The restriction of z^γ to $\Gamma\setminus\{t\}$ will be denoted by $g_{t,\gamma}$. Obviously, $g_{t,\gamma}$ is continuous and nonzero on $\Gamma\setminus\{t\}$. Since $a(t\pm 0) \neq 0$, we can define $\gamma_t = \gamma \in \mathbb{C}$ by the formulas

$$\operatorname{Re}\gamma_t := \frac{1}{2\pi}\arg\frac{a(t-0)}{a(t+0)}, \quad \operatorname{Im}\gamma_t := -\frac{1}{2\pi}\log\left|\frac{a(t-0)}{a(t+0)}\right|, \tag{4.4}$$

where we can take any value of $\arg(a(t-0)/a(t+0))$, which implies that any two choices of $\operatorname{Re}\gamma_t$ differ by an integer only. Clearly, there is a constant $c_t \in \mathbb{C}\setminus\{0\}$ such that $a(t\pm 0) = c_t g_{t,\gamma_t}(t\pm 0)$, which means that a is locally equivalent to $c_t g_{t,\gamma_t}$ at the point $t \in \Gamma$.

For $t \in \Gamma$ and $\gamma \in \mathbb{C}$, consider the weight

$$\varphi_{t,\gamma}(\tau) := |(\tau - t)^\gamma|, \quad t \in \Gamma\setminus\{t\}.$$

From [21, Lemma 7.1] we get the following.

Lemma 4.4. *Let Γ be a Carleson Jordan curve and let $p : \Gamma \to (1,\infty)$ be a continuous function. Suppose $w : \Gamma \to [0,\infty]$ is an arbitrary weight such that the operator S is bounded on $L^{p(\cdot)}(\Gamma, w)$. If, for some $k \in \mathbb{Z}$ and $\gamma \in \mathbb{C}$, the operator $\varphi_{t,k-\gamma} S \varphi_{t,\gamma-k} I$ is bounded on $L^{p(\cdot)}(\Gamma, w)$, then the function $g_{t,\gamma}$ admits a Wiener-Hopf factorization on $L^{p(\cdot)}(\Gamma, w)$.*

4.3. Fredholm criterion

Now we are in a position to prove one of the main results of the paper.

Theorem 4.5. *Let Γ be a Carleson Jordan curve satisfying the logarithmic whirl condition* (1.1) *at each point $t \in \Gamma$ and $p : \Gamma \to (1,\infty)$ be a continuous function satisfying* (1.4)*. Suppose a weight w is given by* (1.3)*, where for $w_1,\dots,w_n \in \mathbb{W}$*

the condition (1.5) *is fulfilled. The operator* $aP+Q$, *where* $a \in PC(\Gamma)$, *is Fredholm on the weighted variable Lebesgue space* $L^{p(\cdot)}(\Gamma, w)$ *if and only if* $a(t \pm 0) \neq 0$ *and*

$$-\frac{1}{2\pi}\arg\frac{a(t-0)}{a(t+0)}+\frac{\delta(t)}{2\pi}\log\left|\frac{a(t-0)}{a(t+0)}\right|+\frac{1}{p(t)}+\theta\mu_t+(1-\theta)\nu_t \notin \mathbb{Z} \tag{4.5}$$

for all $t \in \Gamma$ *and all* $\theta \in [0,1]$, *where*

$$\begin{aligned} &\mu_{t_k} = m(w_k), \quad \nu_{t_k} = M(w_k) \quad \textit{for} \quad k \in \{1,\dots,n\}, \\ &\mu_t = 0, \qquad\quad\;\; \nu_t = 0 \qquad\qquad \textit{for} \quad t \in \Gamma \setminus \{t_1,\dots,t_n\}. \end{aligned} \tag{4.6}$$

Proof. The proof is developed by analogy with the proof of [22, Theorem 3.3].

Necessity. By Theorem 1.1, the operator S is bounded on $L^{p(\cdot)}(\Gamma, w)$. Then from Theorem 2.9 it follows that $w \in A_{p(\cdot)}(\Gamma)$. Therefore, by Lemma 2.10(a), the function $V_t^0(\eta_t^x w)$ is regular and submultiplicative for all $t \in \Gamma$ and all $x \in \mathbb{R}$. Since Γ satisfies (1.1) for all $t \in \Gamma$, from Lemma 2.11 and Theorem 2.8 it follows that the indices of $V_t^0(\eta_t^x w)$ (that is, the indicator functions α_t, β_t of the pair (Γ, w) at $t \in \Gamma$) are calculated by

$$\alpha_{t_k}(x) = m(w_k) + \delta(t_k)x, \quad \beta_{t_k}(x) = M(w_k) + \delta(t_k)x \quad (x \in \mathbb{R}) \tag{4.7}$$

for $k \in \{1,\dots,n\}$ and

$$\alpha_t(x) = \delta(t)x, \quad \beta_t(x) = \delta(t)x \quad (x \in \mathbb{R}) \tag{4.8}$$

for $t \in \Gamma \setminus \{t_1,\dots,t_n\}$. Theorem 4.1 and (4.7)–(4.8) imply (4.5) and $a(t \pm 0)$ for all $t \in \Gamma$ and all $\theta \in [0,1]$.

Sufficiency. If $aP+Q$ is Fredholm, then, by Theorem 3.1, $a(t \pm 0) \neq 0$ for all $t \in \Gamma$. Fix an arbitrary $t \in \Gamma$. Choose $\gamma = \gamma_t \in \mathbb{C}$ as in (4.4). Then the function a is locally equivalent to $c_t g_{t,\gamma_t}$ at the point $t \in \Gamma$, where $c_t \in \mathbb{C} \setminus \{0\}$ is some constant. From (4.5) it follows that there exists an $s_t \in \mathbb{Z}$ such that

$$0 < s_t - \operatorname{Re}\gamma_t - \delta(t)\operatorname{Im}\gamma_t + 1/p(t) + \theta\mu_t + (1-\theta)\nu_t < 1$$

for all $\theta \in [0,1]$. In particular, if $\theta = 1$, then

$$0 < 1/p(t) + \big(s_t - \operatorname{Re}\gamma_t - \delta(t)\operatorname{Im}\gamma_t\big) + \mu_t; \tag{4.9}$$

if $\theta = 0$, then

$$1/p(t) + \big(s_t - \operatorname{Re}\gamma_t - \delta(t)\operatorname{Im}\gamma_t\big) + \nu_t < 1. \tag{4.10}$$

Let $\psi_t(x) := x^{s_t - \operatorname{Re}\gamma_t - \delta(t)\operatorname{Im}\gamma_t}$ for $x \in (0, |\Gamma|]$ and

$$\widetilde{w}(\tau) := \psi_t(|\tau - t|)w(\tau), \quad \tau \in \Gamma.$$

Clearly, the weight $\widetilde{w}$ is of the form (1.3).

If $t \in \Gamma \setminus \{t_1,\dots,t_n\}$, then obviously

$$m(\psi_t) = M(\psi_t) = s_t - \operatorname{Re}\gamma_t - \delta(t)\operatorname{Im}\gamma_t. \tag{4.11}$$

From (4.9)–(4.11) we get

$$0 < 1/p(t) + m(\psi_t) \le 1/p(t) + M(\psi_t) < 1.$$

Combining these inequalities with Theorem 1.1, we conclude that the operator S is bounded on $L^{p(\cdot)}(\Gamma, \widetilde{w})$.

If $t = t_k$ for some $k \in \{1, \dots, n\}$, then one can easily show that

$$m(\psi_{t_k} w_k) = s_{t_k} - \operatorname{Re}\gamma_{t_k} - \delta(t_k)\operatorname{Im}\gamma_{t_k} + m(w_k), \tag{4.12}$$

$$M(\psi_{t_k} w_k) = s_{t_k} - \operatorname{Re}\gamma_{t_k} - \delta(t_k)\operatorname{Im}\gamma_{t_k} + M(w_k). \tag{4.13}$$

From (4.9)–(4.10) and (4.12)–(4.13) we obtain

$$0 < 1/p(t_k) + m(\psi_{t_k} w_k) \le 1/p(t_k) + M(\psi_{t_k} w_k) < 1.$$

These inequalities and Theorem 1.1 yield the boundedness of the operator S on $L^{p(\cdot)}(\Gamma, \widetilde{w})$.

In view of the logarithmic whirl condition (1.1) we have

$$\begin{aligned} \varphi_{t,s_t-\gamma_t}(\tau) &= |\tau - t|^{s_t - \operatorname{Re}\gamma_t} e^{\operatorname{Im}\gamma_t \arg(\tau - t)} \\ &= |\tau - t|^{s_t - \operatorname{Re}\gamma_t} e^{-\operatorname{Im}\gamma_t(\delta(t)\log|\tau-t| + O(1))} \\ &= |\tau - t|^{s_t - \operatorname{Re}\gamma_t - \delta(t)\operatorname{Im}\gamma_t} e^{-\operatorname{Im}\gamma_t O(1)} \end{aligned}$$

as $\tau \to t$. Therefore the operator $\varphi_{t,s_t-\gamma_t} S \varphi_{t,\gamma_t-s_t} I$ is bounded on $L^{p(\cdot)}(\Gamma, w)$. Then, by Lemma 4.4, the function g_{t,γ_t} admits a Wiener-Hopf factorization on $L^{p(\cdot)}(\Gamma, w)$. Due to Theorem 3.3, the operator $g_{t,\gamma_t}P + Q$ is Fredholm. Then the operator $c_t g_{t,\gamma_t} P + Q$ is Fredholm, too. Since the function $c_t g_{t,\gamma_t}$ is locally equivalent to the function a at every point $t \in \Gamma$, the operator $aP + Q$ is Fredholm on $L^{p(\cdot)}(\Gamma, w)$ in view of Theorem 3.2. □

Remark 4.6. In the case of standard Lebesgue spaces a complete description of the set of all $\gamma \in \mathbb{C}$ such that the operator S is bounded on $L^p(\Gamma, \varphi_{t,\gamma} w)$ is known in terms of the indicator functions α_t and β_t (see [4, Sections 3.6–3.7]). This description allowed the authors of [4] to consider the case of Carleson curves that may not satisfy (1.1) and may have distinct spirality indices $\alpha(W_t\eta_t)$ and $\beta(W_t\eta_t)$. However, the only result on the boundedness of S on weighted variable Lebesgue spaces in our disposal is Theorem 1.1, which is not applicable to weights of the form $\varphi_{t,\gamma} w$ unless a Carleson curve is sufficiently nice. If a Carleson curve satisfies (1.1), then $\varphi_{t,\gamma}$ is equivalent to $\psi_{t,\lambda}(\tau) := |\tau - t|^\lambda$ for some $\lambda \in \mathbb{R}$ and one can apply Theorem 1.1 to the weight $\psi_{t,\lambda} w$. Therefore, to treat the case of arbitrary Carleson curves by this method, a more general boundedness result than Theorem 1.1 is needed.

4.4. Criterion for the closedness of the image

Theorem 4.7. *Let Γ be a Carleson Jordan curve satisfying the logarithmic whirl condition* (1.1) *at each point $t \in \Gamma$ and $p : \Gamma \to (1, \infty)$ be a continuous function satisfying* (1.4). *Suppose a weight w is given by* (1.3), *where for $w_1, \dots, w_n \in \mathbb{W}$ the condition* (1.5) *is fulfilled. Suppose $a \in PC(\Gamma)$ has finitely many jumps and $a(t \pm 0) \ne 0$ for all $t \in \Gamma$. Then the image of $aP + Q$ is closed in $L^{p(\cdot)}(\Gamma, w)$ if and only if* (4.5) *holds for all $t \in \Gamma$ and all $\theta \in [0, 1]$.*

Proof. The idea of the proof is borrowed from [4, Proposition 7.16] (see also [24, Theorem 4.3]). The sufficiency part follows from Theorem 4.5.

Let us prove the necessity part. Assume that $a(t \pm 0) \neq 0$ for all $t \in \Gamma$. Since the number of jumps, that is, the points $t \in \Gamma$ at which $a(t-0) \neq a(t+0)$, is finite, it is clear that

$$-\frac{1}{2\pi}\arg\frac{a(t-0)}{a(t+0)} + \frac{\delta(t)}{2\pi}\log\left|\frac{a(t-0)}{a(t+0)}\right| + \frac{1}{1+\varepsilon} \notin \mathbb{Z},$$

$$-\frac{1}{2\pi}\arg\frac{a(t+0)}{a(t-0)} + \frac{\delta(t)}{2\pi}\log\left|\frac{a(t+0)}{a(t-0)}\right| + \frac{1}{1+\varepsilon} \notin \mathbb{Z}$$

for all $t \in \Gamma$ and all sufficiently small $\varepsilon > 0$. By Theorem 4.5, the operators $aP+Q$ and $a^{-1}P+Q$ are Fredholm on the Lebesgue space $L^{1+\varepsilon}(\Gamma)$ whenever $\varepsilon > 0$ is sufficiently small.

If (1.5) holds, then there exists a number $\varepsilon > 0$ such that

$$0 < (1/p(t_k) + m(w_k))(1+\varepsilon) \leq (1/p(t_k) + M(w_k))(1+\varepsilon) < 1$$

for all $k \in \{1, \dots, n\}$. It is easy to see that

$$m(w_k^{1+\varepsilon}) = (1+\varepsilon)m(w_k), \quad M(w_k^{1+\varepsilon}) = (1+\varepsilon)M(w_k).$$

Hence, by Theorem 1.1, the operator S is bounded on $L^{p(\cdot)/(1+\varepsilon)}(\Gamma, w^{1+\varepsilon})$. Taking into account this observation, one can prove as in [24, Lemma 4.1] that there is an $\varepsilon_0 > 0$ such that

$$L^{p(\cdot)}(\Gamma, w) \subset L^{1+\varepsilon_0}(\Gamma), \quad L^{q(\cdot)}(\Gamma, 1/w) \subset L^{1+\varepsilon_0}(\Gamma)$$

and $aP+Q$, $a^{-1}P+Q$ are Fredholm on $L^{1+\varepsilon_0}(\Gamma)$. Then

$$n\big(aP+Q; L^{p(\cdot)}(\Gamma, w)\big) \leq n\big(aP+Q; L^{1+\varepsilon_0}(\Gamma)\big) < \infty, \tag{4.14}$$

and, by duality (see [24, Lemma 3.8]),

$$\begin{aligned} d\big(aP+Q; L^{p(\cdot)}(\Gamma, w)\big) &= n\big(a^{-1}P+Q; L^{q(\cdot)}(\Gamma, 1/w)\big) \\ &\leq n\big(a^{-1}P+Q; L^{1+\varepsilon_0}(\Gamma)\big) < \infty. \end{aligned} \tag{4.15}$$

If (4.5) does not hold, then $aP+Q$ is not Fredholm on $L^{p(\cdot)}(\Gamma, w)$ in view of Theorem 4.5. From this fact and (4.14)–(4.15) we conclude that the image of $aP+Q$ is not closed in $L^{p(\cdot)}(\Gamma, w)$, which contradicts the hypothesis. □

5. Tools for the construction of the symbol calculus

5.1. The Allan-Douglas local principle

Let B be a Banach algebra with identity. A subalgebra Z of B is said to be a *central subalgebra* if $zb = bz$ for all $z \in Z$ and all $b \in B$.

Theorem 5.1 (see [8], Theorem 1.35(a)). *Let B be a Banach algebra with identity e and let Z be a closed central subalgebra of B containing e. Let $M(Z)$ be the maximal ideal space of Z, and for $\omega \in M(Z)$, let J_ω refer to the smallest closed two-sided ideal of B containing the ideal ω. Then an element b is invertible in B if and only if $b + J_\omega$ is invertible in the quotient algebra B/J_ω for all $\omega \in M(Z)$.*

The algebra B/J_ω is referred to as the *local algebra* of B at $\omega \in M(Z)$ and the spectrum of $b + J_\omega$ in B/J_ω is called the *local spectrum* of b at $\omega \in M(Z)$.

5.2. The two projections theorem

Recall that an element p of a Banach algebra is called an *idempotent* (or, somewhat loosely, also a *projection*), if $p^2 = p$.

The following two projections theorem was obtained by Finck, Roch, Silbermann [12] and Gohberg, Krupnik [15] (see also [4, Section 8.3]).

Theorem 5.2. *Let B be a Banach algebra with identity e, let $\mathcal{C}$ be a Banach subalgebra of B which contains e and is isomorphic to $\mathbb{C}^{N\times N}$, and let p and q be two idempotent elements in B such that $cp = pc$ and $cq = qc$ for all $c \in \mathcal{C}$. Let $A = \operatorname{alg}(\mathcal{C}, p, q)$ be the smallest closed subalgebra of B containing $\mathcal{C}, p, q$. Put*

$$x = pqp + (e-p)(e-q)(e-p),$$

denote by $\operatorname{sp} x$ the spectrum of x in B, and suppose the points 0 and 1 are not isolated points of $\operatorname{sp} x$. Then

(a) *for each $z \in \operatorname{sp} x$ the map σ_z of $\mathcal{C} \cup \{p, q\}$ into the algebra $\mathbb{C}^{2N\times 2N}$ of all complex $2N \times 2N$ matrices defined by*

$$\sigma_z c = \begin{bmatrix} c & O \\ O & c \end{bmatrix}, \sigma_z p = \begin{bmatrix} E & O \\ O & O \end{bmatrix}, \sigma_z q = \begin{bmatrix} zE & \sqrt{z(1-z)}E \\ \sqrt{z(1-z)}E & (1-z)E \end{bmatrix},$$

where $c \in \mathcal{C}$, E and O denote the $N \times N$ identity and zero matrices, respectively, and $\sqrt{z(1-z)}$ denotes any complex number whose square is $z(1-z)$, extends to a Banach algebra homomorphism

$$\sigma_z : A \to \mathbb{C}^{2N\times 2N};$$

(b) *every element a of the algebra A is invertible in the algebra B if and only if*

$$\det \sigma_z a \neq 0 \quad \text{for all} \quad z \in \operatorname{sp} x;$$

(c) *the algebra A is inverse closed in B if and only if the spectrum of x in A coincides with the spectrum of x in B.*

A further generalization of the above result to the case of $n \geq 3$ projections is contained in [4, Section 8.4].

6. Algebra of singular integral operators

6.1. Operators of local type

In this section we will suppose that Γ is a Carleson curve satisfying the logarithmic whirl condition (1.1) at each point $t \in \Gamma$, $p : \Gamma \to (1, \infty)$ is a continuous function satisfying (1.4), w is a weight of the form (1.3) with $w_1, \dots, w_n \in \mathbb{W}$ satisfying (1.5). Under these conditions the operator S (defined elementwise) is bounded on $L_N^{p(\cdot)}(\Gamma, w)$, where $N \geq 1$ (see Theorem 1.1).

Let $\mathcal{B} := \mathcal{B}(L_N^{p(\cdot)}(\Gamma, w))$ be the Banach algebra of all bounded linear operators on $L_N^{p(\cdot)}(\Gamma, w)$ and let $\mathcal{K} := \mathcal{K}(L_N^{p(\cdot)}(\Gamma, w))$ be the closed ideal of all compact operators on $L_N^{p(\cdot)}(\Gamma, w)$. We will denote by $\mathcal{B}^\pi$ the Calkin algebra $\mathcal{B}/\mathcal{K}$ and by A^π the coset $A + \mathcal{K}$ for any operator $A \in \mathcal{B}$. An operator $A \in \mathcal{B}$ is said to be of *local type* if $A \operatorname{diag}\{c, \dots, c\}I - \operatorname{diag}\{c, \dots, c\}A \in \mathcal{K}$ for every continuous function c on Γ. This notion goes back to Simonenko [48, 50]. It is easy to see that the set $\mathcal{L}$ of all operators of local type is a Banach subalgebra of $\mathcal{B}$.

Lemma 6.1. *We have*

$$\mathcal{K} \subset \operatorname{alg}(S, PC, L_N^{p(\cdot)}(\Gamma, w)) \subset \mathcal{L}. \tag{6.1}$$

Proof. Let $\operatorname{alg}(S, C, L_N^{p(\cdot)}(\Gamma, w))$ be the smallest closed subalgebra of $\mathcal{B}$ containing the operators of multiplication by continuous matrix functions and the operator S. Obviously,

$$\operatorname{alg}(S, C, L_N^{p(\cdot)}(\Gamma, w)) \subset \operatorname{alg}(S, PC, L_N^{p(\cdot)}(\Gamma, w)). \tag{6.2}$$

On the other hand, by analogy with [17, Lemma 9.1] or [23, Lemma 5.1] one can show that

$$\mathcal{K} \subset \operatorname{alg}(S, C, L_N^{p(\cdot)}(\Gamma, w)). \tag{6.3}$$

Combining (6.2) and (6.3), we arrive at the first embedding in (6.1).

If $a \in PC_{N \times N}(\Gamma)$ and $c \in C(\Gamma)$, then the operators aI and $\operatorname{diag}\{c, \dots, c\}I$ obviously commute. Then aI is of local type. By [21, Lemma 6.5], the operator $\operatorname{diag}\{c, \dots, c\}S - S \operatorname{diag}\{c, \dots, c\}I$ is compact. Thus, S is of local type, too. Since the generators of $\operatorname{alg}(S, PC, L_N^{p(\cdot)}(\Gamma, w))$ are of local type, each element of this algebra is of local type, which proves the second embedding in (6.1). □

Lemma 6.2. *An operator $A \in \mathcal{L}$ is Fredholm if and only if the coset A^π is invertible in the quotient algebra $\mathcal{L}^\pi := \mathcal{L}/\mathcal{K}$.*

The proof is straightforward.

6.2. Localization

From Lemma 6.1 we deduce that the quotient algebras

$$\operatorname{alg}^\pi(S, PC, L_N^{p(\cdot)}(\Gamma, w)) := \operatorname{alg}(S, PC, L_N^{p(\cdot)}(\Gamma, w))/\mathcal{K}$$

and $\mathcal{L}^\pi := \mathcal{L}/\mathcal{K}$ are well defined. We will study the invertibility of an element A^π of $\mathrm{alg}^\pi(S, PC, L_N^{p(\cdot)}(\Gamma, w))$ in the larger algebra $\mathcal{L}^\pi$ by using Theorem 5.1. To this end, consider

$$\mathcal{Z}^\pi := \{(\mathrm{diag}\{c, \dots, c\}I)^\pi : c \in C(\Gamma)\}.$$

From the definition of $\mathcal{L}$ it follows that $\mathcal{Z}^\pi$ is a central subalgebra of $\mathcal{L}^\pi$. The maximal ideal space $M(\mathcal{Z}^\pi)$ of $\mathcal{Z}^\pi$ may be identified with the curve Γ via the Gelfand map $\mathcal{G}$ given by

$$\mathcal{G} : \mathcal{Z}^\pi \to C(\Gamma), \quad (\mathcal{G}(\mathrm{diag}\{c, \dots, c\}I)^\pi)(t) = c(t) \quad (t \in \Gamma).$$

In accordance with Theorem 5.1, for every $t \in \Gamma$ we define $\mathcal{J}_t \subset \mathcal{L}^\pi$ as the smallest closed two-sided ideal of $\mathcal{L}^\pi$ containing the set

$$\{(\mathrm{diag}\{c, \dots, c\}I)^\pi \ : \ c \in C(\Gamma),\ c(t) = 0\}.$$

Consider a function $\chi_t \in PC(\Gamma)$ which is continuous on $\Gamma \setminus \{t\}$ and satisfies $\chi_t(t-0) = 0$ and $\chi_t(t+0) = 1$. For $a \in PC_{N\times N}(\Gamma)$ define $a_t \in PC_{N\times N}(\Gamma)$ by

$$a_t := a(t-0)(1-\chi_t) + a(t+0)\chi_t. \tag{6.4}$$

Clearly $(aI)^\pi - (a_tI)^\pi \in \mathcal{J}_t$. Hence, for any operator $A \in \mathrm{alg}(S, PC, L_N^{p(\cdot)}(\Gamma, w))$, the coset $A^\pi + \mathcal{J}_t$ belongs to the smallest closed subalgebra $\mathcal{A}_t$ of $\mathcal{L}^\pi/\mathcal{J}_t$ containing the cosets

$$p := P^\pi + \mathcal{J}_t, \quad q := (\mathrm{diag}\{\chi_t, \dots, \chi_t\}I)^\pi + \mathcal{J}_t \tag{6.5}$$

and the algebra

$$\mathcal{C} := \{(cI)^\pi + \mathcal{J}_t \ : \ c \in \mathbb{C}^{N\times N}\}. \tag{6.6}$$

The latter algebra is obviously isomorphic to $\mathbb{C}^{N\times N}$, so $\mathcal{C}$ and $\mathbb{C}^{N\times N}$ can (and will) be identified with each other. It is easy to see that

$$p^2 = p, \quad q^2 = q, \quad pc = cp, \quad qc = cq \tag{6.7}$$

for all $c \in \mathcal{C}$. To apply Theorem 5.2 to the algebras $\mathcal{L}^\pi/\mathcal{J}_t$ and $\mathcal{A}_t = \mathrm{alg}(\mathcal{C}, p, q)$, we need to identify the spectrum of the element

$$\begin{aligned} x &:= pqp + (e-p)(e-q)(e-p) \\ &= \big(P\,\mathrm{diag}\{\chi_t, \dots, \chi_t\}P + Q\,\mathrm{diag}\{1-\chi_t, \dots, 1-\chi_t\}Q\big)^\pi + \mathcal{J}_t \end{aligned} \tag{6.8}$$

in the algebra $\mathcal{L}^\pi/\mathcal{J}_t$.

6.3. Spiralic horns

Given two real numbers a, b satisfying $0 < a \le b < 1$, two complex numbers z_1, z_2, and a real number δ, we define the *spiralic horn* between z_1 and z_2 as the set

$$\begin{aligned} &\mathcal{S}(z_1, z_2; \delta; a, b) := \{z_1, z_2\} \cup \\ &\cup \left\{ u \in \mathbb{C} \setminus \{z_1, z_2\} : \frac{1}{2\pi}\left(\arg\frac{u-z_1}{u-z_2} - \delta \log\left|\frac{u-z_1}{u-z_2}\right|\right) \in [a, b] + \mathbb{Z} \right\}. \end{aligned}$$

If $z_1 = z_2$, then $\mathcal{S}(z_1, z_2; \delta; a, b)$ degenerates to the point z_1. Assume $z_1 \ne z_2$. Then the set $\mathcal{S}(z_1, z_2; 0; 1/2, 1/2)$ is the segment between z_1 and z_2. If $a \ne 1/2$, then $\mathcal{S}(z_1, z_2; 0; a, a)$ is the circular arc between z_1 and z_2. If $a < 1/2$ (resp. $a > 1/2$), then one sees the straight line between z_1 and z_2 under the angle $2\pi a$ (resp. $2\pi(1-a)$) and running through the arc from z_1 to z_2 this straight line is located at the left-hand side (resp. right-hand side). The importance of these arcs in the Fredholm theory of singular integral operators was first observed by Widom [52], they were exploited intensively by Gohberg and Krupnik [13], [14, Chap. 9].

If $z_1 \neq z_2$ and $a < b$, then $\mathcal{S}(z_1, z_2; 0; a, b)$ is an ordinary horn bounded by the circular arcs $\mathcal{S}(z_1, z_2; 0; a, a)$ and $\mathcal{S}(z_1, z_2; 0; b, b)$. If $\delta \neq 0$, then $\mathcal{S}(z_1, z_2; \delta; a, a)$ is a logarithmic double spiral. If $a < b$, then $\mathcal{S}(z_1, z_2; \delta; a, b)$ is the union of logarithmic double spirals

$$\mathcal{S}(z_1, z_2; \delta; a, b) = \bigcup_{\lambda \in [a,b]} \mathcal{S}(z_1, z_2; \delta; \lambda, \lambda).$$

Hence this set is the closed set between two logarithmic double spirals. It was introduced in [3] (see also [4, Sections 7.3–7.6] and [6, 5]).

6.4. The local spectrum

Now we are ready to identify the spectrum of x in the algebra $B = \mathcal{L}^\pi / \mathcal{J}_t$.

Lemma 6.3. *Let $\chi_t \in PC(\Gamma)$ be a continuous function on $\Gamma \setminus \{t\}$ such that*

$$\chi_t(t-0) = 0, \quad \chi_t(t+0) = 1,$$
$$\chi_t(\Gamma \setminus \{t\}) \cap \mathcal{S}(0, 1; \delta(t); 1/p(t) + \mu_t, 1/p(t) + \nu_t) = \emptyset,$$

where μ_t and ν_t are defined by (4.6). Then the spectrum of the element x given by (6.8) in the algebra $B = \mathcal{L}^\pi / \mathcal{J}_t$ coincides with $\mathcal{S}(0, 1; \delta(t); 1/p(t) + \mu_t, 1/p(t) + \nu_t)$.

Proof. From Theorem 4.5 we immediately get that the set of all $\lambda \in \mathbb{C}$ such that the operator $(\chi_t - \lambda)P + Q$ is not Fredholm on $L^{p(\cdot)}(\Gamma, w)$ coincides with $\mathcal{S}(0, 1; \delta(t); 1/p(t) + \mu_t, 1/p(t) + \nu_t)$. After this observation the proof can be developed by a literal repetition of the proof of [17, Lemma 9.4] with the Boyd indices α_M and β_M of an Orlicz space considered there replaced by the numbers $1/p(t) + \mu_t$ and $1/p(t) + \nu_t$, respectively. □

6.5. Construction of the symbol calculus

Now we are in a position to prove the main result of the paper.

Theorem 6.4. *Let $N \in \mathbb{N}$, let Γ be a Carleson Jordan curve satisfying the logarithmic whirl condition (1.1) at every point $t \in \Gamma$, let $p : \Gamma \to (1, \infty)$ be a continuous function satisfying (1.4), and let $w : \Gamma \to [0, \infty]$ be a radial oscillating weight given by (1.3) with $w_1, \dots, w_n \in \mathbb{W}$ satisfying (1.5). Put*

$$\mu_{t_k} = m(w_k), \quad \nu_{t_k} = M(w_k) \quad \text{for} \quad k \in \{1, \dots, n\},$$
$$\mu_t = 0, \qquad \nu_t = 0 \qquad \text{for} \quad t \in \Gamma \setminus \{t_1, \dots, t_n\}$$

and define the "spiralic horn bundle" by

$$\mathcal{M} := \bigcup_{t \in \Gamma} \big(\{t\} \times \mathcal{S}(0, 1; \delta(t); 1/p(t) + \mu_t, 1/p(t) + \nu_t) \big).$$

(a) *For each point* $(t,z) \in \mathcal{M}$*, the map*

$$\sigma_{t,z} : \{S\} \cup \{aI \ : \ a \in PC_{N\times N}(\Gamma)\} \to \mathbb{C}^{2N\times 2N}$$

given by

$$\sigma_{t,z}(S) = \begin{bmatrix} E & O \\ O & -E \end{bmatrix},$$

$$\sigma_{t,z}(aI) = \begin{bmatrix} a(t+0)z + a(t-0)(1-z) & (a(t+0)-a(t-0))\sqrt{z(1-z)} \\ (a(t+0)-a(t-0))\sqrt{z(1-z)} & a(t+0)(1-z)+a(t-0)z \end{bmatrix},$$

where E *and* O *denote the* $N \times N$ *identity and zero matrices, respectively, and* $\sqrt{z(1-z)}$ *denotes any complex number whose square is* $z(1-z)$*, extends to a Banach algebra homomorphism*

$$\sigma_{t,z} : \mathrm{alg}(S, PC, L_N^{p(\cdot)}(\Gamma, w)) \to \mathbb{C}^{2N\times 2N}$$

with the property that $\sigma_{t,z}(K)$ *is the* $2N \times 2N$ *zero matrix whenever* K *is a compact operator on* $L_N^{p(\cdot)}(\Gamma, w)$.

(b) *An operator* $A \in \mathrm{alg}(S, PC, L_N^{p(\cdot)}(\Gamma, w))$ *is Fredholm on* $L_N^{p(\cdot)}(\Gamma, w)$ *if and only if*

$$\det \sigma_{t,z}(A) \neq 0 \quad \textit{for all} \quad (t,z) \in \mathcal{M}.$$

(c) *The quotient algebra* $\mathrm{alg}^\pi(S, PC, L_N^{p(\cdot)}(\Gamma, w))$ *is inverse closed in the Calkin algebra* $\mathcal{B}^\pi$*, that is, if a coset* $A^\pi \in \mathrm{alg}^\pi(S, PC, L_N^{p(\cdot)}(\Gamma, w))$ *is invertible in* $\mathcal{B}^\pi$*, then* $(A^\pi)^{-1} \in \mathrm{alg}^\pi(S, PC, L_N^{p(\cdot)}(\Gamma, w))$.

Proof. The idea of the proof is borrowed from [4, Section 8.5]. Here we follow the proof of [22, Theorem 5.1] and [23, Theorem 5.4]. Since the latter two sources may not be readily available, we give a selfcontained proof.

Fix $t \in \Gamma$ and choose a function $\chi_t \in PC(\Gamma)$ as in the assumptions of Lemma 6.3. From (6.7) and Lemma 6.3 we deduce that the algebras $\mathcal{L}^\pi/\mathcal{J}_t$ and $\mathcal{A}_t = \mathrm{alg}(\mathcal{C}, p, q)$, where p, q, and $\mathcal{C}$ are given by (6.5) and (6.6), respectively, satisfy all the assumptions of Theorem 5.2.

(a) By Theorem 5.2(a), for every $z \in \mathcal{S}(0,1;\delta(t);1/p(t)+\mu_t, 1/p(t)+\nu_t)$, the map

$$\sigma_{t,z} = \sigma_z \circ \pi_t : \mathrm{alg}(S, PC, L_N^{p(\cdot)}(\Gamma, w)) \to \mathbb{C}^{2N\times 2N},$$

where σ_z is given in Theorem 5.2(a) and π_t acts by the rule $A \mapsto A^\pi + \mathcal{J}_t$ for every $A \in \mathrm{alg}(S, PC, L_N^{p(\cdot)}(\Gamma, w))$, is a well-defined Banach algebra homomorphism. It is easy to check that

$$\sigma_{t,z}(S) = 2\sigma_z p - \sigma_z e = \begin{bmatrix} E & O \\ O & -E \end{bmatrix}.$$

If $a \in PC_{N\times N}(\Gamma)$, then from (6.4) and $(aI - a_tI)^\pi \in \mathcal{J}_t$ it follows that

$$\sigma_{t,z}(aI) = \sigma_{t,z}(a_tI) = \sigma_z(a(t-0))\sigma_z(e-q) + \sigma_z(a(t+0))\sigma_z q$$
$$= \begin{bmatrix} a(t+0)z + a(t-0)(1-z) & (a(t+0)-a(t-0))\sqrt{z(1-z)} \\ (a(t+0)-a(t-0))\sqrt{z(1-z)} & a(t+0)(1-z)+a(t-0)z \end{bmatrix}.$$

From Lemma 6.1 it follows that $\pi_t(K)$ is correctly defined for every $K \in \mathcal{K}$ and $\pi_t(K) = \mathcal{J}_t$. Hence

$$\sigma_{t,z}(K) = \sigma_z(0) = \begin{bmatrix} O & O \\ O & O \end{bmatrix}.$$

Part (a) is proved.

(b) From Lemma 6.2 it follows that the Fredholmness of an operator A from $\operatorname{alg}(S, PC, L_N^{p(\cdot)}(\Gamma, w))$ is equivalent to the invertibility of $A^\pi \in \mathcal{L}^\pi$. By Theorem 5.1, the latter is equivalent to the invertibility of $\pi_t(A) = A^\pi + \mathcal{J}_t$ in $\mathcal{L}^\pi/\mathcal{J}_t$ for every $t \in \Gamma$. By Theorem 5.2(b), this is equivalent to

$$\det \sigma_{t,z}(A) = \det \sigma_z \pi_t(A) \neq 0 \quad \text{for all} \quad (t,z) \in \mathcal{M}. \tag{6.9}$$

Part (b) is proved.

(c) Since the compact simply connected set $\mathcal{S}(0,1;\delta(t);1/p(t)+\mu_t, 1/p(t)+\nu_t)$ does not separate the complex plane, it follows that the spectra of the element x given by (6.8) in the algebras $\mathcal{L}^\pi/\mathcal{J}_t$ and $\mathcal{A}_t$ coincide. Suppose A belongs to $\operatorname{alg}(S, PC, L_N^{p(\cdot)}(\Gamma, w))$. If A^π is invertible in $\mathcal{B}^\pi$, then (6.9) is fulfilled. Consequently, by Theorem 5.2(b), (c), $\pi_t(A) = A^\pi + \mathcal{J}_t$ is invertible in $\mathcal{A}_t$ for every $t \in \Gamma$. Applying the Allan-Douglas local principle (Theorem 5.1) to the algebra $\operatorname{alg}^\pi(S, PC, L_N^{p(\cdot)}(\Gamma, w))$, its central subalgebra $\mathcal{Z}^\pi$ and the ideals $\mathcal{J}_t$, we obtain that A^π is invertible in $\operatorname{alg}^\pi(S, PC, L_N^{p(\cdot)}(\Gamma, w))$. Thus, $\operatorname{alg}^\pi(S, PC, L_N^{p(\cdot)}(\Gamma, w))$ is inverse closed in $\mathcal{B}^\pi$. □

For constant p, Khvedelidze weights, and piecewise Lyapunov curves, Theorem 6.4 was obtained in [13] by a different method. For constant p, arbitrary Muckenhoupt weights, and arbitrary Carleson curves, an analogue of this result is contained in [4, Section 8.5]. In the case of variable exponent $p : \Gamma \to (1, \infty)$ and Khvedelidze weights, an earlier version of this theorem for Lyapunov curves or Radon curves without cusps is in [23, Theorem 5.4] and for Carleson curves satisfying (1.1) is in [22, Theorem 5.1].

6.6. Final remarks

Remark 6.5 (On index formulas). We do not consider formulas for the index of an operator in the algebra $\operatorname{alg}(S, PC, L_N^{p(\cdot)}(\Gamma, w))$ since the approach to the study of Banach algebras of SIOs based on the Allan-Douglas local principle and the two projections theorem does not allow us to get formulas for the index of an arbitrary operator in the Banach algebra of SIOs with piecewise continuous coefficients. These formulas can be obtained similarly to the classical situation considered by

Gohberg and Krupnik [13] (see also [4, Chap. 10]). For reflexive Orlicz spaces over Carleson curves with logarithmic whirl points this was done by the author [19]. In the case of variable Lebesgue spaces with radial oscillating weights weights over Carleson curves with logarithmic whirl points, the index formulas are almost the same as in [19]. It is only necessary to replace the Boyd indices α_M and β_M of an Orlicz space L^M by the numbers $1/p(t_k) + m(w_k)$ and $1/p(t_k) + M(w_k)$ (at the nodes t_k of the weight), respectively, or both Boyd indices by $1/p(t)$ (at all other points t) in corresponding index formulas.

Remark 6.6 (On semi-Fredholm operators). Since one has Theorem 4.7 at hands, by using ideas of Spitkovsky [51], one can prove that if $a, b \in PC_{N\times N}(\Gamma)$, then $aP + bQ$ is semi-Fredholm if and only if it is Fredholm. With the help of the linear dilation procedure (see, e.g., [13]) this fact can be extended to the operators of the form $\sum_i \prod_j (a_{ij}P + b_{ij}Q)$ with $a_{ij}, b_{ij} \in PC_{N\times N}(\Gamma)$. Since the property of an operator to be Fredholm or semi-Fredholm is stable under small perturbations, we finally arrive at the following result. Its proof, given for Khvedelidze weights in [24], works also for weights considered in this paper.

Theorem 6.7. *Suppose that all the hypotheses of Theorem* 6.4 *are fulfilled. If an operator in the algebra* $\mathrm{alg}(S, PC; L^{p(\cdot)}_N(\Gamma, w))$ *is semi-Fredholm, then it is Fredholm.*

Remark 6.8 (On coefficients beyond $PC(\Gamma)$). The approach of this paper is based on the Wiener-Hopf factorization, local principles, and the two-projection theorem. Note that there is also another approach to study SIOs on standard Lebesgue spaces $L^p(\Gamma, w)$ with slowly oscillating (i.e., not arbitrary!) Muckenhoupt weights over slowly oscillating (i.e., not all!) Carleson curves. It is based based on the technique of pseudodifferential operators and limit operators developed by Rabinovich and his coauthors (see, e.g., [42] and [6, 7]). This approach allows one to study SIOs not only with piecewise continuous coefficients, but also with slowly oscillating coefficients. Notice that a satisfactory theory of pseudodifferential operators in the setting of variable Lebesgue spaces did not exist a couple of years ago. Very recently Rabinovich and S. Samko [44] have started to develop such a theory, however they do not consider in that paper SIOs on curves.

Remark 6.9 (On more general variable exponents). It is well known that the boundedness of S is guaranteed as soon as a suitable maximal function is bounded on $L^{p(\cdot)}(\Gamma, w)$. Lerner [34], among other things, observed that if $p(x) = \alpha + \sin(\log\log(1/|x|)\chi_E(x))$, where $\alpha > 2$ is some constant and χ_E is the characteristic function of the ball $E := \{x \in \mathbb{R}^n : |x| \le 1/e\}$, then the Hardy-Littlewood maximal function is bounded on $L^{p(\cdot)}(\mathbb{R}^n)$. Clearly, the exponent p in this example is discontinuous at the origin, so it does not satisfy (an $\mathbb{R}^n$ analog of) the condition (1.4). This exponent belongs to the class of pointwise multipliers for BMO (the space of functions of bounded mean oscillation). Kapanadze and Kopaliani [16] proved that if p belongs to $VMO^{1/|\log|}$, a weighted space of functions with vanishing mean oscillation, then the Hardy-Littlewood maximal function is bounded on variable Lebesgue space $L^{p(\cdot)}(\Omega)$ over a bounded domain Ω. Thus, I believe that necessary and sufficient conditions for the boundedness of the Cauchy singular integral operator (and other singular integrals and maximal functions) on weighted variable Lebesgue spaces $L^{p(\cdot)}(\Gamma, w)$ should be formulated in terms of integral means of the exponent p (i.e., in BMO terms), but not in pointwise terms like (1.4). It is natural that the number $1/p(t)$ in (4.5) under eventual more general boundedness hypotheses for S should be replaced by a pair of indices of a suitable submultiplicative function (see, e.g., [21, Section 4.4]). I believe that, in general, these indices may be different and the following is true.

Conjecture 6.10. *Suppose $\mathbb{T}$ is the unit circle and $a : \mathbb{T} \to \mathbb{C}$ be a continuous on $\mathbb{T} \setminus \{1\}$ function such that $a(1-0) = 0$ and $a(1+0) = 1$. There is a measurable function $p : \Gamma \to (1, \infty)$ such that the operator S is bounded on $L^{p(\cdot)}(\mathbb{T})$ and the essential spectrum*

$$\sigma_{\mathrm{ess}}(aP + Q) := \{\lambda \in \mathbb{C} : (aP + Q) - \lambda I \text{ is not Fredholm on } L^{p(\cdot)}(\mathbb{T})\}$$

of the operator $aP + Q$ is massive, that is, it has a nonzero plane measure.

If $p : \mathbb{T} \to (1, \infty)$ is a measurable function such that

$$1 < \operatorname*{ess\,inf}_{\tau \in \mathbb{T}} p(\tau), \quad \operatorname*{ess\,sup}_{\tau \in \mathbb{T}} p(\tau) < \infty \tag{6.10}$$

and S is bounded on $L^{p(\cdot)}(\mathbb{T})$, then from [21, Theorem 7.3] it follows that

$$a(\mathbb{T} \setminus \{1\}) \cup \mathcal{S}(0, 1; 0; \alpha, \beta) \subset \sigma_{\mathrm{ess}}(aP + Q)$$

where α and β are the lower and upper indices of the following submultiplicative function

$$Q(x) := \limsup_{R \to 0} \frac{\|\chi_{\Delta(xR)}\|_{p(\cdot)} \|\chi_{\Delta(R)}\|_{q(\cdot)}}{|\Delta(R)|}, \quad x \in (0, \infty),$$

where $\Delta(R) := \Gamma(1, R) \setminus \Gamma(1, R/2)$. Notice that if p satisfies (1.4), then $\alpha = \beta = 1/p(1)$ by [21, Lemma 5.8]. If $0 < \alpha < \beta < 1$, then the horn $\mathcal{S}(0, 1; 0; \alpha, \beta)$ has a nonzero plane measure. Thus, to confirm Conjecture 6.10, it is sufficient to construct an exponent $p : \mathbb{T} \to (1, \infty)$ such that (6.10) holds, S is bounded on $L^{p(\cdot)}(\mathbb{T})$, and $0 < \alpha < \beta < 1$.

Remark 6.11 (On arbitrary Carleson curves). Very recently the author [25] has obtained sufficient conditions for the boundedness of the Cauchy singular integral operator S on variable Lebesgue spaces $L^{p(\cdot)}(\Gamma, w)$ with special weights $w(\tau) = |(\tau - t)^\gamma|$, where γ is a complex number, over arbitrary Carleson Jordan curves. This allows him to obtain analogues of Theorems 4.5 and 6.4 for the case of nonweighted variable Lebesgue spaces $L^{p(\cdot)}(\Gamma)$ over arbitrary Carleson curves (that is, without condition (1.1)). As in the present paper, local spectra of singular integral operators can be massive, but the reason for that is different. In the present paper this effect is due to oscillation of weights, but in the forthcoming paper [25] this effect is due to oscillation of arbitrary Carleson curves.

References

[1] V.D. Aslanov and Yu.I. Karlovich, *One-sided invertibility of functional operators in reflexive Orlicz spaces.* Akad. Nauk Azerbaidzhan. SSR Dokl. **45** (1989), no. 11–12, 3–7 (in Russian).

[2] C. Bennett and R. Sharpley, *Interpolation of Operators.* Academic Press, Boston, 1988.

[3] A. Böttcher and Yu.I. Karlovich, *Toeplitz and singular integral operators on Carleson curves with logarithmic whirl points.* Integral Equations Operator Theory **22** (1995), 127–161.

[4] A. Böttcher and Yu.I. Karlovich, *Carleson Curves, Muckenhoupt Weights, and Toeplitz Operators.* Birkhäuser, Basel, 1997.

[5] A. Böttcher and Yu.I. Karlovich, *Cauchy's singular integral operator and its beautiful spectrum.* In: "Systems, approximation, singular integral operators, and related topics" (Bordeaux, 2000). Operator Theory: Advances and Applications **129** (2001), 109–142.

[6] A. Böttcher, Yu.I. Karlovich, and V.S. Rabinovich, *Emergence, persistence, and disappearance of logarithmic spirals in the spectra of singular integral operators.* Integral Equations Operator Theory **25** (1996), 406–444.

[7] A. Böttcher, Yu.I. Karlovich, and V.S. Rabinovich, *The method of limit operators for one-dimensional singular integrals with slowly oscillating data.* J. Operator Theory **43** (2000), 171–198.

[8] A. Böttcher and B. Silbermann, *Analysis of Toeplitz Operators.* 2nd edition. Springer-Verlag, Berlin, 2006.

[9] D.W. Boyd, *Indices for the Orlicz spaces.* Pacific J. Math. **38** (1971), 315–323.

[10] K.P. Clancey and I. Gohberg, *Factorization of Matrix Functions and Singular Integral Operators.* Operator Theory: Advances and Applications **3**. Birkhäuser, Basel, 1981.

[11] G. David, *Opérateurs intégraux singuliers sur certaines courbes du plan complexe.* Ann. Sci. École Norm. Super. **17** (1984), 157–189.

[12] T. Finck, S. Roch, and B. Silbermann, *Two projections theorems and symbol calculus for operators with massive local spectra.* Math. Nachr. **162** (1993), 167–185.

[13] I. Gohberg and N. Krupnik, *Singular integral operators with piecewise continuous coefficients and their symbols.* Math. USSR Izvestiya **5** (1971), 955–979.

[14] I. Gohberg and N. Krupnik, *One-Dimensional Linear Singular Integral Equations.* Vols. 1 and 2. Operator Theory: Advances and Applications **53–54**. Birkhäuser, Basel, 1992.

[15] I. Gohberg and N. Krupnik, *Extension theorems for Fredholm and invertibility symbols.* Integral Equations Operator Theory **16** (1993), 514–529.

[16] E. Kapanadze and T. Kopaliani, *A note on maximal operator on $L^{p(t)}(\Omega)$ spaces.* Georgian Math. J. **15** (2008), 307–316.

[17] A.Yu. Karlovich, *Algebras of singular integral operators with piecewise continuous coefficients on reflexive Orlicz spaces.* Math. Nachr. **179** (1996), 187–222.

[18] A.Yu. Karlovich, *Singular integral operators with piecewise continuous coefficients in reflexive rearrangement-invariant spaces.* Integral Equations Operator Theory **32** (1998), 436–481.

[19] A.Yu. Karlovich, *The index of singular integral operators in reflexive Orlicz spaces.* Math. Notes **64** (1998), 330–341.

[20] A.Yu. Karlovich, *Algebras of singular integral operators with PC coefficients in rearrangement-invariant spaces with Muckenhoupt weights.* J. Operator Theory **47** (2002), 303–323.

[21] A.Yu. Karlovich, *Fredholmness of singular integral operators with piecewise continuous coefficients on weighted Banach function spaces.* J. Integr. Equat. Appl. **15** (2003), 263–320.

[22] A.Yu. Karlovich, *Algebras of singular integral operators on Nakano spaces with Khvedelidze weights over Carleson curves with logarithmic whirl points.* Ivzestija VUZov. Severo-Kavkazskii region. Estestvennie nauki, Special issue "Pseudodifferential Equations and Some Problems of Mathematical Physics", dedicated to the 70th birthday of Prof. I.B. Simonenko, Rostov-on-Don, 2005, 135–142. Preprint is available at arXiv:math/0507312.

[23] A.Yu. Karlovich, *Algebras of singular integral operators with piecewise continuous coefficients on weighted Nakano spaces.* In: "The Extended Field of Operator Theory". Operator Theory: Advances and Applications **171** (2006), 171–188.

[24] A.Yu. Karlovich, *Semi-Fredholm singular integral operators with piecewise continuous coefficients on weighted variable Lebesgue spaces are Fredholm.* Operators and Matrices **1** (2007), 427–444.

[25] A.Yu. Karlovich, *Singular integral operators on variable Lebesgue spaces over arbitrary Carleson curves.* In: "Topics in Operator Theory". Operator Theory: Advances and Applications **202**, to appear. Preprint, arXiv:0810.3110, 2008.

[26] B.V. Khvedelidze, *Linear discontinuous boundary problems in the theory of functions, singular integral equations and some of their applications.* Trudy Tbiliss. Mat. Inst. Razmadze **23** (1956), 3–158 (in Russian).

[27] V. Kokilashvili, V. Paatashvili, and S. Samko, *Boundary value problems for analytic functions in the class of Cauchy type integrals with density in* $L^{p(\cdot)}(\Gamma)$. Boundary Value Problems **1** (2005), 43–71.

[28] V. Kokilashvili, V. Paatashvili, and S. Samko, *Boundedness in Lebesgue spaces with variable exponent of the Cauchy singular operator on Carleson curves.* In: "Modern Operator Theory and Applications. The Igor Borisovich Simonenko Anniversary Volume". Operator Theory: Advances and Applications **170** (2006), 167–186.

[29] V. Kokilashvili, N. Samko, and S. Samko, *Singular operators in variable spaces* $L^{p(\cdot)}(\Omega, \rho)$ *with oscillating weights.* Math. Nachr. **280** (2007), 1145–1156.

[30] V. Kokilashvili and S. Samko, *Singular integral equations in the Lebesgue spaces with variable exponent.* Proc. A. Razmadze Math. Inst. **131** (2003), 61–78.

[31] O. Kováčik and J. Rákosník, *On spaces* $L^{p(x)}$ *and* $W^{k,p(x)}$. Czechoslovak Math. J. **41(116)** (1991), 592–618.

[32] S.G. Krein, Ju.I. Petunin, and E.M. Semenov, *Interpolation of Linear Operators.* AMS Translations of Mathematical Monographs **54**, Providence, RI, 1982.

[33] N. Krupnik, *Banach Algebras with Symbol and Singular Integral Operators.* Operator Theory: Advances and Applications **26**. Birkhäuser, Basel, 1987.

[34] A.K. Lerner, *Some remarks on the Hardy-Littlewood maximal function on variable L^p spaces.* Math. Z. **251** (2005), 509–521.

[35] G.S. Litvinchuk and I.M. Spitkovsky, *Factorization of Measurable Matrix Functions.* Operator Theory: Advances and Applications **25**. Birkhäuser, Basel, 1987.

[36] L. Maligranda, *Indices and interpolation.* Dissert. Math. **234** (1985), 1–49.

[37] L. Maligranda, *Orlicz Spaces and Interpolation.* Sem. Math. 5, Dep. Mat., Universidade Estadual de Campinas, Campinas SP, Brazil, 1989.

[38] W. Matuszewska and W. Orlicz, *On certain properties of φ-functions.* Bull. Acad. Polon. Sci. Ser. Sci. Math. Astronom. Phys. **8** (1960), 439–443. Reprinted in: W. Orlicz, *Collected Papers*, PWN, Warszawa, 1988, 1112–1116.

[39] W. Matuszewska and W. Orlicz, *On some classes of functions with regard to their orders of growth.* Studia Math. **26** (1965), 11–24. Reprinted in: W. Orlicz, *Collected Papers*, PWN, Warszawa, 1988, 1217–1230.

[40] J. Musielak, *Orlicz Spaces and Modular Spaces.* Lecture Notes in Mathematics **1034**. Springer-Verlag, Berlin, 1983.

[41] H. Nakano, *Modulared Semi-Ordered Linear Spaces.* Maruzen Co., Ltd., Tokyo, 1950.

[42] V.S. Rabinovich, *Algebras of singular integral operators on composed contours with nodes that are logarithmic whirl points.* Izv. Math. **60** (1996), 1261–1292.

[43] V. Rabinovich, N. Samko, and S. Samko, *Local Fredholm spectra and Fredholm properties of singular integral operators on Carleson curves acting on weighted Hölder spaces.* Integral Equations Operator Theory **56** (2006), 257–283.

[44] V. Rabinovich and S. Samko, *Boundedness and Fredholmness of pseudodifferential operators in variable exponent spaces.* Integral Equations Operator Theory **60** (2008), 507–537.

[45] S. Roch and B. Silbermann, *Algebras of Convolution Operators and Their Image in the Calkin Algebra.* Report R-Math-05/90, Karl Weierstrass Inst. F. Math., Berlin, 1990.

[46] N. Samko, *Singular integral operators in weighted spaces of continuous functions with oscillating continuity moduli and oscillating weights.* In: "The Extended Field of Operator Theory". Operator Theory: Advances and Applications **171** (2006), 323–347.

[47] I.B. Simonenko, *The Riemann boundary value problem for n pairs functions with measurable coefficients and its application to the investigation of singular integral operators in the spaces L^p with weight.* Izv. Akad. Nauk SSSR, Ser. Matem. **28** (1964), 277–306 (in Russian).

[48] I.B. Simonenko, *A new general method of investigating linear operator equations of singular integral equations type.* Part I: Izv. Akad. Nauk SSSR, Ser. Matem. **29** (1965), 567–586 (in Russian); Part II: Izv. Akad. Nauk SSSR, Ser. Matem. **29** (1965), 757–782 (in Russian).

[49] I.B. Simonenko, *Some general questions in the theory of the Riemann boundary value problem.* Math. USSR Izvestiya **2** (1968), 1091–1099.

[50] I.B. Simonenko and Chin Ngok Min, *Local Method in the Theory of One-Dimensional Singular Integral Equations with Piecewise Continuous Coefficients. Noetherity.* Rostov-on-Don State Univ., Rostov-on-Don, 1986 (Russian).

[51] I.M. Spitkovsky, *Singular integral operators with PC symbols on the spaces with general weights.* J. Funct. Anal. **105** (1992), 129–143.

[52] H. Widom, *Singular integral equations on L^p.* Trans. Amer. Math. Soc. **97** (1960), 131–160.

Alexei Yu. Karlovich
Departamento de Matemática
Faculdade de Ciências e Tecnologia
Universidade Nova de Lisboa
Quinta da Torre
2829–516 Caparica, Portugal
e-mail: oyk@fct.unl.pt

Operator Theory:
Advances and Applications, Vol. 195, 213–230

Reverse Martingales in Riesz Spaces

Mareli Korostenski, Coenraad C.A. Labuschagne
and Bruce A. Watson

Abstract. Reverse martingales in the measure-free setting of Dedekind complete Riesz spaces are considered. The space of such reverse martingales is shown to be a Dedekind complete Riesz space with respect to its natural ordering. By means of a reverse upcrossing theorem, we characterize the convergent reverse (sub, super) martingales in Dedekind complete Riesz spaces with weak order units.

Mathematics Subject Classification (2000). Primary 06F20: Secondary 46A40, 47B60, 60G48, 60G42.

Keywords. Reverse filtration, reverse martingales, Riesz space

1. Introduction

Reverse martingales play an important role in stochastic processes, in particular when considering Brownian motion, see for example [14, p. 19], and laws of large numbers, see [15, p. 150]. Various aspects of stochastic processes have been considered in the measure-free setting of Riesz spaces in [2, 7, 8, 9, 10, 16, 17, 18, 20] where optional stopping theorems, sub and super martingale convergence and ergodicity were considered. More recently amarts were considered in this setting, see [11].

Here we develop decomposition, upcrossing and convergence theorems for reverse martingales in this measure-free Riesz spaces setting. In particular the reverse martingales are shown to form a Riesz space homeomorphically Riesz-isomorphic to a Dedekind complete Riesz subspaces of the original space. Finally the convergent reverse martingales are characterized.

In Section 2 some basic Riesz space theory is given and the definition of a conditional expectation on a Dedekind complete Riesz space with weak order unit is recalled from [7] for the readers convenience – justification for this definition

BAW supported in part by the Centre for Applicable Analysis and Number Theory and by the NRF grant number FA2007041200006. CCAL supported in part by the NRF.

along with some of the resulting consequences can be found in [7, 10], with which we assume the reader to be familiar. Reverse filtrations, reverse adapted sequences and martingales on Riesz spaces with a weak order unit are defined as well as the space of reverse adapted sequences and, its subspace, the reverse martingales.

Analogues of the Krickeberg and Riesz decompositions are derived for reverse martingales in Section 3, see [4, 5, 6], for classical versions of these decompositions, and [10], for the martingale version in Riesz spaces. It is shown that there is an order continuous Riesz isomorphism with order continuous inverse between the Riesz space of reverse martingales and a Dedekind complete Riesz subspace (with infinite suprema in the subspace coinciding with those in the larger space) of the original Riesz space.

In Section 4 we give a Riesz space upcrossing theorem for reverse martingales in a Riesz spaces with weak order unit, see [8] for the analogous martingale upcrossing theorem in Riesz spaces.

By means of the reverse martingale upcrossing theorem of Section 4, in Section 5 we are able to give reverse martingale convergence theorems in the Riesz space setting. In particular if the reverse martingale is bounded, it is shown to be convergent. If suitable universal completeness conditions are placed on the Riesz space, it is shown that all reverse martingales are convergent and hence bounded, see [10] for the martingale analogue in Riesz spaces. Thus when the bounded reverse martingales and the convergent reverse martingales coincide is characterized.

2. Preliminaries

We assume the reader to be familiar with the basic terminology of Riesz spaces as can be found in [21]. However, we recall from [21, p. 17–18] the following notation. If E is a Riesz space and $x, y \in E$, then we denote by $x \vee y$ and $x \wedge y$ respectively the supremum and infimum of x and y. In terms of these operations, the positive part of x is $x^+ = x \vee 0$, the negative part of x is $x^- = (-x) \vee 0$ and $|x| := x^+ + x^-$ is the absolute value of x. Consistent with this we have $x \vee y = (x-y)^+ + y$ and $x \wedge y = x + y - x \vee y$.

If E is a Riesz space, then $e \in E_+$ is a weak order unit for E, if $f \wedge ne \uparrow_{n \in \mathbb{N}} f$ for all $f \in E_+$, where E_+ denotes the positive cone of E, see [21] for more details.

The concepts of conditional expectations, filtrations (stochastic bases), adapted sequences, (sub and super) martingales, stopping times and stopped processes were considered in Riesz spaces with weak order units in [7, 8, 9, 10]. We recall for the reader's convenience the following definitions from [7, 9].

Definition 2.1. *Let E be a Riesz space with a weak order unit e.*

(a) *A conditional expectation operator, T, on E, is a positive linear order continuous projection on E which maps weak order units to weak order units and has $\mathcal{R}(T)$ as a Dedekind complete Riesz subspace of E (by this we mean that $\mathcal{R}(T)$ is a Riesz subspace of E, and that it is Dedekind complete with* *(infinite) suprema in $\mathcal{R}(T)$ of subsets of $\mathcal{R}(T)$ coinciding with their suprema in E).*

(b) *A filtration (or stochastic basis) on E is a sequence of conditional expectations, $(T_i)_{i\in\mathbb{N}}$ on E with $T_iT_j = T_i = T_jT_i$ for all $i \le j$, $i, j \in \mathbb{N}$.*

(c) *Adapted sequence, $(f_i, T_i)_{i\in\mathbb{N}}$, on E is a sequence of pairs, such that $(T_i)_{i\in\mathbb{N}}$ is a filtration on E and $f_i \in \mathcal{R}(T_i)$ for each $i \in \mathbb{N}$.*

(d) *A (sub, super) martingale, $(f_i, T_i)_{i\in\mathbb{N}}$, on E is an adapted sequence on E, with $f_i(\le, \ge) = T_if_j$ for all $i \le j$, $i, j \in \mathbb{N}$.*

It is easily seen, since a conditional expectation operator is a projection and maps weak order units to weak order units, that if $(T_i)_{i\in\mathbb{N}}$ is a filtration and w is a weak order unit, then $e := T_1w$ is a weak order unit invariant under $T_i, i \in \mathbb{N}$.

Replacing the index set $\mathbb{N}$ by $-\mathbb{N}$, the negative integers, in (b), (c) and (d) of Definition 2.1, we arrive at the corresponding reverse objects, see [4, pages 26 and 213] and [13, pages 115–119], as below.

Definition 2.2. *Let E be a Riesz space with a weak order unit e.*

(a) *A reverse filtration on E is a sequence of conditional expectations, $(T_i)_{i\in-\mathbb{N}}$ on E with $T_iT_j = T_i = T_jT_i$ for all $i \le j$, $i, j \in -\mathbb{N}$, and $e = T_ie, i \in -\mathbb{N}$, for some weak order unit e.*

(b) *Reverse adapted sequence, $(f_i, T_i)_{i\in-\mathbb{N}}$, on E is a sequence of pairs, such that $(T_i)_{i\in-\mathbb{N}}$ is a filtration on E and $f_i \in \mathcal{R}(T_i)$ for each $i \in -\mathbb{N}$.*

(c) *A reverse (sub, super) martingale, $(f_i, T_i)_{i\in-\mathbb{N}}$, on E is a reverse adapted sequence on E, with $f_i(\le, \ge) = T_if_j$ for all $i \le j$, $i, j \in -\mathbb{N}$.*

We remind the reader of the connection between the Riesz space theory of stochastic processes and the classical $\mathcal{L}^1(\Omega, \Sigma, \mu)$ theory of stochastic processes, where μ is a probability measure and Σ is a σ-algebra of subsets of Ω. If one takes E to be the Riesz space $\mathcal{L}^1(\Omega, \Sigma, \mu)$ and $e = 1$, the constant 1 function, then the Riesz space conditional expectation operators leaving e invariant and classical conditional expectation operators coincide as do the classical and Riesz space sub and super martingales, see [9] for details. It is easily seen that the same is true for reverse sub and super martingales.

It is useful in the sequel to have notations for the reverse adapted sequences and reverse martingales relative to a given reverse filtration (stochastic basis).

Definition 2.3. *Let E be a Riesz space with a weak order unit, e, and reverse filtration $(T_i)_{i\in-\mathbb{N}}$ with $e = T_ie$ for all $i \in -\mathbb{N}$.*

We denote by $\mathcal{RA}(E, T_i)$ the set of all reverse adapted sequences $(f_i, T_i)_{i\in-\mathbb{N}}$ in E and by $\mathcal{RM}(E, T_i)$ the set of all reverse martingales $(f_i, T_i)_{i\in-\mathbb{N}}$ in E.

Here $\mathcal{RA}(E, T_i)$ is an ordered vector space if we define the vector space and order operation componentwise, i.e., for $(f_i, T_i)_{i\in-\mathbb{N}}, (g_i, T_i)_{i\in-\mathbb{N}} \in \mathcal{RA}(E, T_i)$

$$\begin{aligned}
(f_i, T_i)_{i\in-\mathbb{N}} + (g_i, T_i)_{i\in-\mathbb{N}} &= (f_i + g_i, T_i)_{i\in-\mathbb{N}}, \\
\lambda(f_i, T_i)_{i\in-\mathbb{N}} &= (\lambda f_i, T_i)_{i\in-\mathbb{N}}, \quad \text{for each } \lambda \in \mathbb{R}, \\
(f_i, T_i)_{i\in-\mathbb{N}} \le (g_i, T_i)_{i\in-\mathbb{N}} &\iff f_i \le g_i, \quad \text{for all } i \in -\mathbb{N}.
\end{aligned}$$

Since each $\mathcal{R}(T_i), i \in -\mathbb{N}$, is a Dedekind complete Riesz space it follows easily that $\mathcal{R}A(E, T_i)$ is a Dedekind complete Riesz space with weak order unit $(e, T_i)_{i\in-\mathbb{N}}$. Clearly $\mathcal{R}M(E, T_i)$ is an ordered vector subspace of $\mathcal{R}A(E, T_i)$. It is also a Riesz space, as will be shown later, but is not in general a Riesz subspace of $\mathcal{R}A(E, T_i)$.

If $(f_i, T_i)_{i\in-\mathbb{N}}$ is a reverse martingale, then $T_i f_{-1} = f_i$ for all $i \in -\mathbb{N}$. Hence

$$(f_i, T_i)_{i\in-\mathbb{N}} = (T_i f_{-1}, T_i)_{i\in-\mathbb{N}}$$

and

$$\mathcal{R}M(E, T_i) = \{(T_i f, T_i)_{i\in-\mathbb{N}} \mid f \in \mathcal{R}(T_{-1})\}. \tag{2.1}$$

3. Riesz and Krickeberg decompositions

In this section we show that $\mathcal{R}M(E, T_i)$ is a Dedekind complete Riesz space under the ordering inherited as a subset of $\mathcal{R}A(E, T_i)$, it is however not a Riesz subspace. In addition to this it is homeomorphically Riesz isomorphic to $\mathcal{R}(T_{-1})$. This is closely linked to the Krickeberg and Riesz decompositions of sub and super martingales which will also be given here.

Theorem 3.1. Krickeberg decomposition
Let E be a Dedekind complete Riesz space with a weak order unit, e, and $(T_i)_{i\in-\mathbb{N}}$ a reverse filtration on E with $T_i e = e$ for all $i \in -\mathbb{N}$.

If $(f_i, T_i)_{i\in-\mathbb{N}}$ is a reverse sub martingale in E, then

$$(g_i, T_i)_{i\in-\mathbb{N}} := (T_i f_{-1}^+, T_i)_{i\in-\mathbb{N}} \tag{3.1}$$

is the minimal positive reverse martingale above $(f_i, T_i)_{i\in-\mathbb{N}}$ and

$$(h_i, T_i)_{i\in-\mathbb{N}} := (T_i f_{-1}^+ - f_i, T_i)_{i\in-\mathbb{N}} \tag{3.2}$$

is a positive reverse super martingale such that

$$(f_i, T_i)_{i\in-\mathbb{N}} = (g_i, T_i)_{i\in-\mathbb{N}} - (h_i, T_i)_{i\in-\mathbb{N}}. \tag{3.3}$$

Proof. As $(T_i)_{i\in-\mathbb{N}}$ is a reverse filtration, $(g_i, T_i)_{i\in-\mathbb{N}}$, as given in (3.1), is a positive reverse martingale. Since $(f_i, T_i)_{i\in-\mathbb{N}}$ is a reverse sub martingale,

$$f_i \le T_i f_{-1} \le T_i f_{-1}^+ = g_i, \quad \text{for all} \quad i \in -\mathbb{N},$$

giving that $(g_i, T_i)_{i\in-\mathbb{N}}$ is above $(f_i, T_i)_{i\in-\mathbb{N}}$. Observe that, since $(f_i, T_i)_{i\in-\mathbb{N}}$ is a reverse sub martingale,

$$h_i = g_i - f_i \ge g_i - T_i f_{-1} \ge g_i - T_i f_{-1}^+ = 0, \quad \text{for all} \quad i \in -\mathbb{N},$$

and

$$T_j h_i = T_j g_i - T_j f_i \le g_j - f_j = h_j, \quad \text{for} \quad j \le i, i, j \in -\mathbb{N},$$

making $(h_i, T_i)_{i\in-\mathbb{N}}$ a positive super martingale. Equation (3.3) holds because of (3.2).

We now consider the minimality. Let $(p_i, T_i)_{i\in -\mathbb{N}}$ be any positive reverse martingale above $(f_i, T_i)_{i\in -\mathbb{N}}$, then $p_{-1} \geq 0$ and $p_{-1} \geq f_{-1}$, giving $p_{-1} \geq f_{-1}^+ = g_{-1}$. Now

$$p_i = T_i p_{-1} \geq T_i g_{-1} = g_i, \quad i \in -\mathbb{N},$$

hence proving the minimality of $(g_i, T_i)_{i\in -\mathbb{N}}$. □

The following composition theorem shows that the reverse martingales not only form a ordered vector spaces, but also a Riesz space. It is easily seen that, other than in some trivial instances, the reverse martingales do not form a Riesz subspace of the reverse adapted sequences.

By the positive cone of $\mathcal{RM}(E, T_i)$ we mean

$$\mathcal{RM}^+(E, T_i) := \{(f_i, T_i)_{i\in -\mathbb{N}} \in \mathcal{RM}(E, T_i) \mid f_i \geq 0 \quad \forall i \in -\mathbb{N}\}.$$

Theorem 3.2. Riesz decomposition
Let E be a Dedekind complete Riesz space with a weak order unit, e, and $(T_i)_{i\in -\mathbb{N}}$ a reverse filtration on E with $T_i e = e$ for all $i \in -\mathbb{N}$, then $\mathcal{RM}(E, T_i)$ is a Dedekind complete Riesz space with weak order unit $(e, T_i)_{i\in -\mathbb{N}}$. In particular, if $(f_i, T_i)_{i\in -\mathbb{N}} \in \mathcal{RM}(E, T_i)$, then

$$(f_i, T_i)^+_{i\in -\mathbb{N}} = (g_i, T_i)_{i\in -\mathbb{N}}, \tag{3.4}$$

$$(f_i, T_i)^-_{i\in -\mathbb{N}} = (h_i, T_i)_{i\in -\mathbb{N}}, \tag{3.5}$$

where $(g_i, T_i)_{i\in -\mathbb{N}}$ and $(h_i, T_i)_{i\in -\mathbb{N}}$ are as given in Theorem 3.1.

Proof. As already noted $\mathcal{RM}(E, T_i)$ is an ordered vector space, so in order to show that it is a Riesz spaces, we need only show the existence of $(f_i, T_i)^+_{i\in -\mathbb{N}}$ and $(f_i, T_i)^-_{i\in -\mathbb{N}}$ for each $(f_i, T_i)_{i\in -\mathbb{N}} \in \mathcal{RM}(E, T_i)$. But, from Theorem 3.1, $(g_i, T_i)_{i\in -\mathbb{N}}$ is the minimal positive reverse martingale above $(f_i, T_i)_{i\in -\mathbb{N}}$, making $(f_i, T_i)^+_{i\in -\mathbb{N}} = (g_i, T_i)_{i\in -\mathbb{N}}$. Since $(f_i, T_i)_{i\in -\mathbb{N}}$ is a reverse martingale $(h_i, T_i)_{i\in -\mathbb{N}}$ of Theorem 3.1 is a positive reverse martingale and since $h_{-1} = g_{-1} - f_{-1} = f_{-1}^-$, it follows that $(h_i, T_i)_{i\in -\mathbb{N}}$ is the minimal positive reverse martingale above $-(f_i, T_i)_{i\in -\mathbb{N}}$ giving $(f_i, T_i)^-_{i\in -\mathbb{N}} = (h_i, T_i)_{i\in -\mathbb{N}}$. From Theorem 3.1,

$$(f_i, T_i)^+_{i\in -\mathbb{N}} - (f_i, T_i)^-_{i\in -\mathbb{N}} = (g_i, T_i)_{i\in -\mathbb{N}} - (h_i, T_i)_{i\in -\mathbb{N}} = (f_i, T_i)_{i\in -\mathbb{N}},$$

and $\mathcal{RM}(E, T_i)$ is a Riesz space.

We now prove that $\mathcal{RM}(E, T_i)$ is Dedekind complete. Let $((f_i^\alpha, T_i)_{i\in -\mathbb{N}})_\alpha$ be an increasing net (in α) of elements from $\mathcal{RM}^+(E, T_i)$ bounded above by $(f_i, T_i)_{i\in -\mathbb{N}} \in \mathcal{RM}^+(E, T_i)$. From the Dedekind completeness of $\mathcal{R}(T_i)$ we have that there exists $g_i \in \mathcal{R}(T_i)$ with $f_i^\alpha \uparrow_\alpha g_i \leq f_i$ for each $i \in -\mathbb{N}$. By the order continuity of T_j, for $j \leq i$ we have

$$T_j g_i \leftarrow_\alpha T_j f_i^\alpha = f_j^\alpha \uparrow_\alpha g_j,$$

showing that $(g_i, T_i)_{i\in -\mathbb{N}}$ is a reverse martingale. Hence $\mathcal{RM}(E, T_i)$ is Dedekind complete.

If $(f_i, T_i)_{i\in-\mathbb{N}} \in \mathcal{R}M^+(E, T_i)$ then

$$\begin{aligned}(f_i, T_i)_{i\in-\mathbb{N}} \wedge n(e, T_i)_{i\in-\mathbb{N}} &= (T_i f_{-1}, T_i)_{i\in-\mathbb{N}} \wedge (ne, T_i)_{i\in-\mathbb{N}} \\ &= (T_i(f_{-1} \wedge ne), T_i)_{i\in-\mathbb{N}}.\end{aligned}$$

Since $f_{-1} \wedge ne \uparrow_n f_{-1}$ and each T_i is order continuous, it follows that

$$(f_i, T_i)_{i\in-\mathbb{N}} \wedge n(e, T_i) \uparrow_n (f_i, T_i)_{i\in-\mathbb{N}},$$

and that (e, T_i) is a weak order unit for $\mathcal{R}M(E, T_i)$. □

From the theorem it follows directly, in $\mathcal{R}M(E, T_i)$, that

$$\begin{aligned}|(f_i, T_i)_{i\in-\mathbb{N}}| &= (T_i|f_{-1}|, T_i)_{i\in-\mathbb{N}}, \\ (f_i, T_i)_{i\in-\mathbb{N}} \vee (g_i, T_i)_{i\in-\mathbb{N}} &= (T_i(f_{-1} \vee g_{-1}), T_i)_{i\in-\mathbb{N}}, \\ (f_i, T_i)_{i\in-\mathbb{N}} \wedge (g_i, T_i)_{i\in-\mathbb{N}} &= (T_i(f_{-1} \wedge g_{-1}), T_i)_{i\in-\mathbb{N}}.\end{aligned}$$

As a consequence of Theorems 3.1 and 3.2, we get the following connection between $\mathcal{R}M(E, T_i)$ and $\mathcal{R}(T_{-1})$.

Corollary 3.3. *Let E be a Dedekind complete Riesz space with a weak order unit and $(T_i)_{i\in-\mathbb{N}}$ a reverse filtration on E. Then $J : \mathcal{R}(T_{-1}) \to \mathcal{R}M(E, T_i)$, defined by*

$$J(f) = (T_i f, T_i)_{i\in-\mathbb{N}}, \quad \text{for all} \quad f \in \mathcal{R}(T_{-1}),$$

is a continuous Riesz isomorphism with continuous inverse.

Proof. From Theorem 3.2, we get that J is a Riesz homomorphism with inverse

$$J^{-1}(f_i, T_i)_{i\in-\mathbb{N}} = f_{-1}, \quad \text{for all} \quad (f_i, T_i)_{i\in-\mathbb{N}} \in \mathcal{R}M(E, T_i).$$

That J^{-1} is order continuous is immediate, while the order continuity of J follows from each $T_i, i \in -\mathbb{N}$, being order continuous. □

4. Reverse upcrossing

Let E be a Dedekind complete Riesz space with a weak order unit and $f \in E^+$. Throughout this section P_f will denote the band projection in E onto the band generated by f. For further details on bands and band projections, we refer the reader to [21], and for commutation relations between band projections and conditional expectations we recommend [9] and [10] to the reader.

Upcrossing theorems play a fundamental role in both martingale and reverse martingale convergence theory, see for example [10, 13, 19]. We recall the following upcrossing theorem for martingales on Riesz spaces from [8, Theorem 4.3].

Theorem 4.1. Upcrossing theorem
Let (f_i, T_i) be a sub (super) martingale in a Riesz space E with weak order unit

and let $g, f \in \mathcal{R}(T_1)$ *with* $g \le f$. *Define the upcrossing yield of* (f_i, T_i), *from time* 1 *to time* N, *across the order interval* $[g, f]$ *to be*

$$Y_N(g,f) = \sum_{i=1}^{N} Q_N^i(f-g),$$

where $S_0^m = 0 = Q_0^m$, $S_n^1 = P_{\bigvee_{i=1}^n (g-f_i)^+}$ *and*

$$\begin{aligned} Q_n^m &= \sum_{j=2}^{n} P_{\bigvee_{i=j}^n (f_i-f)^+}(S_{j-1}^m - S_{j-2}^m) \\ S_n^{m+1} &= \sum_{j=2}^{n} P_{\bigvee_{i=j}^n (g-f_i)^+}(Q_{j-1}^m - Q_{j-2}^m). \end{aligned}$$

Then $T_1 Y_N(g,f) \le T_1(f_N - g)^+$ (*resp.* $T_1(f_N - f)^-$).

As a simple corollary to the above upcrossing theorem, we obtain the following upcrossing theorem for reverse sub (super) martingales.

Corollary 4.2. Reverse upcrossing theorem
Let $(f_i, T_i)_{i \in -\mathbb{N}}$ *be a reverse sub (super) martingale in a Riesz space* E *with weak order unit* $e = T_i e, i \in -\mathbb{N}$, *and let* $g, f \in \mathcal{R}(T_N)$ *with* $g \le f$. *Define the upcrossing yield of* $(f_i, T_i)_{i \in -\mathbb{N}}$, *from time* N *to* M, $N \le M \le -1$, *across the order interval* $[g, f]$ *to be*

$$Y_{N,M}(g,f) = \sum_{k=1}^{M-N+1} Q_{N,M}^k(f-g),$$

where,

$$S_{N,k}^m = 0 = Q_{N,k}^m, \quad -1 \ge N > k,$$

and for $N \le k \le -1$,

$$\begin{aligned} S_{N,k}^1 &= P_{\bigvee_{i=N}^k (g-f_i)^+}, \\ Q_{N,k}^m &= \sum_{j=N+1}^{k} P_{\bigvee_{i=j}^k (f_i-f)^+}(S_{N,j-1}^m - S_{N,j-2}^m), \\ S_{N,k}^{m+1} &= \sum_{j=N+1}^{k} P_{\bigvee_{i=j}^k (g-f_i)^+}(Q_{N,j-1}^m - Q_{N,j-2}^m). \end{aligned}$$

Then

$$T_N Y_{N,M}(g,f) \le T_N(f_M - g)^+ \ (\textit{resp. } T_N(f_M - f)^-).$$

Proof. Let $N \le -1$ be held fixed. Let $\tilde{f}_i = f_{i+N-1}$ and $\tilde{T}_i = T_{i+N-1}$ for $i = 1, \dots, -N$. Then $(\tilde{f}_i, \tilde{T}_i)_{i=1,\dots,-N}$ is a sub (super) martingale and as such Theorem 4.1 is applicable. In particular $g, f \in \mathcal{R}(\tilde{T}_1)$. Let $\tilde{S}_n^m, \tilde{Q}_n^m$ and $\tilde{Y}_n$ be S_n^m, Q_n^m

and Y_n of Theorem 4.1 as applied to $(\tilde{f}_i, \tilde{T}_i)_{i=1,\dots,-N}$. Then the upcrossing yield from N to M across the order interval $[g, f]$ is $Y_{N,M}(g, f) := \tilde{Y}_{M-N+1}(g, f)$ and

$$\tilde{T}_1\tilde{Y}_{M-N+1}(g, f) \le \tilde{T}_1(\tilde{f}_{M-N+1} - g)^+ \text{ (resp. } \tilde{T}_1(\tilde{f}_{M-N+1} - f)^-),$$

which translates to

$$T_N Y_{N,M}(g, f) \le T_N(f_M - g)^+ \text{ (resp. } T_N(f_M - f)^-).$$

Here

$$\tilde{Y}_{M-N+1}(g, f) = \sum_{i=1}^{M-N+1} \tilde{Q}^i_{M-N+1}(f - g),$$

where $\tilde{S}^m_0 = 0 = \tilde{Q}^m_0$,

$$\tilde{S}^1_n = P_{\bigvee_{i=1}^n (g-\tilde{f}_i)^+} = P_{\bigvee_{i=N}^{n+N-1}(g-f_i)^+}$$

and

$$\begin{aligned}
\tilde{Q}^m_n &= \sum_{j=2}^{n} P_{\bigvee_{i=j}^n (\tilde{f}_i - f)^+}(\tilde{S}^m_{j-1} - \tilde{S}^m_{j-2}) \\
&= \sum_{j=N+1}^{n+N-1} P_{\bigvee_{i=j}^{n+N-1}(f_i - f)^+}(\tilde{S}^m_{j-N} - \tilde{S}^m_{j-N-1}) \\
\tilde{S}^{m+1}_n &= \sum_{j=2}^{n} P_{\bigvee_{i=j}^n (g-\tilde{f}_i)^+}(\tilde{Q}^m_{j-1} - \tilde{Q}^m_{j-2}) \\
&= \sum_{j=N+1}^{n+N-1} P_{\bigvee_{i=j}^{n+N-1}(g-f_i)^+}(\tilde{Q}^m_{j-N} - \tilde{Q}^m_{j-N-1}).
\end{aligned}$$

Setting $S^m_{N,k} = \tilde{S}^m_{k-N+1}$ and $Q^m_{N,k} = \tilde{Q}^m_{k-N+1}$, $N - 1 \le k \le -1$ we obtain

$$S^m_{N,N-1} = 0 = Q^m_{N,N-1},$$

$$S^1_{N,k} = P_{\bigvee_{i=N}^k (g-f_i)^+}, \quad N \le k \le -1,$$

and

$$\begin{aligned}
Q^m_{N,k} &= \sum_{j=N+1}^{k} P_{\bigvee_{i=j}^k (f_i - f)^+}(\tilde{S}^m_{j-N} - \tilde{S}^m_{j-N-1}) \\
&= \sum_{j=N+1}^{k} P_{\bigvee_{i=j}^k (f_i - f)^+}(S^m_{N,j-1} - S^m_{N,j-2}), \\
S^{m+1}_{N,k} &= \sum_{j=N+1}^{k} P_{\bigvee_{i=j}^k (g-f_i)^+}(\tilde{Q}^m_{j-N} - \tilde{Q}^m_{j-N-1}) \\
&= \sum_{j=N+1}^{k} P_{\bigvee_{i=j}^k (g-f_i)^+}(Q^m_{N,j-1} - Q^m_{N,j-2}).
\end{aligned}$$

Finally

$$Y_{N,M}(g,f) = \sum_{i=1}^{M-N+1} \tilde{Q}^i_{M-N+1}(f-g) = \sum_{i=1}^{M-N+1} Q^i_{N,M}(f-g)$$

concluding the proof. □

The following lemma from [10] is used in the Lemma 4.4, which forms the first step towards discussing order convergence of reverse martingales in the Riesz space setting.

Lemma 4.3. *Let $m, M \in E$ with $M > m$ where E is a Dedekind complete Riesz space with weak order unit e. Then there exist $s < t$ such that*

$$(M - te)^+ \wedge (se - m)^+ > 0.$$

We briefly look at some aspects of order convergence on Riesz spaces. Let E be a Dedekind complete Riesz space and let $(f_i)_{i\in-\mathbb{N}}$ be a sequence in E which is order bounded, i.e., there exists $0 \leq g \in E$ such that $-g \leq f_i \leq g$ for all $i \in -\mathbb{N}$. If $u_n := \sup\{f_n, f_{n-1}, \dots\}, n \in -\mathbb{N}$, then u_n exists in E by the Dedekind completeness of E and the fact that $(f_i)_{i\in-\mathbb{N}}$ is order bounded. Furthermore, $(u_n)_{n\in-\mathbb{N}}$ is order bounded and hence $\inf\limits_{n\in-\mathbb{N}} u_n = \inf\limits_{n\in-\mathbb{N}} \sup\{f_n, f_{n-1}, \dots\}$ exists and will be denoted by $\lim\sup\limits_{i\to-\infty} f_i$. Similarly, if $l_n := \inf\{f_n, f_{n-1}, \dots\}$, we denote $\sup\limits_{n\in-\mathbb{N}} l_n = \sup\limits_{n\in-\mathbb{N}} \inf\{f_n, f_{n-1}, \dots\}$ by $\lim\inf\limits_{i\to-\infty} f_i$.

That both $\lim\sup\limits_{i\to-\infty} f_i$ and $\lim\inf\limits_{i\to-\infty} f_i$ exist is equivalent to requiring that $(f_i)_{i\in-\mathbb{N}}$ is order bounded. From [12, Proposition 1.1.10], $(f_i)_{i\in-\mathbb{N}}$ is order convergent if and only if $\lim\sup\limits_{i\to-\infty} f_i = \lim\inf\limits_{i\to-\infty} f_i$.

The following, intuitively obvious but technical, lemma shows that if a sequence in a Dedekind complete Riesz space with weak order unit has distinct limes supremum and limes infimum then the sequence must oscillate an infinite number of times. The sense in which the sequence has this behaviour is also made more precise in the lemma.

Lemma 4.4. *Let $(f_n)_{n\in-\mathbb{N}}$ be a sequence in a Dedekind complete Riesz space E with weak order unit e. If $q \in \mathbb{N}$ and*

$$M := \lim\sup_{j\to-\infty} ((qe \wedge f_j) \vee -qe) > \lim\inf_{j\to-\infty} ((qe \wedge f_j) \vee -qe) =: m$$

let $-q < s < t < q$ be such that

$$h := (M - te)^+ \wedge (se - m)^+ > 0.$$

Then the following limits exist and

$$P_h \leq \lim_{r\to-\infty} S^p_{r,k}, \tag{4.1}$$

$$P_h \leq \lim_{r\to-\infty} Q^p_{r,k}, \tag{4.2}$$

for $p, -k \in \mathbb{N}$, *where*

$$S^p_{N,k} = 0 = Q^p_{N,k}, \quad -1 \geq N > k,$$

and for $N \leq k \leq -1$,

$$\begin{aligned} S^1_{N,k} &= P_{\bigvee_{i=N}^k (se-f_i)^+}, \\ Q^p_{N,k} &= \sum_{j=N+1}^{k} P_{\bigvee_{i=j}^k (f_i - te)^+}(S^p_{N,j-1} - S^p_{N,j-2}), \\ S^{p+1}_{N,k} &= \sum_{j=N+1}^{k} P_{\bigvee_{i=j}^k (se-f_i)^+}(Q^p_{N,j-1} - Q^p_{N,j-2}). \end{aligned}$$

Proof. The existence of the limits in (4.1) and (4.2) follows from the observation that

$$0 \leq S^p_{i,j} \leq S_{r,s} \quad \text{and} \quad 0 \leq Q^p_{i,j} \leq Q_{r,s}$$

for all $r \leq i \leq j \leq s \leq -1$. Also note that

$$\begin{aligned} h_k &:= \bigvee_{i \leq k} [se - ((qe \wedge f_i) \vee -qe)]^+ \geq h, \\ H_k &:= \bigvee_{i \leq k} [((qe \wedge f_i) \vee -qe) - te]^+ \geq h, \end{aligned}$$

for $k \in -\mathbb{N}$. Here $H_{-1} \leq H_{-2} \leq H_{-3} \leq \dots$ with $H_k \to (M - te)^+$ as $k \to -\infty$ and $h_{-1} \leq h_{-2} \leq h_{-3} \leq \dots$ with $h_k \to (se - m)^+$ as $k \to -\infty$. The remainder of the proof is carried out in three inductive steps.

Step I *We prove that* $\lim_{j \to -\infty} S^1_{j,k} \geq P_h$, *for each* $k \in -\mathbb{N}$.

Here

$$\begin{aligned} \lim_{j \to -\infty} S^1_{j,k} &= \sum_{j=-\infty}^{k+1} (S^1_{j-1,k} - S^1_{j,k}) \\ &= \sum_{j=-\infty}^{k+1} (P_{\bigvee_{i=j-1}^k (se-f_i)^+} - P_{\bigvee_{i=j}^k (se-f_i)^+}) \\ &= \lim_{j \to -\infty} P_{\bigvee_{i=j-1}^k (se-f_i)^+} = P_{h_k} \geq P_h. \end{aligned}$$

Step II *Assuming* $\lim_{j \to -\infty} S^p_{j,k} \geq P_h$, *for all* $k \in -\mathbb{N}$, *we prove that* $\lim_{j \to -\infty} Q^p_{j,k} \geq P_h$, *for each* $k \in -\mathbb{N}$.

Here

$$\begin{aligned}\lim_{r\to-\infty} Q^p_{r,k} &= \sum_{r=-\infty}^{k+1} (Q^p_{r-1,k} - Q^p_{r,k}) \\ &= \sum_{r=-\infty}^{k+1} \Bigg[\sum_{j=r}^{k} P_{\bigvee_{i=j}^{k}(f_i-te)^+}(S^p_{r-1,j-1} - S^p_{r-1,j-2}) \\ &\qquad - \sum_{j=r+1}^{k} P_{\bigvee_{i=j}^{k}(f_i-te)^+}(S^p_{r,j-1} - S^p_{r,j-2})\Bigg] \\ &= \sum_{r=-\infty}^{k+1}\sum_{j=r}^{k} \Big[P_{\bigvee_{i=j}^{k}(f_i-te)^+}[(S^p_{r,j-2} - S^p_{r-1,j-2}) - (S^p_{r,j-1} - S^p_{r-1,j-1})\Big]\end{aligned}$$

since $S^p_{r,r-1} - S^p_{r-1,r-2} = 0$. Splitting into two summations and reindexing we obtain

$$\begin{aligned}\lim_{r\to-\infty} Q^p_{r,k} &= \sum_{r=-\infty}^{k+1} \Bigg[\sum_{j=r-2}^{k-2} P_{\bigvee_{i=j+2}^{k}(f_i-te)^+}(S^p_{r,j} - S^p_{r-1,j}) \\ &\qquad - \sum_{j=r-1}^{k-1} P_{\bigvee_{i=j+1}^{k}(f_i-te)^+}(S^p_{r,j} - S^p_{r-1,j})\Bigg].\end{aligned}$$

Since $S^p_{r,r-2} - S^p_{r-1,r-2} = 0$ and $P_{\bigvee_{i=k+1}^{k}(f_i-te)^+} = 0$,

$$\begin{aligned}\lim_{r\to-\infty} Q^p_{r,k} &= \sum_{r=-\infty}^{k+1}\sum_{j=r-1}^{k-1} [P_{\bigvee_{i=j+2}^{k}(f_i-te)^+} - P_{\bigvee_{i=j+1}^{k}(f_i-te)^+}](S^p_{r,j} - S^p_{r-1,j}) \\ &= \sum_{r=-\infty}^{k}\sum_{j=r-1}^{k-1} [P_{\bigvee_{i=j+1}^{k}(f_i-te)^+} - P_{\bigvee_{i=j+2}^{k}(f_i-te)^+}](S^p_{r-1,j} - S^p_{r,j}).\end{aligned}$$

As all terms in the above summation are non-negative, the summation can be reordered to give

$$\begin{aligned}\lim_{r\to-\infty} Q^p_{r,k} &= \sum_{j=-\infty}^{k-1} \left[(P_{\bigvee_{i=j+1}^{k}(f_i-te)^+} - P_{\bigvee_{i=j+2}^{k}(f_i-te)^+}) \sum_{r=-\infty}^{j+1} (S^p_{r-1,j} - S^p_{r,j})\right] \\ &= \sum_{j=-\infty}^{k-1} \left[(P_{\bigvee_{i=j+1}^{k}(f_i-te)^+} - P_{\bigvee_{i=j+2}^{k}(f_i-te)^+}) \lim_{r\to-\infty} S^p_{r,j}\right].\end{aligned}$$

Now by the assumptions of Step II,

$$\lim_{r\to-\infty} Q^p_{r,k} \geq \sum_{j=-\infty}^{k-1} (P_{\bigvee_{i=j+1}^{k}(f_i-te)^+} - P_{\bigvee_{i=j+2}^{k}(f_i-te)^+})P_h = P_{H_k}P_h$$

giving $\lim_{r\to-\infty} Q^p_{r,k} \geq P_h$.

Step III *Assuming* $\lim_{j\to-\infty} Q^p_{j,k} \geq P_h$, *for each* $k \in -\mathbb{N}$, *we prove that* $\lim_{j\to-\infty} S^{p+1}_{j,k} \geq P_h$, *for all* $k \in -\mathbb{N}$.

Replacing $(f_i - te)^+, Q^p_{r,k}, S^p_{r,k}, H_k$ respectively, in the proof of Step II by $(se - f_i)^+, S^{p+1}_{r,k}, Q^p_{r,k}, h_k$ gives immediately the proof of Step III. □

5. Convergence of reverse martingales

We begin by considering a local form of convergence which is the foundation for all the later convergence results.

Lemma 5.1. Local convergence of reverse martingales
Let $(f_i, T_i)_{i\in-\mathbb{N}}$ *be a reverse sub (super) martingale in a Dedekind complete Riesz space* E *with weak order unit* $e = T_i e$ *and* $T_i, i \in -\mathbb{N}$, *strictly positive. If there is a conditional expectation* T *with* $TT_N = T = T_N T$ *for all* $N \in -\mathbb{N}$, *then, for each* $n \in \mathbb{N}$, $(ne \wedge f_i \vee (-ne))_{i\in-\mathbb{N}}$ *is order convergent with order limit* $F_n \in E$ *given by*

$$\lim_{i\to-\infty}\sup ((ne \wedge f_i \vee)(-ne)) = F_n = \lim_{i\to-\infty}\inf ((ne \wedge f_i) \vee (-ne)).$$

Proof. Let $n \in \mathbb{N}$, then

$$M := \lim_{i\to-\infty}\sup ((ne \wedge f_i) \vee (-ne)) \geq \lim_{i\to-\infty}\inf ((ne \wedge f_i) \vee (-ne)) =: m.$$

We show that $M = m$. Suppose this equality to be false, then $M > m$. By Lemma 4.3, there are $s, t \in \mathbb{R}$ with $-n < s < t < n$ such that

$$h := (M - te)^+ \wedge (se - m)^+ > 0.$$

Consider the upcrossing of $(f_i, T_i)_{i\in-\mathbb{N}}$ over the order interval $[se, te]$. Let P_h denote the band projection onto the band generated by h. Let $Q^k_{N,M}$ be as defined in the reverse upcrossing theorem, Corollary 4.2, then by Lemma 4.4,

$$\bigvee_{N\leq-1} Q^k_{N,-1} e \geq P_h e, \quad \text{for all} \quad k \in \mathbb{N}.$$

Thus,

$$H_k := \bigvee_{N\leq-1} \left[((t-s)P_h e) \wedge \frac{1}{k} Y_{N,-1}(se, te)\right] = (t-s)P_h e > 0, \quad \text{for all} \quad k \in \mathbb{N}.$$

From the reverse upcrossing theorem,

$$TY_{N,-1}(se, te) = TT_N Y_{N,-1}(se, te) \leq TT_N |f_{-1} - se| \leq T|f_{-1}| + |s|e,$$

for all $N \leq -1$.

Now, for each $k \in \mathbb{N}$,

$$\left((t-s)P_h e \wedge \frac{1}{k} Y_{-j,-1}(se, te)\right)_{j\in\mathbb{N}}$$

and

$$\left(T[(t-s)P_h e \wedge \frac{1}{k}Y_{-j,-1}(se,te)]\right)_{j\in\mathbb{N}}$$

are increasing sequences bounded above by $(t-s)P_h e$ and $(t-s)TP_h e$ respectively, it thus follows that

$$\lim_{j\to-\infty}(t-s)P_h e \wedge \frac{1}{k}Y_{j,-1}(se,te) = H_k,$$

and

$$\lim_{j\to-\infty} T\left[(t-s)P_h e \wedge \frac{1}{k}Y_{j,-1}(se,te)\right] = TH_k.$$

Thus

$$\bigvee_{N\le -1} T\left(((t-s)P_h e) \wedge \frac{1}{k}Y_{N,-1}(se,te)\right) = TH_k,$$

for each $k \in \mathbb{N}$.

Hence,

$$0 < (t-s)TP_h e = TH_k = \bigvee_{n\le -1} T\left(((t-s)P_h e) \wedge \frac{1}{k}Y_{n,-1}(se,te)\right) \le \frac{g}{k}$$

for all $k \in \mathbb{N}$, which is not possible as E is an Archimedian Riesz space, proving that

$$ne \ge \limsup_i (ne \wedge f_{-i} \vee (-ne)) = \liminf_i (ne \wedge f_{-i} \vee (-ne)) \ge -ne$$

for all $n \in \mathbb{N}$. □

Using the above lemma, we now give a reverse sub (super) martingale convergence theorem for order bounded reverse martingales.

Theorem 5.2. Convergence of order bounded reverse martingales
Let $(f_i, T_i)_{i\in-\mathbb{N}}$ be a reverse sub (super) martingale in a Dedekind complete Riesz space E with weak order unit $e = T_i e$ and T_i strictly positive for all $i \in -\mathbb{N}$. Suppose that there is a conditional expectations T on E with $T_i T = T = TT_i$ for all $i \in -\mathbb{N}$. If there exists $g \in E^+$ such that $|f_i| \le g$ for all $i \in -\mathbb{N}$, then $(f_i)_{i\in-\mathbb{N}}$ is order convergent and has order limit $f_{-\infty} \in E$ given by

$$\lim_{i\to-\infty}\sup f_i = f_{-\infty} = \lim_{i\to-\infty}\inf f_i.$$

Proof. Applying Lemma 5.1, $((ne \wedge f_i) \vee (-ne))_{i\in-\mathbb{N}}$ is order convergent for each $n \in \mathbb{N}$, and has order limit $F_n \in E$ is given by

$$\lim_{i\to-\infty}\sup\,((ne \wedge f_i) \vee (-ne)) = F_n = \lim_{i\to-\infty}\inf\,((ne \wedge f_i) \vee (-ne)).$$

Observe that $(|F_n|)_{n\in\mathbb{N}}$ is an increasing sequence in E and that

$$|(ne \wedge f_i) \vee (-ne)| \le |f_i| \le g, \quad \text{for all} \quad i \in -\mathbb{N}.$$

Hence $|F_n| \le g$, for all $n \in \mathbb{N}$, and thus, from the Dedekind completeness of E, $|F_n| \uparrow F$ for some F in E. In particular

$$\lim_{i\to-\infty}(|f_i| \wedge ne) \le F, \quad \text{for all} \quad n \in \mathbb{N}.$$

Thus $F^S := \lim\sup_{i\to-\infty} f_i$ and $F^I := \lim\inf_{i\to-\infty} f_i$ exist. Now, from Lemma 5.1,

$$\begin{aligned}(ne \wedge F^S) \vee (-ne) &= \lim\sup_{i\to-\infty}((ne \wedge f_i) \vee (-ne)) \\ &= \lim\inf_{i\to-\infty}((ne \wedge f_i) \vee (-ne)) \\ &= (ne \wedge F^I) \vee (-ne)\end{aligned}$$

for all $n \in \mathbb{N}$. Taking the positive and negative parts of the above equality gives

$$ne \wedge F^{S\pm} = ne \wedge F^{I\pm}.$$

Since e is a weak order unit for E, letting $n \to \infty$ in the above equation gives $F^{S\pm} = F^{I\pm}$ and hence $F^S = F^I$. □

The following notation will be used in the remainder of this paper.

Definition 5.3. *Let E be a Dedekind complete Riesz space with weak order unit e and $(T_i)_{i\in-\mathbb{N}}$ be a reverse filtration on E with $T_ie = e$ for all $i \in -\mathbb{N}$. Denote by $\mathcal{RC}(E,T_i)$ be the set of order convergent reverse martingales, i.e., the set of $(f_i,T_i)_{i\in-\mathbb{N}} \in \mathcal{RM}(E,T_i)$ with $(f_i)_{i\in-\mathbb{N}}$ an order convergent sequence in E, and by $\mathcal{RB}(E,T_i)$ the set of order bounded reverse martingales, i.e., the set of reverse martingales $(f_i,T_i)_{i\in-\mathbb{N}} \in \mathcal{RM}(E,T_i)$ for which there exists $g \in E^+$ with $|f_i| \le g$ for all $i \in -\mathbb{N}$.*

It is easily verified that $\mathcal{RC}(E,T_i)$ and $\mathcal{RB}(E,T_i)$ are Riesz subspaces of $\mathcal{RM}(E,T_i)$.

Corollary 5.4. *If E is a Dedekind complete Riesz space with strictly positive conditional expectations T and weak order unit $e = Te$. If $(T_i)_{i\in-\mathbb{N}}$ is a reverse filtration on E with $T_iT = T = TT_i$ for all $i \in -\mathbb{N}$, then $\mathcal{RC}(E,T_i) = \mathcal{RB}(E,T_i)$.*

Proof. From [12, Proposition 1.1.10], we have that $\mathcal{RC}(E,T_i) \subset \mathcal{RB}(E,T_i)$. Theorem 5.2 gives the reverse containment. □

In order to obtain convergence theorems for sub (super) martingales on Riesz spaces analogous to those for norm bounded martingales in $\mathcal{L}^1$, as was seen in [10], the concept of T-universal completeness is required. This will also be needed for the reverse sub (super) martingale case.

Definition 5.5. *Let E be a Dedekind complete Riesz space and T a strictly positive conditional expectation on E. The space E is said to be T-universally complete if for each increasing net (f_α) in E^+ with (Tf_α) order bounded, we have that (f_α) is order convergent.*

For T-universally complete Riesz spaces, we have the following refinement of Theorem 5.2, an analogue of the convergence theorem for norm bounded reverse sub (super) martingales in L^1-spaces.

Theorem 5.6. Reverse martingale convergence – universally complete *Let $(f_i, T_i)_{i\in-\mathbb{N}}$ be a reverse sub (or super) martingale in a T-universally complete Riesz space E with weak order unit $e = Te$ in which the conditional expectation T is strictly positive and has $TT_N = T = T_NT$ for all $N \in -\mathbb{N}$. If there exists $g \in E$ with $T|f_N| \le g$ for all $N \in -\mathbb{N}$, then $(f_i)_{i\in-\mathbb{N}}$ is order convergent.*

Proof. With the above assumptions, the conditions of Lemma 5.1 are met. Thus, $((ne \wedge f_i) \vee (-ne))_{i\in-\mathbb{N}}$ is order convergent, for each $n \in \mathbb{N}$, to the order limit, say $F_n \in E$, given by

$$\lim_{i\to-\infty}\sup\,((ne \wedge f_i) \vee (-ne)) = F_n = \lim_{i\to-\infty}\inf\,((ne \wedge f_i) \vee (-ne)).$$

It should be noted that $(|F_n|)$ is an increasing sequence in E and that

$$|ne \wedge f_i \vee (-ne)| \le |f_i|.$$

Hence

$$T|ne \wedge f_i \vee (-ne)| \le T|f_i| \le g.$$

The order continuity of T now gives $T|F_n| \le g$ for all $n \in \mathbb{N}$. This, along with E being T-universally complete, gives that $|F_n| \uparrow F$ for some F in E. In particular,

$$\lim_{i\to-\infty}(|f_i| \wedge ne) \le F,$$

for all $n \in \mathbb{N}$, and thus $\limsup_{i\to-\infty}|f_i| \le F$ making $(f_i, T_i)_{i\in-\mathbb{N}}$ an order bounded. Theorem 5.2 can now be applied to give that $(f_i, T_i)_{i\in-\mathbb{N}}$ is order convergent. □

As a direct consequence of the above theorem, we have, under suitable T-universal completeness conditions, that $\mathcal{RC}(E, T_i)$ and $\mathcal{RM}(E, T_i)$ are equal.

Theorem 5.7. *Let E be a Dedekind and T-universally complete Riesz space E with weak order unit, where T is a strictly positive conditional expectation on E. Let $(T_i)_{i\in-\mathbb{N}}$ be a reverse filtration on E, with $TT_N = T = T_NT$ for all $N \in -\mathbb{N}$. Then $\mathcal{RC}(E, T_i) = \mathcal{RM}(E, T_i)$.*

Proof. Since $TT_i = T = T_iT$ for all $i \in -\mathbb{N}$, T is strictly positive and E is T-universally complete it follows, for each reverse martingale $(f_i, T_i)_{i\in-\mathbb{N}}$, that

$$T|f_i| = T|T_if_{-1}| \le TT_i|f_{-1}| = T|f_{-1}|, \quad \text{for all} \quad i \in -\mathbb{N}.$$

Theorem 5.6 now gives that $(f_i, T_i)_{i\in-\mathbb{N}}$ is convergent and hence $\mathcal{RC}(E, T_i) = \mathcal{RM}(E, T_i)$. □

In the light of Theorem 5.6, every reverse martingale compatible with T on a T-universally complete Riesz spaces is order convergent. Hence the limit map which takes each such reverse martingale to its order limit can be defined, as below.

Theorem 5.8. *Let E be a Dedekind complete Riesz space with strictly positive conditional expectation T and weak order unit $e = Te$. Let $(T_i)_{i\in-\mathbb{N}}$ be a reverse filtration on E with $T_iT = T = TT_i$ for all $i \in -\mathbb{N}$. If $\mathcal{RM}(E,T_i) = \mathcal{RC}(E,T_i)$, define the limit map $L : \mathcal{RC}(E,T_i) \to E$, by setting $L((f_i,T_i)_{i\in-\mathbb{N}})$ equal to the order limit of the reverse martingale $(f_i,T_i)_{i\in-\mathbb{N}}$. Then L is positive, linear and order continuous and the maps LJ and LJT_{-1} are a conditional expectations on $\mathcal{R}(T_{-1})$ and E respectively, both with range $\bigcap_{i\in-\mathbb{N}} \mathcal{R}(T_i)$, where J is as defined in Corollary 3.3.*

Proof. That L is positive and linear, follows from properties of limits in Riesz spaces.

We now show that L is order continuous. Let $((f_i^\alpha,T_i)_{i\in-\mathbb{N}})_\alpha \subset \mathcal{RM}^+(E,T_i)$ be a net with $(f_i^\alpha,T_i)_{i\in-\mathbb{N}} \downarrow_\alpha (0,T_i)_{i\in-\mathbb{N}}$, in particular $f_{-1}^\alpha \downarrow_\alpha 0$. Let $g_\alpha := L((f_i^\alpha,T_i)_{i\in-\mathbb{N}})$. Then from the positivity of L, $g_\alpha \downarrow_\alpha h \geq 0$.

Here $g_\alpha \in \bigcap_{i\in-\mathbb{N}} \mathcal{R}(T_i)$, so $T_ig_\alpha = g_\alpha$ for all $i \in -\mathbb{N}$. But T_i is order continuous and so $T_ih = h$, for all $i \in -\mathbb{N}$. Hence $(h,T_i)_{i\in-\mathbb{N}} \in \mathcal{RM}^+(E,T_i)$ with

$$\lim_\alpha \lim_{i\to-\infty} T_if_{-1}^\alpha = h \geq 0.$$

Since T is order continuous,

$$0 = T\lim_\alpha f_{-1}^\alpha = \lim_\alpha Tf_{-1}^\alpha = \lim_\alpha \lim_{i\to-\infty} TT_iF_{-1}^\alpha = T\lim_\alpha \lim_{i\to-\infty} T_iF_{-1}^\alpha = Th \geq 0.$$

Thus $Th = 0$, and since T is strictly positive and $h \geq 0$, this yields $h = 0$ making L order continuous.

Under the assumptions of the above theorem, every reverse martingale in $\mathcal{RM}(E,T_i)$ is convergent. Thus LJ is defined everywhere on $\mathcal{R}(T_{-1})$. If

$$f \in \bigcap_{i\in-\mathbb{N}} \mathcal{R}(T_i),$$

then $T_if = f$ for each $i \in -\mathbb{N}$ and hence $LJf = f$. Thus $f \in \mathcal{R}(LJ)$ and

$$\bigcap_{i\in-\mathbb{N}} \mathcal{R}(T_i) \subset \mathcal{R}(LJ).$$

If $g \in \mathcal{R}(LJ)$, then $g = LJf$ for some $f \in \mathcal{R}(T_{-1})$. Since $T_jf = T_iT_jf$ for all $j \leq i$, taking the limit as $j \to -\infty$ yields $g = T_ig \in \mathcal{R}(T_i)$ for all $i \in -\mathbb{N}$. Hence $g \in \bigcap_{i\in-\mathbb{N}} \mathcal{R}(T_i)$ and consequently $\mathcal{R}(LJ) \subset \bigcap_{i\in-\mathbb{N}} \mathcal{R}(T_i)$. Hence $\mathcal{R}(LJ) = \bigcap_{i\in-\mathbb{N}} \mathcal{R}(T_i)$ and LJ is a projection with $LJ(e) = e$.

Since each of the maps L and J are positive, linear and order continuous, see Corollary 3.3 and Lemma 5.8, it follows that LJ is a positive linear order continuous projection preserving e.

Since $\bigcap_{i \in -\mathbb{N}} \mathcal{R}(T_i)$ is an intersection of Dedekind complete Riesz subspaces of E, it is a Dedekind complete Riesz subspace of E and thus a Dedekind complete Riesz subspace of $\mathcal{R}(T_{-1})$. Hence LJ is a conditional expectation on $\mathcal{R}(T_{-1})$.

For LJT_{-1}, merely observe that $\mathcal{R}(LJT_{-1}) = \mathcal{R}(LJ) \subset \mathcal{R}(T_{-1})$ giving $(LJT_{-1})^2 = LJLJT_{-1} = LJT_{-1}$ and $LJT_{-1}e = LJe = e$. That LJT_{-1} is positive and order continuous follows from both LJ and T_{-1} being positive and order continuous. □

References

[1] P. Billingsley, *Probability and Measure*, John Wiley and Sons, 1995.

[2] R. De Marr, A martingale convergence theorem in vector lattices, *Canad. J. Math.*, **18** (1966), 424–432.

[3] J. Diestel, J.J. Uhl, Jr., *Vector measures*, American Mathematical Society, 1977.

[4] G.A. Edgar, L. Sucheston, *Stopping times and directed processes*, Cambridge University Press, 1992.

[5] L. Egghe, *Stopping time techniques for analysts and probabilists*, Cambridge University Press, 1984.

[6] K. Krickeberg, Convergence of martingales with a directed index set, *Trans. Amer. Math. Soc.*, **83** (1956), 313–337.

[7] W.-C. Kuo, C.C.A. Labuschagne, B.A. Watson, Discrete time stochastic processes on Riesz spaces, *Indag. Math. N.S.*, **15** (2004), 435–451.

[8] W.-C. Kuo, C.C.A. Labuschagne, B.A. Watson, Riesz space and fuzzy upcrossing theorems, in *Soft Methodology and Random Information Systems: Advances in Soft Computing*, Springer Verlag, Berlin, 2004, 101–108.

[9] W.-C. Kuo, C.C.A. Labuschagne, B.A. Watson, Conditional expectation operators on Riesz spaces, *J. Math. Anal. Appl.*, **303** (2005), 509–521.

[10] W.-C. Kuo, C.C.A. Labuschagne, B.A. Watson, Convergence of martingales in Riesz spaces, *Indag. Math. N.S.*, **17** (2006), 271–283.

[11] W.-C. Kuo, C.C.A. Labuschagne, B.A. Watson, Amarts on Riesz spaces, *Acta Mathematica Sinica (English Series)*, to appear.

[12] P. Meyer-Nieberg, *Banach lattices*, Springer Verlag, 1991.

[13] J. Neveu, *Discrete-parameter martingales*, North Holland, 1975.

[14] P.E. Protter, *Stochastic Integration and Differential Equations, Second Edition*, Springer, 2005.

[15] L.C.G. Rogers, D. Williams, *Diffusions, Markov processes and martingales, Vol 1*, Cambridge University Press, 2001.

[16] G. Stoica, Martingales in vector lattices, *Bull. Math. Soc. Sci. Math. Roumanie. (N.S.)*, **34(82)** (1990), 357–362.

[17] G. Stoica, Martingales in vector lattices II, *Bull. Math. Soc. Sci. Math. Roumanie. (N.S.)*, **35(83)** (1991), 155–157.

[18] G. Stoica, The structure of stochastic processes in normed vector lattices, *Stud. Cerc. Mat.* **46** (1994), 477–486.

[19] D.W. STROOCK, *Lectures on Stochastic Analysis: Diffusion Theory,* Cambridge University Press, 1987.

[20] V. TROITSKY, Martingales in Banach lattices, *Positivity,* **9** (2005), 437–456.

[21] A.C. ZAANEN, *Introduction to Operator Theory in Riesz Space,* Springer Verlag, 1997.

Mareli Korostenski
School of Mathematics
University of the Witwatersrand
Private Bag 3, P O WITS 2050, South Africa
e-mail: `marelikd@gmail.com`

Coenraad C.A. Labuschagne
School of Computational and Applied Mathematics
University of the Witwatersrand
Private Bag 3, P O WITS 2050, South Africa
e-mail: `Coenraad.Labuschagne@wits.ac.za`

Bruce A. Watson
School of Mathematics and
Centre for Applicable Analysis and Number Theory,
University of the Witwatersrand
Private Bag 3, P O WITS 2050, South Africa
e-mail: `b.alastair.watson@gmail.com`

Operator Theory:
Advances and Applications, Vol. 195, 231–244

Contractivity of the H^∞-calculus and Blaschke Products

Christoph Kriegler and Lutz Weis

Abstract. It is well known that a densely defined operator A on a Hilbert space is accretive if and only if A has a contractive H^∞-calculus for any angle bigger than $\frac{\pi}{2}$. A third equivalent condition is that $\|(A-w)(A+\overline{w})^{-1}\| \leq 1$ for all $\operatorname{Re} w \geq 0$. In the Banach space setting, accretivity does not imply the boundedness of the H^∞-calculus any more. However, we show in this note that the last condition is still equivalent to the contractivity of the H^∞-calculus in all Banach spaces. Furthermore, we give a sufficient condition for the contractivity of the H^∞-calculus on $\mathbb{C}_+$, thereby extending a Hilbert space result of Sz.-Nagy and Foiaş to the Banach space setting.

Mathematics Subject Classification (2000). 47A60, 47B44, 30D50.

Keywords. H^∞-calculus, accretive operators, Blaschke products

1. Introduction

It is well known that the Cayley transform $A \mapsto T = (A-1)(A+1)^{-1}$ provides, in a Hilbert space, a one-to-one correspondence between accretive operators A (i.e., negative generators of contractive semigroups) and bounded contractive operators T which do not have 1 as an eigenvalue (cf. [NF] Theorem 4.1 in IV.4). If $\theta > \frac{\pi}{2}$, then this, combined with von Neumann's inequality for contractions, can be used to construct a contractive $H^\infty(\Sigma_\theta)$-calculus (see Section 2 for the definition) for any accretive operator A (cf. [ADM]).

In a Banach space setting, this correspondence between accretive operators and contractions breaks down and von Neumann's inequality does not necessarily hold [Foi]. Thus in order to construct a contractive H^∞-calculus for Banach space operators we need stronger assumptions than accretivity. In this note we take a hint from one of the many known proofs of von Neumann's inequality (see [Dru]) and use classical approximations of bounded analytic functions on $\Sigma_{\frac{\pi}{2}}$ or $\mathbb{D}$ by Blaschke products. This enables us to connect the holomorphic functional calculus of a negative generator A of a semigroup and its Cayley transform T, thereby proving its contractivity.

In Section 4 we show that a sectorial operator A of type $\frac{\pi}{2}$ on a Banach space X has a contractive $H^\infty(\Sigma_\theta)$-calculus for all $\theta > \frac{\pi}{2}$ if and only if for all $\lambda \in \Sigma_{\frac{\pi}{2}}$ and $x \in D(A)$

$$\|(A-\lambda)x\| \leq \|(A+\overline{\lambda})x\|. \tag{1.1}$$

Condition (1.1) is stronger than accretivity in a general Banach space X but it is easily seen that it is equivalent to accretivity if X is a Hilbert space. Hence our result extends the Hilbert space result quoted earlier and provides a relatively short proof for it. In Theorem 4.4, we extend this functional calculus to functions in $H^\infty(\Sigma_{\frac{\pi}{2}})$ with a continuous boundary function on $i\mathbb{R} \cup \{\infty\}$, in the case that A is only a $\frac{\pi}{2}$-sectorial operator.

In general, (1.1) does not guarantee a bounded $H^\infty(\Sigma_{\frac{\pi}{2}})$-calculus, even in a Hilbert space. However, in Section 5, we show that if we assume in addition that there exists a dense subset D of X such that

$$\int_0^\infty |\langle e^{-tA}x, x'\rangle|^2 dt \leq C_x \|x'\|^2 \text{ for } x \in D,\ x' \in X', \tag{1.2}$$

then A has a contractive $H^\infty(\Sigma_{\frac{\pi}{2}})$-calculus. Condition (1.2) is of importance in scattering theory and, in the Hilbert space setting, was studied by E.B. Davies in [Da1] and [Da2]. He shows in [Da1], Theorem 6.26, that a completely non-unitary semigroup on a Hilbert space satisfies (1.2). Therefore our result generalizes the known fact that a completely non-unitary semigroup has an $H^\infty(\Sigma_{\frac{\pi}{2}})$ functional calculus, which can be derived from [NF], Section III.8. We derive our Banach space result via the Cayley transform from a corresponding result for contractions T on X (see Corollary 5.3) where we assume that

$$\sum_{n=0}^\infty |\langle T^n x, x'\rangle|^2 \leq C_x \|x'\|^2 \text{ for } x \in D,\ x' \in X'. \tag{1.3}$$

Clearly, every sectorial operator A on a Banach space X with a bounded $H^\infty(\Sigma_\theta)$-calculus has a contractive functional calculus in the equivalent norm

$$|\!|\!| x |\!|\!| = \sup\{\|\phi(A)x\| : \phi \in H^\infty(\Sigma_\theta),\ |\phi(\lambda)| \leq 1 \text{ for } \lambda \in \Sigma_\theta\}.$$

However, in Section 6 we show that for many common examples of semigroup generators, $|\!|\!| \cdot |\!|\!|$ is not the natural norm of X.

In Sections 2 and 3, we recall some facts on the H^∞-calculus and Blaschke products that are essential for our argument.

2. Preliminaries on the H^∞-calculus

Notation. Let $\theta \in (0,\pi)$. As usual we shall set

$$\Sigma_\theta = \{re^{i\phi} : r > 0,\ |\phi| < \theta\}$$

and

$$H^\infty(\Sigma_\theta) = \{f : \Sigma_\theta \to \mathbb{C} : f \text{ is analytic and bounded}\}.$$

This space is complete when equipped with the norm $\|f\|_{\infty,\theta} = \sup_{\lambda\in\Sigma_\theta} |f(\lambda)|$. Put

$$H_0^\infty(\Sigma_\theta) = \{f \in H^\infty(\Sigma_\theta) : \exists\, \epsilon, C > 0 \text{ s.th. } |f(\lambda)| \leq C \min(|\lambda|^\epsilon, |\lambda|^{-\epsilon})\}$$

and

$$R(\Sigma_\theta) = \operatorname{span}(H_0^\infty(\Sigma_\theta) \cup \{1, (1 + (\cdot))^{-1}\}).$$

Note that $R(\Sigma_\theta)$ is a subalgebra of $H^\infty(\Sigma_\theta)$ which contains all rational functions of non-positive degree and poles outside Σ_θ.
Let X be a Banach space and $\theta \in (0, \pi)$. An operator $A : D(A) \subseteq X \to X$ is called *θ-sectorial,* if

1. $D(A)$ is dense in X.
2. The spectrum $\sigma(A)$ is contained in $\overline{\Sigma_\theta}$.
3. For all $\omega > \theta$ there is a $C_\omega > 0$ such that $\|\lambda R(\lambda, A)\| \leq C_\omega$ for all $\lambda \in \overline{\Sigma_\omega}^c$.

For such an operator, one can construct for every $\omega > \theta$ a linear and multiplicative mapping $\Phi_\omega : R(\Sigma_\omega) \longrightarrow B(X)$ such that for $\mu \in \mathbb{C}\backslash\overline{\Sigma_\omega}$ and $f \in H_0^\infty(\Sigma_\omega)$

$$\Phi_\omega(1) = \mathrm{Id}_X,\ \Phi_\omega((\mu - (\cdot))^{-1}) = (\mu - A)^{-1} \text{ and } \Phi_\omega(f) = \frac{1}{2\pi i}\int_{\partial\Sigma_{(\omega+\theta)/2}} f(\lambda) R(\lambda, A)\, d\lambda.$$

We call Φ_ω the $H^\infty(\Sigma_\omega)$-calculus of A. For $\omega_1 > \omega_2$, Φ_{ω_1} and Φ_{ω_2} coincide on $R(\Sigma_{\omega_1})$. If in addition A has dense range then Φ_ω can be extended to

$$\begin{aligned} H_A^\infty(\Sigma_\omega) = \{f \in H^\infty(\Sigma_\omega) :\ & \exists\, (f_n)_n \subseteq R(\Sigma_\omega) \text{ s.th.} \\ & f_n \to f \text{ pointwise and } \sup_n \|f_n(A)\| + \|f_n\|_{\infty,\omega} < \infty\}. \end{aligned}$$

For $f \in H_A^\infty(\Sigma_\omega)$ and associated $(f_n)_n \subseteq R(\Sigma_\omega)$, $\Phi_\omega(f)x = \lim_n \Phi_\omega(f_n)x$ for all $x \in X$. This implies clearly that $\|\Phi_\omega(f)\| \leq \liminf_n \|\Phi_\omega(f_n)\|$. The map

$$\Phi_\omega : H_A^\infty(\Sigma_\omega) \longrightarrow B(X)$$

is still linear and multiplicative. Further, $H_A^\infty(\Sigma_\omega)$ equals $H^\infty(\Sigma_\omega)$ iff there exists $C > 0$ such that for all $f \in H_0^\infty(\Sigma_\omega)$ one has $\|f(A)\| \leq C\|f\|_{\infty,\omega}$. In this case, we say that A has a *bounded $H^\infty(\Sigma_\omega)$-calculus.* For more information on the H^∞-calculus, see for example [CDMY], especially Section 2 and [KW], especially Chapter 9. We denote $\Phi_\omega(f)$ by $f(A)$.

3. Preliminaries on Blaschke products

In this section, we develop the necessary background on Blaschke products (see also [Gar] for more information on this topic).

Notation. We put $\mathbb{D} = \{z \in \mathbb{C} : |z| < 1\}$ and consider the two spaces

$$H^\infty(\mathbb{D}) = \{f : \mathbb{D} \to \mathbb{C} \text{ analytic and bounded}\}$$

and

$$A(\mathbb{D}) = \{f \in H^\infty(\mathbb{D}) : f \text{ has a continuous extension to } \overline{\mathbb{D}}\}.$$

They are equipped with the norm $\|f\|_{\infty,\mathbb{D}} = \|f\|_\infty = \sup_{z\in\mathbb{D}} |f(z)|$, for which they are complete. For $\mu \in \mathbb{D}$, we define

$$f_\mu(z) = \frac{z-\mu}{1-\overline{\mu}z}.$$

This function is called a *Blaschke factor on* $\mathbb{D}$ and is clearly analytic on the neighborhood $\frac{1}{\mu}\mathbb{D}$ of $\mathbb{D}$. For $\mu \in \mathbb{D}$, $f_\mu(\partial\mathbb{D}) \subseteq \partial\mathbb{D}$, since for $\theta \in \mathbb{R}$

$$|f_\mu(e^{i\theta})| = \left|\frac{e^{i\theta}-\mu}{1-\overline{\mu}e^{i\theta}}\right| = \left|e^{i\theta}\frac{1-\mu e^{-i\theta}}{1-\overline{\mu e^{-i\theta}}}\right| = 1.$$

Thus by the maximum principle, $\|f_\mu\|_\infty = 1$.
A function $f \in A(\mathbb{D})$ is called a *(finite) Blaschke product on* $\mathbb{D}$ if it is of the form

$$f(z) \equiv e^{i\theta} \quad \text{or} \quad f(z) = e^{i\theta}\prod_{k=1}^n f_{\mu_k}(z)$$

for some $n \in \mathbb{N}$, $\mu_k \in \mathbb{D}$ and $\theta \in \mathbb{R}$. A function $f \in H^\infty(\mathbb{D})$ is a finite Blaschke product if and only if

1. f is analytic on some $a\,\mathbb{D}$ with $a > 1$.
2. $f(\partial\mathbb{D}) \subseteq \partial\mathbb{D}$.

Indeed, a finite Blaschke product clearly has the two properties. Suppose now that f satisfies 1. and 2. If f had infinitely many zeros in $\mathbb{D}$ then there would exist an accumulation point of them in $\overline{\mathbb{D}}$. By 1. and the identity theorem, $f \equiv 0$, which contradicts 2. Hence f has only finitely many zeros and there exists a finite Blaschke product b which has the same zeros counted with their multiplicity. Then f/b and b/f also satisfy 1. and 2. Therefore, for $z \in \mathbb{D}$, $|f(z)/b(z)| \leq 1$ and $|b(z)/f(z)| \leq 1$, and thus $|f(z)/b(z)| = 1$, which implies that $f(z)/b(z)$ is constant. Hence $f(z) = e^{i\theta}b(z)$ for some $\theta \in \mathbb{R}$.
If $T \in B(X)$ with spectrum $\sigma(T) \subseteq \overline{\mathbb{D}}$, then for $f(z) = \sum a_n z^n$ holomorphic in $a\,\mathbb{D}$ for some $a > 1$, we define $f(T) = \sum a_n T^n$, which is the Dunford calculus of T. In particular, $f_\mu(T) = (T-\mu)(1-\overline{\mu}T)^{-1}$.
We also need the Blaschke factors on $\Sigma_{\frac{\pi}{2}}$. We define the conformal mappings

$$\zeta : \mathbb{D} \to \Sigma_{\frac{\pi}{2}},\ z \mapsto -\frac{z+1}{z-1} \quad \text{and} \quad \tau = \zeta^{-1} : \Sigma_{\frac{\pi}{2}} \to \mathbb{D},\ \lambda \mapsto \frac{\lambda-1}{\lambda+1}.$$

Then $H^\infty(\Sigma_{\frac{\pi}{2}}) = \{f\circ\tau : f \in H^\infty(\mathbb{D})\}$. We put $A(\Sigma_{\frac{\pi}{2}}) = \{f\circ\tau : f \in A(\mathbb{D})\} = \{f \in H^\infty(\Sigma_{\frac{\pi}{2}}) : f$ has a continuous extension to $\overline{\Sigma_{\frac{\pi}{2}}}$ and at $\infty\}$. The spaces $H^\infty(\Sigma_{\frac{\pi}{2}})$

and $A(\Sigma_{\frac{\pi}{2}})$ are again equipped with the infinity norm. Note that for $\theta > \frac{\pi}{2}$, $H^\infty(\Sigma_\theta) \not\subset A(\Sigma_{\frac{\pi}{2}})$, but $R(\Sigma_\theta) \subset A(\Sigma_{\frac{\pi}{2}})$. For $w \in \Sigma_{\frac{\pi}{2}}$, we define

$$b_w(\lambda) = \frac{\lambda - w}{\lambda + \overline{w}}, \quad \lambda \in \Sigma_{\frac{\pi}{2}}$$

which we call a *Blaschke factor on* $\Sigma_{\frac{\pi}{2}}$. For $\mu \in \mathbb{D}$, the Blaschke factor f_μ on $\mathbb{D}$ is related to the Blaschke factor $b_{\zeta(\mu)}$ on $\Sigma_{\frac{\pi}{2}}$ by the identity

$$f_\mu \circ \tau = f_\mu(1) \cdot b_{\zeta(\mu)}.$$

As above we call $f \in A(\Sigma_{\frac{\pi}{2}})$ a *(finite) Blaschke product on* $\Sigma_{\frac{\pi}{2}}$, if it is of the form

$$f(\lambda) \equiv e^{i\theta} \quad \text{or} \quad f(\lambda) = e^{i\theta} \prod_{k=1}^{n} b_{w_k}(\lambda)$$

for some $n \in \mathbb{N}$, $w_k \in \Sigma_{\frac{\pi}{2}}$ and $\theta \in \mathbb{R}$. We denote by $B_{\mathbb{D}}$ (resp. $B_{\Sigma_{\frac{\pi}{2}}}$) the set of finite Blaschke products, which is contained in the unit ball of $A(\mathbb{D})$ (resp. $A(\Sigma_{\frac{\pi}{2}})$).

The next lemma and its proof is essentially taken from [Gar]. Since these results are essential for us we include a proof for the convenience of the reader.

Lemma 3.1.

1. (*Carathéodory*). *For every f in the unit ball of $H^\infty(\mathbb{D})$ (resp. $H^\infty(\Sigma_{\frac{\pi}{2}})$), there exists a sequence in $B_{\mathbb{D}}$ (resp. $B_{\Sigma_{\frac{\pi}{2}}}$) which converges pointwise on $\mathbb{D}$ (resp. $\Sigma_{\frac{\pi}{2}}$) to f.*
2. (*Bernard*). co $B_{\mathbb{D}}$, *the convex hull of $B_{\mathbb{D}}$, is norm dense in the unit ball of $A(\mathbb{D})$. Hence also* co $B_{\Sigma_{\frac{\pi}{2}}}$ *is norm dense in the unit ball of $A(\Sigma_{\frac{\pi}{2}})$.*

Proof. 1. It clearly suffices to prove the statement for $\mathbb{D}$. Write $f(z) = \sum_{k=0}^\infty c_k z^k$. It suffices to find for every $n \in \mathbb{N}$ a finite Blaschke product $B_{n,f}(z) = \sum_{k=0}^\infty d_k z^k$ such that $d_k = c_k$ for $k \le n-1$. Indeed, by the Cauchy integral formula, $|c_k|, |d_k| \le 1$ for all $k \in \mathbb{N}$. Then for $|z| < 1$ fixed,

$$|f(z) - B_{n,f}(z)| \le \sum_{k=n}^\infty (|c_k| + |d_k|)|z|^k \le 2|z|^n(1 - |z|)^{-1} \to 0.$$

If $|c_0| = 1$, then by the maximum principle, f is constant and $B_{n,f}(z) \equiv c_0$ suffices. So suppose from now on that $|c_0| < 1$. We proceed by induction. For $n = 1$, $B_{1,f}(z) = f_{-c_0}(z) = \frac{z+c_0}{1+\overline{c_0}z}$ suffices. Suppose that for a given $n \in \mathbb{N}$ and all h in the unit ball of $H^\infty(\mathbb{D})$, there exists $B_{n-1,h}$ such that $B_{n-1,h} - h$ has a zero of multiplicity $n-1$ at 0. Let

$$h(z) = \frac{1}{z}\frac{f(z) - c_0}{1 - \overline{c_0} f(z)} = \frac{1}{z} f_{c_0}(f(z)).$$

If $|z| < 1$ then for all $r \in (|z|, 1)$ we have $|h(z)| \le \frac{1}{r}|f_{c_0}(f(z))| \le \frac{1}{r}$ and hence h is in the unit ball of $H^\infty(\mathbb{D})$. We will construct $B_n = B_{n,f}$ by means of $B_{n-1} = B_{n-1,h}$. Put

$$B_n(z) = \frac{zB_{n-1}(z) + c_0}{1 + \overline{c_0} z B_{n-1}(z)} = f_{-c_0}(zB_{n-1}(z)).$$

Then B_n is analytic on $a\mathbb{D}$ for some $a > 1$ and satisfies $B_n(\partial\mathbb{D}) \subseteq \partial\mathbb{D}$. Hence B_n is a finite Blaschke product.

$$f(z) - B_n(z) = f_{-c_0}(zh(z)) - f_{-c_0}(zB_{n-1}(z)) = \frac{(1 - |c_0|^2)z(h(z) - B_{n-1}(z))}{(1 + \overline{c_0}zh(z))(1 + \overline{c_0}zB_{n-1}(z))}.$$

Here we see that $f - B_n$ has a zero of multiplicity n at 0. The induction is complete.
2. Let $p = \sum_{n=0}^N a_n z^n$ be a polynomial in $A(\mathbb{D})$ with $\|p\|_\infty < 1$. It clearly suffices to show that such a p can be uniformly approximated by a finite convex combination of finite Blaschke products. Let $q(z) = \sum_{n=0}^N \overline{a_n} z^{N-n} = z^N \overline{p(z)}$, so that $q \in A(\mathbb{D})$ and $|q(z)| < 1$ on $\overline{\mathbb{D}}$. For $z, \xi \in \overline{\mathbb{D}}$ put

$$r(z,\xi) = \frac{p(z) + \xi z^N}{1 + \xi q(z)} = f_{-p(z)}(\xi z^N).$$

For every $\xi \in \overline{\mathbb{D}}$, $r(\cdot, \xi)$ can be analytically continued on $a\mathbb{D}$ for some $a > 1$. Also for every $z \in \overline{\mathbb{D}}$, $r(z, \cdot)$ can be analytically continued on $\|q\|_\infty^{-1}\mathbb{D}$. Since $|r(z, e^{it})| = |f_{-p(z)}(e^{it}z^N)| = 1$ for all $z \in \partial\mathbb{D}$, $r(\cdot, e^{it})$ is a finite Blaschke product. By the mean value property for $r(z, \cdot)$,

$$p(z) = r(z, 0) = \int_0^{2\pi} r(z, e^{it}) \frac{dt}{2\pi},$$

where $[0, 2\pi] \ni t \mapsto r(z, e^{it}) \in A(\mathbb{D})$ is continuous. So p can be approximated in $A(\mathbb{D})$ by a sequence of finite convex combinations of finite Blaschke products. □

4. Contractivity of the H^∞-calculus

Our result is motivated by the following observation about Hilbert space operators.

Lemma 4.1. *Let H be a Hilbert space and let $A : D(A) \subseteq H \to H$ be an operator. The following are equivalent:*

1. *A is accretive, i.e., for all $x \in D(A)$ we have* $\operatorname{Re}\langle Ax, x\rangle \geq 0$.
2. *For all $w \in \mathbb{C}$ with* $\operatorname{Re} w > 0$ *and all $x \in D(A)$,* $\|(A - w)x\| \leq \|(A + \overline{w})x\|$.
3. *There exists a $w \in \mathbb{C}$ with* $\operatorname{Re} w > 0$ *such that for all $x \in D(A)$,* $\|(A - w)x\| \leq \|(A + \overline{w})x\|$.
4. *For every finite Blaschke product $b \in B_{\Sigma_{\frac{\pi}{2}}}$,* $\|b(A)\| \leq 1$.

Proof. The following inequalities are equivalent for $x \in D(A)$:

$$\begin{aligned}
\|(A - w)x\| &\leq \|(A + \overline{w})x\| \\
\langle (A - w)x, (A - w)x\rangle &\leq \langle (A + \overline{w})x, (A + \overline{w})x\rangle \\
\langle Ax, Ax\rangle - 2\operatorname{Re}[\overline{w}\langle Ax, x\rangle] + |w|^2\langle x, x\rangle &\leq \langle Ax, Ax\rangle + 2\operatorname{Re}[w\langle Ax, x\rangle] + |w|^2\langle x, x\rangle \\
0 &\leq \operatorname{Re}[(w + \overline{w})\langle Ax, x\rangle] \\
0 &\leq \operatorname{Re}\langle Ax, x\rangle.
\end{aligned}$$

So if 3. holds for some $w \in \Sigma_{\frac{\pi}{2}}$, then 1. holds, which in turn implies that 2. holds. That 2. implies 3. is evident and the equivalence of 2. and 4. holds since for every $w \in \Sigma_{\frac{\pi}{2}}$, $A + \overline{w}$ is a bijection $D(A) \to X$. □

In a general Banach space, the equivalence of 1. and 4. above breaks down. As was shown in [KW, Section 10], there exist accretive operators without a bounded H^∞-calculus. Condition 2. above is however still equivalent to the boundedness of the H^∞-calculus in the Banach space context.

Theorem 4.2. *Let X be a Banach space, $\theta < \frac{\pi}{2}$ and $A : D(A) \subseteq X \to X$ be θ-sectorial with dense range. Then A satisfies*

$$\|(A - w)x\| \le \|(A + \overline{w})x\| \text{ for all } \operatorname{Re} w > 0,\ x \in D(A) \tag{4.1}$$

if and only if A has a bounded $H^\infty(\Sigma_{\frac{\pi}{2}})$-calculus and

$$\|f(A)\| \le \|f\|_{\infty,\frac{\pi}{2}} \quad (f \in H^\infty(\Sigma_{\frac{\pi}{2}})).$$

Proof. The condition (4.1) implies that $b_w(A) = (A-w)(A+\overline{w})^{-1}$ is a contraction. Thus, $b(A)$ is a contraction for every finite Blaschke product b. Suppose now that $f \in H^\infty(\Sigma_{\frac{\pi}{2}})$ with $\|f\|_{\infty,\frac{\pi}{2}} \le 1$. Then by Lemma 3.1, there exists a sequence f_n of finite Blaschke products converging to f pointwise on $\Sigma_{\frac{\pi}{2}}$. Since $\|f_n\|, \|f_n(A)\| \le 1$ for all $n \in \mathbb{N}$, f belongs to $H_A^\infty(\Sigma_{\frac{\pi}{2}})$ and $\|f(A)\| \le \liminf_n \|f_n(A)\| \le 1$. The necessity of (4.1) is clear, since it is equivalent to $\|b_w(A)\| \le \|b_w\|_{\infty,\frac{\pi}{2}} = 1$ for all $\operatorname{Re} w > 0$. □

Of course, the last proof also shows that the boundedness of the $H^\infty(\Sigma_{\frac{\pi}{2}})$-calculus in the situation of theorem 4.2 can be characterized by the uniform boundedness of the Blaschke products:

$$\|b(A)\| \le C \text{ for some } C < \infty \text{ and every finite Blaschke product } b \text{ on } \Sigma_{\frac{\pi}{2}}. \tag{4.2}$$

If one compares the boundedness of the H^∞-calculus to the boundedness of the semigroup, then (4.2) corresponds to the Hille-Yosida condition $\|\lambda^n(\lambda+A)^n\| \le C$ for all $\lambda > 0$ and (4.1) corresponds to the accretivity of the operator A. Note that if the $H^\infty(\Sigma_{\frac{\pi}{2}})$-calculus of A is bounded, then it is always contractive in an equivalent norm on X, given for example by

$$\|\!|x|\!\| = \sup\{\|f(A)x\| : f \in H^\infty(\Sigma_{\frac{\pi}{2}}) : \|f\|_{\infty,\frac{\pi}{2}} \le 1\}.$$

Since θ-sectoriality for $\theta < \frac{\pi}{2}$ already implies that $-A$ generates an analytic semigroup, it is worthwhile to state in addition a somewhat weaker result for $\frac{\pi}{2}$-sectorial operators which covers all negative generators of a bounded semigroup.

To this end we need the following lemma.

Lemma 4.3. *Let $A : D(A) \subseteq X \to X$ be $\frac{\pi}{2}$-sectorial. Suppose that (4.1) holds. For $r \in (0,1)$, let*

$$A_r = (1 - r + A(1+r))(1 + r + A(1-r))^{-1} = \zeta \circ \psi_r \circ \tau(A) \in B(X),$$

where ζ and τ are the conformal mappings as in Section 3 and $\psi_r : \mathbb{D} \to \mathbb{D}$, $z \mapsto rz$.

Then

1. $\sigma(A_r) \subset \Sigma_{\frac{\pi}{2}}$.
2. A_r *is again* $\frac{\pi}{2}$*-sectorial, and this uniformly in* r*, in the sense that for all* $\omega \in (\frac{\pi}{2}, \pi)$ *there exists* $C_\omega > 0$ *such that for all* $r \in (0,1)$ *and* $\lambda \in \mathbb{C}\backslash\overline{\Sigma_\omega}$ *we have* $\|\lambda R(\lambda, A_r)\| \leq C_\omega$.
3. *For* $\operatorname{Re}\lambda < 0$, $R(\lambda, A_r) \to R(\lambda, A)$ *in* $B(X)$ *as* $r \to 1$.
4. (4.1) *holds for* A_r *in place of* A.

Proof. 1. By (4.1), $\tau(A)$ is a contraction, and hence $\sigma(\psi_r \circ \tau(A)) \subset \mathbb{D}$. By the spectral mapping theorem, $\sigma(A_r) = \sigma(\zeta \circ \psi_r \circ \tau(A)) \subset \Sigma_{\frac{\pi}{2}}$.

2. Fix some $\omega \in (\frac{\pi}{2}, \pi)$. Let $\lambda \in \mathbb{C}\backslash\overline{\Sigma_\omega}$. Then $\lambda(\lambda - A_r)^{-1}$

$$= \{\lambda(1+r) + A\lambda(1-r)\}\left[A(\lambda(1-r) - (1+r)) + \lambda(1+r) - (1-r)\right]^{-1}. \quad (4.3)$$

We split this expression into two summands (I) and (II).

$$\begin{aligned}(\mathrm{I}) &= \lambda(1+r)[\ldots]^{-1} \\ &= \frac{\lambda(1+r)}{\lambda(1-r) - (1+r)}\left[A + \frac{\lambda(1+r) - (1-r)}{\lambda(1-r) - (1+r)}\right]^{-1} \\ &= \frac{\lambda(1+r)}{\lambda(1+r) - (1-r)}\mu(A+\mu)^{-1},\end{aligned}$$

with $\mu = \frac{\lambda(1+r)-(1-r)}{\lambda(1-r)-(1+r)}$. Note that $\left|\frac{\lambda(1+r)}{\lambda(1+r)-(1-r)}\right| \leq 1$. Further, $\mu = \frac{\lambda - a^{-1}}{\lambda - a}a$ with $a = \frac{1+r}{1-r} > 1$, so $|\arg\mu| = |\arg(\lambda - a^{-1}) - \arg(\lambda - a)| \leq |\arg(\lambda - a^{-1})| < \pi - \omega$, and thus $\|\mu(A+\mu)^{-1}\| \leq C_\omega = \sup\{\|\nu R(\nu, A)\| : \nu \in \mathbb{C}\backslash\overline{\Sigma_\omega}\}$.

$$(\mathrm{II}) = A\lambda(1-r)[\ldots]^{-1} = \frac{\lambda(1-r)}{\lambda(1-r) - (1+r)}A[A+\mu]^{-1}.$$

Now $\left|\frac{\lambda(1-r)}{\lambda(1-r)-(1+r)}\right| \leq 1$. Further, $\|A(A+\mu)^{-1}\| \leq 1 + \|\mu(A+\mu)^{-1}\| \leq 1 + C_\omega$, since $\mu \in \Sigma_{\pi-\omega}$.

3. This can be deduced using (4.3) and the continuity of the map $\lambda \mapsto AR(\lambda, A)$ from $\mathbb{C}\backslash\overline{\Sigma_\omega}$ to $B(X)$.

4. Let $\operatorname{Re} w > 0$, put $\mu = \tau(w) \in \mathbb{D}$ and $T = \tau(A)$. $b_w(A_r) = b_w \circ \zeta \circ \psi_r \circ \tau(A) = f_\mu(1)^{-1} f_\mu(rT)$. We show that $f_\mu(rT) = (rT - \mu)(1 - \overline{\mu}rT)^{-1}$ is a contraction, which implies that $b_w(A_r)$ is one.

$$f_\mu(z) = \frac{z - \mu}{1 - \overline{\mu}z} = -\mu + \frac{1 - |\mu|^2}{\overline{\mu}}\frac{\overline{\mu}z}{1 - \overline{\mu}z}.$$

This shows that $f_\mu(rz) = \alpha f_{r\mu}(z) + \beta$ with $\alpha = \frac{(1-|\mu|^2)r}{1-|r\mu|^2}$ and $\beta = \mu r\frac{(1-|\mu|^2)r}{1-|r\mu|^2} - \mu$. So, $f_\mu(rT) = \alpha f_{r\mu}(T) + \beta \operatorname{Id}_X$. Now $f_{r\mu}(T) = f_{r\mu}(1) b_{\zeta(r\mu)}(A)$ is a contraction, so that it suffices to show that $|\alpha| + |\beta| \leq 1$.

$$\begin{aligned}
|\alpha| + |\beta| &= \frac{(1-|\mu|^2)r}{1-|r\mu|^2} + |\mu| \left| \frac{r^2(1-|\mu|^2)}{1-|r\mu|^2} - 1 \right| \\
&= \frac{(1-|\mu|^2)r}{1-|r\mu|^2} + |\mu| \frac{r^2-1}{1-|r\mu|^2} \\
&= \frac{(1-a^2)r + a(1-r^2)}{1-a^2r^2},
\end{aligned}$$

where $a = |\mu| \in (0,1)$. So $|\alpha| + |\beta| \leq 1$ iff $r^2(a-a^2) + r(a^2-1) + 1 - a \geq 0$. The left-hand side of the last inequality equals 0 for $r = 1$, and its derivative with respect to r is $2r(a-a^2) + a^2 - 1 = 2(a-a^2)(r-1) - (a-1)^2 \leq 0$ for $r \leq 1$. So the inequality is indeed fulfilled. □

Theorem 4.4. *Let $A : D(A) \subseteq X \to X$ be $\frac{\pi}{2}$-sectorial. Suppose that*

$$\|(A-w)x\| \leq \|(A+\overline{w})x\| \text{ for all } w \in \Sigma_{\frac{\pi}{2}} \text{ and } x \in D(A).$$

Then for all $\theta > \frac{\pi}{2}$ and all $f \in H_0^\infty(\Sigma_\theta)$,

$$\|f(A)\| \leq \|f\|_{\infty,\frac{\pi}{2}},$$

and if A has dense range, this holds for all $f \in H^\infty(\Sigma_\theta)$. Further, there exists a unique contractive algebra homomorphism $\Phi : A(\Sigma_{\frac{\pi}{2}}) \to B(X)$, which coincides with the H^∞-calculus for A on $\bigcup_{\theta > \frac{\pi}{2}} R(\Sigma_\theta)$.

Proof. For $r \in (0,1)$, let A_r be as in Lemma 4.3. By Lemma 4.3 (4), $\|b_w(A_r)\| \leq 1$ for all $w \in \Sigma_{\frac{\pi}{2}}$. Therefore, $f(A_r)$ is a contraction for all $f \in B_{\Sigma_{\frac{\pi}{2}}}$ and thus for all $f \in \operatorname{co} B_{\Sigma_{\frac{\pi}{2}}}$. Let $f \in A(\Sigma_{\frac{\pi}{2}})$ such that $\|f\|_{\infty,\frac{\pi}{2}} \leq 1$. By Lemma 3.1, there exists a sequence $(f_n)_n \subset \operatorname{co} B_{\Sigma_{\frac{\pi}{2}}}$ with $\|f_n - f\|_{\infty,\frac{\pi}{2}} \to 0$. Since $A_r \in B(X)$ with $\sigma(A_r) \subset \Sigma_{\frac{\pi}{2}}$, by the Dunford calculus $f_n(A_r) \to f(A_r)$ and hence $\|f(A_r)\| \leq 1$. (The a priori continuity of $A(\Sigma_{\frac{\pi}{2}}) \to B(X)$, $f \mapsto f(A_r)$ was the reason for the introduction of the A_r.) If now $f = f_0 + a + b(1+(\cdot))^{-1} \in R(\Sigma_\theta)$ for some $\theta > \omega > \frac{\pi}{2}$, then for $r \to 1$: $f(A_r) = (2\pi i)^{-1} \int_{\partial\Sigma_\omega} f(\lambda)R(\lambda, A_r)d\lambda + a \operatorname{Id}_X + b(1+A_r)^{-1} \to (2\pi i)^{-1} \int_{\partial\Sigma_\omega} f(\lambda)R(\lambda, A)d\lambda + a \operatorname{Id}_X + b(1+A)^{-1} = f(A)$ by the convergence of the resolvents (Lemma 4.3 (3)), the uniform sectoriality (Lemma 4.3 (2)) and Lebesgue's theorem. This shows $\|f(A)\| \leq \|f\|_{\infty,\frac{\pi}{2}}$. Existence and uniqueness of Φ now follow from the density of $\bigcup_{\theta > \frac{\pi}{2}} R(\Sigma_\theta)$ in $A(\Sigma_{\frac{\pi}{2}})$.
If A has dense range, then for $f \in H^\infty(\Sigma_\theta)$, $\|f(A)\| \leq \liminf_n \|f\rho_n\|_{\infty,\frac{\pi}{2}} = \|f\|_{\infty,\frac{\pi}{2}}$, where $\rho_n(\lambda) = \left(\lambda(1+\lambda)^{-2}\right)^{\frac{1}{n}}$. □

Combining Theorems 4.2 and 4.4 with lemma 4.1 enables one to deduce the following well-known Hilbert space result. However, our proof is different from the ones given in [NF] or [ADM].

Corollary 4.5. *Let H be a Hilbert space. Suppose that A is a θ-sectorial operator on H with dense range such that A is accretive.*

1. *If $\theta < \frac{\pi}{2}$, then A has a contractive H^∞-calculus for the angle $\frac{\pi}{2}$.*
2. *If $\theta = \frac{\pi}{2}$, then A has a contractive H^∞-calculus for any angle $> \frac{\pi}{2}$.*

Remark 4.6. The key argument in both theorem 4.2 and 4.4 is to show first the contractivity of Blaschke operators and then use a density argument. We point out that in [vN], von Neumann uses a similar approach in the original proof of the inequality named after him.

5. Square integrable vectors and the calculus on $\Sigma_{\frac{\pi}{2}}$

It is not always possible to extend the holomorphic functional calculus for contractions in Hilbert spaces as constructed in [NF] to all of $H^\infty(\mathbb{D})$. Likewise, one cannot expect an $H^\infty(\Sigma_{\frac{\pi}{2}})$-calculus for all operators satisfying condition (4.1). In the Hilbert space case, it can be derived from [NF], Sections III.2 and III.8, that an accretive operator A has an $H^\infty(\Sigma_{\frac{\pi}{2}})$- calculus if the unitary part of A (in the sense of [NF], III.8) has a uniformly continuous spectral measure. In particular, A has an $H^\infty(\Sigma_{\frac{\pi}{2}})$-calculus if it is completely non-unitary (see [NF] III.8). It is also known that in these circumstances A has a dense set D of "square integrable vectors" in the sense that

$$\int_0^\infty |\langle e^{-tA}x, x'\rangle|^2 dt \le C_x \|x'\|^2 \text{ for all } x \in D \text{ and } x' \in X' \tag{5.1}$$

(see [Da1] Section 6.5). In the Banach space setting, one has to find a replacement for the notions of "absolutely continuous spectral measure" and "completely non-unitary operators". Davies' result convinced us that condition (5.1) may be a good replacement for these notions. It also allows us to prove an extension of the Hilbert space result quoted above in the Banach space setting. But first we connect square integrable vectors of negative generators of semigroups to square summable vectors of its Cayley transform.

Lemma 5.1. *Let $-A$ be the generator of a c_0-semigroup e^{-tA} on X. Let $x' \in X'$, $x \in D(A)$ and $y = \frac{1}{2}(1+A)x$. Suppose that $\int_0^\infty |\langle e^{-tA}y, x'\rangle|^2 dt < \infty$. Then for $T = (A-1)(A+1)^{-1} \in B(X)$:*

$$\sum_{n=0}^\infty |\langle T^n x, x'\rangle|^2 = 2\int_0^\infty |\langle e^{-tA}y, x'\rangle|^2 dt.$$

Proof. Put $f(\lambda) = \langle (\lambda + \frac{1}{2}A)^{-1}y, x'\rangle$. Then $f(\lambda) = \int_0^\infty e^{-\lambda t} g(t)dt$ with $g(t) = \langle e^{-\frac{t}{2}A}y, x'\rangle$ and $\|g\|_{L^2(0,\infty)} < \infty$. Further, $f^{(k)}(\lambda) = k!(-1)^k \langle (\lambda + \frac{1}{2}A)^{-k-1}y, x'\rangle$

for $k \in \mathbb{N}_0$. Now for $n \in \mathbb{N}_0$

$$\begin{aligned}\langle T^n x, x'\rangle &= \langle [(A-1)(A+1)^{-1}]^n x, x'\rangle = \langle (\mathrm{Id}_X - 2(A+1)^{-1})^n x, x'\rangle \\ &= \sum_{k=0}^{n} \binom{n}{k} (-1)^k \left\langle \left(\frac{1}{2} + \frac{1}{2}A\right)^{-k} x, x' \right\rangle \\ &= \sum_{k=0}^{n} \binom{n}{k} \frac{1}{k!} k! (-1)^k \left\langle \left(\frac{1}{2} + \frac{1}{2}A\right)^{-k-1} y, x' \right\rangle = \sum_{k=0}^{n} \binom{n}{k} \frac{1}{k!} f^{(k)}\left(\frac{1}{2}\right) =: q_n.\end{aligned}$$

By [Sho] (see also [Roo] Theorem 1 for the case $\nu = 0$), the fact that f is the Laplace transform of a function in $L^2(0,\infty)$ implies that

$$2\int_0^\infty |\langle e^{-tA} y, x'\rangle|^2 dt = \int_0^\infty |g(t)|^2 dt = \sum_{n=0}^\infty |q_n|^2 = \sum_{n=0}^\infty |\langle T^n x, x'\rangle|^2.$$

□

Theorem 5.2. *Let X be a Banach space and $D \subset X$ a dense subset. Let A be a $\frac{\pi}{2}$-sectorial operator on X satisfying (4.1) and (5.1), i.e.,*

1. $\|(A-w)x\| \le \|(A+\overline{w})x\|$ *for all* $x \in D(A)$ *and* $\mathrm{Re}\, w > 0$.
2. $\forall x \in D\, \exists C > 0\, \forall x' \in X' : \int_0^\infty |\langle e^{-tA} x, x'\rangle|^2 dt \le C\|x'\|^2$.

Then there is a unique extension of the $A(\Sigma_{\frac{\pi}{2}})$-calculus in Theorem 4.4 to a linear, multiplicative and contractive $\Phi : H^\infty(\Sigma_{\frac{\pi}{2}}) \to B(X)$ with the following convergence property:
If $f, f_1, f_2, \ldots \in H^\infty(\Sigma_{\frac{\pi}{2}})$ with $\sup_{n\in\mathbb{N}} \|f_n\|_{\infty,\frac{\pi}{2}} < \infty$ and $f_n(\lambda) \to f(\lambda)$ for a.a. $\lambda \in i\mathbb{R}$, then $\Phi(f_n)x \to \Phi(f)x$ for all $x \in X$.

Proof. For $r \in (0,1)$ and $f \in H^\infty(\mathbb{D})$, we put $f_r(\lambda) = f(r\lambda)$, so that $f_r \in A(\mathbb{D})$. Let $T = (A-1)(A+1)^{-1}$. We claim:

$$\text{For } f(z) = \sum_{n=0}^\infty a_n z^n \in H^\infty(\mathbb{D}) \text{ and } x \in X,\ f_r(T)x \text{ converges in } X \text{ for } r \to 1. \tag{5.2}$$

Let first $x \in (\frac{1}{2} + \frac{1}{2}A)^{-1}(D)$ and $x' \in X'$. By Lemma 5.1, the sequence $\langle T^n x, x'\rangle$ is in l^2 with $\|(\langle T^n x, x'\rangle)_n\|_{l^2} \le C\|x'\|$. So for $r, s \in (0,1)$, $r > s$,

$$\begin{aligned}|\langle f_r(T)x - f_s(T)x, x'\rangle| &\le \sum_{n=0}^\infty |a_n(r^n - s^n)\langle T^n x, x'\rangle| \\ &\le (1 - \frac{s^N}{r^N}) \sum_{n=0}^N |a_n|\, |\langle T^n x, x'\rangle| + 2\|(a_n)_n\|_{l^2(\{N+1,N+2,\ldots\})}\, \|\left(\langle T^n x, x'\rangle\right)_n\|_{l^2}.\end{aligned}$$

Since the a_n are the Fourier coefficients of $f \in H^\infty(\mathbb{D}) \subset L^2(\partial\mathbb{D})$, we have $(a_n)_n \in l^2$. Now, for a given $\epsilon > 0$, choose first N large, and then r, s sufficiently near to 1 in order to dominate the expression by ϵ, uniformly for $\|x'\| \le 1$. This gives (5.2) for $x \in (\frac{1}{2} + \frac{1}{2}A)^{-1}(D)$. If $x \in X$ is arbitrary, we can choose $x_\epsilon \in (\frac{1}{2} + \frac{1}{2}A)^{-1}(D)$ such that $\|x - x_\epsilon\| \le \epsilon$. Indeed, since D is dense in X and $(\frac{1}{2} + \frac{1}{2}A)^{-1}$ is an isomorphism $(X, \|\cdot\|_X) \to (D(A), \|\cdot\|_X + \|A\cdot\|_X)$, $(\frac{1}{2}+\frac{1}{2}A)^{-1}(D)$ is dense in $D(A)$ with respect to $\|\cdot\|_X + \|A\cdot\|_X$, and thus also with respect to $\|\cdot\|_X$. Finally, $D(A)$ is dense in X. Then

$$\begin{aligned}\|f_r(T)x - f_s(T)x\| &\le \|f_r(T)(x-x_\epsilon)\| + \|f_s(T)(x-x_\epsilon)\| + \|f_r(T)x_\epsilon - f_s(T)x_\epsilon\| \\ &\le 2\epsilon \sup_{t\in(0,1)} \|f_t(T)\| + \epsilon\end{aligned}$$

for r, s close to 1. Note that $\|f_t(T)\| \le \|f_t\|_{\infty,\mathbb{D}}$. Indeed, the assumption 1. implies by theorem 4.4 that $\|p(T)\| = \|p\circ\tau(A)\| \le \|p\circ\tau\|_{\infty,\frac{\pi}{2}} = \|p\|_{\infty,\mathbb{D}}$ for any polynomial p. By means of the density of the polynomials in $A(\mathbb{D})$, we conclude that $\|f_t(T)\| \le \|f_t\|_{\infty,\mathbb{D}} \le \|f\|_{\infty,\mathbb{D}}$ and finally get (5.2). Now define

$$\text{for } f \in H^\infty(\mathbb{D}) : \Psi(f)x = \lim_{r\to 1} f_r(T)x.$$

Then $\Psi(p) = p(T)$ for any polynomial p, since $\|p_r(T) - p(T)\| \le \|p_r - p\|_{\infty,\mathbb{D}} \to 0$ for $r \to 1$. It is clear that $\Psi : H^\infty(\mathbb{D}) \to B(X)$ is linear and contractive. To show the multiplicativity, let $f, g \in H^\infty(\mathbb{D})$. Then $\Psi(fg)x = \lim_r (fg)_r(T)x = \lim_r f_r(T)g_r(T)x = \lim_r f_r(T)(\lim_s g_s(T)x) = \Psi(f)\Psi(g)x$, where the penultimate equality follows from $\sup_{r\in(0,1)} \|f_r(T)\| < \infty$. Now we pull back Ψ to $H^\infty(\Sigma_{\frac{\pi}{2}})$ and put

$$\Phi : H^\infty(\Sigma_{\frac{\pi}{2}}) \to B(X),\ \Phi(f) = \Psi(f\circ\zeta),$$

with $\zeta(z) = -\frac{z+1}{z-1}$ as in Section 3. For a polynomial p, $\Phi(p\circ\tau) = \Psi(p) = p(T) = p\circ\tau(A)$, so that Φ coincides with Φ from Theorem 4.4 on the set $\{p\circ\tau : p \text{ polynomial}\}$, which is dense in $A(\Sigma_{\frac{\pi}{2}})$. Therefore, Φ coincides on $A(\Sigma_{\frac{\pi}{2}})$ with the Φ from Theorem 4.4.
We now show the convergence property of the calculus. Let $f, f_1, f_2, \ldots$ be as in the assumption. Put $g = f\circ\zeta$ and $g_n = f_n\circ\zeta$ and denote a_k and $a_k^{(n)}$ their Fourier coefficients. Then by Lebesgue's theorem, $g_n \to g$ in $L^2(\partial\mathbb{D})$. If $x \in (\frac{1}{2}+\frac{1}{2}A)^{-1}(D)$ and $x' \in X'$,

$$\begin{aligned}|\langle\Phi(f_n - f)x, x'\rangle| &= |\langle\Psi(g_n - g)x, x'\rangle| = |\lim_r \sum (a_k^{(n)} - a_k)r^k\langle T^k x, x'\rangle| \\ &= |\sum (a_k^{(n)} - a_k)\langle T^k x, x'\rangle| \le \|(a_k^{(n)} - a_k)_k\|_{l^2} C_x \|x'\| \to 0.\end{aligned}$$

The statement for arbitrary $x \in X$ follows again from the density of $(\frac{1}{2}+\frac{1}{2}A)^{-1}(D)$ and $\|\Psi(g_n)\| \le \|g_n\|_\infty \le C$.
The uniqueness of the constructed functional calculus follows from this convergence property and another appeal to (4.1). □

The proof shows of course the following counterpart for contractions:

Corollary 5.3. *Let $T \in B(X)$ have a contractive $A(\mathbb{D})$-calculus and satisfy*

$$\sum_{k=0}^{\infty} |\langle T^k x, x'\rangle|^2 \le C_x \|x'\|^2 \text{ for } x' \in X' \text{ and } x \text{ in a dense subset of } X.$$

Then $f(T)x = \lim_{r\to 1} f_r(T)x$ *extends the* $A(\mathbb{D})$*-calculus to a contractive algebra homomorphism* $H^\infty(\mathbb{D}) \to B(X)$ *which has the obvious analogue of the convergence property in Theorem* 5.2.

Remark 5.4. To define an H^∞-calculus for a sectorial operator A on all of $H^\infty(\Sigma_{\frac{\pi}{2}})$, one usually needs the assumption $\overline{R(A)} = X$. In Theorem 5.2, no such assumption is made explicitly. Nevertheless, any A satisfying (4.1) and (5.1) has dense image. Indeed, let $f_n(\lambda) = \frac{n\lambda}{1+n\lambda}$ and $f(\lambda) \equiv 1$. Then $f_n(\lambda) \to f(\lambda)$ for $\lambda \in i\mathbb{R}\backslash\{0\}$, and therefore, by Theorem 5.2, $\Phi(f_n)x \to \Phi(f)x$ for every $x \in X$. Since $f, f_n \in R(\Sigma_\theta)$ for any $\theta \in (\frac{\pi}{2}, \pi)$, we know from Theorem 4.4 that $\Phi(f), \Phi(f_n)$ are given by the H^∞-calculus. Thus, $\Phi(f_n)x = -AR(-\frac{1}{n}, A)x \in R(A)$ and $\Phi(f)x = x$. This shows $\overline{R(A)} = X$.

6. An example

The proof of Theorem 4.2 shows that a θ-sectorial operator A on a Banach space X with $\theta < \frac{\pi}{2}$ has a contractive $H^\infty(\Sigma_{\frac{\pi}{2}})$-calculus if and only if for all Blaschke factors $b_w(\lambda) = (\lambda - w)(\lambda + \overline{w})^{-1}$ with $\operatorname{Re} w > 0$, we have $\|b_w(A)\| \le 1$, or, equivalently,

$$\| \operatorname{Id}_X + 2 \operatorname{Re} w R(-\overline{w}, A)\| \le 1 \text{ for all } \operatorname{Re} w > 0. \tag{6.1}$$

This condition allows one to show that many operators which are accretive in a scale of $L^p(K,\mu)$-spaces cannot have a contractive H^∞-calculus on the whole scale. Indeed if K is a compact metric space, we say that $T \in B(C(K))$ satisfies the Daugavet equation (see for example [WW]), if

$$\| \operatorname{Id}_X + T\| = 1 + \|T\|. \tag{6.2}$$

Hence if $X = C(K)$ and $T = -2\operatorname{Re}\lambda R(\lambda, A)$ satisfies (6.2) for some λ with $\operatorname{Re}\lambda < 0$, then (6.1) cannot hold. This is the case for a large class of operators A, for example when $R(\lambda, A)$ is weakly compact or an integral operator

$$Tf(y) = \int_K k(x,y) f(y) dy$$

with a measurable kernel. For these and weaker conditions for (6.2), see [WW].

References

[ADM] D. Albrecht, X. Duong, A. McIntosh, *Operator theory and harmonic analysis.* Proc. Centre Math. Appl. Austral. Nat. Univ. 34(pt.3), 77–136 (1996).

[CDMY] M. Cowling, I. Doust, A. McIntosh, A. Yagi, *Banach space operators with a bounded H^∞ functional calculus.* J. Austral. Math. Soc., Ser. A 60, No.1, 51–89 (1996).

[Da1] E.B. Davies, *One parameter semigroups.* London Mathematical Society, Monographs, No.15, 230p. (1980).

[Da2] E.B. Davies, *Non-unitary scattering and capture. I: Hilbert space theory.* Comm. Math. Phys. 71, 277–288 (1980).

[Dru] S. Drury, *Remarks on von Neumann's inequality.* Proc. Spec. Year Analysis, Univ. Conn. 1980–81, Lecture Notes in Math. 995, 12–32 (1983).

[Foi] C. Foiaş, *Sur certains théorèmes de J. von Neumann concernant les ensembles spectraux.* Acta Sci. Math. Szeged 18, 15–20 (1957).

[Gar] J.B. Garnett, *Bounded analytic functions. Revised* 1*st ed.* Graduate Texts in Mathematics 236. Springer-Verlag. xiv, 460 p. (2006).

[KW] P.C. Kunstmann, L. Weis, *Maximal* L_p*-regularity for parabolic equations, Fourier multiplier theorems and* H^∞*-functional calculus.* Lecture Notes in Math. 1855, 65–311 (2004).

[LM] C. Le Merdy, H^∞*-functional calculus and applications to maximal regularity.* Publ. Math. UFR Sci. Tech. Besançon. 16, 41–77 (1998).

[NF] B. Sz.-Nagy, C. Foiaş, *Harmonic analysis of operators on Hilbert space.* North-Holland Publishing Co. xiii, 387 p. (1970).

[vN] J. von Neumann, *Eine Spektraltheorie für allgemeine Operatoren eines unitären Raumes.* Math. Nachr. 4, 258–281 (1951).

[Roo] P.G. Rooney, *Laplace transforms and generalized Laguerre polynomials.* Canad. J. Math. 10, 177–182 (1958).

[Sho] J. Shohat, *Laguerre polynomials and the Laplace transform.* Duke Math. J. 6, 615–626 (1940).

[WW] L. Weis, D. Werner, *The Daugavet equation for operators not fixing a copy of* $C[0,1]$. J. Operator Theory 39, No.1, 89–98 (1998).

Christoph Kriegler and Lutz Weis
Institut für Analysis
Universität Karlsruhe
Englerstraße 2
D-76128 Karlsruhe, Germany
e-mail: `kriegler@math.uni-karlsruhe.de`
`weis@math.uni-karlsruhe.de`

Operator Theory:
Advances and Applications, Vol. 195, 245–256

Invertibility in the Algebra of τ-measurable Operators

Isaac Daniel Tembo

Abstract. Let $\mathcal{M}$ be a semi-finite von Neumann algebra in a Hilbert space H and τ be a faithful normal semi-finite trace on $\mathcal{M}$. The algebra $\widetilde{\mathcal{M}}$ of τ-measurable operators affiliated with $\mathcal{M}$, equipped with the topology of convergence in measure, is a topological$*$-algebra. We show that inversion is continuous on Q, the set of invertible elements in $\widetilde{\mathcal{M}}$, and that Q is open if and only if $\widetilde{\mathcal{M}} = \mathcal{M}$.

Mathematics Subject Classification (2000). Primary 47L60; Secondary 46B5.

Keywords. von Neumann algebra, τ-measurable operator, invertibility

1. Introduction

In a Banach algebra the set $\mathcal{Q}$ of invertible elements is open and inversion is continuous on $\mathcal{Q}$. A topological algebra in which the set of invertible elements is open is called a $\mathcal{Q}$-algebra. Every Banach algebra and some commutative locally m-convex algebras are $\mathcal{Q}$-algebras, but this is not the case for topological algebras in general. The purpose of this paper is to determine when the algebra of τ-measurable operators affiliated with a semi-finite von Neumann algebra with semi-finite faithful normal trace τ is a $\mathcal{Q}$-algebra, and to investigate continuity of inversion in such algebras. The next section consists of preliminaries needed to understand and prove the main results, which are presented in the final section.

The author wishes to thank Jurie Conradie of the Department of Mathematics and Applied Mathematics of the University of Cape Town for useful suggestions regarding the proofs of the main results of this paper.

2. The algebra $\widetilde{\mathcal{M}}$

The algebra of bounded linear operators from a Hilbert space H into itself will be denoted by $\mathcal{B}(H)$. Throughout the paper $\mathcal{M}$ will be a semi-finite von Neumann algebra on the Hilbert space H, equipped with a faithful semi-finite normal trace τ. Denote by $\mathcal{M}^p$ the lattice of self-adjoint projections in $\mathcal{M}$, and by $\mathcal{M}^+$ the set of positive elements of $\mathcal{M}$. I denotes the identity of $\mathcal{M}$, and $\|\cdot\|$ the C*-norm on $\mathcal{M}$. We say that a closed densely defined operator T, with domain denoted by $D(T)$, is affiliated with a von Neumann algebra $\mathcal{M}$ (and we write $T\eta\mathcal{M}$) when $U^*TU = T$ for each unitary operator $U \in \mathcal{M}'$, the commutant of $\mathcal{M}$.

Definition 2.1. ([5], Definition 2.1) *Let E be a subspace of H. We say that E is τ-dense in H if for every $\delta > 0$, there exists a projection $P \in \mathcal{M}^p$ such that $P(H) \subseteq E$ and $\tau(I - P) < \delta$.*

Definition 2.2. *We define the set of τ-measurable operators, denoted by $\widetilde{\mathcal{M}}$, as the set of all closed densely defined operators S affiliated with $\mathcal{M}$ with a τ-dense domain $D(S)$.*

Proposition 2.3. ([13], 1.21) *Suppose S is a closed densely defined operator affiliated with $\mathcal{M}$, and $S = v|S|$ is the polar decomposition of S.*
The following are equivalent:
- (a) $S \in \widetilde{\mathcal{M}}$
- (b) $|S| \in \widetilde{\mathcal{M}}$
- (c) *there exists a $t > 0$ such that $\tau(E_{(t,\infty)}(|S|)) < \infty$.*

The set $\widetilde{\mathcal{M}}$ is a *-algebra with sum and product operations defined as the respective closures of the algebraic sum and product, and usual scalar multiplication. The sets

$$\widetilde{\mathcal{M}}(\epsilon, \delta) = \{S \in \widetilde{\mathcal{M}}\ ;\ \exists P \in \mathcal{M}^p :\ P(H) \subseteq D(S), \|SP\| \leq \epsilon, \tau(I - P) \leq \delta\}$$

for $\epsilon > 0, \delta > 0$, form a neighbourhood basis at 0 for a metrisable vector topology on $\widetilde{\mathcal{M}}$, called the *topology of convergence in measure* (τ_m). $\widetilde{\mathcal{M}}$ with the topology τ_m, is a complete metrisable topological *-algebra in which $\mathcal{M}$ is dense. We write $S_i \to_{\tau_m} S$ in case a net $(S_i)_{i\in I} \subset \widetilde{\mathcal{M}}$ converges to $S \in \widetilde{\mathcal{M}}$ for the measure topology on $\widetilde{\mathcal{M}}$.

Example 2.4. *If $\mathcal{M}$ is a commutative von Neumann algebra, then $\mathcal{M}$ is *-isomorphic to $L_\infty(X, \Sigma, \mu)$ for some localizable measure space (X, Σ, μ), where $L_\infty(X, \Sigma, \mu)$ is the space of equivalence classes of a.e. complex-valued essentially bounded measurable functions on (X, Σ, μ). With trace the integral, it is well known that $\widetilde{\mathcal{M}}$ is the set of all complex measurable functions that are bounded except on a set of finite measure.*

Suppose $S \in \widetilde{\mathcal{M}}$ and $S = v|S|$ is the polar decomposition of S. We define the *distribution function* of S, denoted by $d_t(S)$, by

$$d_t(S) = \tau(E_{(t,\infty)}(|S|)),$$

and the *generalised singular function* of S by

$$\mu_t(S) = \inf\{s \geq 0 :\ d_s(S) \leq t\} = \inf\{s \geq 0 :\ \tau(E_{(s,\infty)}(|S|)) \leq t\}.$$

For the properties of the generalised singular functions, the reader can consult [5]. The generalised singular function and the topology of convergence in measure are related in the following manner:

$$S \in \widetilde{\mathcal{M}}(\epsilon, t) \Leftrightarrow \mu_t(S) \leq \epsilon.$$
$$S_i \rightarrow_{\tau_m} S \Leftrightarrow \mu_t(S_i - S) \rightarrow 0 \text{ for all } t > 0.$$

Suppose $S \in \widetilde{\mathcal{M}}$. Let $\lim_{t\to\infty} \mu_t(S)$ be denoted by $\mu_\infty(S)$. We define $\widetilde{\mathcal{M}}_0$ to be the kernel of μ_∞, i.e., $\widetilde{\mathcal{M}}_0 = \{S \in \widetilde{\mathcal{M}} : \mu_\infty(S) = 0\}$.

Proposition 2.5. [11] *The following are equivalent:*

(a) $\widetilde{\mathcal{M}} = \mathcal{M}$
(b) $\inf\{\tau(P) : P \in \mathcal{M}^p : \tau(P) \neq 0\} > 0$
(c) $\widetilde{\mathcal{M}}_0 = \mathcal{M}_0$, *where* $\widetilde{\mathcal{M}}_0 \cap \mathcal{M} = \mathcal{M}_0$.

An operator $S \in \widetilde{\mathcal{M}}$ is said to be invertible in $\widetilde{\mathcal{M}}$ if there exists an element $T \in \widetilde{\mathcal{M}}$ such that $\overline{ST} = \overline{TS} = I$, where the bar indicates closures.

Definition 2.6. *For $S \in \widetilde{\mathcal{M}}$, define*

$$\rho_{\widetilde{\mathcal{M}}}(S) = \{\lambda \in \mathbb{C} : (\lambda I - S)^{-1} \in \widetilde{\mathcal{M}}\},$$

that is to say there exists an operator $T \in \widetilde{\mathcal{M}}$ such that

$$\overline{T(\lambda I - S)} = \overline{(\lambda I - S)T} = I \text{ and we write } T = (\lambda I - S)^{-1}.$$

We then define the spectrum of $S \in \widetilde{\mathcal{M}}$ relative to $\mathcal{M}$ by

$$\sigma_{\widetilde{\mathcal{M}}}(S) = \mathbb{C} \backslash \rho_{\widetilde{\mathcal{M}}}(S).$$

Define the point spectrum of S, denoted by $\sigma_p(S)$, by

$$\sigma_p(S) = \{\lambda \in \mathbb{C} : \exists x \in H, x \neq 0 \text{ such that } Sx = \lambda x\}.$$

Denote by π the canonical quotient map $\pi : \widetilde{\mathcal{M}} \rightarrow \widetilde{\mathcal{M}}/\widetilde{\mathcal{M}}_0$. Since $\widetilde{\mathcal{M}}_0$ is an ideal, we have that $\widetilde{\mathcal{M}}/\widetilde{\mathcal{M}}_0$ is an algebra with identity. Then we define the essential spectrum of S, denoted by $\sigma_e(S)$, as

$$\sigma_e(S) = \sigma_{\widetilde{\mathcal{M}}/\widetilde{\mathcal{M}}_0}(\pi(S)).$$

It is immediate that if $S \in \widetilde{\mathcal{M}}_0$ then $\sigma_e(S) = \{0\}$. We shall need the following type of spectral representation first introduced by Ovchinnikov in [9] and extended in [2].

Proposition 2.7. [2] *Let $0 \leq S \in \widetilde{\mathcal{M}}$, with spectral family $\{E_\lambda(S) : \lambda \geq 0\}$. Let $\mu_t(S)$ be the decreasing rearrangement of S. Let*

$$\lambda_0 = \lim_{t\to\infty} \mu_t(S) \quad \text{and} \quad \alpha_0 = \inf\{t > 0 : \mu_t(S) = \lambda_0\}.$$

and $S_0 = (I - E_{\lambda_0}(S))S$. Then

$$S_0 = \int_0^\infty \mu_t(S_0) d\tilde{E}_t(S),$$

where

$$\tilde{E}_t(S) = \begin{cases} 0 & t < 0 \\ (I - E_{\lambda_0}(S))(I - E_{\mu_t(S)_-}(S)) & 0 \le t < \alpha_0 \\ I - E_{\lambda_0}(S) & t \ge \alpha_0 \end{cases}$$

This representation of S is called the Schmidt spectral decomposition.

3. The main results

We shall need the following elementary lemmas.

Lemma 3.1. *Suppose $S, T \in \widetilde{\mathcal{M}}$ are such that $\overline{ST} = \overline{TS} = I$. Then* $\operatorname{ran}(S)$ *is τ-dense in H and $T = S^{-1}$ where $S^{-1} : \operatorname{ran}(S) \to D(S)$ is defined by $S^{-1}y = x$ where $y = Sx \in \operatorname{ran}(S)$, for some $x \in D(S)$.*

Proof. We have $STx = x$ for all $x \in D(ST)$. Hence if $x \in D(ST)$, $x = STx \in \operatorname{ran}(S)$ and therefore $D(ST) \subseteq \operatorname{ran}(S)$. But by ([13], 1.20), $D(ST)$ is τ-dense. Hence $\operatorname{ran}(S)$ is τ-dense.
Also we have $TSx = x$ for all $x \in D(TS)$, where $D(TS) = \{x \in D(S) : \; Sx \in D(T)\}$. Hence $Ty = x$ if $y = Sx \in \operatorname{ran}(S)$ and $y \in D(T)$. Let $D = \operatorname{ran}(S) \cap D(T)$. Then D is τ-dense, since for all $\delta > 0$, there exist projections $P_1, P_2 \in \mathcal{M}^p$ such that $P_1(H) \subseteq \operatorname{ran}(S)$ and $\tau(I - P_1) \le \frac{\delta}{2}$, and $P_2(H) \subseteq D(T)$ and $\tau(I - P_2) \le \frac{\delta}{2}$. Let $P = P_1 \wedge P_2$. Then $P(H) \subseteq D$ and $\tau(I - P) = \tau(I - P_1 \wedge P_2) = \tau((I - P_1) \vee (I - P_2) \le \tau(I - P_1) + \tau(I - P_2) \le \delta$, showing that D is τ-dense. We also have that $T|_D \subseteq S^{-1}$. Now S^{-1} is closed and by Lemma 3.2, $S^{-1}\eta\mathcal{M}$. Since from what we have shown above, $\operatorname{ran}(S)$ is τ-dense, we have that $S^{-1} \in \widetilde{\mathcal{M}}$. By the uniqueness of extensions in $\widetilde{\mathcal{M}}$ ([13], 1.15), $T = S^{-1}$. □

Lemma 3.2. *Let S be a closed, densely defined injective normal operator affiliated with a von Neumann algebra $\mathcal{M}$. Then its inverse $S^{-1} :$ $\operatorname{ran}(S) \to D(S)$, defined by $S^{-1}(Sx) = x$ for every $x \in D(S)$, is also affiliated with $\mathcal{M}$.*

Proof. Since S is normal, $SS^* = S^*S$ so that $|S| = |S^*|$. Then we have $\ker(S) = \ker(|S|) = \ker(|S^*|) = \ker(S^*)$. Since S is injective we have that S^* is also injective so that $\operatorname{ran}(S)$ is dense in H, so that S^{-1} is densely defined. Since S is closed, S^{-1} is closed as well. Since $S\eta\mathcal{M}$, we have $US = SU$ for every unitary $U \in \mathcal{M}'$, that is

$$D(S) = D(SU) \quad \text{and} \quad USx = SUx \quad \text{for all } x \in D(S). \tag{1}$$

We want to show that for every $U \in \mathcal{M}'$, $US^{-1} = S^{-1}U$. Let $x \in \operatorname{ran}(S)$, then $x = Sz$ for some $z \in D(S) = D(SU)$. Then we have

$$Ux = USz = SUz \in \operatorname{ran}(S) = D(S^{-1}), \qquad \text{from equation (1).} \tag{2}$$

Hence $x \in D(S^{-1}U)$. Thus $\operatorname{ran}(S) \subseteq D(S^{-1}U)$.

Conversely, if $x \in D(S^{-1}U)$, then $Ux \in D(S^{-1}) = \operatorname{ran}(S)$ so that $x = U^*Ux \in U^*(\operatorname{ran}(S))$. Now $x = U^*Ux = U^*Sz = SU^*z \in \operatorname{ran}(S)$, for some $z \in D(S)$ from equation (1) Hence $D(S^{-1}U) \subseteq \operatorname{ran}(S)$, and so, $D(S^{-1}U) = \operatorname{ran}(S) = D(S^{-1})$.

Now if $x \in D(S^{-1}) = \text{ran}(S)$, $x = Sz$, for some $z \in D(S)$. Now $S^{-1}Ux = S^{-1}USz = S^{-1}SUz = Uz = US^{-1}x$. Hence we have that $S^{-1}\eta\mathcal{M}$. □

Proposition 3.3. *Suppose $S \in \widetilde{\mathcal{M}}^{sa}$ and $\{E_\lambda(S)\}_{\lambda\in\mathbb{R}}$ is the spectral family for S. S is invertible in $\widetilde{\mathcal{M}}$ if and only if S is injective and there exists a $t > 0$ such that*

$$\tau(E_{(-t,\, t)}(S)) < \infty.$$

In this case $S^{-1} = h(S)$, where $h : \mathbb{R} \to \mathbb{R}$ is defined by

$$h(\lambda) = \begin{cases} \frac{1}{\lambda}, & \lambda \neq 0 \\ 0, & \lambda = 0 \end{cases}$$

Proof. Suppose S is injective and there exists a $t > 0$ such that $\tau(E_{(-t,t)}(S)) < \infty$. Since S is injective, it has an inverse $S^{-1} : \text{ran}(S) \to D(S)$, and since S is self-adjoint, ran(S) is dense in H since $R(S) = I - N(S^*) = I - N(S) = I$. By Lemma 3.2, $S^{-1}\eta\mathcal{M}$ and $S^{-1} = h(S)$. Let h be as above and $g(\lambda) = |\lambda|$. By the functional calculus it follows that, for $\lambda > 0$,

$$\begin{aligned} E_\lambda(|h(S)|) &= E_{((g\circ h)^{-1}(-\infty,\lambda])}(S) \\ &= E_{(-\infty,\, -\frac{1}{\lambda}]}(S) + E_{[\frac{1}{\lambda},\, \infty)}(S). \end{aligned}$$

Thus we have, for $\lambda > 0$,

$$\begin{aligned} I - E_\lambda(|h(S)|) &= I - E_{(-\infty,\, -\frac{1}{\lambda}]}(S) - E_{[\frac{1}{\lambda},\, \infty)}(S) \\ &= E_{(-\frac{1}{\lambda},\frac{1}{\lambda})}(S). \end{aligned}$$

Now $h(S)$ is a closed densely defined operator affiliated with $\mathcal{M}$, by ([6], 5.6). And by assumption there exists a $t > 0$ such that

$$\tau(I - E_{\frac{1}{t}}(|h(S)|)) = \tau(E_{(-t,\, t)}(S)) < \infty.$$

Thus $h(S) \in \widetilde{\mathcal{M}}$.

By the functional calculus, we have $\overline{Sh(S)} = \overline{h(S)S} = \int_{-\infty}^{\infty} \lambda h(\lambda) dE_\lambda(S) = \int_{-\infty}^{0^-} dE_\lambda(S) + \int_{0^+}^{\infty} dE_\lambda(S) = I - (E_0(S) - E_{0^-}(S)) = I$ since $\{E_\lambda(S)\}$ is right continuous and since S is injective, and using ([7], 9.11-1), which can be extended to the unbounded case. Thus S is invertible in $\widetilde{\mathcal{M}}$ with inverse $h(S)$.

Conversely, suppose S is invertible in $\widetilde{\mathcal{M}}$. Then there is a $T \in \widetilde{\mathcal{M}}$ such that $\overline{ST} = \overline{TS} = I$. Suppose $Sx = 0$. Then $Sx \in D(T)$. Now since $TS \subseteq \overline{TS}$, we have $0 = TSx = \overline{TS}x = Ix = x$, and so S is injective. By Lemma 3.2, S has an inverse S^{-1} which is affiliated with $\mathcal{M}$, and $S^{-1} = h(S)$. It can then be shown that $S^{-1} = T \in \widetilde{\mathcal{M}}$. Thus there exists a $t > 0$ such that

$$\infty > \tau(I - E_{\frac{1}{t}}(|h(S)|)) = \tau(E_{(-t,\, t)}(S)).$$ □

Proposition 3.4. *Suppose $S \in \widetilde{\mathcal{M}}$, $S = U|S|$ the polar decomposition for S. Then*

1. *if S is invertible in $\widetilde{\mathcal{M}}$ then $|S|$ is invertible in $\widetilde{\mathcal{M}}$ and $|S|^{-1} = S^{-1}U$.*
2. *if S is normal and $|S|$ is invertible in $\widetilde{\mathcal{M}}$ then S is invertible in $\widetilde{\mathcal{M}}$.*

Proof. (1) Suppose S is invertible in $\widetilde{\mathcal{M}}$. Then there exists a $T \in \widetilde{\mathcal{M}}$ such that $\overline{ST} = \overline{TS} = I$. Also S invertible implies that S is injective and so $\mathrm{ran}(S^*)$ is dense in H. Furthermore $\mathrm{ran}(S)$ is τ-dense by Lemma 3.1. Hence U is a unitary, since it is a partial isometry with initial space the closure of $\mathrm{ran}(S^*)$ and final space the closure of $\mathrm{ran}(S)$. Thus $U^*U = UU^* = I$. We then have

$$\begin{aligned}
& \overline{ST} = \overline{TS} = I \\
\Rightarrow \quad & \overline{U|S|T} = \overline{TU|S|} = I \\
\Rightarrow \quad & I = \overline{U^*\overline{U|S|T}U} = \overline{U^*\overline{U|S|TU}} \\
& = \overline{U^*U|S|TU} = \overline{|S|TU} \\
\Rightarrow \quad & \overline{|S|TU} = \overline{TU|S|} = I,
\end{aligned}$$

by uniqueness of extensions ([13], 1.15), since $\overline{U^*\overline{U|S|T}U}$ and $\overline{U^*\overline{U|S|TU}}$ are both extensions of $U^*U|S|TU$. Since T is closed and U is bounded, $TU = \overline{TU} \in \widetilde{\mathcal{M}}$ is the inverse of $|S|$ in $\widetilde{\mathcal{M}}$.

(2) Suppose $|S|$ is invertible in $\widetilde{\mathcal{M}}$ and S is normal, then there is a $T \in \widetilde{\mathcal{M}}$ such that $\overline{T|S|} = I = \overline{|S|T}$. Since S is normal, $SS^* = S^*S$ so that $|S| = |S^*|$. Then we have $\ker(S) = \ker(|S|) = \ker(|S^*|) = \ker(S^*)$. Since S is injective then we have that S^* is also injective so that $\mathrm{ran}(S)$ and $\mathrm{ran}(S^*)$ are dense in H. Hence U is a unitary since U is a partial isometry with initial space the closure of $\mathrm{ran}(S^*)$ and final space the closure of $\mathrm{ran}(S)$. Since S is normal, it is affiliated with some abelian von Neumann subalgebra $\mathcal{A}$ of $\mathcal{M}$, by ([6], 5.6.18). From this we have that $U \in \mathcal{A}$ and $|S|\eta\mathcal{A}$, by ([10], 9.29). Now $|S|$ affiliated with $\mathcal{A}$ implies its inverse T is affiliated with $\mathcal{A}$, by Lemma 3.2. Thus $UTU^* = T$ since $U \in \mathcal{A} \subset \mathcal{A}'$. Now

$$\overline{TU^*S} = \overline{TU^*U|S|} = \overline{T|S|} = I.$$

Also

$$\begin{aligned}
\overline{STU^*} &= \overline{|S^*|UTU^*} \\
&= \overline{|S^*|T} \quad \text{since } UTU^* = T \\
&= \overline{|S|T} \quad \text{since } |S| = |S^*| \text{ when } S \text{ is normal} \\
&= I.
\end{aligned}$$

Thus we have that $TU^* \in \widetilde{\mathcal{M}}$ is the inverse of S in $\widetilde{\mathcal{M}}$. □

Proposition 3.5. [11] *Suppose $S \in \widetilde{\mathcal{M}}$ and S is self-adjoint, then*

$$\sigma_e(S) = \{\mu \in \mathbb{R} : \tau(E_{(\mu-\frac{1}{n},\, \mu+\frac{1}{n})}(S)) = \infty \quad \textit{for all}\, n \in \mathbb{N}\}.$$

Proposition 3.6. *Suppose $S \in \widetilde{\mathcal{M}}$ and S is self-adjoint*

$$\sigma_{\widetilde{\mathcal{M}}}(S) = \sigma_p(S) \cup \sigma_e(S)$$

Proof. For $\mu \in \mathbb{R}$, $\mu \notin \sigma_{\widetilde{\mathcal{M}}}(S)$ if and only if $S - \mu I$ is invertible in $\widetilde{\mathcal{M}}$. Let $E_\lambda(S)$ be the spectral family for S. Set $k(\lambda) = \lambda - \mu$. Then $k(S) = S - \mu I$. By the change of measure principle, $E_\lambda(k(S)) = E_{k^{-1}(-\infty,\lambda]}(S) = E_{(-\infty,\lambda+\mu]}(S)$ and so

$\{E_{\lambda+\mu}(S) : \lambda \in \mathbb{R}\}$ is the spectral family for $S - \mu I$. Now by Proposition 3.3

$$\lambda \notin \sigma_{\widetilde{\mathcal{M}}}(S) \Leftrightarrow S - \lambda I \text{ is injective and there exists a } t > 0 \text{ such that}$$
$$\tau(E_{(\lambda-t,\lambda+t)}(S)) = \tau(E_{(-t,t)}(S - \lambda I)) < \infty$$
$$\lambda \in \sigma_{\widetilde{\mathcal{M}}}(S) \Leftrightarrow S - \lambda I \text{ is not injective or } \tau(E_{(\lambda-t,\lambda+t)}(S)) = \infty \text{ for all } t > 0.$$
$$\Leftrightarrow \lambda \in \sigma_p(S) \text{ or } \lambda \in \sigma_e(S).$$

Thus we have that $\sigma_{\widetilde{\mathcal{M}}}(S) = \sigma_p(S) \cup \sigma_e(S)$. □

Lemma 3.7. *Suppose $0 \le S \in \widetilde{\mathcal{M}}_0$, $\alpha_0 < \infty$ and $\tau(I) = \infty$. Then $0 \in \sigma_p(S)$.*

Proof. If $S \in \widetilde{\mathcal{M}}_0$, then $\lambda_0 = 0$. If $\alpha_0 < \infty$ then

$$\tau(I - E_0(S)) = \alpha_o < \infty.$$

If $\tau(I) = \infty$, then $E_0(S) \neq 0$ so that $\ker(S) = E_0(S)(H) \neq \{0\}$. Hence $0 \in \sigma_p(S)$. □

Proposition 3.8. *Suppose $0 \le S \in \widetilde{\mathcal{M}}_0$ and let m be Lebesgue measure on the positive real line. Then*

$$\{\lambda > 0 : m\{t > 0 : \mu_t(S) = \lambda\} > 0\} \subseteq \sigma_p(S),$$

and

$$\sigma_p(S) \subseteq \{\lambda > 0 : m\{t > 0 : \mu_t(S) = \lambda\} > 0\} \cup \{0\}.$$

If furthermore $\tau(I) = \infty$ and $\alpha_0 < \infty$, then

$$\sigma_p(S) = \{\lambda \ge 0 : m\{t > 0 : \mu_t(S) = \lambda\} > 0\}.$$

Proof. Suppose $\lambda_1 > 0$ and $m\{t > 0 : \mu_t(S) = \lambda_1\} > 0$, then the set $\{t > 0 : \mu_t(S) = \lambda_1\}$ is an interval of constancy for $\mu_t(S)$. Thus λ_1 is a point of discontinuity for $d_\lambda(S)$. This implies that

$$\lim_{\lambda \uparrow \lambda_1} \tau(E_{(\lambda,\infty)}(S)) = d_{\lambda_1 - 0}(S) \neq d_{\lambda_1+0}(S) = d_{\lambda_1}(S), \tag{3}$$

using the right continuity of $d_\lambda(S)$. Now we show that $E_{\lambda_1-0}(S) = \lim_{\lambda\uparrow\lambda_1} E_\lambda(S) \neq E_{\lambda_1}(S)$. For each $0 < \lambda < \lambda_1$, $d_\lambda(S) < \infty$ since $S \in \widetilde{\mathcal{M}}_0$. Now suppose, to the contrary that $E_{\lambda_1-0}(S) = E_{\lambda_1}(S)$, then $E_\lambda(S) \uparrow E_{\lambda_1}(S)$ as $\lambda \uparrow \lambda_1$, so that $E_{(\lambda,\infty)}(S) \downarrow E_{(\lambda_1,\infty)}$ as $\lambda \uparrow \lambda_1$. By the normality of the trace and the fact that $d_\lambda(S) < \infty$, we have that

$$d_\lambda(S) = \tau(E_{(\lambda,\infty)}(S)) \downarrow \tau(E_{(\lambda_1,\infty)}(S)) = d_{\lambda_1}(S), \text{ as } \lambda \uparrow \lambda_1,$$

a contradiction to equation (3). Hence $E_{\lambda_1-0}(S) \neq E_{\lambda_1}(S)$, and by ([7], 9.11-1), we have

$$\ker(S - \lambda_1 I) = (E_{\lambda_1}(S) - E_{\lambda_1-0}(S))(H) \neq \{0\}.$$

Hence $\lambda_1 \in \sigma_p(S)$.
Now suppose $\lambda_1 > 0$ and $m\{t > 0 : \mu_t(S) = \lambda_1\} = 0$. Then $d_\lambda(S)$ is continuous at λ_1. Thus we have

$$\lim_{\lambda \uparrow \lambda_1} \tau(E_{(\lambda,\infty)}(S)) = \tau(E_{(\lambda_1,\infty)}(S)) < \infty, \text{ since } S \in \widetilde{\mathcal{M}}_0. \tag{4}$$

Suppose to the contrary that $I - E_{\lambda_1 - 0}(S) \neq E_{(\lambda_1,\infty)}(S)$, then there exists a projection Q such that

$$E_{(\lambda,\infty)}(S) \downarrow Q > E_{(\lambda_1,\infty)}(S) \quad \text{as } \lambda \uparrow \lambda_1.$$

Now since $S \in \widetilde{\mathcal{M}}_0$, just as in the first part of the proof, the normality and faithfulness of the trace implies that

$$\tau(E_{(\lambda,\infty)}(S)) \downarrow \tau(Q) > \tau(E_{(\lambda_1,\infty)}(S)) \quad \text{as } \lambda \uparrow \lambda_1.$$

From this it follows that $\lim_{\lambda \uparrow \lambda_1} \tau(E_{(\lambda,\infty)}(S)) \neq \tau(E_{(\lambda_1,\infty)}(S))$, a contradiction to equation (4). Thus $I - E_{\lambda_1 - 0}(S) = E_{(\lambda_1,\infty)}(S)$, from which we have that $E_{\lambda_1 - 0}(S) = E_{\lambda_1}(S)$. Hence, again by ([7], 9.11-1)

$$\ker(S - \lambda_1 I) = (E_{\lambda_1}(S) - E_{\lambda_1 - 0}(S))(H) = \{0\}.$$

Hence $\lambda_1 \notin \sigma_p(S)$, and therefore,

$$\sigma_p(S) \subseteq \{\lambda > 0 : \; m\{t > 0 : \; \mu_t(S) = \lambda\} > 0\} \cup \{0\}.$$

Now if $\alpha_0 < \infty$, then $m\{t > 0 : \; \mu_t(S) = 0\} > 0$ and by Lemma 3.7, $0 \in \sigma_p(S)$. Thus, together with what we have already proved above, we have that

$$\sigma_p(S) = \{\lambda \geq 0 : \; m\{t > 0 : \; \mu_t(S) = \lambda\} > 0\}. \qquad \square$$

We now show that inversion is continuous on the set $\mathcal{Q}$ of invertible elements in $\widetilde{\mathcal{M}}$ with respect to the topology of convergence in measure.

Theorem 3.9. [3] *Let $S \in \widetilde{\mathcal{M}}$, (S_n) be a sequence in $\widetilde{\mathcal{M}}$, and $S_n \to_{\tau_m} S$. Then*

$$|S_n| \to_{\tau_m} |S|.$$

Proposition 3.10. *Let $\mathcal{Q}$ be the set of invertible elements in $\widetilde{\mathcal{M}}$, and (S_n) a sequence in $\mathcal{Q}$ such that $S_n \to_{\tau_m} I$. Then*

$$S_n^{-1} \to_{\tau_m} I,$$

that is to say inversion is τ_m-continuous on $\mathcal{Q}$.

Proof. We first prove the result for the case when the S_n are positive. So suppose $S_n > 0$ for all $n \geq 1$. From the equality

$$I - S_n^{-1} = (S_n - I)S_n^{-1}$$

we have, for $t > 0$,

$$\begin{aligned} \mu_t(I - S_n^{-1}) &= \mu_t((S_n - I)S_n^{-1}) \\ &\leq \mu_{\frac{t}{2}}(S_n - I)\mu_{\frac{t}{2}}(S_n^{-1}), \text{ by ([6], 2.5).} \end{aligned}$$

Now since $S_n \to_{\tau_m} I$, it follows that $\mu_t(S_n - I) \to 0$ as $n \to \infty$ for all $t > 0$. It will suffice to show that for fixed $t > 0$, the $\mu_t(S_n^{-1})$ have a common upper bound for all $n \in \mathbb{N}$. By Proposition 3.3, we have that $S_n^{-1} = h(S_n)$ where

$$h(\lambda) = \begin{cases} \frac{1}{\lambda}, & \lambda \neq 0 \\ 0, & \lambda = 0 \end{cases}$$

and that for $\beta > 0$, $\tau(E_{(\beta,\infty)}(|h(S_n)|)) = \tau(E_{(0,\frac{1}{\beta})}(S_n))$. Hence

$$\mu_t(S_n^{-1}) = \inf\{\beta > 0 : \tau(E_{(0,\frac{1}{\beta})}(S_n)) < t\} = \inf\Big\{\frac{1}{\alpha} : \tau(E_{(0,\alpha)}(S_n)) < t\Big\}$$
$$= \frac{1}{\sup\{\alpha > 0 : \tau(E_{(0,\alpha)}(S_n)) < t\}}$$

We have that

$$\mu_t(S_n - I) = \inf\{\theta \geq 0 : \tau(E_{(\theta,\infty)}(|S_n - I|)) < t\}$$

and $\mu_t(S_n - I) \to_n 0$ for each $t > 0$, by assumption, since $S_n \to_{\tau_m} I$. Let $t > 0$. There exists an $n_0 \in \mathbb{N}$ such that for all $n \geq n_0$, $\inf\{\theta \geq 0 : \tau(E_{(\theta,\infty)}(|S_n - I|)) < t\} = \mu_t(S_n - I) < \frac{1}{2}$. Hence $\tau(E_{(\frac{1}{2},\infty)}(|S_n - I|)) < t$ for all $n \geq n_0$. Now, by the functional calculus, $|S_n - I| = k(S_n)$ where $k(\lambda) = |\lambda - 1|$. Thus

$$E_{(\theta,\infty)}(|S_n - I|) = E_{(\theta,\infty)}(k(S_n)) = E_{k^{-1}(\theta,\infty)}(S_n),$$

the last equality following by the change of measure principle. Now

$$\begin{aligned} k^{-1}(\theta,\infty) &= \{\lambda \in \mathbb{R}^+ : \theta < k(\lambda) < \infty\} \\ &= (-\infty, 1-\theta) \cup (\theta + 1, \infty) \end{aligned}$$

With $\theta = \frac{1}{2}$, and for all $n \geq n_0$, we have

$$t > \tau(E_{(\frac{1}{2},\infty)}(|S_n - I|)) = \tau[E_{(0,\frac{1}{2})}(S_n) + E_{(\frac{3}{2},\infty)}(S_n)] \geq \tau(E_{(0,\frac{1}{2})}(S_n)).$$

Thus $\sup\{\alpha > 0 : \tau(E_{(0,\alpha)}(S_n)) < t\} \geq \frac{1}{2}$ so that for all $n \geq n_0$, $\mu_t(S_n^{-1}) \leq 2$. Now let $q = \max\{\mu_t(S_n^{-1}) : n = 1, 2, \ldots, n_0 - 1\}$ and $K = \max\{q, 2\}$. Then for all $n \in \mathbb{N}$, $\mu_t(S_n^{-1}) \leq K$. Thus

$$\mu_t(S_n^{-1} - I) \leq K\mu_{\frac{t}{2}}(S_n - I) \to_n 0.$$

Since this holds for all $t > 0$, $S_n^{-1} \to_{\tau_m} I$.

Now let (S_n) be an arbitrary sequence in $\mathcal{Q}$ with $S_n \to_{\tau_m} I$. Let $S_n = U_n|S_n|$ be the polar decompositions for S_n. By Proposition 3.4(1), $|S_n| \in \mathcal{Q}$, $|S_n|^{-1} = S_n^{-1}U_n$ and $S_n^{-1} = |S_n|^{-1}U_n^*$. By Theorem 3.9, we have that $|S_n| \to_{\tau_m} I$. By the first part of the proof, $|S_n|^{-1} \to_{\tau_m} I$.

Now we have

$$\begin{aligned} S_n \to I \quad &\Rightarrow \quad U_n|S_n| \to_{\tau_m} I \\ &\Rightarrow \quad U_n|S_n||S_n|^{-1} \to_{\tau_m} I, \text{ by continuity of multiplication} \\ &\Rightarrow \quad U_n \to_{\tau_m} I \\ &\Rightarrow \quad U_n^* \to_{\tau_m} I^* = I, \text{ by continuity of adjunction} \\ &\Rightarrow \quad S_n^{-1} = |S_n|^{-1}U_n^* \to_{\tau_m} I.I = I, \text{ by continuity of multiplication.} \end{aligned}$$

Hence inversion is τ_m-continuous. □

Remark. In [14], Theorem 2.4, it is shown that if f is a real-valued Borel function on the reals bounded on bounded subsets and continuous at the spectrum points of a self-adjoint element T of the metrisable $*$-algebra of τ- measurable operators with respect to τ, then the function $\mapsto f(T)$ is τ_m-continuous. This result can not be applied in Proposition 3.10 above, since the inversion function of Proposition 3.3 is not bounded on bounded subsets of the reals.

Definition 3.11. *A subalgebra $\mathcal{A}$ of a semi-finite von Neumann algebra $\mathcal{M}$ with a faithful semi-finite normal trace τ is said to be proper if the restriction of τ to $\mathcal{A}$ is a semi-finite trace on $\mathcal{A}$.*

Proposition 3.12. *Suppose $0 < S \in \widetilde{\mathcal{M}}_0$ and let the projections $\tilde{E}_t(S)$, $t \in \mathbb{R}$ be defined as in Proposition 2.7. Then the commutative subalgebra $\mathcal{M}_S$ of $\mathcal{M}$ generated by the projections $\{\tilde{E}_t(S) : t \in \mathbb{R}\}$ is proper.*

Proof. This follows immediately from the fact that $\tau(I - E_\lambda(S)) < \infty$ for all $\lambda > 0$, since $S \in \widetilde{\mathcal{M}}_0$. □

We now show that $\widetilde{\mathcal{M}}$ is a $\mathcal{Q}$-algebra if and only if $\widetilde{\mathcal{M}} = \mathcal{M}$.
The following is a special case of a result which appears in the proof of ([2], Theorem 3.5).

Lemma 3.13. ([2], 3.5) *Let $0 \leq S \in \widetilde{\mathcal{M}}_0$ and denote by $\mathcal{K}$ the collection of the intervals of constancy of $\mu_t(S)$. Put $C = \cup\{F : F \in \mathcal{K}\}$. Let M_S be the von Neumann subalgebra of $L_\infty(\mathbb{R}^+)$ generated by the characteristic functions of members of $\mathcal{K}$ and the Lebesgue measurable subsets of $\mathbb{R}^+ \backslash C$, and let $\mathcal{M}_S$ be the proper abelian subalgebra of $\mathcal{M}$ defined in Proposition 3.12. For each $f \in \widetilde{M}_S$, let $J_S(f) = \int_0^\infty f(t) d\tilde{E}_t(S)$, where $\tilde{E}_t(S)$ is as defined in Proposition 2.7.*
Then $J_S(f) \in \widetilde{\mathcal{M}}_S$ and $\mu_t(J_S(f)) = \mu_t(f)$.

Now we construct an unbounded function in $\widetilde{M}_S$ that we shall use in what follows to characterize those $\widetilde{\mathcal{M}}$ which are $\mathcal{Q}$-algebras.

Proposition 3.14. *Suppose $0 < S \in \widetilde{\mathcal{M}}_0 \backslash \mathcal{M}$ and let $\widetilde{\mathcal{M}}_S$ be defined as in Proposition 3.12. Then there exists an unbounded operator $T \in \widetilde{\mathcal{M}}_S \cap \widetilde{\mathcal{M}}_0$ such that $\sigma_{\widetilde{\mathcal{M}}}(T)$ is unbounded.*

Proof. We show that we can define $T = J_S(f)$ for an appropriate $f \in \widetilde{M}_S$, where M_S is as in Lemma 3.13. Let C be defined as in Lemma 3.13. We consider two cases.
Case 1: There exists a $t_0 > 0$ such that $(0, t_0) \cap C = \emptyset$. This implies that there exists an $n_0 \in \mathbb{N}^+$, say, such that $(0, \frac{1}{n_0}) \cap C = \emptyset$. Then in this case define

$$f(t) = \begin{cases} \sum_{n=n_0}^{\infty} n \chi_{(\frac{1}{n+1}, \frac{1}{n}]}(t) & t < \frac{1}{n_0} \\ 0 & t \geq \frac{1}{n_0} \end{cases}$$

Case 2: For all $t_0 > 0$, $(0, t_0) \cap C \neq \emptyset$. This implies that for each $t_0 > 0$ there exists an interval $F_{t_0} \in \mathcal{K}$ such that $(0, t_0) \cap F_{t_0} \neq \emptyset$, with $(0, t_0) \cap F_{t_0}$ an interval of constancy of $\mu_t(S)$. Now since S is unbounded, and so $\mu_t(S) \uparrow \infty$ as $t \downarrow 0$, we can find a_1, b_1 such that $0 < a_1 < b_1 \leq 1$ and $(0, 1) \cap F_1 = (a_1, b_1)$. Put $K_1 = (a_1, b_1)$.

Now by assumption there exists an $F_{a_1} \in \mathcal{K}$ such that $(0, a_1) \cap F_{a_1} \neq \emptyset$, and as before we can find $0 < a_2 < b_2 \leq a_1$ such that $(0, a_1) \cap F_{a_1} = (a_2, b_2)$. Put $K_2 = (a_2, b_2)$. Continuing in this way, we get a sequence of intervals $\{K_k\}_{k=1}^{\infty}$ such that $m(K_k) > 0$ for all k, the K_k are mutually disjoint, and $m(K_k) \downarrow 0$ as $k \uparrow \infty$, where

$$K_k = (0, a_{k-1}) \cap F_{a_{k-1}} = (a_k, b_k), \text{ with } 0 < a_k < b_k \leq a_{k-1}.$$

Let $G_k = [b_{k+1}, a_k]$.
Now in this case, define the function

$$f(t) = \begin{cases} \sum_{k=1}^{\infty} [k\chi_{K_k}(t) + k\chi_{G_k}(t)] & t < b_1 \\ 0 & t \geq b_1 \end{cases}$$

Then in both cases 1 and 2, we have that $f \in \widetilde{M_S}$, since by construction, f is bounded except on a set of m-measure less than 1, and f is affiliated with M_S since it is generated by characteristic functions from M_S. Now with the function f the one constructed above and using Lemma 3.13, define the operator

$$T = J_S(f) = \int_0^{\infty} f(t) d\tilde{E}_t(S).$$

We have that $T \in \widetilde{\mathcal{M}_S} \subseteq \widetilde{\mathcal{M}}$, $\mu_t(T) = \mu_t(f) \to 0$ as $t \to \infty$, by the construction of f, so that in fact $0 < T \in \widetilde{\mathcal{M}_0}$. Then we have

$$\begin{aligned} \sigma_{\widetilde{\mathcal{M}}}(T) &= \sigma_p(T) \cup \sigma_e(T) \text{ from Proposition 3.5} \\ &= \{\lambda \in \mathbb{R}^+ : m\{t > 0 : \mu_t(T) = \lambda\} > 0\} \cup \{0\} \text{ from Proposition 3.8} \\ &= \{\lambda \in \mathbb{R}^+ : m\{t > 0 : \mu_t(f) = \lambda\} > 0\} \cup \{0\} \\ &= \mathbb{N}, \end{aligned}$$

since by the construction of f, for each $n \in \mathbb{N}^+$ there is an interval of constancy for $\mu_t(f)$ on which $\mu_t(f) = n$. This shows that $\sigma_{\widetilde{\mathcal{M}}}(T)$ is unbounded. □

Proposition 3.15. *$\widetilde{\mathcal{M}}$ is a $\mathcal{Q}$-algebra if and only if $\widetilde{\mathcal{M}} = \mathcal{M}$.*

Proof. If $\widetilde{\mathcal{M}} = \mathcal{M}$, then $\widetilde{\mathcal{M}}$ is a Banach algebra and it follows that $\widetilde{\mathcal{M}}$ is a $\mathcal{Q}$-algebra.

To show that $\widetilde{\mathcal{M}}$ a $\mathcal{Q}$-algebra implies that $\widetilde{\mathcal{M}} = \mathcal{M}$, we suppose that $\widetilde{\mathcal{M}} \neq \mathcal{M}$ and then show that $\widetilde{\mathcal{M}}$ is not a $\mathcal{Q}$-algebra. Suppose $\widetilde{\mathcal{M}} \neq \mathcal{M}$, then by Proposition 2.5, we have that $\widetilde{\mathcal{M}_0} \neq \mathcal{M}_0$. So let $0 < S \in \widetilde{\mathcal{M}_0} \backslash \mathcal{M}_0$, that is S is a positive unbounded operator in $\widetilde{\mathcal{M}_0}$. With this S define $\widetilde{\mathcal{M}_S}$ as before, as in Proposition 3.12. Now by Proposition 3.14, there is a $T \in \widetilde{\mathcal{M}}$ such that $\sigma_{\widetilde{\mathcal{M}}}(T)$ is unbounded. By ([1], 4.8-3.), $\widetilde{\mathcal{M}}$ is not a $\mathcal{Q}$-algebra. Thus we have that if $\widetilde{\mathcal{M}}$ is a $\mathcal{Q}$-algebra, then $\widetilde{\mathcal{M}} = \mathcal{M}$. This completes the proof. □

Corollary 3.16. *$\widetilde{\mathcal{M}}$ is a $\mathcal{Q}$-algebra if and only if* $\inf\{\tau(P) : P \in \mathcal{M}^p : \tau(P) \neq 0\} > 0$.

Proof. Follows from Proposition 2.5 □

References

[1] Bechenstein.E., Narici L. and Suffel C. Topological Algebras. 1977.

[2] Dodds, P.G. Dodds, T.K. and de Pagter, B. *Fully Symmetric Operator Spaces*, Int. Equat. Oper. Theory, **15** 1992, 943–972.

[3] Dodds, P.G. Dodds, T.K. and de Pagter, B. *Lipschitz Continuity of the Absolute Value and Riesz Projections in Symmetric Operator Spaces*, J. Funct. Anal., **148**, 1997, 28–69.

[4] Dunford, N. and Schwartz, J.T. Spectral Theory. Part 2: Self Adjoint Operators in Hilbert Space. Interscience Publishers, New York, 1971.

[5] Fack, T. and Kosaki, H. *Generalised S-numbers of τ-measurable operators.* Pacific J. Math; **123**, 1986, 269–300.

[6] Kadison, R.V. and Ringrose, J.R. Fundamentals of the theory of operator algebras. Volume I. Pure and Applied Mathematics. Academic Press, London, 1983.

[7] Kreyszig, E. Introductory Functional Analysis with Applications. John Wiley and Sons, 1978.

[8] Nelson, E. *Notes on Non-Commutative Integration.* J. Funct. Anal. **15**, 1974, 103–116.

[9] Ovchinnikov V.I. *s-Numbers of Measurable Operators.* Funktsional'nyi Analiz i Ego Prilozheniya, **4**, 1970, 78–85.

[10] Stratila S. and Zsido L. Lectures on von Neumann algebras. Abacus Press, Tunridge Wells, 1979.

[11] Ströh, A. and West, G. P. *τ-compact operators affiliated with a semifinite von Neumann algebra.* Proc. Irish. Sect. A. **93**, 1993, 73–83.

[12] Takesaki, M. Theory of Operator Algebras 1. Springer-Verlag, New York, 1979.

[13] Terp, M. *L^p-spaces associated with von Neumann Algebras.* Notes, Copenhagen Univ. (1981)

[14] Tikhonov, O.E. *Continuity of operator functions in topologies connected with a trace on a von Neumann algebra.* (Russian) Izv. Vyssh. Uchebn. Zaved. Mat. **1** 1987, 77–79.

[15] West, G. P.*Non-commutative Banach Function Spaces*, MSc Thesis, University of Cape Town, South Africa, 1990.

Isaac Daniel Tembo
Department of Mathematics and Statistics
University of Zambia
Box 32379
Lusaka, Zambia
e-mail: isaac_tembo1@yahoo.co.uk
isaactembo@unza.zm

Operator Theory:
Advances and Applications, Vol. 195, 257–271

Analyticity of Semigroups Related to a Class of Block Operator Matrices

Carsten Trunk

Abstract. We derive various properties, e.g., analyticity of the associated semigroup and existence of a Riesz basis consisting of eigenfunctions, of the operator matrix $\mathcal{A} = \begin{bmatrix} 0 & I \\ -A_0 & -D \end{bmatrix}$. Here the entries A_0 and D are unbounded operators. Such operator matrices are associated with second-order problems of the form $\ddot{z}(t) + A_0 z(t) + D\dot{z}(t) = 0$ which are used as models for small motions of some hydrodynamical systems.

Mathematics Subject Classification (2000). Primary 47D03; Secondary 34G10, 46C20, 47A70, 46C05.

Keywords. Operator matrices, second-order equations, spectrum, Riesz basis, analytic semigroup

1. Introduction

The aim of this paper is the study of second-order equations of the form

$$\ddot{z}(t) + A_0 z(t) + D\dot{z}(t) = 0. \tag{1.1}$$

Here the operator A_0 is a possibly unbounded positive operator on a Hilbert space H and is assumed to be boundedly invertible. It is assumed that the operator D is an unbounded operator such that, for some $\theta > 1$, the operator $A_0^{-\theta/2} D A_0^{-\theta/2}$ is bounded and non-negative in H. Problems of such type arise in hydrodynamics, see [20] and references therein. Moreover, we mention that systems of the form

$$\begin{aligned} -\frac{\partial v}{\partial t} &= \frac{\partial^2 u}{\partial t^2} - \frac{\partial^2 u}{\partial r^2}, \\ v &= \frac{\partial^4 u}{\partial r^4}, \end{aligned}$$

with some appropriate boundary conditions, can be written as a second-order equation (1.1) by setting $A_0 := \frac{\partial^2}{\partial r^2}$ and $D := A_0^2$.

The second-order equation (1.1) is equivalent to the standard first-order equation $\dot{x}(t) = \mathcal{A}x(t)$, where $\mathcal{A} : \operatorname{dom}\mathcal{A} \subset \operatorname{dom}(A_0^{1/2}) \times H \to \operatorname{dom}(A_0^{1/2}) \times H$, is given by

$$\mathcal{A} = \begin{bmatrix} 0 & I \\ -A_0 & -D \end{bmatrix}, \tag{1.2}$$

$$\operatorname{dom}\mathcal{A} = \left\{ \left(\begin{smallmatrix} z \\ w \end{smallmatrix} \right) \in \operatorname{dom}(A_0^{1/2}) \times \operatorname{dom}(A_0^{1/2}) \mid A_0 z + Dw \in H \right\}.$$

In this paper we focus on two properties of the operator $\mathcal{A}$: Analyticity of the generated semigroup and the Riesz basis property in the phase space $H_{1/2} \times H$. In a first step, we show under the assumption that $A_0^{-\theta}D$ is boundedly invertible in $H_{1/2}$, where $H_{1/2} = \operatorname{dom}(A_0^{1/2})$ is equipped with the norm $\|x\|_{1/2} = \|A_0^{1/2}x\|_H$, that $\mathcal{A}$ generates a C_0-semigroup of contractions in $H_{1/2} \times H$ and thus the spectrum of $\mathcal{A}$ is located in the closed left half-plane. Moreover, under the additional assumption that A_0 has a compact resolvent, we show that the essential spectrum of $\mathcal{A}$ consists of the point zero only. In this case we are able to show that $\mathcal{A}$ is the generator of an analytic semigroup and we develop some conditions for the existence of a Riesz basis in the phase space $H_{1/2} \times H$ consisting of eigenvectors of $\mathcal{A}$.

The block operator matrix in (1.2) has been studied in the literature for more than 20 years. Interest in this particular model is motivated by various problems such as stabilization, see for example [6, 27, 28, 30], solvability of the Riccati equations [11], minimum-phase property [16] and compensator problems with partial observations [12].

The case $\theta \leq 1$ is very well studied. We mention here only a few results. In this case, it is well known that $\mathcal{A}$ generates a C_0-semigroup of contractions in $H_{1/2} \times H$. This goes back to [3] and [26], see also [4, 8]. Analyticity has been studied in [3, 4, 8, 9, 13, 14]. The Riesz basis property has been shown in [22] in the situation where A_0^{-1} is a compact operator, D is of the form $D = \alpha A_0 + B$, for some $\alpha \geq 0$ with a symmetric operator B and $-1/\alpha \notin \sigma_p(\mathcal{A})$, if $\alpha \neq 0$ (and with some additional assumptions in the case $\alpha = 0$). Similar results were obtained [9, Appendix A] in a more special situation. All these assumptions guarantee that the essential spectrum of $\mathcal{A}$ consists at most of one point. For the case that the essential spectrum of $\mathcal{A}$ consists of infinitely many points we refer to [17].

We proceed as follows. In Section 2 we provide some useful results on the spectrum of operators in Krein spaces. In particular, we recall the notion of spectral points of positive and negative type and of type π_+ and type π_-. One main tool of this paper is to show that certain spectral points of $\mathcal{A}$ are of positive/negative type or of type π_+/π_-. In Section 3 we give the precise definition of the operator $\mathcal{A}$ and prove some of its properties. In particular we show that $\mathcal{A}$ generates a C_0-semigroup of contractions and, if A_0^{-1} is compact, that zero is the only point in the essential spectrum. The main results of this paper are contained in Section 4 where we always assume that that A_0^{-1} is a compact operator and that $A_0^{-\theta}D$ is boundedly invertible in $H_{1/2}$. We show that $\mathcal{A}$ generates an analytic strongly continuous semigroup. Further, we show that ∞ is a spectral point of negative type and that every real spectral point is of type π_+. As a consequence we obtain that $\mathcal{A}$ is definitizable and that the non-real spectrum of $\mathcal{A}$ consists of at most finitely many points belonging to the point spectrum of $\mathcal{A}$. Under additional weak conditions there exists a Riesz basis of $H_{1/2} \times H$ consisting of eigenvalues of $\mathcal{A}$.

2. Spectrum of operators in Krein spaces

We briefly recall that a complex linear space $\mathcal{H}$ with a Hermitian nondegenerate sesquilinear form $[\cdot\,,\cdot]$ is called a *Krein space* if there exists a so-called *fundamental decomposition*

$$\mathcal{H} = \mathcal{H}_+ \oplus \mathcal{H}_- \tag{2.1}$$

with subspaces $\mathcal{H}_\pm$ being orthogonal to each other with respect to $[\cdot\,,\cdot]$ such that $(\mathcal{H}_\pm, \pm[\cdot\,,\cdot])$ are Hilbert spaces. In the following all topological notions are understood with respect to some Hilbert space norm $\|\cdot\|$ on $\mathcal{H}$ such that $[\cdot\,,\cdot]$ is $\|\cdot\|$-continuous. Any two such norms are equivalent. For the basic theory of Krein space and operators acting therein we refer to [7] and [1].

Let A be a closed operator in $\mathcal{H}$. We define the extended spectrum $\sigma_e(A)$ of A by $\sigma_e(A) := \sigma(A)$ if A is bounded and $\sigma_e(A) := \sigma(A) \cup \{\infty\}$ if A is unbounded. The resolvent set of A is denoted by $\rho(A)$. The operator A is called *Fredholm* if the dimension of the kernel of A and the codimension of the range of A are finite. The set

$$\sigma_{\text{ess}}(A) := \{\lambda \in \mathbb{C} \mid A - \lambda I \text{ is not Fredholm}\}$$

is called the *essential spectrum* of A. We say that $\lambda \in \mathbb{C}$ belongs to the *approximate point spectrum of* A, denoted by $\sigma_{ap}(A)$, if there exists a sequence $(x_n) \subset \operatorname{dom}(A)$ with $\|x_n\| = 1$, $n = 1, 2, \ldots$, such that

$$\|x_n\| = 1 \quad \text{and} \quad \lim_{n\to\infty} \|Ax_n - \lambda x_n\| = 0.$$

Obviously, the continuous and the point spectrum of a closed operator are subsets of the approximate point spectrum. Moreover, we have the following, cf. [10, §IV 1.10].

Remark 2.1. The boundary points of $\sigma(A)$ in $\mathbb{C}$ belong to $\sigma_{ap}(A)$.

Let A be a self-adjoint operator in the Krein space $(\mathcal{H}, [\cdot\,,\cdot])$, i.e., A coincides with its adjoint A^+ with respect to the indefinite inner product $[\cdot,\cdot]$. Then all real spectral points of A belong to $\sigma_{ap}(A)$ (see, e.g., Corollary VI.6.2 in [7]).

The indefiniteness of the scalar product on $\mathcal{H}$ leads to the definition of several subsets of the spectrum of an operator. The following definition was given in [18], [21], [25] and [2].

Definition 2.2. For a self-adjoint operator A in the Krein space $(\mathcal{H}, [\cdot\,,\cdot])$ a point $\lambda_0 \in \sigma(A)$ is called a spectral point of *positive (negative) type of A* if $\lambda_0 \in \sigma_{ap}(A)$ and for every sequence $(x_n) \subset \operatorname{dom}(A)$ with $\|x_n\| = 1$ and $\|(A - \lambda_0 I)x_n\| \to 0$ as $n \to \infty$ we have

$$\liminf_{n\to\infty} [x_n, x_n] > 0 \quad \big(\text{resp. } \limsup_{n\to\infty} [x_n, x_n] < 0\big).$$

The point ∞ is said to be of *positive (negative) type of A* if A is unbounded and for every sequence $(x_n) \subset \operatorname{dom}(A)$ with $\lim_{n\to\infty} \|x_n\| = 0$ and $\|Ax_n\| = 1$ we have

$$\liminf_{n\to\infty} [Ax_n, Ax_n] > 0 \quad \big(\text{resp. } \limsup_{n\to\infty} [Ax_n, Ax_n] < 0\big).$$

We denote the set of all points of $\sigma_e(A)$ of positive (negative) type by $\sigma_{++}(A)$ (resp. $\sigma_{--}(A)$).

The sets $\sigma_{++}(A)$ and $\sigma_{--}(A)$ are contained in $\overline{\mathbb{R}}$. Indeed, for $\lambda \in \sigma_{++}(A) \setminus \{\infty\}$ and (x_n) as in the first part of Definition 2.2 we have $-(\operatorname{Im}\lambda)[x_n, x_n] = \operatorname{Im}[(A - \lambda)x_n, x_n] \to 0$ for $n \to \infty$ which implies $\operatorname{Im}\lambda = 0$. Here $\overline{\mathbb{R}}$ denotes the set $\mathbb{R} \cup \{\infty\}$ and $\overline{\mathbb{C}}$ the set $\mathbb{C} \cup \{\infty\}$, each equipped with the usual topology.

In a similar way as above we define subsets $\sigma_{\pi_+}(A)$ and $\sigma_{\pi_-}(A)$ of $\sigma_e(A)$ containing $\sigma_{++}(A)$ and $\sigma_{--}(A)$, respectively (cf. Definition 5 in [2]). They will play an important role in the following.

Definition 2.3. For a self-adjoint operator A in $\mathcal{H}$ a point $\lambda_0 \in \sigma(A)$ is called a spectral point of *type π_+ (type π_-) of A* if $\lambda_0 \in \sigma_{ap}(A)$ and if there exists a linear submanifold $\mathcal{H}_0 \subset \mathcal{H}$ with $\operatorname{codim}\mathcal{H}_0 < \infty$ such that for every sequence $(x_n) \subset \mathcal{H}_0 \cap \operatorname{dom}(A)$ with $\|x_n\| = 1$ and $\|(A - \lambda_0 I)x_n\| \to 0$ as $n \to \infty$ we have

$$\liminf_{n\to\infty} [x_n, x_n] > 0 \quad \big(\text{resp. } \limsup_{n\to\infty} [x_n, x_n] < 0\big).$$

The point ∞ is said to be of *type π_+ (type π_-) of A* if A is unbounded and if there exists a linear submanifold $\mathcal{H}_0 \subset \mathcal{H}$ with $\operatorname{codim}\mathcal{H}_0 < \infty$ such that for every sequence $(x_n) \subset \mathcal{H}_0 \cap \operatorname{dom}(A)$ with $\lim_{n\to\infty} \|x_n\| = 0$ and $\|Ax_n\| = 1$ we have

$$\liminf_{n\to\infty} [Ax_n, Ax_n] > 0 \quad \big(\text{resp. } \limsup_{n\to\infty} [Ax_n, Ax_n] < 0\big).$$

We denote the set of all points of $\sigma_e(A)$ of type π_+ (type π_-) of A by $\sigma_{\pi_+}(A)$ (resp. $\sigma_{\pi_-}(A)$).

For example, if $A - \lambda_0$ is a Fredholm operator, then $\lambda_0 \in \sigma_{\pi_+}(A)$ (and $\lambda_0 \in \sigma_{\pi_-}(A)$).

We will collect some properties of spectral point of positive/negative type and of type π_+/π_- in the following proposition.

Proposition 2.4. *Let λ_0 be a point of $\sigma_{\pi_+}(A)$ ($\sigma_{\pi_-}(A)$, respectively). Then there exists an open neighbourhood $\mathcal{U}$ in $\overline{\mathbb{C}}$ of λ_0 such that the following holds.*

(i) *We have*

$$\mathcal{U} \setminus \overline{\mathbb{R}} \subset \sigma_p(A).$$

Moreover, if $\lambda_0 \in \mathbb{C}$ *is non-real then the operator* $A - \lambda_0$ *has a closed range and* $\dim \ker (A - \lambda_0) < \infty$.

(ii) $\mathcal{U} \cap \sigma_{ap}(A) \subset \sigma_{\pi_+}(A)$ *(resp.* $\mathcal{U} \cap \sigma_{ap}(A) \subset \sigma_{\pi_-}(A)$*).*

(iii) *If* $\lambda_0 = \infty$ *then* $\infty \in \sigma_{++}(A)$*. If* $\infty \in \sigma_{\pi_-}(A)$ *then* $\infty \in \sigma_{--}(A)$*.*

(iv) *Assume, in addition, that* $\lambda_0 \in \mathbb{R}$ *and that there is* $[a,b] \subset \mathcal{U}$, $\lambda_0 \in [a,b]$, *such that each point of* $[a,b]$ *is an accumulation point of* $\rho(A)$*. Then there exists an open neighbourhood* $\mathcal{V}$ *in* $\mathbb{C}$ *of* $[a,b]$ *such that* $\mathcal{V} \setminus \mathbb{R} \subset \rho(A)$ *and either* $\mathcal{V} \cap \sigma(A) \cap \mathbb{R} \subset \sigma_{++}(A)$ *or there exists a finite number of points* $\lambda_1, \ldots, \lambda_n \in \sigma_{\pi_+}(A) \cap \sigma_p(A)$ *such that*

$$(\mathcal{V} \cap \sigma(A) \cap \mathbb{R}) \setminus \{\lambda_1, \ldots, \lambda_n\} \subset \sigma_{++}(A).$$

Moreover, in this case there exist numbers $m \geq 1$ *and* $M > 0$ *such that*

$$\|(A-\lambda)^{-1}\| \leq \frac{M}{|\mathrm{Im}\,\lambda|^m} \quad \text{for all } \lambda \in \mathcal{V} \setminus \overline{\mathbb{R}}.$$

For a proof of Proposition 2.4 we refer to [2] and [5]. Recall that a Krein space $(\mathcal{H}, [\cdot\,,\cdot])$ is called a *Pontryagin space* if one of the spaces $\mathcal{H}_+, \mathcal{H}_-$ in (2.1) is finite dimensional. Moreover, we will call a Krein space $(\mathcal{H}, [\cdot\,,\cdot])$ an *anti Hilbert space* if $(\mathcal{H}, -[\cdot\,,\cdot])$ is a Hilbert space.

Theorem 2.5. *Let* A *be a self-adjoint operator in* $(\mathcal{H}, [\cdot\,,\cdot])$*. Let* A *satisfy*

$$\sigma_e(A) = \sigma_{++}(A) \quad (\text{resp. } \sigma_e(A) = \sigma_{--}(A)). \tag{2.2}$$

Then $(\mathcal{H}, [\cdot\,,\cdot])$ *is a Hilbert space (anti-Hilbert space, respectively).*

If A *satisfies instead of* (2.2) *the following condition*

$$\sigma_e(A) = \sigma_{++}(A) \cup \sigma_{--}(A),$$

then A *is similar to a self-adjoint operator in a Hilbert space.*

The proof of Theorem 2.5 will be given together with Theorem 2.6 below.

Observe, that in the case of an unbounded operator A condition (2.2) implies also $\infty \in \sigma_{++}(A)$.

Recall that a self-adjoint operator A in a Krein space $(\mathcal{H}, [\cdot\,,\cdot])$ is called *definitizable* if $\rho(A) \neq \emptyset$ and if there exists a rational function $p \neq 0$ having poles only in $\rho(A)$ such that $[p(A)x, x] \geq 0$ for all $x \in \mathcal{H}$. Then the spectrum of A is real or its non-real part consists of a finite number of points, cf. [23, 24]. Moreover, A has a spectral function $E(\cdot)$ defined on the ring generated by all connected subsets of $\overline{\mathbb{R}}$ whose endpoints do not belong to some finite set which is contained in $\{t \in \mathbb{R} : p(t) = 0\} \cup \{\infty\}$ (see [23, 24]).

We will use the following theorem in Section 4.

Theorem 2.6. *Let* A *be a self-adjoint operator in* $(\mathcal{H}, [\cdot\,,\cdot])$ *with* $\rho(A) \neq \emptyset$ *satisfying*

$$\sigma_{\mathrm{ess}}(A) \subset \mathbb{R} \quad \text{and} \quad \sigma_e(A) = \sigma_{\pi_+}(A) \quad (\text{resp. } \sigma_e(A) = \sigma_{\pi_-}(A)). \tag{2.3}$$

Then $(\mathcal{H},[\cdot,\cdot])$ *is a Pontryagin space and the space* $\mathcal{H}_-$ *(resp.* $\mathcal{H}_+$*) in the fundamental decomposition* (2.1) *is of finite dimension. Moreover, the set* $\sigma(A)\setminus\mathbb{R}$ *consists of at most finitely many eigenvalues with finite-dimensional root subspaces, i.e.,*

$$\sigma(A)\setminus\mathbb{R}\subset\sigma_p(A)\setminus\sigma_{\mathrm{ess}}(A)$$

If A *with* $\rho(A)\neq\emptyset$ *satisfies instead of* (2.3) *the following condition*

$$\sigma_{\mathrm{ess}}(A)\subset\mathbb{R}\quad \textit{and}\quad \sigma_e(A)=\sigma_{\pi_+}(A)\cup\sigma_{\pi_-}(A), \tag{2.4}$$

then the non-real spectrum of A *consists of at most finitely many points which belong to* $\sigma_p(A)\setminus\sigma_{\mathrm{ess}}(A)$*. Moreover, the operator* A *is definitizable.*

Proof of the Theorems 2.5 *and* 2.6. Let A be a self-adjoint operator in the Krein space $(\mathcal{H},[\cdot,\cdot])$ with $\rho(A)\neq\emptyset$ satisfying (2.4). The resolvent set of a self-adjoint operator in a Krein space is symmetric with respect to the real axis (cf. [7, VI.6.3]), hence there are points from $\rho(A)$ in the upper and in the lower half-plane. This and $\sigma_{\mathrm{ess}}(A)\subset\mathbb{R}$ imply that $\sigma(A)\setminus\mathbb{R}$ consists only of isolated eigenvalues with finite algebraic multiplicity (see Chapter IV, §5.6 in [19]). In particular, each point in $\overline{\mathbb{R}}$ is an accumulation point of $\rho(A)$ and Proposition 2.4 implies that the non-real spectrum of A cannot accumulate to a real point. Moreover, from (2.4) and Proposition 2.4 (iii) we conclude $\infty\in\sigma_{++}(A)\cup\sigma_{--}(A)\cup\rho(A)$. Therefore the non-real spectrum of A is bounded and consists of at most finitely many points which belong to $\sigma_p(A)\setminus\sigma_{\mathrm{ess}}(A)$.

Relation (2.4), Theorem 23 in [2] and Theorem 4.7 in [18] imply that A is a definitizable operator and the second part of Theorem 2.6 is proved. In order to show the first part of Theorem 2.6 we assume without loss of generality

$$\sigma_{\mathrm{ess}}(A)\subset\mathbb{R}\quad\text{and}\quad\sigma_e(A)=\sigma_{\pi_+}(A). \tag{2.5}$$

It remains to show that $(\mathcal{H},[\cdot,\cdot])$ is a Pontryagin space. As $\infty\in\sigma_{++}(A)\cup\rho(A)$, there are at most only finitely many points in $\sigma_e(A)$ which belong to $\sigma_{\pi_+}(A)\setminus\sigma_{++}(A)$, cf. Proposition 2.4. Then, with Proposition 25, Theorem 26 of [2] and the properties of the spectral function of a definitizable operator (see [23, 24]), the assertions of Theorem 2.6 hold true.

Theorem 2.5 is now a consequence of Theorem 2.6: Assume without loss of generality

$$\sigma_e(A)=\sigma_{++}(A).$$

Then Theorem 2.6 implies that $(\mathcal{H},[\cdot,\cdot])$ is a Pontryagin space and the space $\mathcal{H}_-$ in the fundamental decomposition (2.1) is of finite dimension. If $\mathcal{H}_-\neq 0$, then there exists at least one non-positive eigenvector of A (see §12 in [15]) for some eigenvalue λ_0. This implies $\lambda_0\notin\sigma_{++}(A)$, a contradiction. Hence $\mathcal{H}_-=0$ and $(\mathcal{H},[\cdot,\cdot])$ is a Hilbert space. The second part of Theorem 2.5 follows from the fact that the sets $\sigma_{++}(A)$ and $\sigma_{--}(A)$ have a positive distance, hence, with the help of the spectral function of the definitizable operator A, $\mathcal{H}$ can be decomposed into the orthogonal sum of a A-invariant Hilbert and a A-invariant anti-Hilbert space. □

3. Framework and first results

Throughout this paper we make the following assumptions.

(I) The operator $A_0 : \mathrm{dom}A_0 \subset H \to H$ is a self-adjoint, positive definite linear operator on a Hilbert space H such that zero is in the resolvent set of A_0, i.e., A_0 is uniformly positive. Here $\mathrm{dom}A_0$ denotes the domain of A_0.

Since A_0 is self-adjoint and positive definite, A_0^α is well defined for $\alpha \geq 0$. A scale of Hilbert spaces H_α is defined as follows: For $\alpha \geq 0$, we define

$$H_\alpha = \mathrm{dom}(A_0^\alpha)$$

equipped with the norm induced by the inner product

$$\langle x, y\rangle_{H_\alpha} = \langle A_0^\alpha x, A_0^\alpha y\rangle_H, \qquad x, y \in H_\alpha$$

and $H_{-\alpha} = H_\alpha^*$. Here the duality is taken with respect to the pivot space H, that is, equivalently, $H_{-\alpha}$ is the completion of H with respect to the norm

$$\|z\|_{H_{-\alpha}} = \|A_0^{-\alpha} z\|_H.$$

Thus A_0 extends (restricts) to $A_0 : H_\alpha \to H_{\alpha-1}$ for $\alpha \in \mathbb{R}$. We use the same notation A_0 to denote this extension (restriction).

We denote the inner product on H by $\langle\cdot,\cdot\rangle_H$ or $\langle\cdot,\cdot\rangle$, and the duality pairing on $H_{-\alpha} \times H_\alpha$ by $\langle\cdot,\cdot\rangle_{H_{-\alpha}\times H_\alpha}$. Note that for $(z', z) \in H \times H_\alpha$, $\alpha > 0$, we have

$$\langle z', z\rangle_{H_{-\alpha}\times H_\alpha} = \langle z', z\rangle_H.$$

(II) For some $\theta > 1$ the operator $D : H_{\frac{\theta}{2}} \to H_{-\frac{\theta}{2}}$ is a bounded operator such that $A_0^{-\theta/2} D A_0^{-\theta/2}$ is a bounded non-negative self-adjoint operator in H, that is,

$$\langle Dz, z\rangle_{H_{-\frac{\theta}{2}}\times H_{\frac{\theta}{2}}} \geq 0, \qquad z \in H_{\frac{\theta}{2}}.$$

In particular, we have that $A_0^{-\theta} D$ is a bounded self-adjoint operator in $H_{\frac{\theta}{2}}$.

(III) The operator $A_0^{-\theta} D$ admits a bounded extension to an operator in H. We use the same notation $A_0^{-\theta} D$ to denote this extension. Note, that (by complex interpolation), the operator $A_0^{-\theta} D$ admits a bounded extension to an operator in $H_{\frac{1}{2}}$, which we also denote by $A_0^{-\theta} D$.

Example 3.1. Let $A_0 : \mathrm{dom}A_0 \subset H \to H$ be a uniformly positive operator on a Hilbert space H. Choose

$$D := A_0^\theta \quad \text{for some } \theta > 1.$$

Then, obviously, we have $A_0^{-\theta/2} D A_0^{-\theta/2} = A_0^{-\theta} D = I$ and (I)–(III) are valid.

The equation (1.1) is equivalent to the following standard first-order equation

$$\dot{x}(t) = \mathcal{A}x(t),$$

where $\mathcal{A} : \operatorname{dom}\mathcal{A} \subset H_{\frac{1}{2}} \times H \to H_{\frac{1}{2}} \times H$, is given by

$$\mathcal{A} = \begin{bmatrix} 0 & I \\ -A_0 & -D \end{bmatrix},$$

$$\operatorname{dom}\mathcal{A} = \left\{ \left(\begin{smallmatrix} z \\ w \end{smallmatrix} \right) \in H_{\frac{1}{2}} \times H_{\frac{1}{2}} \mid A_0 z + D w \in H \right\}.$$

Proposition 3.2. *The operator $\mathcal{A}$ is a closed operator in $H_{\frac{1}{2}} \times H$.*

Proof. Let $\left(\begin{smallmatrix} z_n \\ w_n \end{smallmatrix} \right) \subset \operatorname{dom}\mathcal{A}$ be a sequence and let $x, z \in H_{\frac{1}{2}}$, $y, w \in H$ such that $z_n \to z$ in $H_{\frac{1}{2}}$ and $w_n \to w$ in H as $n \to \infty$ and

$$\mathcal{A} \left(\begin{smallmatrix} z_n \\ w_n \end{smallmatrix} \right) \to \left(\begin{smallmatrix} x \\ y \end{smallmatrix} \right) \quad \text{in } H_{\frac{1}{2}} \times H \quad \text{as } n \to \infty. \tag{3.1}$$

Hence, $w_n \to x$ in $H_{\frac{1}{2}}$ and $w = x$, $x \in H_{\frac{1}{2}}$. As $A_0^{-\theta} D$ is bounded in $H_{\frac{1}{2}}$,

$$A_0^{-\theta} D w_n \to A_0^{-\theta} D x \quad \text{in } H_{\frac{1}{2}} \quad \text{as } n \to \infty. \tag{3.2}$$

Moreover, by (3.1),

$$-A_0 z_n - D w_n \to y \text{ in } H$$

and hence

$$-A_0^{-\theta+1} z_n - A_0^{-\theta} D w_n \to A_0^{-\theta} y \text{ in } H \text{ as } n \to \infty.$$

From (3.2) we conclude

$$-A_0^{-\theta+1} z_n \to A_0^{-\theta} D x + A_0^{-\theta} y \quad \text{in } H \quad \text{as } n \to \infty.$$

Furthermore, $-A_0^{-\theta+1} z_n \to -A_0^{-\theta+1} z$ in H as $n \to \infty$ and we have

$$-A_0^{-\theta+1} z = A_0^{-\theta} D x + A_0^{-\theta} y,$$

hence $-A_0 z = Dx + y$. This shows $A_0 z + Dx = -y \in H$ and $\mathcal{A}$ is a closed operator. □

For $\left(\begin{smallmatrix} x_1 \\ y_1 \end{smallmatrix} \right), \left(\begin{smallmatrix} x_2 \\ y_2 \end{smallmatrix} \right) \in H_{\frac{1}{2}} \times H$ we define an indefinite inner product on $H_{\frac{1}{2}} \times H$ by

$$\left[\begin{pmatrix} x_1 \\ y_1 \end{pmatrix}, \begin{pmatrix} x_2 \\ y_2 \end{pmatrix} \right] := \left\langle J \begin{pmatrix} x_1 \\ y_1 \end{pmatrix}, \begin{pmatrix} x_2 \\ y_2 \end{pmatrix} \right\rangle = \langle x_1, x_2 \rangle_{H_{\frac{1}{2}}} - \langle y_1, y_2 \rangle, \tag{3.3}$$

where

$$J = \begin{bmatrix} I & 0 \\ 0 & -I \end{bmatrix} \tag{3.4}$$

is a self-adjoint operator in the Hilbert space $H_{\frac{1}{2}} \times H$. Then $(H_{\frac{1}{2}} \times H, [\cdot\,,\cdot])$ is a Krein space and we obtain the following theorem.

Theorem 3.3. *Assume, in addition to our standard assumptions* (I)–(III), *that the operator $A_0^{-\theta} D$ is boundedly invertible in $H_{\frac{1}{2}}$. Then $\mathcal{A}$ is a self-adjoint operator in the Krein space $(H_{\frac{1}{2}} \times H, [\cdot\,,\cdot])$, its spectrum is contained in the closed left half-plane and lies symmetric with respect to the real line. The operator $\mathcal{A}$ is the generator of a strongly continuous semigroup of contractions on $H_{\frac{1}{2}} \times H$.*

Proof. Observe, that the operator $A_0^{-\theta}D$ is a uniformly positive operator in the Hilbert space $H_{\frac{1}{2}}$. The operator $A_0^{-\theta+1}+A_0^{-\theta}$ is a bounded non-negative operator in the Hilbert space $H_{\frac{1}{2}}$, therefore, for $\lambda > 0$, the operator

$$V(\lambda) := A_0^{-\theta+1} + \lambda A_0^{-\theta}D + \lambda^2 A_0^{-\theta} \tag{3.5}$$

is a bounded, uniformly positive operator in the Hilbert space $H_{\frac{1}{2}}$. In particular, $V(\lambda)$ is boundedly invertible in $H_{\frac{1}{2}}$ for all $\lambda > 0$. Let $x \in H_{\frac{1}{2}}$ and $y \in H$. We set

$$z := -V(1)^{-1}\left(A_0^{-\theta}y + A_0^{-\theta}Dx + A_0^{-\theta}x\right) \quad \text{and} \quad w := x + z.$$

Then $z, w \in H_{\frac{1}{2}}$ and we have

$$V(1)z = A_0^{-\theta+1}z + A_0^{-\theta}Dz + A_0^{-\theta}z = -\left(A_0^{-\theta}y + A_0^{-\theta}Dx + A_0^{-\theta}x\right),$$

hence

$$A_0 z + Dz + z = -y - Dx - x.$$

This implies

$$A_0 z + Dw = A_0 z + Dx + Dz = -y - x - z = -y - w \in H,$$

that is, $\binom{z}{w} \in \operatorname{dom}\mathcal{A}$. Moreover, we have

$$-z + w = x \quad \text{and} \quad -A_0 z - Dw - w = y$$

and, thus, $\mathcal{A} - 1$ is surjective. Assume that 1 is an eigenvalue of $\mathcal{A}$ with an corresponding eigenvector $\binom{z}{w}$. Then $z = w$ follows and $A_0 z + Dz + z = 0$. Then

$$A_0^{-\theta+1}z + A_0^{-\theta}Dz + A_0^{-\theta}z = V(1)z = 0.$$

This gives $z = w = 0$, a contradiction. From Proposition 3.2 we conclude

$$1 \in \rho(\mathcal{A}).$$

It is easy to see that $J\mathcal{A}$, where J is defined as in (3.4), is a symmetric operator in the Hilbert space $H_{\frac{1}{2}} \times H$. Then $J(\mathcal{A}-1)$ is a symmetric and boundedly invertible operator in the Hilbert space $H_{\frac{1}{2}} \times H$. Thus $J(\mathcal{A}-1)$ and $J\mathcal{A}$ are self-adjoint operators in the Hilbert space $H_{\frac{1}{2}} \times H$ and $\mathcal{A}$ is a self-adjoint operator in the Krein space $(H_{\frac{1}{2}} \times H, [\cdot\,,\cdot])$ with

$$\mathcal{A}^* = J\mathcal{A}J, \qquad \text{with } \operatorname{dom}(\mathcal{A}^*) = J\operatorname{dom}(\mathcal{A}).$$

For $\binom{z}{w} \in \operatorname{dom}\mathcal{A}$

$$\operatorname{Re}\left\langle \mathcal{A}\begin{pmatrix} z \\ w \end{pmatrix}, \begin{pmatrix} z \\ w \end{pmatrix} \right\rangle = \operatorname{Re}\left\langle \begin{pmatrix} w \\ -A_0 z - Dw \end{pmatrix}, \begin{pmatrix} z \\ w \end{pmatrix} \right\rangle = -\langle Dw, w\rangle_{H_{-\frac{\theta}{2}} \times H_{\frac{\theta}{2}}} \le 0$$

and, by a similar argument,

$$\operatorname{Re}\left\langle \mathcal{A}^*\begin{pmatrix} z \\ w \end{pmatrix}, \begin{pmatrix} z \\ w \end{pmatrix} \right\rangle \le 0$$

for all $\binom{z}{w} \in \operatorname{dom}\mathcal{A}^*$. Hence, $\mathcal{A}$ is the generator of a strongly continuous semigroup of contractions on the state space $H_{\frac{1}{2}} \times H$. □

The next example shows that, in general, $\mathcal{A}$ is not a boundedly invertible operator.

Example 3.4. Let $A_0 : \mathrm{dom} A_0 \subset H \to H$ be a uniformly positive operator on a Hilbert space H. Choose $D := A_0^2$. Then (cf. Example 3.1) (I)–(III) are valid. Assume $y_0 \in H_{\frac{1}{2}} \setminus H_1$. For all $x \in H_{\frac{1}{2}}$ we have

$$-A_0 x - A_0^2 y_0 \in H_{-\frac{3}{2}} \setminus H_{-1},$$

that is $-A_0 x - A_0^2 y_0 \neq 0$. Hence

$$\mathcal{A}\begin{pmatrix} x \\ y_0 \end{pmatrix} = \begin{pmatrix} y_0 \\ -A_0 x - A_0^2 y_0 \end{pmatrix} \neq \begin{pmatrix} y_0 \\ 0 \end{pmatrix}$$

for all $x \in H_{\frac{1}{2}}$ and $\mathcal{A}$ is not a boundedly invertible operator.

If the operator A_0 has a compact resolvent, then we have the following description of the essential spectrum of $\mathcal{A}$.

Proposition 3.5. *Assume, in addition to our standard assumptions* (I)–(III), *that* A_0^{-1} *is a compact operator in* H *and the operator* $A_0^{-\theta} D$ *is boundedly invertible in* $H_{\frac{1}{2}}$. *Then*

$$\sigma_{\mathrm{ess}}(\mathcal{A}) = \{0\}. \tag{3.6}$$

Proof. The operator $V(1)$ (see (3.5)) is a bounded, uniformly positive operator in the Hilbert space $H_{\frac{1}{2}}$. It is easily seen, that

$$(\mathcal{A} - 1)^{-1} = \begin{bmatrix} V(1)^{-1} A^{-\theta+1} - I & -V(1)^{-1} A_0^{-\theta} \\ V(1)^{-1} A_0^{-\theta+1} & -V(1)^{-1} A_0^{-\theta} \end{bmatrix}. \tag{3.7}$$

The left upper corner of the matrix of the right-hand side of (3.7) is the difference between a compact operator and the identity, the other entries of this matrix are compact operators. Hence,

$$\sigma_{\mathrm{ess}}\left((\mathcal{A} - 1)^{-1}\right) = \{0\} \cup \sigma_{\mathrm{ess}}(-I) = \{0\} \cup \{-1\}$$

and $\sigma_{\mathrm{ess}}(\mathcal{A}) = \{0\}$ follows. □

4. Analyticity and expansions in eigenfunctions

The following result is the main result of this paper.

Theorem 4.1. *Assume, in addition to our standard assumptions* (I)–(III), *that* A_0^{-1} *is a compact operator in* H *and the operator* $A_0^{-\theta} D$ *is boundedly invertible in* $H_{\frac{1}{2}}$. *Then* $\mathcal{A}$ *generates an analytic semigroup on* $H_{1/2} \times H$.

The proof of Theorem 4.1 will be given after Theorem 4.2 below.

Theorem 4.2. *Assume, in addition to our standard assumptions* (I)–(III), *that* A_0^{-1} *is a compact operator in* H *and the operator* $A_0^{-\theta}D$ *is boundedly invertible in* $H_{\frac{1}{2}}$. *Then the operator* $\mathcal{A}$ *is definitizable and*

$$\infty \in \sigma_{--}(\mathcal{A}), \quad 0 \in \sigma_{++}(\mathcal{A}) \quad \textit{and} \quad \mathbb{R} \subset \sigma_{\pi_+}(\mathcal{A}) \cup \rho(\mathcal{A}). \tag{4.1}$$

Moreover, there exists a neighbourhood $\mathcal{U}$ *of* ∞ *in* $\overline{\mathbb{C}}$ *and constants* $M > 0$, $m \in \mathbb{N}$ *and* $\eta > 0$ *such that*

$$\mathcal{U} \setminus \overline{\mathbb{R}} \subset \rho(\mathcal{A}) \quad \textit{and} \quad \mathcal{U} \cap \mathbb{R} \subset \sigma_{--}(\mathcal{A}) \cup \rho(\mathcal{A}) \tag{4.2}$$

and

$$\|(\mathcal{A} - \lambda I)^{-1}\| \le \tfrac{M}{|\operatorname{Im}\lambda|} \quad \textit{for all } \lambda \in \mathcal{U} \setminus \overline{\mathbb{R}}, \tag{4.3}$$

$$\|(\mathcal{A} - \lambda I)^{-1}\| \le \tfrac{M}{|\operatorname{Im}\lambda|^m} \quad \textit{for all } \lambda \in \rho(A) \setminus \mathbb{R} \textit{ with } |\operatorname{Im}\lambda| \le \eta. \tag{4.4}$$

Further, the non-real spectrum of $\mathcal{A}$ *consists of at most finitely many points.*

Proof. In the first three steps we show that (4.1) holds.

Step 1. Let $\left(\begin{smallmatrix} x_n \\ y_n \end{smallmatrix}\right) \subset \operatorname{dom}\mathcal{A}$ be a sequence with $\|\left(\begin{smallmatrix} x_n \\ y_n \end{smallmatrix}\right)\|^2_{H_{\frac{1}{2}} \times H} = \|x_n\|^2_{H_{\frac{1}{2}}} + \|y_n\|^2 \to 0$ as $n \to \infty$ and $\|\mathcal{A}\left(\begin{smallmatrix} x_n \\ y_n \end{smallmatrix}\right)\|_{H_{\frac{1}{2}} \times H} = 1$. Then

$$\|y_n\|^2_{H_{\frac{1}{2}}} + \|A_0 x_n + D y_n\|^2 = 1, \tag{4.5}$$

We have $\|y_n\| \to 0$ as $n \to \infty$. Assume that there exists a subsequence of (y_n) which does not converge to zero in $H_{\frac{1}{2}}$. Without loss of generality, let us assume that $\|y_n\|_{H_{\frac{1}{2}}} \ge \gamma > 0$ for $n \in \mathbb{N}$. By (4.5), $A_0^{-\frac{1}{2}}(A_0 x_n + D y_n)$ is bounded in H. As (x_n) converges to zero in $H_{\frac{1}{2}}$, the sequence $\left(A_0^{-\frac{1}{2}} D y_n\right)$ is a bounded sequence in H, hence it has a subsequence $\left(A_0^{-\frac{1}{2}} D y_{n_k}\right)$ which converges weakly in H to some $y_0 \in H$ for $k \to \infty$. Then $A_0^{-\theta} D y_{n_k}$ converges in H to $A_0^{-\theta+\frac{1}{2}} y_0$ as $k \to \infty$. As $A_0^{-\theta}D$ is a bounded operator in H and (y_n) tends to zero in H, $(A_0^{-\theta} D y_{n_k})$ converges to zero in H as $k \to \infty$ and we have

$$A_0^{-\theta+\frac{1}{2}} y_0 = 0,$$

hence $y_0 = 0$. Therefore, $(A_0^{-\frac{1}{2}} D y_{n_k})$ converges weakly to zero in H. It follows from the compactness of the operator $A_0^{-\theta+1}$ that

$$A_0^{-\theta+1} A_0^{-\frac{1}{2}} D y_{n_k} = A_0^{\frac{1}{2}} A_0^{-\theta} D y_{n_k} \to 0 \quad \text{in } H \quad \text{as } k \to \infty$$

and

$$A_0^{-\theta} D y_{n_k} \to 0 \quad \text{in } H_{\frac{1}{2}} \quad \text{as } k \to \infty.$$

By assumption, the operator $A_0^{-\theta}D$ is boundedly invertible in $H_{\frac{1}{2}}$, that is

$$y_{n_k} \to 0 \quad \text{in } H_{\frac{1}{2}} \quad \text{as } k \to \infty.$$

This is a contradiction, therefore we have

$$y_n \to 0 \quad \text{in } H_{\frac{1}{2}} \quad \text{as } k \to \infty.$$

By (4.5)

$$\limsup_{n\to\infty} [\mathcal{A}\left(\begin{smallmatrix} x_n \\ y_n \end{smallmatrix}\right), \mathcal{A}\left(\begin{smallmatrix} x_n \\ y_n \end{smallmatrix}\right)] = \limsup_{n\to\infty} \left(\|y_n\|^2_{H_{\frac{1}{2}}} - \|A_0 x_n + D y_n\|^2 \right) = -1, \qquad (4.6)$$

that is, $\infty \in \sigma_{--}(\mathcal{A})$.

Step 2. Proposition 3.5 implies $0 \in \sigma(\mathcal{A})$. As all real spectral points of $\mathcal{A}$ belong to the approximative spectrum of $\mathcal{A}$ (cf. [7, Corollary VI.6.2]) we have

$$0 \in \sigma_{ap}(\mathcal{A}).$$

Let $(\left(\begin{smallmatrix} x_n \\ y_n \end{smallmatrix}\right)) \subset \operatorname{dom}\mathcal{A}$ be a sequence with $\|\left(\begin{smallmatrix} x_n \\ y_n \end{smallmatrix}\right)\|^2_{H_{\frac{1}{2}} \times H} = \|x_n\|^2_{H_{\frac{1}{2}}} + \|y_n\|^2 = 1$ and $\mathcal{A}\left(\begin{smallmatrix} x_n \\ y_n \end{smallmatrix}\right) \to 0$ as $n \to \infty$. Then

$$\|y_n\|^2_{H_{\frac{1}{2}}} \to 0 \quad \text{and} \quad \|x_n\|^2_{H_{\frac{1}{2}}} \to 1 \quad \text{as } n \to \infty.$$

This gives

$$\liminf_{n\to\infty} [\left(\begin{smallmatrix} x_n \\ y_n \end{smallmatrix}\right), \left(\begin{smallmatrix} x_n \\ y_n \end{smallmatrix}\right)] = \liminf_{n\to\infty} \left(\|x_n\|^2_{H_{\frac{1}{2}}} - \|y_n\|^2 \right) = 1,$$

that is, $0 \in \sigma_{++}(\mathcal{A})$.

Step 3. We now choose $\mu \in (-\infty, 0)$ and

$$G_\mu := \operatorname{span}\{x \in H_{\frac{1}{2}} \mid A_0 x = \nu x, \nu \le \mu^2\}.$$

Then G_μ is a finite-dimensional subspace of $H_{\frac{1}{2}}$. For every sequence $\left(\begin{smallmatrix} x_n \\ y_n \end{smallmatrix}\right)$ in $\operatorname{dom}\mathcal{A} \cap (G_\mu \times G_\mu)^\perp$, where $\perp$ denotes the orthogonal complement in $H_{\frac{1}{2}} \times H$ with respect to the usual Hilbert space product in $H_{\frac{1}{2}} \times H$, with $\|\left(\begin{smallmatrix} x_n \\ y_n \end{smallmatrix}\right)\|^2_{H_{\frac{1}{2}} \times H} = 1$ and $(\mathcal{A} - \mu I)\left(\begin{smallmatrix} x_n \\ y_n \end{smallmatrix}\right) \to 0$ as $n \to \infty$ we have

$$\|y_n - \mu x_n\|_{H_{\frac{1}{2}}} \to 0 \quad \text{and} \quad \|A_0 x_n + D y_n + \mu y_n\| \to 0 \quad \text{as } n \to \infty.$$

This gives

$$\begin{aligned} \liminf_{n\to\infty} [\left(\begin{smallmatrix} x_n \\ y_n \end{smallmatrix}\right), \left(\begin{smallmatrix} x_n \\ y_n \end{smallmatrix}\right)] &= \liminf_{n\to\infty} \left(\langle x_n, x_n \rangle_{H_{\frac{1}{2}}} - \langle y_n, y_n \rangle \right) \\ &= \liminf_{n\to\infty} \left(\langle A_0 x_n, x_n \rangle - \mu^2 \langle x_n, x_n \rangle \right) > 0, \end{aligned}$$

where the last inequality follows from the fact that $x_n \in G_\mu^\perp$, $n \in \mathbb{N}$. Therefore $\mathbb{R} \subset \sigma_{\pi_+}(\mathcal{A})$ and (4.1) is valid.

Step 4. By Proposition 3.5 the essential spectrum of $\mathcal{A}$ is real. Moreover, for $\lambda \in \mathbb{C} \setminus \mathbb{R}$, $A - \lambda$ is a Fredholm operator and, therefore, $\mathbb{C} \setminus \mathbb{R} \subset \sigma_{\pi_+}(\mathcal{A}) \cup \rho(\mathcal{A})$, cf. [2]. Together with (4.1) we obtain

$$\sigma_e(\mathcal{A}) = \sigma_{\pi_+}(\mathcal{A}) \cup \sigma_{\pi_-}(\mathcal{A}).$$

Then, by Theorem 2.6, $\mathcal{A}$ is definitizable. From this and (4.1) the remaining assertions of Theorem 4.2 follow (see, e.g., [24, Proposition II.2.1] or [2, Proposition 3 and Theorem 20]). □

Proof of Theorem 4.1. Since $\mathcal{A}$ is the generator of a strongly continuous semigroup, estimate (4.3) shows immediately that $\mathcal{A}$ generates an analytic semigroup. □

In the sequel we assume the Hilbert space H to be separable. An at most countable set $\mathcal{M}$ of elements of a Hilbert space is said to be a *Riesz basis* if there exists an isomorphic mapping $\mathcal{M}$ onto an orthonormal basis, cf. [29, Lecture VI].

Theorem 4.3. *Assume, in addition to our standard assumptions* (I)–(III), *that* H *is separable,* A_0^{-1} *is a compact operator in* H *and the operator* $A_0^{-\theta}D$ *is boundedly invertible in* $H_{\frac{1}{2}}$. *Then there exists a subspace of* $H_{\frac{1}{2}} \times H$ *of at most finite codimension which has a Riesz basis consisting of eigenvectors of* $\mathcal{A}$ *and there exists a Riesz basis of* $H_{\frac{1}{2}} \times H$ *consisting of eigenvectors and finitely many associated vectors of* $\mathcal{A}$. *Moreover assume, in addition, that for all* $\mu \in \sigma_{\pi_+}(\mathcal{A}) \setminus \sigma_{++}(\mathcal{A})$ *there exists no non-zero* $\binom{y}{\mu y} \in \ker(\mathcal{A} - \mu I)$ *such that*

$$\langle y, y\rangle_{H_{\frac{1}{2}}} = \mu^2 \langle y, y\rangle. \tag{4.7}$$

Then $\mathcal{A}$ *has no associated vectors, i.e., there are no Jordan chains of length greater than one, and there exists a Riesz basis of* $H_{\frac{1}{2}} \times H$ *consisting of eigenvectors of* $\mathcal{A}$.

Proof. The first part of Theorem 4.3 follows from the properties of the spectral function of $\mathcal{A}$ and Theorem 4.2. Relation (4.7) implies that there are no neutral eigenvectors. Hence the spectrum of $\mathcal{A}$ is real and there are no Jordan chains of $\mathcal{A}$ corresponding to the eigenvalue μ of length greater than one. Moreover,

$$\sigma(\mathcal{A}) \subset \sigma_{++}(\mathcal{A}) \cup \sigma_{--}(\mathcal{A}) \subset \mathbb{R}.$$

Then, by Theorem 2.5, the operator $\mathcal{A}$ is similar to a self-adjoint operator in a Hilbert space, hence there exists a Riesz basis of $H_{\frac{1}{2}} \times H$ consisting of eigenvectors of $\mathcal{A}$. □

References

[1] T.Ya. Azizov and I.S. Iokhvidov, *Linear Operators in Spaces with an Indefinite Metric*, John Wiley & Sons, 1989.

[2] T.Ya. Azizov, P. Jonas, and C. Trunk, *Spectral points of type π_+ and π_- of self-adjoint operators in Krein spaces*, J. Funct. Anal. **226** (2005), 114–137.

[3] H.T. Banks and K. Ito, *A unified framework for approximation in inverse problems for distributed parameter systems*, Control Theory Adv. Tech. **4** (1988), 73–90.

[4] H.T. Banks, K. Ito, and Y. Wang, *Well-posedness for damped second-order systems with unbounded input operators, Differential Integral Equations* **8** (1995), 587–606.

[5] J. Behrndt, F. Philipp, and C. Trunk, *Properties of the spectrum of type π_+ and type π_- of self-adjoint operators in Krein spaces*, Methods Funct. Anal. Topology **12** (2006), 326–340.

[6] C.D. Benchimol, *A note on weak stabilizability of contraction semigroups*, SIAM J. Control Optimization **16** (1978), 373–379.

[7] J. Bognar, *Indefinite Inner Product Spaces*, Springer, 1974.

[8] S. Chen, K. Liu, and Z. Liu, *Spectrum and stability for elastic systems with global or local Kelvin-Voigt damping*, SIAM J. Appl. Math. **59** (1999), 651–668.

[9] S. Chen and R. Triggiani, *Proof of extensions of two conjectures on structural damping for elastic systems*, Pacific J. Math. **136** (1989), 15–55.

[10] K.-J. Engel and R. Nagel, *One-Parameter Semigroups for Linear Evolution Equations*, Springer Verlag, New York, 2000.

[11] E. Hendrickson and I. Lasiecka, *Numerical approximations and regularizations of Riccati equations arising in hyperbolic dynamics with unbounded control operators*, Comput. Optim. Appl. **2** (1993), 343–390.

[12] E. Hendrickson and I. Lasiecka, *Finite-dimensional approximations of boundary control problems arising in partially observed hyperbolic systems*, Dynam. Contin. Discrete Impuls. Systems **1** (1995), 101–142.

[13] R.O. Hryniv and A.A. Shkalikov, *Operator models in the theory of elasticity and in hydrodynamics, and associated analytic semigroups*, Moscow Univ. Math. Bull. **54** (1999), 1–10.

[14] F. Huang, *On the mathematical model for linear elastic systems with analytic damping*, SIAM J. Control Optim. **26** (1988) 714–724.

[15] I.S. Iohvidov, M.G. Krein, and H. Langer, *Introduction to the Spectral Theory of Operators in Spaces with an Indefinite Metric*, Akademie-Verlag, Berlin, 1982.

[16] B. Jacob, K. Morris, and C. Trunk, *Minimum-phase infinite-dimensional second-order systems*, IEEE Transactions on Automatic Control, **52** (2007), 1654–1665.

[17] B. Jacob, C. Trunk, and M. Winklmeier, *Analyticity and Riesz basis property of semigroups associated to damped vibrations*, Journal of Evolution Equations, **8** (2008), 263–281.

[18] P. Jonas, *On locally definite operators in Krein spaces*, in: Spectral Theory and Applications, Theta Ser. Adv. Math. **2** (2003), Theta, Bucharest, 95–127.

[19] T. Kato, *Perturbation Theory for Linear Operators*, Second Edition, Springer, 1976.

[20] N.D. Kopachevsky, R. Mennicken, Ju.S. Pashkova, and C. Tretter, *Complete second-order linear differential operator equations in Hilbert space and applications in hydrodynamics*, Trans. Am. Math. Soc. **356** (2004), 4737–4766.

[21] P. Lancaster, A.S. Markus, and V.I. Matsaev, *Definitizable operators and quasihyperbolic operator polynomials*, J. Funct. Anal. **131** (1995), 1–28.

[22] P. Lancaster, A. Shkalikov, *Damped vibrations of beams and related spectral problems*, Canadian Applied Mathematics Quarterly **2** (1994), 45–90.

[23] H. Langer, *Spektraltheorie linearer Operatoren in J-Räumen und einige Anwendungen auf die Schar* $L(\lambda) = \lambda^2 + \lambda B + C$, Habilitationsschrift, Technische Universität Dresden, 1965.

[24] H. Langer, *Spectral functions of definitizable operators in Krein spaces*, Lect. Notes Math. **948** (1982), 1–46.

[25] H. Langer, A.S. Markus, and V.I. Matsaev, *Locally definite operators in indefinite inner product spaces*, Math. Ann. **308** (1997), 405–424.

[26] I. Lasiecka, *Stabilization of wave and plate-like equations with nonlinear dissipation on the boundary*, J. Differential Equations **79** (1989), 340–381.

[27] I. Lasiecka and R. Triggiani, *Uniform exponential energy decay of wave equations in a bounded region with $L_2(0,\infty;L_2(\Gamma))$-feedback control in the Dirichlet boundary condition*, J. Differential Equations **66** (1987), 340–390.

[28] N. Levan, *The stabilizability problem: A Hilbert space operator decomposition approach*, IEEE Trans. Circuits and Systems **25** (1978), 721–727.

[29] N.K. Nikolskii, *Treatise on the Shift Operator*, Springer 1986.

[30] M. Slemrod, *Stabilization of boundary control systems*, J. Differential Equation **25** (1976), 402–415.

Carsten Trunk
Technische Universität Ilmenau
Institut für Mathematik
Postfach 100565
D-98684 Ilmenau, Germany
e-mail: carsten.trunk@tu-ilmenau.de

Operator Theory:
Advances and Applications, Vol. 195, 273–286

Derivations of τ-measurable Operators

Martin Weigt

Abstract. It is well known that all derivations on a C*-algebra are continuous and that all derivations on a von Neumann algebra are inner. The aim of this work is to provide some extensions of these results to the algebra of τ-measurable operators affiliated with a von Neumann algebra equipped with a faithful semifinite normal trace τ.

Mathematics Subject Classification (2000). 46H35, 46L51.

Keywords. Atomic projection lattice, inner derivation, derivation, τ-measurable operators, faithful semifinite normal trace, topology of convergence in measure, von Neumann algebra

1. Introduction

Let $\mathcal{A}$ be an algebra. A *derivation* on $\mathcal{A}$ is a linear map $D : \mathcal{A} \to \mathcal{A}$ such that $D(xy) = xD(y) + D(x)y$ for all $x, y \in \mathcal{A}$. A derivation on $\mathcal{A}$ is said to be *inner* if there exists an $a \in \mathcal{A}$ such that $D(x) = ax - xa$ for every $x \in \mathcal{A}$. The study of derivations on C*-algebras is well developed ([28], [12] and [29]). For instance, it is known that all derivations on a C*-algebra are continuous ([28], Lemma 4.1.3) and that all derivations on a von Neumann algebra are inner ([28], Theorem 4.1.6).

The study of non-commutative integration, initiated by Segal in [30], brought about some interesting classes of *-algebras consisting of unbounded operators affiliated with a von Neumann algebra. Of special importance here is the *-algebra of measurable operators affiliated with a von Neumann algebra. Several other notions of measurability have arisen since Segal published his paper. Of special importance to us is the *-algebra $\widetilde{\mathcal{M}}$ of τ-measurable operators affiliated with a (semifinite) von Neumann algebra $\mathcal{M}$ equipped with a faithful semifinite normal trace τ. Equipped with the topology of convergence in measure, $\widetilde{\mathcal{M}}$ is a complete metrizable topological *-algebra.

All of this naturally leads to questions about derivations on $\widetilde{\mathcal{M}}$. We ask whether derivations on $\widetilde{\mathcal{M}}$ are continuous and inner. These questions have recently been answered in the negative in [6] and [7], namely that if $\mathcal{M}$ is commutative and the projection lattice of $\mathcal{M}$ is not atomic, then $\widetilde{\mathcal{M}}$ admits at least one discontinuous (hence non-inner) derivation.

Motivated by this result, we prove that if the projection lattice of $\mathcal{M}$ is atomic, then all derivations on $\widetilde{\mathcal{M}}$ are continuous (Theorem 3.5). Conversely, we show that if $\mathcal{M}$ is a finite type I von Neumann algebra and all derivations on $\widetilde{\mathcal{M}}$ are continuous, then the projection lattice of $\mathcal{M}$ is atomic (Theorem 3.12).

In light of the above, we also consider the problem as to when all continuous derivations on $\widetilde{\mathcal{M}}$ are inner. This problem is still open and Sh.A. Ayupov et al. have recently made some progress on this problem in [1]. We show that if the range of a derivation on $\widetilde{\mathcal{M}}$ is contained inside a non-commutative L_p-space, then it is inner (Theorem 3.16).

This paper consists of three sections, including the introduction. Section 2 consists of preliminaries needed to understand and prove the main results, the subject of Section 3.

2. The algebra $\widetilde{\mathcal{M}}$ and non-commutative L_p-spaces

Throughout the paper, unless stated otherwise, $\mathcal{M}$ denotes a (semifinite) von Neumann algebra equipped with a faithful semifinite normal trace τ on a Hilbert space $\mathcal{H}$.

2.1. The algebra $\widetilde{\mathcal{M}}$

Throughout, $\mathcal{D}(x)$ denotes the domain of an unbounded operator x. The notation $\mathcal{U}(\mathcal{M})$ denotes the set of closed densely defined operators affiliated with $\mathcal{M}$. If $\mathcal{M}$ is finite, then $\mathcal{U}(\mathcal{M})$ is a *-algebra ([11]). The *-algebra of measurable operators affiliated with a von Neumann algebra $\mathcal{M}$, as defined in [30], will be denoted by $\mathcal{S}(\mathcal{M})$. If $\mathcal{M}$ is commutative, then $\mathcal{M} \cong L_\infty(X, \Sigma, \mu)$ and $\mathcal{S}(\mathcal{M}) \cong L_0(X, \Sigma, \mu)$ for some localizable measure space (X, Σ, μ) ([30]).

The projection lattice of a von Neumann algebra $\mathcal{M}$ is denoted by $\mathcal{M}_p$. We denote the centre of $\mathcal{M}$ by $Z(\mathcal{M})$.

Definition 2.1. ([18], Definition 2.1) Let E be a subspace of $\mathcal{H}$. We say that E is τ-dense in $\mathcal{H}$ if for every $\delta > 0$, there exists a projection $p \in \mathcal{M}$ such that $p\mathcal{H} \subset E$ and $\tau(1-p) < \delta$.

An unbounded operator x is said to be τ-measurable if $x \in \mathcal{U}(\mathcal{M})$ and $\mathcal{D}(x)$ is τ-dense in $\mathcal{H}$.

The set of τ-measurable operators affiliated with $\mathcal{M}$ will be denoted by $\widetilde{\mathcal{M}}$.

Proposition 2.2. ([18], *p.* 271; [30], *Corollary* 4.1) *If* $\tau(1) < \infty$, *then* $\widetilde{\mathcal{M}} = \mathcal{U}(\mathcal{M})$. *For any finite von Neumann algebra* $\mathcal{M}$, *one has* $S(\mathcal{M}) = \mathcal{U}(\mathcal{M})$.

Let $\epsilon, \delta > 0$ and let $\widetilde{\mathcal{M}}(\epsilon, \delta)$ denote the set

$$\{x \in \widetilde{\mathcal{M}} : \text{ there is a } p \in \mathcal{M}_p \text{ such that } p\mathcal{H} \subset \mathcal{D}(x), \|xp\| \leq \epsilon, \tau(1-p) < \delta\}.$$

The sets $\{\widetilde{\mathcal{M}}(\epsilon,\delta) : \epsilon,\delta > 0\}$ form a system of basic neighbourhoods of zero for a topology γ_{cm} on $\widetilde{\mathcal{M}}$, called the *topology of convergence in measure* on $\widetilde{\mathcal{M}}$ ([18], p. 272).

Furthermore, $\widetilde{\mathcal{M}}$ is a *-algebra under strong sum, strong product, scalar multiplication and ordinary adjunction (the *strong sum* (resp. *strong product*) of unbounded operators x and y is defined to be $\overline{x+y}$ (resp. $\overline{xy}$), where $\overline{x}$ denotes the closure of the operator x). When equipped with the topology of convergence in measure, $\widetilde{\mathcal{M}}$ is a complete metrizable topological *-algebra ([18], p. 272). Lastly, $\mathcal{M}$ is dense in $\widetilde{\mathcal{M}}$ with respect to the topology of convergence in measure ([18], p. 272).

In the sequel, unless stated otherwise, if $x, y \in \widetilde{\mathcal{M}}$, we will write $x+y$ and xy to mean the strong sum and strong product respectively of x and y.

Example. ([35]) Let $\mathcal{M}$ be a commutative von Neumann algebra. Then $\mathcal{M} \cong L_\infty(X,\Sigma,\mu)$ for some localizable measure space (X,Σ,μ), and $\mathcal{M}$ is equipped with the trace defined by $\tau(f) = \int_X f \, d\mu$ for all (almost everywhere) positive $f \in L_\infty(X,\Sigma,\mu)$. The topologies of convergence in measure on $L_\infty(X,\Sigma,\mu)$ and $\widetilde{L}_\infty(X,\Sigma,\mu)$ are the restrictions of the topology of convergence in measure on $L_0(X,\Sigma,\mu)$ to $L_\infty(X,\Sigma,\mu)$ and $\widetilde{L}_\infty(X,\Sigma,\mu)$ respectively. By taking completions, we find that $\widetilde{\mathcal{M}}$ is topologically *-isomorphic to $\widetilde{L}_\infty(X,\Sigma,\mu)$.

Theorem 2.3. ([31], *Theorem* 2.3(i)) *Let $p \in \mathcal{M}_p$ and $\tau_p = \tau|_{(p\mathcal{M}p)^+}$. Then τ_p is a faithful semifinite normal trace on $p\mathcal{M}p$. Furthermore, $\widetilde{p\mathcal{M}p} = p\widetilde{\mathcal{M}}p$, where $\widetilde{p\mathcal{M}p}$ is the algebra of τ_p-measurable operators affiliated with $p\mathcal{M}p$.*

Proposition 2.4. ([31], *Examples* 2.2(3)) *The following statements are equivalent.*

(i) $\widetilde{\mathcal{M}} = \mathcal{M}$.
(ii) $\inf\{\tau(p) : 0 \neq p \in \mathcal{M}_p\} > 0$.
(iii) *The topology of convergence in measure coincides with the norm topology.*

Let $x \in \widetilde{\mathcal{M}}$. The generalized singular function ([18], Definition 2.1) of x, denoted by $\mu_t(x)$, is defined by

$$\mu_t(x) = \inf\{\|xp\| : p \in \mathcal{M}_p, p\mathcal{H} \subset \mathcal{D}(x), \text{ and } \tau(1-p) \leq t\}, \ t > 0.$$

Proposition 2.5. ([18], *Lemma* 2.5) *Let $x, y, z \in \widetilde{\mathcal{M}}$. The following statements hold for every $t > 0$.*

(i) $\mu_t(xyz) \leq \|x\|\mu_t(y)\|z\|$, *where the cases $\|x\| = \infty$ and $\|z\| = \infty$ are allowed.*
(ii) $\mu_t(x^*) = \mu_t(x)$.

Theorem 2.6. ([36], *Theorem* 2.1) *Let $\mathcal{I}$ be a γ_{cm}-closed two-sided ideal of $\widetilde{\mathcal{M}}$. Then $\mathcal{I} \cap \mathcal{M}$ is a norm closed two-sided ideal of $\mathcal{M}$ and $\mathcal{I} = \overline{\mathcal{I} \cap \mathcal{M}}^{\gamma_{cm}}$.*

An *atom* of $\mathcal{M}$ is a nonzero projection in $\mathcal{M}$ having no nonzero proper subprojections in $\mathcal{M}$. We say that $\mathcal{M}_p$ is *atomic* if for every nonzero $p \in \mathcal{M}_p$, there exists an atom $q \in \mathcal{M}_p$ such that $q \leq p$.

2.2. Non-commutative L_p-spaces

For $0 < p < \infty$, we define $L_p(\mathcal{M}, \tau)$ to be the set of all $x \in \widetilde{\mathcal{M}}$ such that

$$\|x\|_p = \left(\int_0^\infty \mu_t(x)^p \, \mathrm{d}t \right)^{\frac{1}{p}} < \infty$$

(a special case of [15], Definition 4.1). In addition, we put $L_\infty(\mathcal{M}, \tau) = \mathcal{M}$ and denote by $\|.\|_\infty$ the usual operator norm defined on $\mathcal{M}$. It is well known that $L_p(\mathcal{M}, \tau)$ is a Banach space under $\|.\|_p$ whenever $1 \leq p \leq \infty$ ([15], Theorem 4.5). Also, whenever $1 \leq p \leq \infty$, $L_p(\mathcal{M}, \tau)$ is a Banach $\mathcal{M}$-module ([37], Proposition 2.5). If $1 < p < \infty$, then L_p is reflexive (this is a special case of [16], Corollary 5.16). The proof of the following lemma follows directly from Proposition 2.5.

Lemma 2.7. *Let $1 \leq p < \infty$ and let $u \in \mathcal{M}$ be a unitary operator. Then $\|u^* x u\|_p = \|x\|_p$ for every $x \in L_p(\mathcal{M}, \tau)$.*

There are other definitions of non-commutative L_p-spaces which we will not discuss, see for example [18], p. 271, and [37]. These definitions are equivalent to the one given above.

3. The main results

The *separating space* $\mathcal{S}(D)$ of a derivation D on $\widetilde{\mathcal{M}}$ is defined to be the set

$$\{y \in \widetilde{\mathcal{M}} : \text{ there is a sequence } (x_n) \text{ in } \widetilde{\mathcal{M}} \text{ such that } x_n \to 0 \text{ and } D(x_n) \to y\}$$

with respect to the topology of convergence in measure on $\widetilde{\mathcal{M}}$. Since $\widetilde{\mathcal{M}}$ is a complete metrizable topological *-algebra in the topology of convergence in measure, the closed graph theorem tells us that D is $\gamma_{cm} - \gamma_{cm}$ continuous if and only if $\mathcal{S}(D) = \{0\}$.

Recall that every derivation on a C*-algebra is continuous and that every derivation on a von Neumann algebra is inner. The following result demonstrates that these results do not carry over to algebras of measurable operators.

Theorem 3.1 (A.F. Ber, V.I. Chilin and F.A. Sukochev). ([6], *Theorem* 3, *and* [7], *Theorem* 3.4) *Let $\mathcal{M}$ be a commutative von Neumann algebra. Then $S(\mathcal{M})$ admits a nonzero derivation if and only if $\mathcal{M}_p$ is not atomic.*

By recalling that every commutative von Neumann algebra is finite, the following corollary follows from Proposition 2.2 and Theorem 3.1.

Corollary 3.2. *Let $\mathcal{M}$ be a commutative von Neumann algebra with a faithful finite normal trace. Then $\widetilde{\mathcal{M}}$ admits a nonzero derivation if and only if $\mathcal{M}_p$ is not atomic.*

Note that if $\mathcal{M}$ is commutative, then a derivation on $\widetilde{\mathcal{M}}$ is $\gamma_{cm} - \gamma_{cm}$ continuous if and only if it is zero ([1], p. 11).

Using Theorem 3.1 and the fact that every finite type I von Neumann algebra is a direct sum of finite matrix algebras over commutative von Neumann algebras ([28], Theorems 2.3.2 and 2.3.3), one can prove the following result.

Theorem 3.3. ([8], *Theorem* 5) *If $\mathcal{M}$ is a finite type I von Neumann algebra and every derivation $D : S(\mathcal{M}) \to \mathcal{S}(\mathcal{M})$ is inner, then $\mathcal{M}_p$ is atomic.*

The following corollary is an immediate consequence of Proposition 2.2 and Theorem 3.3.

Corollary 3.4. *Let $\mathcal{M}$ be a type I von Neumann algebra with a faithful finite normal trace τ. If every derivation $D : \widetilde{\mathcal{M}} \to \widetilde{\mathcal{M}}$ is inner, then $\mathcal{M}_p$ is atomic.*

Our next result was motivated by Corollary 3.2 and the remark immediately thereafter.

Theorem 3.5. *If $\mathcal{M}_p$ is atomic, then all derivations on $\widetilde{\mathcal{M}}$ are $\gamma_{cm} - \gamma_{cm}$ continuous.*

Proof. Let D be a derivation on $\widetilde{\mathcal{M}}$ and q an atomic projection in $\mathcal{M}$. Let $y \in \mathcal{S}(D)$. Then there exists a sequence x_n in $\widetilde{\mathcal{M}}$ with $x_n \to 0$ (γ_{cm}) and $D(x_n) \to y$ (γ_{cm}). Therefore

$$\begin{aligned} D(qx_nq) &= qD(x_nq) + D(q)x_nq \\ &= qx_nD(q) + qD(x_n)q + D(q)x_nq \\ &\to qyq\ (\gamma_{cm}). \end{aligned}$$

By Theorem 2.3, it follows that $q\widetilde{\mathcal{M}}q = \widetilde{q\mathcal{M}q}$. Since q is an atomic projection, $q\mathcal{M}q = \mathbb{C}q$.

We show that $q\mathcal{M}q = \widetilde{q\mathcal{M}q}$. Since q is an atomic projection, it is the only nonzero projection in $q\mathcal{M}q$. Hence

$$\inf\{\tau(p) : p \text{ a nonzero projection in } q\mathcal{M}q\} = \tau(q) > 0,$$

since $q \neq 0$ and the trace τ is faithful. Therefore, by Proposition 2.4, $q\mathcal{M}q = \widetilde{q\mathcal{M}q}$.

Thus $q\widetilde{\mathcal{M}}q = \widetilde{q\mathcal{M}q} = q\mathcal{M}q = \mathbb{C}q$. Since $x_n \to 0$ (γ_{cm}), $qx_nq \to 0$ (γ_{cm}). Since $q\widetilde{\mathcal{M}}q$ is finite-dimensional, $D|_{q\widetilde{\mathcal{M}}q}$ is $\gamma_{cm} - \gamma_{cm}$ continuous, implying that $D(qx_nq) \to 0$ (γ_{cm}). Recall that $D(qx_nq) \to qyq$ (γ_{cm}). Therefore $qyq = 0$. This is true for every atomic projection q in $\mathcal{M}$ and for every $y \in \mathcal{S}(D)$.

We show next that $\mathcal{S}(D)$ has no atomic projections of $\mathcal{M}$. Suppose that $\mathcal{S}(D)$ has at least one atomic projection q_0 of $\mathcal{M}$. Then, from what we have proved above, $q_0q_0q_0 = 0$, i.e., $q_0 = 0$. This is a contradiction since $q_0 \neq 0$. Therefore $\mathcal{S}(D)$ has no atomic projections of $\mathcal{M}$.

For a contradiction, let us assume that $\mathcal{S}(D) \neq \{0\}$. By Theorem 2.6, and the fact that $\mathcal{S}(D)$ is a γ_{cm}-closed two-sided ideal of $\widetilde{\mathcal{M}}$, it follows that $\mathcal{S}(D) \cap \mathcal{M}$ is a norm closed two-sided ideal of $\mathcal{M}$. Furthermore, $\mathcal{S}(D) \cap \mathcal{M} \neq \{0\}$ since, by Theorem 2.6,

$$\overline{\mathcal{S}(D) \cap \mathcal{M}}^{\gamma_{cm}} = \mathcal{S}(D) \neq \{0\}.$$

Therefore $\mathcal{S}(D) \cap \mathcal{M}$ has at least one nonzero projection p. Since $\mathcal{M}_p$ is atomic, there exists an atomic projection q_0 of $\mathcal{M}$ such that $0 < q_0 \leq p$. So $q_0 = q_0 p \in \mathcal{S}(D) \cap \mathcal{M}$ because $q_0 \in \mathcal{M}$, p is in $\mathcal{S}(D)$, and $\mathcal{S}(D) \cap \mathcal{M}$ is a two-sided ideal of $\mathcal{M}$. This is a contradiction since $\mathcal{S}(D)$ has no atomic projections of $\mathcal{M}$. Hence $\mathcal{S}(D) = \{0\}$, and so D is $\gamma_{cm} - \gamma_{cm}$ continuous. □

In light of Corollary 3.2 and the remark thereafter, it would be interesting to know if the converse of Theorem 3.5 holds. We now solve this problem in the affirmative for finite type I von Neumann algebras (Theorem 3.12), thereby extending Corollary 3.4. For this, we need the following four lemmas. The proof of [8], Theorem 5 (i.e., Theorem 3.3), relies strongly on [11], Lemma 2, which itself depends on the fact that $S(\mathcal{M})$ is a regular algebra whenever $\mathcal{M}$ is finite.

In general, $\widetilde{\mathcal{M}}$ is not regular even if $\mathcal{M}$ is commutative: The crucial fact is that a finite von Neumann algebra $\mathcal{M}$ can be enlarged to a regular algebra $\mathcal{U}(\mathcal{M})$, the algebra of closed densely defined operators affiliated with $\mathcal{M}$. By [10], p. 211, $\mathcal{U}(\mathcal{M})$ is also the smallest regular algebra to contain $\mathcal{M}$, in the sense that the only regular subalgebra of $\mathcal{U}(\mathcal{M})$ to contain $\mathcal{M}$ is $\mathcal{U}(\mathcal{M})$ itself. Thus, if $\mathcal{M}$ is commutative and $\widetilde{\mathcal{M}} \neq \mathcal{U}(\mathcal{M})$, then $\widetilde{\mathcal{M}}$ is not regular.

Lemma 3.6. *Let p be a central projection of $\mathcal{M}$ and D a derivation on $\widetilde{p\mathcal{M}p}$. Then D can be extended to a derivation $\overline{D}$ on $\widetilde{\mathcal{M}}$.*

Proof. Let $\overline{D}(x) = D(pxp)$ for every $x \in \widetilde{\mathcal{M}}$. Since p is a central projection, $D(p) = 0$. Using this, it is easily verified that $\overline{D}$ is a derivation on $\widetilde{\mathcal{M}}$ extending D. □

For $p \in \mathcal{M}_p$, we denote by $c(p)$ the least central projection majorizing p, and we call $c(p)$ the *central support* of p. If $p, q \in \mathcal{M}_p$, then $qxp = 0$ for every $x \in \mathcal{M}$ if and only if $c(p)c(q) = 0$ ([32], Corollary V.1.7). This is needed in the proof of the following known result, which we give for completeness.

Lemma 3.7. ([20], *Theorem* 2.1) *If q is an atom in a von Neumann algebra $\mathcal{M}$, then the central support $c(q)$ of q is an atom in $Z(\mathcal{M})$.*

Proof. Let p be a central projection of $\mathcal{M}$ such that $0 < p \leq c(q)$. Then $qp \in \mathcal{M}_p$ and $qp \leq q$. Since q is an atom, it follows that $qp = 0$ or $qp = q$.

We show that $qp \neq 0$, so suppose that $qp = 0$. Then $qxp = qpx = 0$ for every $x \in \mathcal{M}$, and so $c(q)c(p) = 0$, i.e., $pc(q) = c(q)p = 0$. Since $p \leq c(q)$, it follows that $p = 0$. This is a contradiction since $0 \neq p$. Thus $qp \neq 0$.

Therefore $qp = q$, implying that $q = qp \leq p$. Hence $c(q) \leq p$, since p is a central projection majorizing q, and thus $c(q) = p$, implying that $c(q)$ is an atom in $Z(\mathcal{M})$. □

Lemma 3.8. *Let (p_α) be a family of central orthogonal projections in $\mathcal{M}$ such that $\sum_\alpha p_\alpha = 1$. If $(p_\alpha \mathcal{M} p_\alpha)_p$ is atomic for every α, then $\mathcal{M}_p$ is atomic.*

Proof. Suppose $(p_\alpha \mathcal{M} p_\alpha)_p$ is atomic for every α. Let $0 < p \in \mathcal{M}_p$. Since p_α is a central projection for every α, it follows that $pp_\alpha \in \mathcal{M}_p$ for every α. By hypothesis, $\sum_\alpha p_\alpha = 1$, and so $\sum_\alpha pp_\alpha = p$. Since $p \neq 0$, we have that $pp_\alpha \neq 0$ for some α. It is easily verified that $p \geq pp_\alpha$ and that $pp_\alpha \leq p_\alpha$. Therefore, pp_α is a projection in $p_\alpha \mathcal{M} p_\alpha$ and, by hypothesis, there exists an atom q_α in $p_\alpha \mathcal{M} p_\alpha$ such that $p \geq pp_\alpha \geq q_\alpha$. Clearly, q_α is also an atom in $\mathcal{M}$. This completes the proof. □

For the next lemma, we recall the notion of a direct sum of von Neumann algebras ([14], pp. 20–21). For each α, let $\mathcal{A}_\alpha$ be a von Neumann algebra on the Hilbert space $\mathcal{H}_\alpha$, and let $\|\cdot\|_\alpha$ denote the norm on $\mathcal{A}_\alpha$. If $x_\alpha \in \mathcal{A}_\alpha$ for each α, and $\sup_\alpha \|x_\alpha\|_\alpha < \infty$, define a bounded linear operator x on the direct sum $\mathcal{H}$ of the family of Hilbert spaces $(\mathcal{H}_\alpha)$ by $x(\xi) = (x_\alpha(\xi_\alpha))$, where $\xi = (\xi_\alpha) \in \mathcal{H}$. The *direct sum* $\mathcal{A}$ of the von Neumann algebras $\mathcal{A}_\alpha$ is defined to be the set of all such operators x, and we write $\mathcal{A} = \oplus_\alpha \mathcal{A}_\alpha$. It can be verified that $\mathcal{A}$ is a von Neumann algebra on $\mathcal{H}$ with coordinate-wise operations, and norm $x \mapsto \sup_\alpha \|x_\alpha\|_\alpha$.

Lemma 3.9. *Let $\mathcal{M} = \mathcal{A} \overline{\otimes} \mathcal{B}(\mathcal{H})$, where $\mathcal{A}$ is a commutative von Neumann algebra and $\mathcal{H}$ a finite-dimensional Hilbert space (so $\mathcal{M}$ is a type I von Neumann algebra). If $\mathcal{A}_p$ is atomic, so is $\mathcal{M}_p$.*

Proof. Since $\mathcal{A}$ is commutative and $\mathcal{A}_p$ is atomic, $\mathcal{A} \cong L_\infty(X, \Sigma, \mu)$ for some localizable atomic measure space (X, Σ, μ). Therefore we can find a disjoint family of atoms $(A_\lambda : \lambda \in \Lambda)$ satisfying $X = \cup A_\lambda$. Recall that measurable functions on atoms are constant. Hence $\mathcal{A} \cong l_\infty(\Lambda)$, where $l_\infty(\Lambda)$ is the space of all bounded nets in $\mathbb{C}$ indexed by Λ. Hence, by [14], p. 29,

$$\begin{aligned}\mathcal{M} &= \mathcal{A} \overline{\otimes} \mathcal{B}(\mathcal{H}) \\ &\cong l_\infty(\Lambda) \overline{\otimes} \mathcal{B}(\mathcal{H}) \\ &= \oplus_{\alpha\in\Lambda}(\mathbb{C}_\alpha \overline{\otimes} \mathcal{B}(\mathcal{H})) \\ &\cong \oplus_{\alpha\in\Lambda}\mathcal{B}(\mathcal{H}_\alpha),\end{aligned}$$

where $\mathcal{H}_\alpha = \mathcal{H}$ and $\mathbb{C}_\alpha = \mathbb{C}$ for every $\alpha \in \Lambda$. Every $(\mathcal{B}(\mathcal{H}_\alpha))_p$ is atomic, and one can find a family of central orthogonal projections (p_α) in $\mathcal{M}$ such that $\sum_\alpha p_\alpha = 1$ and $p_\alpha \mathcal{M} p_\alpha \cong \mathcal{B}(\mathcal{H}_\alpha)$, namely $p_\alpha = 1_\alpha$, where 1_α denotes the identity operator on $\mathcal{H}_\alpha$, for every $\alpha \in \Lambda$. By Lemma 3.8, $\mathcal{M}_p$ is atomic. □

If $\mathcal{M}$ is a finite von Neumann algebra, then $\mathcal{M}$ can be imbedded into the maximal ring of right quotients $\mathcal{Q}_\mathcal{M}$ of $\mathcal{M}$ ([11]). Furthermore, $\mathcal{U}(\mathcal{M}) \cong \mathcal{Q}_\mathcal{M}$ ([11]). Therefore, if $\mathcal{A}$ and $\mathcal{B}$ are *-isomorphic finite von Neumann algebras, then $\mathcal{U}(\mathcal{A}) \cong \mathcal{U}(\mathcal{B})$. In particular, if $\mathcal{A}$ and $\mathcal{B}$ are equipped with faithful finite normal traces, then $\widetilde{\mathcal{A}} = \mathcal{U}(\mathcal{A}) \cong \mathcal{U}(\mathcal{B}) = \widetilde{\mathcal{B}}$. This is needed in the proof of our next proposition.

If $\mathcal{A}$ is a von Neumann algebra with a faithful finite normal trace, then $\widetilde{M_n(\mathcal{A})} = M_n(\widetilde{\mathcal{A}})$ for every $n \in \mathbb{N}$ ([11], Lemma 2). Let $(a^m_{i,j})$ be a sequence in $M_n(\widetilde{\mathcal{A}})$ and $(a_{i,j}) \in M_n(\widetilde{\mathcal{A}})$. Then $(a^m_{i,j}) \to (a_{i,j})$ in measure as $m \to \infty$ if and only if $a^m_{i,j} \to a_{i,j}$ as $m \to \infty$ for every i,j (this is an immediate consequence of [17], Lemma 2.1).

The proof of the following proposition is a slight modification of the proof [8], Theorem 5, and we give the proof for completeness. It is a special case of Theorem 3.12.

Proposition 3.10. *Suppose that $\mathcal{M}$ is a type I von Neumann algebra with a faithful finite normal trace. If all derivations on $\widetilde{\mathcal{M}}$ are $\gamma_{cm} - \gamma_{cm}$ continuous, then $\mathcal{M}_p$ is atomic.*

Proof. Since $\mathcal{M}$ is a finite type I von Neumann algebra, there exists a sequence of central projections (p_n) such that $\mathcal{M} = \oplus_{n=1}^{\infty} p_n \mathcal{M} p_n$, and $p_n \mathcal{M} p_n \cong M_{k_n}(\mathcal{A}_n)$, where $\mathcal{A}_n$ is a commutative von Neumann algebra for every n ([28], Theorems 2.3.2 and 2.3.3), and the k_n are integers. Assume that $\mathcal{M}_p$ is not atomic. Then, by Lemmas 3.8 and 3.9, it follows that $(\mathcal{A}_r)_p$ is not atomic for some $r \in \mathbb{N}$. By Corollary 3.2, there exists a derivation δ on $\widetilde{\mathcal{A}_r}$ which is not $\gamma_{cm} - \gamma_{cm}$ continuous.

We show that there exists a derivation D_r on $\widetilde{M_{k_r}(\mathcal{A}_r)} = M_{k_r}(\widetilde{\mathcal{A}_i})$ which is not $\gamma_{cm} - \gamma_{cm}$ continuous. Define D_r to be the linear map defined by $D_r((a_{i,j})) = (\delta(a_{i,j}))$ for every $(a_{i,j}) \in M_{k_r}(\widetilde{\mathcal{A}_r})$. It is easily verified that D_r is a derivation. Since δ is not $\gamma_{cm} - \gamma_{cm}$ continuous, there exists a sequence (a_m) in $\widetilde{\mathcal{A}_r}$ such that $a_m \to 0$ in measure and $\delta(a_m)$ does not converge to zero in measure. Let $(a_{i,j}^m)$ be the sequence in $M_{k_r}(\mathcal{A}_r)$ defined as $a_{i,j}^m = a_m$ for every m and for all i,j. By the preceding remarks, $(a_{i,j}^m) \to (b_{i,j})$ as $m \to \infty$, where $b_{i,j} = 0$ for all i,j. Also, by the preceding remarks, $D_r((a_{i,j}^m)) = (\delta(a_{i,j}^m))$ does not converge to $D_r((b_{i,j})) = (\delta(b_{i,j}))$ in measure. Therefore, D_r is not $\gamma_{cm} - \gamma_{cm}$ continuous.

Since $p_r \mathcal{M} p_r \cong M_{k_r}(\mathcal{A}_r)$ and the traces of $p_r \mathcal{M} p_r$ and $M_{k_r}(\mathcal{A}_r)$ are finite, it follows that the algebra $M_{k_r}(\widetilde{\mathcal{A}_r}) = \widetilde{M_{k_r}(\mathcal{A}_r)}$ is isomorphic to $\widetilde{p_r \mathcal{M} p_r}$, and this isomorphism is also $\gamma_{cm} - \gamma_{cm}$ bicontinuous (this follows from [24], Proposition 4.7). Therefore, the derivation D_r can be identified with a derivation on $\widetilde{p_r \mathcal{M} p_r}$ which is not $\gamma_{cm} - \gamma_{cm}$ continuous. By Lemma 3.6, D_r can be extended to a derivation D on $\widetilde{\mathcal{M}}$. It follows that D is not $\gamma_{cm} - \gamma_{cm}$ continuous. This contradicts the hypothesis, implying that $\mathcal{M}_p$ is atomic. □

The following proposition is the first part of the proof of Theorem 3.12.

Proposition 3.11. *Let $\mathcal{M}$ be a finite type I von Neumann algebra. If all derivations on $\widetilde{\mathcal{M}}$ are $\gamma_{cm} - \gamma_{cm}$ continuous, then $(Z(\mathcal{M}))_p$ is atomic.*

Proof. Suppose that $\mathcal{M}$ is a finite type I von Neumann algebra such that all derivations on $\widetilde{\mathcal{M}}$ are $\gamma_{cm} - \gamma_{cm}$ continuous. Since $\mathcal{M}$ is a finite von Neumann algebra, $\tau|_{Z(\mathcal{M})}$ is semifinite ([14], Proposition 10, p. 12).

Let q be nonzero central projection such that $\tau(q) < \infty$. Let D be a derivation on $\widetilde{q\mathcal{M}q}$. By Lemma 3.6, D can be extended to a derivation $\overline{D}$ on $\widetilde{\mathcal{M}}$. By hypothesis, $\overline{D}$ is $\gamma_{cm} - \gamma_{cm}$ continuous, and thus, D is $\gamma_{cm} - \gamma_{cm}$ continuous. Since $q\mathcal{M}q$ is a type I von Neumann algebra (on the Hilbert space $q\mathcal{H}$) with finite trace, it follows from Proposition 3.10 that $(q\mathcal{M}q)_p$ is atomic. Therefore there exists an atom q_0 in $q\mathcal{M}q$ such that $q \geq q_0$. It is clear that q_0 is also an atom in $\mathcal{M}$.

Let $p \in Z(\mathcal{M})_p$. Using again the fact that $\tau|_{Z(\mathcal{M})}$ is semifinite, it follows that there exists a nonzero $p_1 \in Z(\mathcal{M})_p$ such that $\tau(p_1) < \infty$ and $p \geq p_1$. By the previous paragraph, there is an atom p_2 in $\mathcal{M}$ such that $p_1 \geq p_2$. Hence every central projection majorizes an atom of $\mathcal{M}$. By Lemma 3.7, $c(p_2)$ is an atom in $Z(\mathcal{M})$. Since $p \in Z(\mathcal{M})_p$, it follows that $p \geq c(p_2)$. Thus $Z(\mathcal{M})_p$ is atomic. □

Theorem 3.12. *Let $\mathcal{M}$ be a finite type I von Neumann algebra. If all derivations on $\widetilde{\mathcal{M}}$ are $\gamma_{cm} - \gamma_{cm}$ continuous, then $\mathcal{M}_p$ is atomic.*

Proof. Since $\mathcal{M}$ is a finite type I von Neumann algebra, we can write $\mathcal{M} \cong \oplus_{n=1}^{\infty} \mathcal{M}_n$, where every $\mathcal{M}_n$ is a type I_n algebra such that $\mathcal{M}_n \cong Z(\mathcal{M}_n) \mathbin{\overline{\otimes}} \mathcal{B}(\mathcal{H}_n)$, where $\mathcal{H}_n$ is a finite-dimensional Hilbert space for each n ([28], Theorems 2.3.2 and 2.3.3). To be more precise, one can find a sequence of central projections (p_n) in $\mathcal{M}$ such that, for every n, $\mathcal{M}_n \cong p_n \mathcal{M} p_n$ and $\sum_{n=1}^{\infty} p_n = 1$. By Theorem 2.3, $\widetilde{\mathcal{M}_n} = p_n \widetilde{\mathcal{M}} p_n$ for every n.

Consider now a fixed n and let D_n be a derivation on $\widetilde{\mathcal{M}_n}$. It follows from Lemma 3.6 that D_n can be extended to a derivation $\overline{D_n}$ on $\widetilde{\mathcal{M}}$. By hypothesis, $\overline{D_n}$ is $\gamma_{cm} - \gamma_{cm}$ continuous, and so the same holds for D_n. Hence all derivations on $\widetilde{\mathcal{M}_n}$ are $\gamma_{cm} - \gamma_{cm}$ continuous. This holds for every n. By Proposition 3.11, $Z(\mathcal{M}_n)_p$ is atomic for every n. It is an immediate consequence of Lemma 3.9, and the fact that $\mathcal{M}_n \cong Z(\mathcal{M}_n) \mathbin{\overline{\otimes}} \mathcal{B}(\mathcal{H})$ for every n, that every $(\mathcal{M}_n)_p$ is atomic. Finally, by Lemma 3.8, $\mathcal{M}_p$ is atomic. □

The following example demonstrates that the converse of Theorem 3.5 does in general not hold.

Example. Let $\mathcal{A} = L_\infty([0,1])$, $\mathcal{H}$ an infinite-dimensional Hilbert space, and $\mathcal{M} = \mathcal{A} \mathbin{\overline{\otimes}} \mathcal{B}(\mathcal{H})$. Then $\mathcal{M}$ is a type I_∞ von Neumann algebra.

Now

$$\begin{aligned} Z(\mathcal{M}) &= Z(\mathcal{A} \mathbin{\overline{\otimes}} \mathcal{B}(\mathcal{H})) \\ &= Z(\mathcal{A}) \mathbin{\overline{\otimes}} Z(\mathcal{B}(\mathcal{H})) \\ &= \mathcal{A} \mathbin{\overline{\otimes}} \mathbb{C}1 \\ &\cong \mathcal{A} \\ &= L_\infty([0,1]). \end{aligned}$$

Therefore $Z(\mathcal{M})$ has no atoms. Therefore, by Lemma 3.7, $\mathcal{M}$ has no atoms, implying that $\mathcal{M}_p$ is not atomic. Recently, F.A. Sukochev, A.F. Ber and B. de Pagter proved that all derivations on $\widetilde{\mathcal{M}}$ are $\gamma_{cm} - \gamma_{cm}$ continuous ([9]). This result has also been proved in [3].

We have already seen that derivations on $\widetilde{\mathcal{M}}$ need not be continuous and not inner. Therefore, the following problem presents itself.

Problem. ([5], Problem 5) Is every $\gamma_{cm} - \gamma_{cm}$ continuous derivation on $\widetilde{\mathcal{M}}$ inner?

We now give results which provide some affirmative answers to this question. Recall that if $\mathcal{M}$ is commutative, then every $\gamma_{cm} - \gamma_{cm}$ continuous derivation on $\widetilde{\mathcal{M}}$ is inner, i.e., the zero derivation is the only $\gamma_{cm} - \gamma_{cm}$ continuous derivation on $\widetilde{\mathcal{M}}$. The following lemma is elementary and we give a proof for completeness.

Lemma 3.13. ([5]) *If D is a $\gamma_{cm} - \gamma_{cm}$ continuous derivation on $\widetilde{\mathcal{M}}$ such that $D(\mathcal{M}) \subset \mathcal{M}$, then there exists $a \in \mathcal{M}$ such that $D(x) = ax - xa$ for all $x \in \widetilde{\mathcal{M}}$, i.e., D is inner.*

Proof. Let D be a derivation as in the hypothesis. Then, since $D(\mathcal{M}) \subset \mathcal{M}$, there exists $a \in \mathcal{M}$ such that $D(x) = ax - xa$ for every $x \in \mathcal{M}$ ([28], Theorem 4.1.6). Since D is continuous and $\mathcal{M}$ is dense in $\widetilde{\mathcal{M}}$, it follows that $D(x) = ax - xa$ for every $x \in \widetilde{\mathcal{M}}$, i.e., D is inner. □

In 2006, the author presented Theorem 3.5 at the conference "Great Plains Operator Theory Symposium" in Iowa City, USA ([34]). He then communicated with Sh.A. Ayupov, who then, with S. Albeverio and K.K. Kudaybergenov, extended Theorem 3.5 in [1]. Derivations on $\widetilde{\mathcal{M}}$ have recently also been studied in [2] and [3].

Proposition 3.14. *Let $D : \widetilde{\mathcal{M}} \to \widetilde{\mathcal{M}}$ be a derivation. If $D|_{\mathcal{M}}$ is $\gamma_{cm} - \gamma_{cm}$ continuous, then D is $\gamma_{cm} - \gamma_{cm}$ continuous.*

Proof. Every derivation on a *-algebra $\mathcal{A}$ can be written as a linear combination of two *-derivations ([12], p. 229). Therefore, we may assume without loss of generality that D is a *-derivation. It suffices to show that $\mathcal{S}(D) = \{0\}$. Let $y \in \mathcal{S}(D)$. Then there is a sequence (x_n) in $\widetilde{\mathcal{M}}$ such that $x_n \to 0$ (γ_{cm}) and $D(x_n) \to y$ (γ_{cm}). Since $x_n(1 + x_n^* x_n)^{-1}$ is affiliated with $\mathcal{M}$ for every n, it follows that $x_n(1 + x_n^* x_n)^{-1} \in \mathcal{M}$ for every n. Since inversion is continuous on $\widetilde{\mathcal{M}}$ in the measure topology ([33]) and $1 + x_n^* x_n \to 1$ (γ_{cm}), it is immediate that $x_n(1 + x_n^* x_n)^{-1} \to 0$ (γ_{cm}). Observe that

$$D\Big(x_n(1 + x_n^* x_n)^{-1}\Big) = x_n D\Big((1 + x_n^* x_n)^{-1}\Big) + D(x_n)(1 + x_n^* x_n)^{-1},$$

and

$$\begin{aligned} 0 = D(1) &= D\Big((1 + x_n^* x_n)(1 + x_n^* x_n)^{-1}\Big) \\ &= (1 + x_n^* x_n) D\Big((1 + x_n^* x_n)^{-1}\Big) + D(1 + x_n^* x_n)(1 + x_n^* x_n)^{-1}. \end{aligned}$$

Therefore

$$(1 + x_n^* x_n) D\Big((1 + x_n^* x_n)^{-1}\Big) = -D(1 + x_n^* x_n)(1 + x_n^* x_n)^{-1},$$

implying that

$$\begin{aligned} D\Big((1+x_n^*x_n)^{-1}\Big) &= -(1+x_n^*x_n)^{-1}D(1+x_n^*x_n)(1+x_n^*x_n)^{-1} \\ &= -(1+x_n^*x_n)^{-1}D(x_n^*x_n)(1+x_n^*x_n)^{-1} \\ &= -(1+x_n^*x_n)^{-1}\Big(x_n^*D(x_n)+D(x_n)^*x_n\Big)(1+x_n^*x_n)^{-1} \\ &\to -1^{-1}(0.y+y^*.0)1^{-1}(\gamma_{cm}) \\ &= 0 \end{aligned}$$

by continuity of inversion. Hence $D\Big(x_n(1+x_n^*x_n)^{-1}\Big) \to y\ (\gamma_{cm})$. Let $y_n = x_n(1+x_n^*x_n)^{-1}$ for every n. Then (y_n) is a sequence in $\mathcal{M}$ such that $y_n \to 0\ (\gamma_{cm})$ and $D(y_n) \to y\ (\gamma_{cm})$. By hypothesis, $y = 0$, implying that $\mathcal{S}(D) = \{0\}$. □

In what follows, we will need the following result.

Theorem 3.15 (Ryll-Nardzewski fixed point theorem). ([25], p. 444) *Suppose that X is a locally convex Hausdorff space and $\emptyset \neq K \subset X$ a weakly compact convex subset of X. Let $\mathcal{F} : K \to K$ be a non-contracting semigroup of weakly continuous affine maps (here, non-contracting means that for every $x, y \in K$ with $x \neq y$, there exists a seminorm p such that $\inf_{\phi\in\mathcal{F}} p(\phi(x)-\phi(y)) > 0$). Then there exists $x \in K$ such that x is a fixed point of $\mathcal{F}$.*

The proof of our next result is similar to that of [21], Lemma 4.1, and [19], Theorem 1. Recall from Section 5.1 the notion of the non-commutative L_p-spaces $L_p(\mathcal{M},\tau)$, $0 < p < \infty$, and that every $L_p(\mathcal{M},\tau)$, $p \geq 1$, is a Banach $\mathcal{M}$-module with the norm $\|.\|_p$ as defined in that section.

Theorem 3.16. *Let D be a derivation on $\widetilde{\mathcal{M}}$ such that $D(\widetilde{\mathcal{M}}) \subset L_p(\mathcal{M},\tau)$ for some $1 < p < \infty$. Then there exists an $a \in L_p(\mathcal{M},\tau)$ such that $D(x) = ax - xa$ for every $x \in \widetilde{\mathcal{M}}$.*

Proof. We denote the unitary group of $\mathcal{M}$ by $\mathcal{M}_u$. Let $\mathcal{K} = \{u^*D(u) : u \in \mathcal{M}_u\}$ and $\mathcal{L}$ be the $\sigma(L_p,(L_p)^*)$-closed convex hull of $\mathcal{K}$ (here $L_p = L_p(\mathcal{M},\tau)$). Let $D_b = D|_{\mathcal{M}}$. Then D_b is a derivation on $\mathcal{M}$ into L_p. By [26], Theorem 2, D_b is norm-norm continuous.

Therefore $\mathcal{L}$ is a bounded subset of (the Banach space) L_p: $\|u^*D(u)\|_p \leq \|u^*\|\|D(u)\|_p = \|D(u)\|_p$. Now $\mathcal{M}_u$ is a norm bounded subset of $\mathcal{M}$. Since D_b is (norm-norm) continuous, $\{D(u) : u \in \mathcal{M}_u\}$ is a norm bounded subset of L_p. So $\sup_{u\in\mathcal{M}_u}\|D(u)\|_p < \infty$. Hence $\sup_{u\in\mathcal{M}_u}\|u^*D(u)\|_p < \infty$, meaning that $\mathcal{K}$, and thus $\mathcal{L}$, is a norm bounded subset of L_p.

Since L_p is reflexive, it follows from the Banach-Alaoglu theorem that $\mathcal{L}$ is $\sigma(L_p,(L_p)^*)$-compact. For each $u \in \mathcal{M}_u$, define the affine map $A_u(x) = u^*xu + u^*D(u)$ for all $x \in L_p$. Since L_p is a Banach $\mathcal{M}$-module, $A_u(L_p) \subset L_p$ for every $u \in \mathcal{M}_u$. Since $\|u^*xu\|_p = \|x\|_p$ for every $u \in \mathcal{M}_u$ and every $x \in L_p$ (Lemma

2.7), it follows easily that $A_u : L_p \to L_p$ is norm continuous for every $u \in \mathcal{M}_u$. A standard result tells us that A_u is $\sigma(L_p, (L_p)^*)$-continuous for every $u \in \mathcal{M}_u$. Let $u, v \in \mathcal{M}_u$. Then an easy computation shows that $A_v(u^*D(u)) = (uv)^*D(uv)$. Now $uv \in \mathcal{M}_u$. Thus $A_v(\mathcal{K}) \subset \mathcal{K}$. Therefore, since A_v is $\sigma(L_p, (L_p)^*)$-continuous, $A_v(\mathcal{L}) \subset \mathcal{L}$. Since $A_u A_v(x) = A_{vu}(x)$ for every $x \in L_p$, it follows that $\{A_u : u \in \mathcal{M}_u\}$ is a semigroup.

Now let $x, y \in \mathcal{L}$ with $x \neq y$. Then, by Lemma 2.7,

$$\begin{aligned} \|A_u(x) - A_u(y)\|_p &= \|u^*xu + u^*D(u) - u^*yu - u^*D(u)\|_p \\ &= \|u^*xu - u^*yu\|_p \\ &= \|u^*(x-y)u\|_p \\ &= \|x - y\|_p \end{aligned}$$

for all $u \in \mathcal{M}_u$ and for all $1 < p < \infty$. Therefore

$$\inf_{u \in \mathcal{M}_u} \|A_u(x) - A_u(y)\|_p > 0$$

for every $1 < p < \infty$. Hence $\{A_u : u \in \mathcal{M}_u\}$ is non-contracting. By the Ryll-Nardzewski fixed point theorem, there exists $a_0 \in \mathcal{L}$ such that $A_u(a_0) = a_0$ for every $u \in \mathcal{M}_u$. Therefore $u^*a_0u + u^*D(u) = a_0$ for every $u \in \mathcal{M}_u$. Let $a = -a_0$. It follows that $D(u) = au - ua$ for every $u \in \mathcal{M}_u$.

Let $x \in \mathcal{M}$. Then it is well known that $x = \sum_{i=1}^4 \lambda_i u_i$, where $\lambda_i \in \mathbb{C}$ and $u_i \in \mathcal{M}_u$ for all $1 \leq i \leq 4$. Once again, an easy calculation shows that $D(x) = ax - xa$. This holds for every $x \in \mathcal{M}$. Thus $D|_\mathcal{M}$ is $\gamma_{cm} - \gamma_{cm}$ continuous. By Proposition 3.14, D is $\gamma_{cm} - \gamma_{cm}$ continuous. Therefore, since $\mathcal{M}$ is dense in $\widetilde{\mathcal{M}}$ with respect to the measure topology, it follows that $D(x) = ax - xa$ for every $x \in \widetilde{\mathcal{M}}$, implying that D is inner. □

In the proof of [28], Theorem 4.1.6, a similar argument was used for the case where $\mathcal{M}$ is a countably decomposable finite von Neumann algebra: Sakai introduced the maps $T_u(x) = uxu^* + D(u)u^*$ for all $x \in \mathcal{M}$ and for all $u \in \mathcal{M}_u$. He showed, by using Zorn's Lemma, instead of the Ryll-Nardzewski fixed point theorem, that the maps T_u have a fixed point.

Acknowledgement

1. The financial assistance of the Department of Labour (DoL) towards this research is hereby acknowledged. Opinions expressed and conclusions arrived at, are those of the author and are not necessarily to be attributed to the DoL.
2. The author wishes to thank Jurie Conradie of the Department of Mathematics and Applied Mathematics of the University of Cape Town for useful suggestions regarding the proofs of the main results of this paper.

References

[1] S. Albeverio, Sh.A. Ayupov, K.K. Kudaybergenov, Derivations on the algebra of measurable operators affiliated with a type I von Neumann algebra, *arXiv:math.OA/0703171*, 7 August 2008.

[2] S. Albeverio, Sh.A. Ayupov, K.K. Kudaybergenov, Structure of derivations on various algebras of measurable operators for type I von Neumann algebras, *arXiv:math.OA/0808.0149*, 1 August 2008.

[3] S. Albeverio, Sh.A. Ayupov, K.K. Kudaybergenov, Description of derivations on measurable operator algebras of type I, *arXiv:math.OA/0710.3344*, 17 October 2007.

[4] Sh.A. Ayupov, Derivations on algebras of measurable operators, *Dokl. Akad. Nauk RUz.* **3**, 2000, 14–17.

[5] Sh.A. Ayupov, Derivations on unbounded operator algebras, Abstracts of the international conference "Operator algebras and quantum probability", Tashkent, September 7–19, 2005.

[6] A.F. Ber, V.I. Chilin and F.A. Sukochev, Derivations in commutative regular algebras, *Math. Notes* **75**, 2004, 418–419.

[7] A.F. Ber, V.I. Chilin and F.A. Sukochev, Derivations in commutative regular algebras, *Extracta Math.* **21**, 2006, 107–147.

[8] A. Ber, F. Sukochev, Derivations on the algebras of measurable operators, Manuscript presented at the international conference "Operator spaces, non-commutative L^p spaces and applications", 25–29 June 2007, Marseilles, France.

[9] A.F. Ber, B. de Pagter and F.A. Sukochev, private communication, May 2008.

[10] S.K. Berberian, *Baer *-rings*, Springer-Verlag, Berlin-Heidelberg New York, 1972.

[11] S.K. Berberian, The maximal ring of quotients of a finite von Neumann algebra, *Rocky Mountain J. Math.* **12**, 1982, 149–164.

[12] O. Bratteli and D.W. Robinson, *Operator algebras and quantum statistical mechanics I*, Springer-Verlag, New York, 1979.

[13] H.G. Dales, *Banach algebras and automatic continuity*, Clarendon Press, Oxford, 2000.

[14] J. Dixmier, *Von Neumann algebras*, North-Holland Mathematical Library, Amsterdam – New York – Oxford, 1981.

[15] P.G. Dodds, T.K. Dodds and B. de Pagter, Non-commutative Banach function spaces, *Math. Z.* **201**, 1989, 583–597.

[16] P.G. Dodds, T.K. Dodds and B. de Pagter, Non-commutative Köthe duality, *Trans. Amer. Math. Soc.* **339**, 1993, 717–750.

[17] P.G. Dodds, T.K. Dodds, B. de Pagter and F.A. Sukochev, Lipschitz continuity of the absolute value and Riesz projections in symmetric operator spaces, *J. Func. Anal.* **148**, 1997, 28–69.

[18] T. Fack and H. Kosaki, Generalised s-numbers of τ-measurable operators, *Pacific J. Math.* **123**, 1986, 269–300.

[19] T.B. Hoover, Derivations, homomorphisms and operator ideals, *Proc. Amer. Math. Soc.* **62**, 1977, 293–298.

[20] J. Hou and M. Gao, Elements of rank one and multiplications on von Neumann algebras, *Chinese J. Contemp. Math.* **15**, 1994, 367–374.

[21] A. Inoue, S. Ôta, Derivations on algebras of unbounded operators, *Trans. Amer. Math. Soc.* **261**, 1980, 567–577.

[22] R.V. Kadison and J.R. Ringrose, *Fundamentals of the theory of operator algebras*, Academic Press, New York, 1996.

[23] J.L. Kelley and I. Namioka, *Linear Topological Spaces*, Van Nostrand, Toronto-New York-London, 1963.

[24] L.E. Labuschagne, Composition operators on non-commutative L_p-spaces, *Expo. Math.* 17, 1999, 429–468.

[25] I. Namioka and E. Asplund, A geometric proof of Ryll-Nardzewski's fixed point theorem, *Bull. Amer. Math. Soc.* **73**, 1967, 443–445.

[26] J.R. Ringrose, Automatic continuity of derivations of operator algebras, *J. London Math. Soc.* **5**, 1972, 432–438.

[27] W. Rudin, *Functional analysis*, McGraw-Hill, New York, 1973.

[28] S. Sakai, *C^*-algebras and W^*-algebras*, Springer-Verlag, Berlin-Heidelberg-New York, 1971.

[29] S. Sakai, *Operator algebras in dynamical systems: the theory of unbounded derivations in C^*-algebras*, Cambridge University Press, Cambridge, 1991.

[30] I.E. Segal, A non-commutative extension of abstract integration, *Ann. of Math.* **57**, 1953, 401–457.

[31] A. Stroh and G. West, τ-compact operators affiliated to a semifinite von Neumann algebra, *Proc. Roy. Irish Acad. Sect. A,* **93**, 1993, 73–86.

[32] M. Takesaki, *Theory of operator algebras I*, Springer-Verlag, Berlin-Heidelberg-New York, 1979.

[33] I. Tembo, Invertibility in the algebra of measurable operators, Manuscript presented at the international conference "The eighteenth international workshop on operator theory and applications", 2–6 July 2007, Potchefstroom, South Africa.

[34] M. Weigt, On automatic continuity of derivations on algebras of measurable operators, Abstracts of the international conference "Great Plains Operator Theory Symposium", 23–28 May 2006, Iowa City, USA.

[35] G.P. West, *Non-commutative Banach Function Spaces*, MSc Thesis, University of Cape Town, South Africa, 1990.

[36] G. West, Ideals of τ-measurable operators, *Quaestiones Math.* **18**, 1995, 333–344.

[37] F.J. Yeadon, Non-commutative L^p-spaces, *Math. Proc. Camb. Phil. Soc.* **77**, 1975, 91–102.

Martin Weigt
Department of Mathematics and Applied Mathematics
University of Cape Town
Rondebosch 7701, South Africa
e-mail: Martin.Weigt@uct.ac.za
weigt.martin@gmail.com

Operator Theory:
Advances and Applications, Vol. 195, 287–291

On Hyperinvariant Subspaces for Linear Maps of Real Vector Spaces

W. Żelazko

Abstract. Let X be a real vector space and T a linear map of X into itself. We give necessary and sufficient conditions for existence of a proper hyperinvariant subspace for T. These conditions are not identical with those for a complex vector space.

Mathematics Subject Classification (2000). 47A15, 15A04.

Keywords. Real vector space; linear map; hyperinvariant subspace.

Let X be a vector space and $L(X)$ – the algebra of all linear maps (operators) from X into itself. If $T \in L(X)$, then T' denotes the commutant of T, i.e., the set $\{S \in L(X) : ST = TS\}$. A vector subspace $X_0 \subset X$ is said *hyperinvariant* for T if it is invariant for all operators S in T' ($TX_0 \subset X_0$). It is said *proper* if $(0) \neq X_0 \neq X$. In [3] we have obtained the following result.

Theorem A. *Let X be an infinite-dimensional complex vector space and let T be a non-constant operator in $L(X)$. Then T has a proper hyperinvariant subspace if and only if its spectrum $\sigma(T)$ is non-void.*

In the above T is non-constant if $T \neq \lambda I$, $\lambda \in \mathbf{C}$, and I is the identity operator on X, while $\sigma(T) = \{\lambda \in \mathbf{C} : T - \lambda I \notin L(X)^{-1}\}$, where $L(X)^{-1}$ is the set (group) of all invertible elements in $L(X)$.

In this paper we shall obtain an analogous result for a real space X (Theorem 1) and use it for characterizing continuous operators on the locally convex space (s) possessing proper closed hyperinvariant subspaces (Theorem 2).

Observe first that the real version of Theorem A fails to be true, if we replace spectrum $\sigma(T)$ by the real spectrum

$$\sigma_R(T) = \{\lambda \in \mathbf{R} : T - \lambda I \notin L(X)^{-1}\}.$$

Example. Let $X = C[0,1]$ be the vector space of all complex-valued continuous functions on the closed unit interval, treated as a real vector space. Put

$$\varphi(t) = \begin{cases} 1, & \text{if } 0 \leq t \leq 1/2; \\ 1/2 + t, & \text{if } 1/2 \leq t \leq 1, \end{cases}$$

and define T in $L(X)$ as the multiplication operator

$$(Tx)(t) = i\varphi(t)x(t), \quad x \in X.$$

Clearly T has a void real spectrum, nevertheless it has a proper hyperinvariant subspace. In fact, consider the operator $T_1 = T^2 + I$ and put

$$X_0 = \ker T_1 = \{x \in X : \operatorname{supp}(x) \subset [0, 1/2]\},$$

where $\operatorname{supp}(x)$ denotes the support of x, i.e., the closure of the set $\{t \in [0,1] : x(t) \neq 0\}$. If $S \in T_1'$ and $x_o \in X_0$, then $0 = ST_1x_0 = T_1Sx_0$, so that $SX_0 \subset X_0$ and X_0 is hyperinvariant for T_1, and so also for T^2. But $T' \subset (T^2)'$, and so X_0 is hyperinvariant for T. Our claim follows.

Recall that an operator T in $L(X)$ is said *algebraic*, if there is a real (non-constant) polynomial p such that $p(T) = 0$ (we assume $T^0 = I$), and denote by p_T a polynomial of the lowest possible degree such that this relation holds true (a minimal polynomial for T). A real polynomial is decomposable if it is a product of two real polynomials of lower degree.

Our result reads now

Theorem 1. *Let X be a real vector space. A non-constant operator T in $L(X)$ has a proper hyperinvariant subspace if and only if either of the following hold true:*

(a) *T is non-algebraic and there exists a non-zero real polynomial p such that the operator $p(T)$ is non-invertible in $L(X)$;*

(b) *T is algebraic, and its minimal polynomial p_T is decomposable.*

Proof. Consider first the case when T is non-algebraic. If $p(T)$ is non-invertible for some non-zero polynomial p, then either $p(T)$ has a non-zero kernel X_1 ($X_1 \neq X$ since T is non-algebraic), or is not a map onto. In the first case X_1 is hyperinvariant for T. In fact, let $S \in T'$ and $x_0 \in X_1$. We have $Sp(T) = p(T)S$, and so $0 = Sp(T)x_0 = p(T)Sx_0$, so that $Sx_0 \in X_1$ and X_1 is hyperinvariant for T. In the second case put $X_1 = p(T)X$. We have $X_1 \neq X$ and $X_1 \neq (0)$, since T is a non-algebraic operator. If $S \in T'$, then $SX_1 = Sp(T)X = p(T)SX \subset p(T)X = X_1$, and again X_1 is hyperinvariant for T.

Suppose now that for each non-zero real polynomial p the operator $p(T)$ is invertible. We have to show that there is no hyperinvariant subspace for T. Denote by Q the field of all real rational functions. If $\varphi \in Q$ then $\varphi(T)$ is a well-defined element of $L(X)$ and it commutes with T. We can also treat X as a vector space over Q if we define

$$\varphi x = \varphi(T)x, \quad x \in X. \tag{1}$$

Denote by $L_Q(X)$ the algebra of all Q-linear operators on X. We have

$$L_Q(X) \subset T' \subset L(X). \tag{2}$$

Assume, towards contradiction, that T has a proper hyperinvariant subspace $X_0 \subset X$ and choose a non-zero element $x_1 \in X_0$. Relation (2) implies

$$X_1 = Qx_1 \subset X_0$$

and X_1 is (an infinite-dimensional) real vector subspace of X_0. Since X_0 is proper, there is an element $x_2 \in X \setminus X_0$. We put $X_2 = Qx_2$. Observe that

$$X_1 \cap X_2 = (0). \tag{3}$$

Otherwise there are $\varphi_1, \varphi_2 \in Q$ such that $\varphi_1 x_1 = \varphi_2 x_2$, which implies $x_2 = \varphi_2^{-1}\varphi_1 x_1 \in X_1 \subset X_0$ – a contradiction by the definition of x_2. Thus x_1 and x_2 are Q-linearly independent and can be imbedded into a Q-Hamel basis for X. It gives a direct sum decomposition

$$X = X_1 + X_2 + Y, \tag{4}$$

where Y is the Q-linear span of remaining elements of the considered Q-Hamel basis (it can happen $Y = (0)$). By (4) every element x in X can be written as

$$x = \varphi_1^{(x)} x_1 + \varphi_2^{(x)} x_2 + y, \quad \varphi_1^{(x)}, \varphi_2^{(x)} \in Q, \quad y \in Y. \tag{5}$$

Define an operator S_0 in $L(X)$ setting for an x of the form (5)

$$S_0 x = \varphi_2^{(x)} x_1 + \varphi_1^{(x)} x_2.$$

Clearly S_0 is a Q-linear operator on X, and so, by the formula (2), it belongs to T'. On the other hand, $S_0 x_1 = x_2$, $x_1 \in X_0$ and $x_2 \notin X_0$, which is in contradiction with the fact that X_0 is invariant for S_0. Our claim follows.

Consider now the case when T is an algebraic operator. Since T is non-constant the degree of p_T is not less than two. If p_T is decomposable, which always happens if its degree is larger than two, then $p_T = p_1 p_2$, with $p_1(T) \neq 0$ and $p_2(T) \neq 0$. Thus $x_1 = p_2(T)x_0 \neq 0$ for some x_0 in X. We have $p_1(T)x_1 = 0$, so that the kernel X_1 of $p_1(T)$ is a proper linear subspace of X. Similarly as in the first part of this proof we show that X_1 is hyperinvariant for T.

If p_T is non-decomposable, it is of order two and is of the form

$$p_T(t) = (t - \alpha)^2 + \beta^2$$

with α, β real and $\beta \neq 0$. Put $T_1 = \beta^{-1}(T - \alpha I)$ so that

$$T_1^2 = -I. \tag{6}$$

By the means of T_1 we can treat X as a complex vector space, setting

$$(\alpha + \beta i)x = \alpha x + \beta T_1 x, \quad \alpha, \beta \in \mathbf{R}, \quad x \in X. \tag{7}$$

Denote by $L_{\mathbf{C}}(X)$ the set of all complex linear operators on X in the sense of complex scalar multiplication given by the formula (7). Clearly elements of $L_{\mathbf{C}}(X)$, treated as real operators, are in $L(X)$ and belong to T'. On the other hand, for any $x, y \in X$, $x \neq 0$, there is an operator $S \in L_{\mathbf{C}}(X)$ with $Sx = y$ (the algebra $L_{\mathbf{C}}(X)$ is transitive on X), so that there is no common invariant subspace for all operators in $L_{\mathbf{C}}(X)$ and so in T'. The conclusion follows. □

Theorem A was used in [4] for obtaining a new proof of the fact that all non-constant operators on the complex locally convex space (s) have proper closed hyperinvariant subspaces (the result was originally proved in [3]). Using the above Theorem 1, we shall obtain a similar (but different) fact concerning the real space

(s) (Theorem 2 below). However first we recall some properties of the space (s) useful in the sequel. The space (s) (it is Banach's notation [1], Köthe [2] and German school denote this space by ω), consists of all scalar-valued sequences $x = (\xi_k(x))_1^\infty$ provided with the coordinatewise convergence, which can be given by the increasing sequence of seminorms $||x||_k = \max\{|\xi_1(x)|, \dots, |\xi_k(x)|\}$. With this topology (s) is a complete metrizable locally convex space. The dual space $(s)'$ of all its continuous (scalar-valued) linear functionals consists of all eventually zero sequences $f = (\eta_k(f))_1^\infty$ ($\eta_k(f) = 0$ for $k > k(f)$), provided with the topology of uniform convergence on bounded subsets of (s). It is known that this topology coincides with the maximal locally convex topology $\tau_{\max}^{LC}$, which is given by means of all seminorms. Under this topology all linear subspaces of $(s)'$ are closed an all its linear operators are continuous. Note that each Hamel basis for $(s)'$ is countable. We have $f(x) = \sum_0^\infty \xi_i(x)\eta_i(f)$. It turns out, that the space (s) is reflexive, i.e., the natural imbedding $(s) \subset (s)''$ is onto and the topology of $(s)''$ coincides with the topology of (s) For proofs of these facts the reader is referred to Köthe [2], pp. 287–292, also §18.5, p. 214. Denote by $L((s))$ the algebra of all continuous endomorphisms of the space (s) and for such an endomorphism T a subspace $X_0 \subset (s)$ is called *hyperinvariant*, if it is invariant for all operators S in $L((s))$ which commute with T. As a corollary to our Theorem 1 we obtain the following

Theorem 2. *A non-constant operator T in $L((s))$ for a real space (s) has a proper closed hyperinvariant subspace if and only if it does not satisfy relation of the form*

$$(T - \alpha I)^2 + \beta^2 I = 0, \quad \alpha, \beta \in \mathbf{R}, \quad \beta \neq 0. \tag{8}$$

Proof. We shall perform the proof for the dual space $(s)'$. Since all its linear subspaces are closed and all linear operators are continuous, the problem becomes purely algebraic. If $T \in L((s))$ then its dual operator T^* is given by the formula

$$(T^* f)(x) = f(Tx), \quad f \in (s)', \quad x \in (s).$$

If T has a proper closed hyperinvariant subspace $X_0 \subset (s)$, then $X_0^\perp = \{f \in (s)' : f(x) = 0 \text{ for all } x \in X_0\}$ is a proper hyperinvariant subspace for T^*, and conversely, if T^* has a proper (closed) hyperinvariant subspace $Y \subset (s)'$, then the subspace $Y^\perp = \{x \in (s) : f(x) = 0 \text{ for all } f \in Y\} \subset (s)$ is a proper closed hyperinvariant subspace for T (easy proofs of both facts are left to the reader). Moreover, T satisfies relation (8) if and only if T^* satisfies the same relation. Thus we have to show that T^* has a proper closed hyperinvariant subspace if and only if it does not satisfy relation (8). If T^* is algebraic, then by Theorem 1, T^* has a proper (closed) hyperinvariant subspace in $(s)'$ if and only if it does not satisfy (8). It remains to be shown that if T^* is non-algebraic, then it has a proper hyperinvariant subspace. In view of Theorem 1 it is sufficient to show that $p(T^*)$ is non-invertible for some (non-zero) real polynomial p. Assume, towards contradiction that all operators $p(T^*)$ are invertible. We shall show that in this case T^* also has a proper hyperinvariant subspace, which will be the desired contradiction. Consider the family $(T^* - \lambda I^*)^{-1}$, $\lambda \in \mathbf{R}$ of elements of $L((s)')$, where I^* is the unity operator on $(s)'$, and a non-zero element $f_0 \in (s)'$. Since the set

$(T^* - \lambda I^*)^{-1} f_0$, $\lambda \in \mathbf{R}$ is uncountable, it is linearly dependent and there are real non-zero coefficients $\alpha_1, \ldots, \alpha_k$ and real numbers $\lambda_1, \ldots, \lambda_k$, $\lambda_i \neq \lambda_j$ such that

$$\sum_{i=1}^{k} \alpha_i (T^* - \lambda_i I^*)^{-1} f_0 = 0. \tag{9}$$

We can rewrite (9) as

$$q(T^*)^{-1} p(T^*) f_0 = 0, \tag{10}$$

where $q(t) = (t - \lambda_1) \ldots (t - \lambda_k)$ and p is a suitable real polynomial. We have $p \neq 0$ since the rational functions $(t - \lambda_1)^{-1}, \ldots, (t - \lambda_k)^{-1}$ are linearly independent. Relation (9) implies $p(T^*) f_0 = 0$, so that the kernel Y of $p(T^*)$ is a proper linear subspace of $(s)'$ ($f_0 \in Y$ and T^* is non-algebraic), which, as in the proof of Theorem 1, is hyperinvariant for T^*. The contradiction proves our assertion and the conclusion follows. □

An example of a continuous operator in $L((s))$ without a proper hyperinvariant subspace can be obtained as follows. Denote by $(e_i)_0^\infty$ the Schauder basis of unit vectors in (s) ($e_i = (\delta_{ij})_{j=1}^\infty$) and define T setting $Te_{2k} = e_{2k-1}$, $Te_{2k-1} = -e_{2k}$, $k = 1, 2, \ldots$. Then T satisfies $T^2 = -I$ and by Theorem 2 has no proper closed hyperinvariant subspace (in fact, no hyperinvariant subspace).

References

[1] S. Banach, *Théorie des Opérations Linéaires*, Warszawa, 1932.

[2] G. Köthe, *Topological Vector Spaces I*, Springer, 1971.

[3] W. Żelazko, *On existence of hyperinvariant subspaces for linear maps*, Banach J. Math. Anal. **3** (2009), no. 1, 143–148.

[4] W. Żelazko, *Operators on locally convex spaces*, in: Operator Theory: Advances and Applications, Vol. 187, 237–247, 2008 Birkhäuser Verlag Basel.

W. Żelazko
Institute of Mathematics
Polish Academy of Sciences
Śniadeckich 8, P.O.B. 21
00-956 Warszawa, Poland
e-mail: `W.Zelazko@impan.pl`

List of Participants: IWOTA 2007

Attumbu, Henry M	Ghana
Ball, Joseph A	USA
Behrndt, Jussi	Germany
Belyi, Sergey	USA
Bhattacharyya, Tirthankar	India
Binding, Paul	Canada
Blasco, Oscar	Spain
Bračič, Janko	Slovenia
Câmara, Cristina	Portugal
Conradie, Jurie	South Africa
Currie, Sonja	South Africa
De Beer, Richard	South Africa
De Kock, Mienie	USA
Diogo, Cristina	Portugal
Dritschel, Michael	UK
Dubi, Chen	Israel
Duvenhage, Rocco	South Africa
El-Sayed Ahmed, Ahmed	Egypt
Emenyu, John	Uganda
Fourie, Jan	South Africa
Garcia-Pacheco, Francisco Javier	USA
Grobler, Jacobus J	South Africa
Groenewald, Gilbert J	South Africa
Hryniv, Rostyslav	Ukraine
Hudson, Robin	UK
Jaftha, Jacob	South Africa
Janse van Rensburg, Dawie	South Africa
Kaashoek, Marinus A	Netherlands
Karlovych, Oleksiy	Portugal
Kuzma, Bojan	Slovenia
Labuschagne, Louis E	South Africa
Langer, Matthias	UK
Lindeboom, Lenore	South Africa

Lindner, Marko	UK
Malheiro, Maria Teresa	Portugal
Martins, Maria do Carmo	Portugal
Möller, Manfred	South Africa
Ran, Andre	Netherlands
Renison, Martin	South Africa
Rodman, Leiba	USA
Strašek, Rok	Slovenia
Ströh, Anton	South Africa
Tembo, Isaac	South Africa
ter Horst, Sanne	Netherlands
Trunk, Carsten	Germany
Venter, Rudolf G	South Africa
Verduyn-Lunel, Sjoerd	Netherlands
Vinnikov, Victor	Israel
Watson, Bruce	South Africa
Weigt, Martin	South Africa
Weis, Lutz	Germany
Woracek, Harald	Austria
Zelazko, Wieslaw	Poland

Operator Theory: Advances and Applications (OT)

Edited by
Israel Gohberg, Tel Aviv University, Israel

BIRKHÄUSER

This series is devoted to the publication of current research in operator theory, with particular emphasis on applications to classical analysis and the theory of integral equations, as well as to numerical analysis, mathematical physics and mathematical methods in electrical engineering.

OT 195: Grobler, J.J. / Labuschagne, L.E. / Möller, M. (eds.), Operator Algebras, Operator Theory and Applications (2009). ISBN 978-3-0346-0173-3

OT 194: González, M. / Martínez-Abejón, A,, Tauberian Operators (2009). ISBN 978-3-7643-8997-0

OT 193: Cialdea, A. / Lanzara, F. / P.E. Ricci (eds.), Analysis, Partial Differential Equations and Applications. The Vladimir Maz'ya Anniversary Volume (2009). ISBN 978-3-7643-9897-2

OT 192: Gohberg, I. / Leiterer, J., Holomorphic Operator Functions of One Variable and Applications. Methods from Complex Analysis in Several Variables (2009). ISBN 978-3-0346-0125-2

OT 190/191: Adamyan, V. / Berezansky, Yu.M. / Gohberg, I. / Gorbachuk, M.L. / Gorbachuk, V. / Kochubei, A.N. / Langer, H. / Popov, G. (eds.), Modern Analysis and Applications. The Mark Krein Centenary Conference (2009).
Volume 1 (OT 190): Operator Theory and Related Topics. ISBN 978-3-7643-9918-4
Volume 2 (OT 191): Differential Operators and Mechanics. ISBN 978-3-7643-9920-7
Set Volumes 1/2: ISBN 978-3-7643-9924-5

OT 189: Rodino, L. / Wong, M.W. (eds.), New Developments in Pseudo-Differential Operators (2008). ISBN 978-3-7643-8968-0

OT 188: Behrndt, J. / Förster, K.-H. / Langer, H. / Trunk, C. (eds.), Spectral Theory in Inner Product Spaces and Applications (2008). ISBN 978-3-7643-8910-9

OT 187: Ando, T. / Curto, R.E. / Jung, I.B. / Lee, W.Y. (eds.), Recent Advances in Operator Theory and Applications (2008). ISBN 978-3-7643-8892-8

OT 186: Janas, J. / Kurasov, P. / Laptev, A. / Naboko, S. / Stolz, G. (eds.), Methods of Spectral Analysis in Mathematical Physics (2008). ISBN 978-3-7643-8754-9

OT 185: Vasilevski, N., Commutative Algebras of Toeplitz Operators on the Bergman Space (2008). ISBN 978-3-7643-8725-9

OT 184: Qin, Y., Nonlinear Parabolic-Hyperbolic Coupled Systems and Their Attractors (2008). Subseries **A**dvances in **P**artial **D**ifferential **E**quations. ISBN 978-3-7643-8813-3

OT 183: Nazaikinskii, V.E.. / Savin, A.Yu. / Sternin, B.Yu., Elliptic Theory and Noncommutative Geometry (2008). Subseries **A**dvances in **P**artial **D**ifferential **E**quations. ISBN 978-3-7643-8774-7

OT 182: van der Mee, C., Exponentially Dichotomous Operators and Applications (2008). Subseries **L**inear **O**perators and **L**inear **S**ystems. ISBN 978-3-7643-8732-7

OT 181: Bastos, A. / Gohberg, I. / Lebre, A.B. / Speck, F.-O. (eds.), Operator Algebras, Operator Theory and Applications (2008). ISBN 978-3-7643-8683-2

OT 180: Okada, S. / Ricker, W.J. / Sánchez Pérez, E.A., Optimal Domain and Integral Extension of Operators Acting in Function Spaces (2008). ISBN 978-3-7643-8647-4

OT 179: Ball, J.A. / Eidelman, Y. / Helton, J.W. / Olshevsky, V. / Rovnyak, J. (eds.), Recent Advances in Matrix and Operator Theory (2007). ISBN 978-3-7643-8538-5

OT 178: Bart, H. / Gohberg, I. / Kaashoek, M.A. / Ran, A.C.M., Factorization of Matrix and Operator Functions: The State Space Method (2007). ISBN 978-3-7643-8267-4

Operator Theory: Advances and Applications (OT)

Edited by
Israel Gohberg, Tel Aviv University, Israel

BIRKHÄUSER

This series is devoted to the publication of current research in operator theory, with particular emphasis on applications to classical analysis and the theory of integral equations, as well as to numerical analysis, mathematical physics and mathematical methods in electrical engineering.

Kolibri 360 GmbH
Ferdinand-Jühlke-Straße 7
99095 Erfurt, Deutschland
produktsicherheit@kolibri360.de